# Roman Funerary Monuments of South-Western Pannonia in their Material, Social, and Religious Context

Branka Migotti, Marjeta Šašel Kos and Ivan Radman-Livaja

in collaboration with

Bojan Djurić, Ljubica Perinić and Mirko Belak

Archaeopress Roman Archaeology 45

ARCHAEOPRESS PUBLISHING LTD
Summertown Pavilion
18-24 Middle Way
Summertown
Oxford OX2 7LG

www.archaeopress.com

ISBN 978-1-78969-021-7 (paperback)
ISBN 978-1-78969-049-1 (hardback)
ISBN 978-1-78969-022-4 (e-Pdf)



Cover: Stele found in Donji Čehi (page 88).



This book is available direct from Archaeopress or from our website www.archaeopress.com

In memory of our colleague and friend Zoran Gregl (1952–2017)

# Contents

# List of Figures

## Introduction

B. Migotti



## Chapter I Catalogues

B. Migotti and M. Šašel Kos

## Chapter II Discussion of the evidence

B. Migotti



## Chapter III Geology and Quarries

B. Migotti

# Foreword and Acknowledgments

Branka Migotti

This book has come about as a result of the project *Roman Funerary Monuments of South-Western Pannonia in their Material, Social, and Religious Context*, unfolding from 4th May 2015 till 5th May 2018 in the Croatian Academy of Sciences and Arts, and fully supported by the Croatian Science Foundation (project number: IP-2014-09-4632). The Croatian-Slovenian research team included four researches: Marjeta Šašel Kos (The Institute of Archaeology of the Research Centre of the Slovenian Academy of Sciences and Arts, Ljubljana), Ivan Radman-Livaja (The Archaeological Museum in Zagreb), Ljubica Perinić (The Division of Archaeology of the Croatian Academy of Sciences and Arts, Zagreb), and Mirko Belak (The Croatian Geological Institute, Zagreb), with Bojan Djurić (The Faculty of Arts of the University of Ljubljana) as the consultant, and Branka Migotti (The Division of Archaeology of the Croatian Academy of Sciences and Arts, Zagreb) as the principal investigator.

The authors are extremely grateful to many colleagues in the museums and other institutions in Croatia, Slovenia and Bosnia and Herzegovina, whose assistance in compiling the Catalogue(s) and studying the monuments was invaluable in writing this book. Some of them provided technical data or photographs, while others shared with us their opinions or gave advice on various issues during the study of the evidence. Without some of the so-obtained information and data this book would have never been realized, while the remainder much improved its overall quality. It goes without saying that the authors alone take full responsibility for possible misconceptions and mistakes of any kinds. Since we are indebted to so many colleagues in the above-mentioned way, it is not possible to describe their contributions individually. However, I must single out two of them, whose engagement was such as to practically make them collaborators on the project: Rosana Škrgulja (The Museum of Sisak, HR) and Berislav Scheibal (Daruvar, HR). To them Domagoj Perkić (The Dubrovnik Museums, HR) should be added for kindly letting us source from his unpublished thesis, Ursula Rothe (The Faculty of Arts and Social Sciences of the Open University, UK), for her many enlightening remarks on the topic of clothes and for allowing the use of illustrations from her book, and Claudio Mazzoli (The Department of Geosciences of the University of Padua, IT) for generously characterizing two samples gratis. The remainder who also merit our sincere thanks are many: Iskra Baćani (The Museum of Sisak, HR), Siniša Bizjak (The Arts Academy of the University of Split, HR), Kristina Brkić (Petrinja, HR), Zdenko Burkowsky (The Museum of Sisak, HR), Adnan Busuladžić (The Archaeology Department of the Faculty of Philosophy, The University of Sarajevo, BA), Nenad Cambi (The Croatian Academy of Sciences and Art, Zagreb, HR); Nikola Cesarik (The Department of Archaeology of the University of Zadar, HR), Robert Čimin (The Museum of Koprivnica, HR), Lazo Čučković (The Museum of Karlovac, HR), Gordana Divac (The Directorate for the Protection of Cultural Heritage, The Conservation Department in Krapina, HR), Ozren Domiter (The Archaeological Museum in Zagreb, HR), Aleksandar Durman (The Faculty of Humanities and Social Sciences, The University of Zagreb, HR), Krešimir Filipec (The Faculty of Humanities and Social Sciences, The University of Zagreb, HR), Vladimir Goss (The Department of Art History of the Faculty of Humanities and Social Sciences, The University of Rijeka, HR), †Zoran Gregl (The Archaeological Museum in Zagreb, HR), Ortolf Harl (Wien, AU), Jelena Hihlik (The Museum of Pakrac, HR), Hana Ivezić (The Archaeological Museum in Zagreb, HR), Igor Krajcar (The Archaeological Museum in Zagreb, HR), Mirta Jambrović (Zagreb, HR), Goran Jakovljević (The Museum of Bjelovar, HR), Zoran Jovanić (Laktaši, BA), Vjekoslav Jukić (The Ministry of Economy of the Republic of Croatia, Zagreb, HR), Spomenka Jurić (Sisak, HR), Krešimir Karlo (The Directorate for the Protection of Cultural Heritage, The Conservation Department in Bjelovar, HR), Vesna Kolić-Klikić (The Museum of Nova Gradiška, HR), Marijana Korunek (The Directorate for the Protection of Cultural Heritage, The Conservation Department in Varaždin, HR), Gabrielle Kremer (The Institute of the History of Culture of Antiquity of the Austrian Academy of Sciences, Wien, AU), Bruna Kuntić-Makvić (The Department of History of the Faculty of Humanities and Social Sciences, The University of Zagreb, HR), Dora Kušan Špalj (The Archaeological Museum in Zagreb, HR), Tino Leleković (The Division of Archaeology of the Croatian Academy of Sciences and Arts, Zagreb, HR), Mladen Levak (Novigrad Podravski, HR), Mladen Medar (Bjelovar, HR), Marija Mihaljević (The Museum of Nova Gradiška, HR), Zsolt Mráv (The Hungarian National Museum Budapest, HU), Aleksandra Nikoloska (The Macedonian Academy of Sciences and Arts, Skopje, MK), Siniša Njegovan Stárek (Daruvar), Veronika Pilat (Daruvar, HR), Lana Okroša Rožić (The Museum of Križevci), Mirela Pavličić Hein (The Museum of Požega, HR), Darko Periša (The Catholic University of Croatia, Zagreb, HR), Tanja Pintarić (The Museum of Turopolje, Velika Gorica, HR), Caterina Previato

(The University of Padua, IT), Daniela Radosavac (The Museum of Kozara, Prijedor, BA), Hrvoje Raguž (The Directorate for the Protection of Cultural Heritage, The Conservation Department in Karlovac, HR), Igor Rižnar (Ljubljana, SI), Dubravka Sokač-Štimac (Požega, HR), Ljubica Srdić (The Republic Institute for the Protection of Cultural, Historical and Natural Heritage, Banja Luka, BA), Amra Šačić (The History Department of the Faculty of Philosophy, The University of Sarajevo, BA), Marina Šimek (Varaždin, HR), Lovorka Štimac Dedić (The Museum of Varaždin, HR), Ivica Šustić (Petrinja, HR), Bojan Vujinović (The County Museum Gradiška, BA), Vlatka Vukelić (The University Department of the Croatian Studies, The University of Zagreb, HR). Finally, our thanks go to the directors (other than those mentioned above among colleague archaeologists) of several museums for their immediate permission for stone sampling: Sanjin Mihelić (The Archaeological Museum in Zagreb, HR), Vlatko Čakširan (The Museum of Sisak, HR), Maja Žebčević Matić (The Museum of Požega, HR), Miran Bojanić Morandini (The Museum of Varaždin, HR); Milan Pavlović (The Museum of Bjelovar, HR), Mirsad Sijarić (The National Museum of Bosnia and Herzegovina, BA) and Zoran Radonjić (The Museum of Kozara, Prijedor, BA).

Last but not least, it is my duty and privilege to express special gratitude to Bojan Djurić, whose contribution by far exceeded his role as project consultant.

# Introduction

## Branka Migotti

Three starting points have provided a background for the research on the project *Roman Funerary Monuments of Southwestern Pannonia in their Material, Social, and Religious Context*, which has resulted in this book. First is the fact that funerary monuments contain material and spiritual components operating at several interconnected levels: ethno-cultural, social, economic, artistic and ritual-religious, and thus have considerable potential to complement our knowledge of the structures and workings of provincial societies of the Roman Empire. Through historical, art-historical and scientific data on funerary monuments (forms, dimensions, quality of execution, kind of stone used, iconography and epigraphy) it is possible to acquire knowledge of the ethnic-cultural origin, social and financial status, and religious affiliation of the deceased. When properly systematised, such data can offer various insights into the social and cultural identity of the society under investigation: social stratification (concerning military and civil ranks and status), ethnic origin, religious affiliations and the relationship between classical cults and Christianity, artistic trends, as well as production and trading aspects. The provinces of Noricum and Pannonia count among those that did not know stonemasonry before the Roman conquest, at least not at a commercial scale; expectedly, it was the army that introduced the tombstone in the northeastern provinces of the Roman Empire.[1] The only funerary stone tentatively interpreted as of the original or resemantized Celtic provenance from northern Croatia, now walled into the facade of the church in the village of Križovljan (Catalogue, Peripheral areas II. Poetovio, Križovljan), is most probably a portrait niche from a Roman-period funerary building. In view of that, funerary monuments of SW Pannonia make an important medium for the study of the autochthonous substrate and oncoming Romanisation, that is, the reception of Roman culture by the indigenous society and the subsequent integration, with stress on the incipient military role in this process.

Second, the funerary monuments of the selected area have never been studied comprehensively, either by individual kinds or even less as a whole, including the epitaphs that were mostly considered separately in the *CIL* or in modern-day epigraphic Internet databases, such as the *Epigraphic Database Clauss-Slaby* and the *Epigraphic Database Heidelberg*. Therefore, several quite old publications still figure as the main source for the study of Roman funerary monuments of northwest Croatia (Brunšmid 1903–1912; *AIJ* 1938). Although still a valuable basic source in the absence of more up-to-date works, these publications are outdated in terms of the approach and the lack of new evidence. The contributions by Brunšmid comprise inscribed and non-inscribed monuments, while the monograph of Hoffiller and Saria (*AIJ*) only cover inscribed stones. Needless to say, in all of them the stone material was judged by the naked eye, with, expectedly, quite a lot of mistaken assessments. With rare exceptions, in more recent works funerary monuments were mostly published individually or in smaller groups, or as sections within various thematic monographs, but still lacking archaeometric analyses.[2]

The third starting point concerns the fact that researchers have by now became completely aware of the necessity to study inscribed stone monuments in their entirety, that is, by giving equal importance to the inscription and depiction.[3] The same is true of including archaeometric characterizations, yet not as the aim in itself, but as a contribution to a better insight into the politics of the economy of production of, and trade in, stone monuments. Nevertheless, in practice such theoretical approach has often remained a declarative aim, with epigraphy taking precedence by epigraphers and a formal-artistic analysis being favoured by archaeologists and art historians, especially in thematic corpora, and with archaeometric research remaining partial and not fully incorporated. Among several contrary recent examples, a volume of the *CSIR* on the *stelai* and funerary altars of Flavia Solva should be pointed out as an improvement in the series. It emerged as a work produced by two authors, covering epigraphy and iconography equally, with an in-depth social analysis based on both these components.[4] Even the third component – petrographic analyses – was included, but only to a measure enabled by characterizations of the monuments conducted previously through another project, resulting in only a very small percentage of the characterized pieces. Indeed, petrographic analyses are still rare as an obligatory component in comprehensive studies of stone monuments, but such an approach has

[1] O. Harl 1991a: 5–12; Jaeger 2003: 476; Carroll 2006: 17; M. Scholz 2012: 2; Russell 2013: 11; Weber-Hiden 2014: 71; Migotti 2016a.

[2] Cambi 1989; Burkowsky 1996; Gregl 1996; Gregl and Migotti 2000; Cambi 2002; Cambi 2005; Migotti 2005b; Migotti 2007a; Migotti 2008; Migotti 2010; Belaj and Migotti 2011; Migotti 2012; Migotti 2013a; Migotti 2013b.

[3] Cf. Migotti, *The Interdependence* (in press).

[4] Pochmarski and Weber-Hiden 2016.

been evolving over the last three decades. Two of a few positive examples in this regard should be mentioned. The first relates to two complementary papers on the stone monuments from the military camp in Vindonissa, and the second is the project of the Institute of the History of Culture of Antiquity of the Austrian Academy of Sciences, Wien: *Steindenkmäler und Steingewinnung im Raum Carnuntum-Vindobona* (FWF – Projekt P 26368-G21) led by Gabrielle Kremer.[5] Finally, a whole new impetus and a new research protocol have been given to this field of research by Bojan Djurić and his team in the last two decades, with the aim of establishing a course for a more comprehensive research that will cover all aspects of the phenomenon of stone production and use. In order to achieve this goal, the investigation has been focused on the production of monuments at all stages, from the extraction of stone and the primary roughing out of the product at the quarry workshop to its final use, be it near or far away from the starting point. The research protocol introduced and defined by Bojan Djurić comprised two main points: the supply of stone and stone monuments in the territory of study (between the Eastern Alps and the Carpathians), and, consequently, transportation routes and commercial relationships, which turned out to be more complicated than previously considered in relation to Eastern Alpine marbles. With these investigations, the study of Roman stone in the region in question has acquired a completely new dimension, making it much easier for researchers of smaller sections of the whole area to better understand wider procedures and more successfully contribute to the general knowledge of issues that remain contested.[6]

Until recently, the stone provenance and workshop origin of Roman funerary monuments of northwest Croatia were rarely taken into consideration, with hypotheses on the import vs. local production mostly resting on typological, iconographical, and stylistic traits. A decade or so ago, marble characterizations of various monuments, funerary included, have been carried out for Mursa (present-day Osijek) in Pannonia Inferior, as well as a certain number of those kept in the Archaeological Museum in Zagreb; among the latter were also monuments from the three Roman cities of the study area (Andautonia, Siscia, and Aquae Balissae).[7] With the passage of time, it turned out that characterizations conducted by Harald Müller in the Archaeological Museum in Zagreb in 2008 required some updating through more refined methods; such research was indeed conducted within the project which has generated this book. On the other hand, no archaeometric research was carried out on stone other than marble from Andautonia, Siscia and Aquae Balissae, while the origin of stone and its association with possible quarries were looked for in only several examples of recently published monuments.[8] For all the above reasons, the interpretative potential of the evidence discussed here had remained much underused before the above-mentioned project started. Therefore, one of the aims of this book was to present the results of the study of a corpus of stone monuments by equally treating all of its components: structural typology, iconography, inscription and stone fabric.

Because of the multifaceted interpretive potential of the evidence, the area of study has been limited to three southwestern Pannonian Roman cities in the territory of north-western Croatia: *municipium Andautonia* (Ščitarjevo), *colonia Siscia* (Sisak), and *municipium Aquae Balissae* (Daruvar) (Figures 1 and 2). It was hoped that juxtaposing three cities of different dimensions and municipal profiles, and of unequal importance in the wider area, would offer a good opportunity for a meaningful comparison of various components of the study. Before the conquest, Siscia/Segest(ic)a had been an important Pannonian emporium. Since its occupation by the Romans in the year 35 BC, the city became a strategically well-placed legionary fortress, but soon enough also a thriving production, commercial and transport hub, covering an area of some 400,000 m$^2$. The military were present in Siscia even after the town's elevation to the rank of colony (*colonia Flavia*) under Vespasian, while Septimius Severus honoured the city with a renewed colony title: *colonia Septimia Siscia Augusta*. In the 3rd century a mint was established in Siscia, while in Late Antiquity it was the most important city of southwest Pannonia, figuring as the capital of Pannonia Savia and an early Christian bishopric. Throughout, the importance and cosmopolitan profile of the city was based on its favourable strategic and traffic position, as well as iron mines in its wider surroundings.[9] Andautonia came under Roman control in the 2nd half of the 1st century BC. Its development as a Roman town started in the beginning of the 1st century AD, to reach its peak in the 2nd century, when it stretched 1000 m in length and 400 m in width. The city was elevated to the rank of *municipium* by the end of the 1st century; apart from *municipium Andautonia*, it was confirmed as *res publica Andautoniensium* situated on the road between Poetovio (Ptuj, SI) and Siscia, and thus on the shortest route from southwestern Pannonia toward North Italy, was the main source of Andautonia's economic growth. The importance of the city seems to have considerably dropped in Late Antiquity.[10] Aquae Balissae was a tribal centre of the Iasi

[5] Tomasevic Buck 2007; Schuler 2007; Rohatsch *et al.* 2016.

[6] Djurić, in press. See also the list of papers by B. Djurić in the Literature.

[7] Djurić, Müller and Filipović 2010; Djurić 2013.

[8] Migotti 2005b; Belaj and Migotti 2011.

[9] Lolić 2003; Šašel Kos 2017b: 174–175.

[10] Nemeth-Ehrlich and Kušan Špalj 2003.

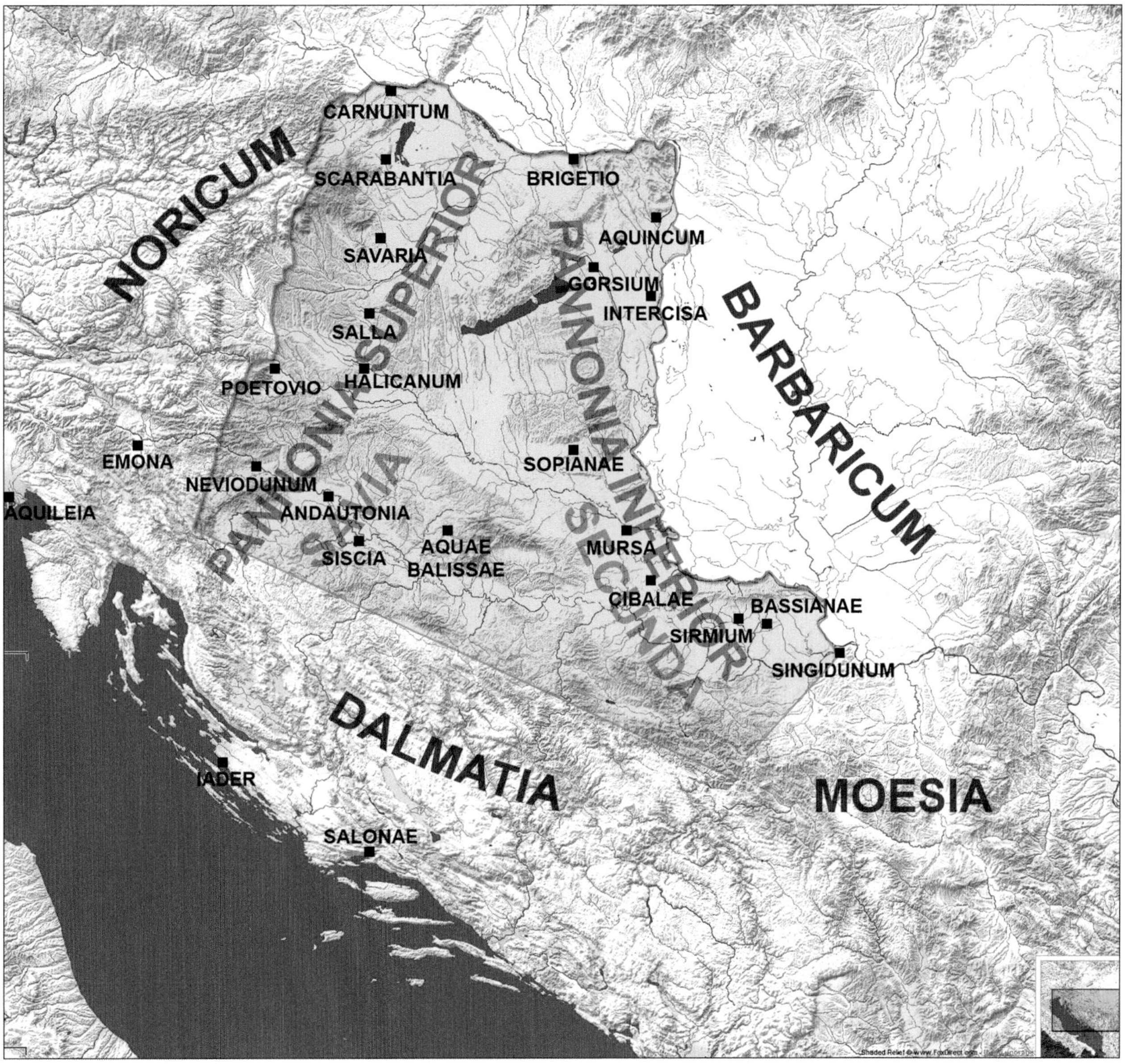


Figure 1. Map of the province of Pannonia (after Migotti 2012c)

(*civitas Iasorum*), which in time acquired the status of autonomous city, attested literally and epigraphically as *res publica Iasorum, Aquae Balissae*, and *municipium Iasorum*; the last-mentioned dates from the reign of the Emperor Hadrian. Due to the lack of archaeological excavations, the perimeter of Aquae Balissae has remained conjectural, but was estimated as covering an area of 100–120 m in diameter. The importance of the city was based on its healing thermal springs and was enhanced by its strategic location in between the Sava and Drava Rivers as the main traffic arteries of the region. Still, the presence of the military either at the beginning of Roman rule or at any later point remains obscure despite epigraphic testimonies of veterans and even active soldiers in the archaeological evidence.[11]

In this book, published and unpublished funerary monuments of all kinds (*stelai*, sarcophagi, ash-chests, *tituli*, altars, medallions, and funerary buildings) dating from the 1st century to the late Roman period are systematically studied through an interdisciplinary approach comprising archaeological, epigraphic, and geological analyses. A total of 200 monuments were considered, of which 143 from Siscia, 30 from Andautonia and 27 from Aquae Balissae, comprising 48 *stelai*, 39 sarcophagi, 54 ash-chests, five *tituli*, five altars, two funerary medallions, 13 funerary buildings, and 30 unclassifiable pieces, some of which no longer exist. There is no doubt that inclusion of all categories of monuments, in addition to funerary, would have widened the interpretative potential of the evidence, but this was not done for two reasons. First is because the Project leader's main research interest is in Roman funerary monuments of northern

[11] Schejbal 2004; Migotti 2017c.

Figure 2. Map of south-western Pannonia (after Migotti 2017a)

Croatia, which resulted in quite a few publications of individual pieces or thematic groups before the start of the Project research. This has enabled a fairly sound overview of the published pieces, providing at the same time the tools to deal with the remainder of the evidence. Also, due to such circumstances enough time remained within the research protocol for the search of unpublished or mostly unknown monuments hidden in museum depots, or out in the open at various places throughout northwest Croatia and northwest Bosnia and Herzegovina. Second, including votive and other monuments would have much strained the research, which was quite limited by the Project's time protocol. Finally, funerary monuments make the most numerous category of Roman stones preserved everywhere in the Roman Empire, the Norico-Pannonian region included, which in a way makes them a representative sample for studying a provincial Roman society.[12] On balance, the research was conceived of as a pilot study in nature, with the possibility for building on it through including other categories of monuments and also widening the territorial scope of study.

[12] O. Harl 1991b: 17; Hope 2001: 1; Carroll 2006: 15; Beltrán Lloris 2015: 95; Hemelrijk 2015: 33; Mitthof and Cenati 2016: 120; Djurić, in press.

# Chapter I
# Catalogues

Branka Migotti and Marjeta Šašel Kos

***Abbreviations of museums, collections and institutions***

AMZ = Arheološki muzej u Zagrebu/The Archaeological Museum in Zagreb
GMB = Gradski Muzej Bjelovar/The Museum of Bjelovar
GMS = Gradski muzej Sisak/The Museum of Sisak
GMK = Gradski muzej Karlovac/The Museum of Karlovac
GMP = Gradski muzej Požega/The Museum of Požega
GMV = Gradski muzej Varaždin/The Museum of Varaždin
HGI = Hrvatski geološki institut/The Croatian Geological Institute, Zagreb
MK = Muzej Kozare/The Museum of Kozara, Prijedor (Bosnia and Herzegovina)
ZMBiH = Zemaljski muzej Bosne i Hercegovine/The National Museum of Bosnia and Herzegovina, Sarajevo
ZJD = Zbirka Janković/The Janković Collection, Daruvar

***Other abbreviations***

Cities and territories: A (Andautonia), AB (Aquae Balissae), S (Siscia), t (the presumed territory), pt (the possible territory); measurements: H (height), L (length), T (thickness), TB (thickness of the bottom), W (width); monument parts: lion pediment (the pediment featuring a lying lion with a ram's head in his paws along each of the sides), lion top (a horizontal parallelepiped with two lions lying back to back by the sides of a central element).

***Catalogue structure***

The territory included comprises the presumed and possible areas of the three municipalities of south-western (SW) Pannonia within the modern Croatian borders, as well as some bordering areas clearly belonging to municipalities other than those three. The latter were included in order to check the evidence that most probably, or even securely, did not belong to the three municipalities discussed in this book. It was hoped that a comparison between these two sets of data would help to observe differences and similarities between them, as a possible indicator of the territorial-administrative affiliation of the areas quite away from the municipal centres, or at least of workshop profiles or influences operating there. Apart from inconclusive territorial affiliations related to outside municipalities, some of the findspots remained ambiguous in terms of belonging to one or other of the three municipalities discussed here. All this will be discussed against the background of the well-known fact that even the presumed territories of Roman cities mostly remain tentatively defined (see Chapter II.1).

SISCIA (Sisak)
The cities (S); the presumed territory of Siscia (S[t]); the possible territory of Siscia (S[pt])

ANDAUTONIA (Ščitarjevo)
The city (A); the presumed territory of Andautonia (A[t]); the possible territory of Andautonia (A[pt])

AQUAE BALISSAE
The city (AB); the presumed territory of Aquae Balissae (AB[t])

PERIPHERAL AREAS

I. Metulum, south of Siscia (the province of Dalmatia)
II. Poetovio, NW of Andautonia (the province of Pannonia)
III. Unknown municipality south of Aquae Balissae (the province of Pannonia)

The types of monuments included are arranged in the following order: I. *Stelai*; II. Sarcophagi; III. Ash-chests; IV. *Tituli*; V. Altars; VI. Medallions; VII. Funerary buildings (*aediculae*/mausoleums, underground chambers); VIII. Unclassifiable. Fragmented monuments without a preserved inscription were included only if they could be reasonably attributed to the funerary category and possibly affiliated with the type (stele, sarcophagus, etc.) within this category. Given that the discussion of the population is one of the research aims, included also were fragmentary inscriptions if they are preserved enough to reveal either the category or some onomastic data, preferably both. Group VIII comprises monuments with no clear determination for two reasons: a fragmentary state concealing the shape, or the shape that could have made a part of various monument kinds. *Tituli* make a fairly evasive category, as it is not always easy to assess whether they were independent inscriptions or parts of larger monuments. Altar shafts are classified as altars. For a more detailed discussion of the controversies of classification, see Chapter II.3.1.

### *Additional explanations of catalogue entries*

The entries comprise the following components: *Provenance*, *Location*, *Stone* (material), *Preservation* (the state of preservation), *Description* (with measurements), *Inscription* (with commentaries), *Date*, *Publication*, *Notes* and *Illustration(s)*. Whenever a component of the catalogue entry is omitted, it means that no data whatsoever exists.

*Provenance:* A small number of finds stem from known (original) archaeological contexts; the majority were recovered from secondary findspots, and are tentatively related with the cemeteries established or presumed with a high degree of certainty in the vicinity. Thus, the term findspot in most cases means a secondary finding place. Distances between places are given as measured in a straight line from the three municipal centres (Siscia/Sisak, Andautonia/Ščitarjevo and Aquae Balissae/Daruvar) respectively, towards the sites in their presumed or possible territories. A considerable number of entries in the Catalogue Siscia concerning the area of Kordun (the '*possible territory of Siscia in the south-west*') was sourced nearly exclusively from Perkić 2012, with the provenance given as the 1997–2001 research, although individual pieces were actually spotted or registered at different points in the given chronological span. The research in question was conducted by Domagoj Perkić under the auspices of the Directorate for the Protection of Cultural Heritage, Conservation Department in Karlovac, in the period 1997–2001.
*Description:* Italic type in measurements denotes the preserved, and not the original, dimensions of fragmentary monuments. However, monuments considered as completely preserved are those which may be fragmentary, but still with all dimensions reconstructable. The *stelai* are considered as complete even when they are found without pedestals; as a matter of fact, the latter are very rarely found (cf. S[t] I.4; A [t] I.6 and 8; (AB[t] I.6). The portraits are defined as busts or half-figures, the latter featuring the arms, which are of necessity missing in busts. The artistic rendering of the monuments was not estimated against a direct comparison with Roman Imperial art, but was judged by provincial standards, although this is not explicitly stated in the descriptions. This was not only because Roman sculpture in Pannonia, as well as in the majority of the provinces, generally does not measure up to the city-of-Rome artistic production. Provincial arts have their own inherent qualities which do not necessarily result from the lack of skill in faithfully copying Roman models; they could have been conditioned by economic limitations, or could have been an expression of autonomous artistic conceptions, something in the line of A. Riegel's *Kunstwollen*.[1] Within such parameters, the majority of the present evidence falls between average and low quality, with rare pieces that stand out. Under the title *Sarcophagi* from Siscia and Andautonia some examples are marked by a superscript star (*). These are chests that cannot be positively classified as either sarcophagi or ash-chest, because of the discrepancy between their measurements indicative of sarcophagi and the considerable thickness of their bottoms, suggestive of ash-chests (see Chapter II.3.1).
*Date*: Within each kind, the monuments are roughly arranged in chronological order, as far as this was possible given the many obstacles to a more precise dating. All centuries refer to AD dating, without this acronym being adduced. The terms *Late Roman* and *Late Antique* are considered as alternative, so the former is used throughout, referring to the 4th and 5th centuries if there are no indications for a more precise date within this time-span. Except in quite rare occurrences of dating based on historical/epigraphic data, such as in S II.2 (*colonia Septimia*) or AB(t) I.5. (*legio IIII Flavia Antoniniana*), dates are mostly based on general parameters, such as monument kind and/or type, portraits and clothes, and, in lesser degree, art-historical analysis of motifs, as well as style and execution techniques. In inscribed monuments epigraphic and palaeographic traits were used as a dating tool, but again, restrictedly. Dating is commented upon in an entry only if this was demanded by specific circumstances, mostly in the cases of opposing dates in previous publications, or in the above-mentioned cases of epigraphic and historical grounds for the chronology. More extensive analyses of dating parameters are given in Chapter II.3.1.
*Publication*: The majority of the finds are mentioned in the literature, but very often just briefly mentioned, and not rarely with incorrect attributions and/or dating. On the other hand, many of them are published here for the first time in terms of a detailed analysis. The literature given in the catalogue entries is not all-inclusive, but is rather based on two criteria: the importance of the works and their recent date, the latter comprising their coverage in earlier works.

SISCIA

### *The town of Sisak (S)*

*I. Stelai*

1. Stele (S I.1)
*Provenance:* Building of the embankment of the Sava River in 1935, at the site of Štakori, approximately 1 km NE of Sisak (possibly N cemetery).
*Location:* AMZ, inv. no. AMZ KS-968.
*Stone*: Upper-Cretaceous biocalcarenite (rudstone-floatstone). A possible origin: the Outer Dinarides, Appenines and southern Alps; a probable origin:

[1] Cf. O. Harl 1991a: 4, 18–19; Ertel 2010: 167–168; Kremer and Kitz 2016: 80.

Figure 1. (S I.1) – stele (AMZ, I. Krajcar)

Figure 2. (S I.1) – stone, microscope (HGI, M. Belak)

the Aurisina quarries, NE Adriatic (M. Belak 2018; C. Mazzoli, 2018).

*Preservation*: Fragmentarily preserved and partly restored. The pediment tip with the hole for a finial and the lower part of the stele below the inscription are missing. The right side moulding of the pediment is chipped and there are diagonal cracks on the inscription panel.

*Description*: *H 130* cm, W 71.5–73.5 cm, T 20.5 cm; type 1. The preserved part of the stele is composed of a moulded pediment with a 12-petal rosette in high relief in its lower middle part, and a plain inscription panel of a sub-rectangular shape, slightly widening towards the bottom. The workmanship is neat and the letters precisely cut.

*Inscription*: *T(itus) Tullius T(iti) f(ilius) / Pup(inia) Terg(este) / Tert(ius), an(norum) LV, / h(ic) s(itus) e(st).*

Elegant letters of uneven height (5.4–12 cm), with punctuation marks between the words. *Litterae longae*: line 1: first and last *T*; line 2: *T*; line 3: *T* and *L*.

Translation: Titus Tullius Tertius, son of Titus, of the voting-tribe *Pupinia*, from Tergeste, 50 years old, lies here.

Commentary: Tertius must have been related to the Tullii Crispini, well documented in Tergeste and in the valley of the Rižana, where they had their property; their stamps are attested on the amphorae of Dressel 6B type, exporting olive oil to Venetia and Danubian provinces (Zaccaria 2015: 286–287). The Tullii, who are attested in several provinces, are particularly well documented in northern Italy, but also in Dalmatia and Pannonia (*OPEL* IV: 132–133). The Latin *cognomen* Tertius is spread everywhere, but it was most popular in northern Italy and Noricum, as well as in Gallia Narbonensis, Dalmatia, and Pannonia (*OPEL* IV: 114–115; cf. Kakoschke 2012: 688–690 for Noricum).

*Date*: First half of the 1st century.

*Publication*: Klemenc 1935; *AE* 1935:162; *AIJ* 575; Mócsy 1959: 212, no. 57/19; Zaccaria 2015: 298 no. 6. Internet databases: *lupa* 3807; *HD*023446; *EDCS*-11301126.

*Notes*: On account of the state of preservation and the findspot away from any burial context, J. Klemenc presumed that the stele was broken during unloading, never having reached the cemetery. Given, however, that its findspot is quite close to the N cemetery, the stele should still be tentatively associated with it.

2. Stele (S I.2)

*Provenance*: Digging of a lime pit in 1874 in the area of the N cemetery; recovered together with a cylindrical urn holding bones and ashes.

*Location*: AMZ, inv. no. AMZ KS-346.

*Stone*: Upper-Cretaceous/Paleogene micritic limestone. A possible origin: the mountains of NW Croatia; a probable origin: the Banovina, S of Sisak (M. Belak 2018).

*Preservation*: Fragmentarily preserved. The lowermost part of the base is missing, and the right corner of the pediment is chipped off and restored. The stele is weathered throughout, with considerable cracks and chips on the pediment, inscription panel, and base.

*Description*: *H 175* cm, W 70 cm, T 21 cm; type 2. The stele is composed of the following segments: a pediment contained within a pentagonal upper part, holding a large eight-petal rosette in high relief, and with tendrils in the spandrels' corners; an inscription panel flanked by spirally fluted columns with Attic bases and composite capitals; a smooth base with relief remains

Figure 3. (S I.2) – stele (AMZ, I. Krajcar)

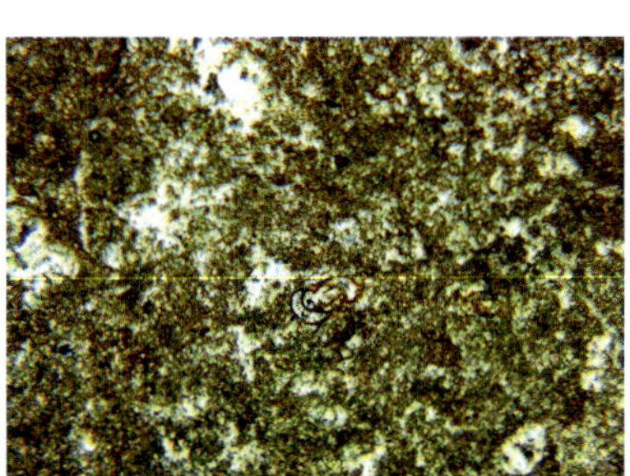

Figure 4. (S I.2) – stone, microscope (HGI, M. Belak)

of a sea monster. The workmanship on the pediment is better than on the somewhat crudely carved columns.

*Inscription*: *M(arcus) Mucius / Hegetor, / medicus / coh(ortis) XXXII vol(untariorum), /$^{5}$ an(norum) XXXXVII / h(ic) s(itus) e(st). / Mucia Corin/thia sibi et / patrono p(osuit).*

Irregular letters of uneven height. Letters, as well as the *ductus* of lines are slightly uneven, lines are not straight. There are punctuation marks between the words.

Translation: Marcus Mucius Hegetor, physician of the 32nd cohort of volunteers, 47 years old, lies here. Mucia Corinthia erected (the monument) for herself and her patron.

Commentary: Hegetor's cohort had presumably been stationed in Siscia in the 1st century AD, until the reign of Vespasian, when it was transferred to Germania (Lőrinz 2001: 44; 57–58; Radman-Livaja 2012: 171). Hegetor was recruited in one of the eastern provinces; his and his freedwoman's cognomina are Greek. The name Mucius also appears on the *tesserae*, on which it may or may not have been a *gentilicium* in a few instances; however, on more than 40 of them, Mucc(i) us is a personal name, resulting in the most frequently attested name on the lead tags (Radman-Livaja 2014: 233–234). He may have been the owner of a fuller's shop or somebody employed in it. Mucius is a rather rare Latin *nomen*, documented several times in Italy, Africa proconsularis, and Numidia, and a few times elsewhere, notably in Dalmatia (Solin and Salomies 1994: 122; Alföldy 1969: 100; *OPEL* III: 89; *Epigraphik-Datenbank* Clauss/Slaby).

*Date*: Mid 1st century. This chronology, suggested by B. Lőrinz on historical grounds, was challenged by J. Spaul (2000: 47–48), who claimed the date at the end of the 2nd century, but with inadequate argumentation. The early date is further substantiated by the stele's structural typology (cf. Djurić 2008: 161–162).

*Publication*: *CIL* III 10854; *ILS* 2601; Brunšmid 1909: 154–155, no. 346; Schober 1923: 55; Gummerus 1932: no. 392; *AIJ* 567; Mócsy 1959: 211, no. 57: 3; Davies 1969: 98 no. 54; Lőrinz 2001: 297, no. 479; Gračanin 2004: 36 no. 28. Internet databases: *lupa* 3073; *HD*072121; *EDCS*-29000476.

3. Stele (S I.3)

*Provenance*: Construction works in 2009 at no. 7 Kvirinova Street, in the area of the N cemetery.

*Location*: GMS, inv. no. GMS-22659.

*Stone*: Badenian biocalcarenite. A probable origin: the Banovina, S of Sisak (M. Belak 2018).

*Preservation*: Fragmentarily preserved, with the lower and upper parts missing. The stele is damaged around the edges, the lower corners broken off, and there is staining and surface pits at places – the smaller ones natural and the bigger ones caused by damage.

*Description*: *H* 95 cm, *W* 77 cm, T 28 cm. The surviving fragment of the stele contains a nearly completely preserved moulded inscription panel and the bottom of a relief niche, featuring the following motifs, from left to right: the lowest segment of a much damaged

Figure 5. (S I.3) – stele (GMS, B. Suntešić)

Figure 6. (S I.3) – stone, microscope (HGI, M. Belak)

smooth column with a Tuscan base, a human figure of which only the lowest part of a bell-shaped tunic and the lower legs in profile striding towards the right survive, a claw-footed round tripod placed centrally in front of a couch and flanked by two large vessels standing on the floor (a one-handled jug on the left, and a square two-handled bottle on the right side), and, at the far right, a baluster-type leg. Although very fragmentary, the scene can be safely restored as representing the deceased reclining on a couch (*klinē*) and a single servant turned towards him. The stele can be reconstructed tentatively as containing a rectangular or a slightly arched relief niche, an inscription panel below, and possibly, although not obligatory, a further relief niche between the inscription panel and the base; the latter is also not obligatory, nor are columns flanking the niche(s) and inscription panel (cf. *lupa* 88, 7071, 7073, 15873).

*Inscription*: *C(aius) Antonius / Sentinus, / vet(eranus) leg(ionis) XIIII / G(eminae) M(artiae) V(ictricis), an(norum) L. /[5] H(eres) f(aciendum) c(uravit). / H(ic) c(onditus) (est).*

Translation: Gaius Antonius Sentinus, veteran of Legion *XIV Gemina* (composed of two) martial (and) victorious, 50 years old. The heir had (this monument) made. He is buried here.

Commentary: To explain *H. C.* in the last line as *heres curavit* (the heir commissioned it) is unlikely, as this would be a repetition of the formula from the previous line. The two letters could perhaps be supplemented as *honoris causa* (cf. Calabi Limentani 1991: 495), although the formula is rare in a funerary context, and is mainly attested in the eastern provinces, but an expanded formula *honoris causa* is documented in an epitaph from Novae in Moesia Inferior (present-day Svištov in Bulgaria: *CIL* III 756).

Sentinus' family name is not particularly characteristic and he was perhaps recruited around the mid 1st century AD in northern Italy or some early Romanized western province, where one of the Celtic languages had been spoken. This could be inferred from his otherwise Latin cognomen (Solin and Salomies 1994: 400), which may have been an assonance name among the Celtic-speaking indigenous inhabitants (Delamarre 2007: 166; on this phenomenon see Raepsaet-Charlier 2012). One Antonius Sido has been attested on a *tessera*; he or perhaps his ancestors originated from one of the eastern provinces of the Empire, where the Antonii are well documented (Radman-Livaja 2014: 163). The Legion *XIV Gemina* had been transferred, most likely in AD 9, from Illyricum to Mogontiacum in Upper Germania, from where it was sent in AD 43 to Britannia. Under Nero, in AD 61, it earned the titles *Martia Victrix* and remained in the province until AD 70, except in AD 68, when it was temporarily sent either to Pannonia or Dalmatia. The legion was again in Mogontiacum between *c.* AD 70 and 92, or slightly afterwards, when it was transferred to Carnuntum in Pannonia, due to the war against the Marcomanni and Quadi, begun by Domitian in AD 96 and ended by Nerva (Franke 2000: 197–198; Strobel 2010: 155; Radman-Livaja 2012: 170–171). Under Domitian, the legion could have been briefly stationed at Mursella near Mursa. Possibly Sentinus served in a vexillation of this legion, which may have been temporarily stationed at Siscia (Radman-Livaja 2012: 170, 175; Dušanić 1983: 20–21 [= 2010: 958–959]; Hoti 1992: 143; Škrgulja and Migotti 2015: 39).

*Date*: Late 1st century or early 2nd century.

*Publication*: Škrgulja and Migotti 2015.

4. Stele (S I.4)

*Provenance:* Unknown; possibly stemming from the E cemetery.

*Location*: GMS, inv. no. GMS-27775.

*Stone*: Most probably, Pohorje marble (W. Prochaska 2017).

*Preservation*: A fragmentarily preserved pediment, broken off on all sides and partly damaged on the surface.

Figure 7. (S I.4) – stele (GMS, B. Suntešić)

*Description*: *W 45* cm, *H 20* cm, T 10 cm. The rectangular upper part of the stele contains a moulded pediment holding a central rosette in the shape of a segmented relief roundel, and a dolphin with down-turned head in each of the spandrels.
*Date*: Later 1st century or 2nd century (probably Flavian).
*Publication*: Škrgulja 2018: 245, no. 268.

5. Stele (S I.5)
*Provenance*: Demolition of a grain warehouse in 1954 in the mid-west part of the town (within the town walls).
*Location*: GMS, inv. no. GMS-14 AZA.
*Stone*: Badenian siliciclastic, slightly sandy biocalcarenite. A probable origin: the Banovina, S of Sisak (M. Belak 2018).
*Preservation*: Fragmentarily preserved, with the upper and lower parts missing. The inscription panel is much weathered and pitted, with the inscription only partly legible, and with chippings around the edges.
*Description*: *H 87* cm, W 56 cm, T 18 cm. The vertical, rectangular inscription field is moulded, with a stump of a smooth panel beneath.
*Inscription*: *---] / et Qu[---] / uxori [..] / ++natu / ++[...] f(ilio?) / M++R++++ / an(norum) XXX. / H(ic) s(iti) s(unt).*
Translation: ... and Qu[], to (his) wife ... 30 years old. They lie here.
Commentary: Possibilities to restore names starting in *Qu* are many; *Quadratus*, *Quartinus*, *Quartus*, and *Quint(us* or *-illus)* have been attested at Siscia to date, including the evidence from the *tesserae* (Radman-Livaja 2014: 250–251).
*Date:* End of the 1st century or 2nd century.
*Publication*: Internet databases: *lupa* 22450; *EDCS*-62500018.

Figure 8. (S I.5) – stele (GMS, B. Suntešić)

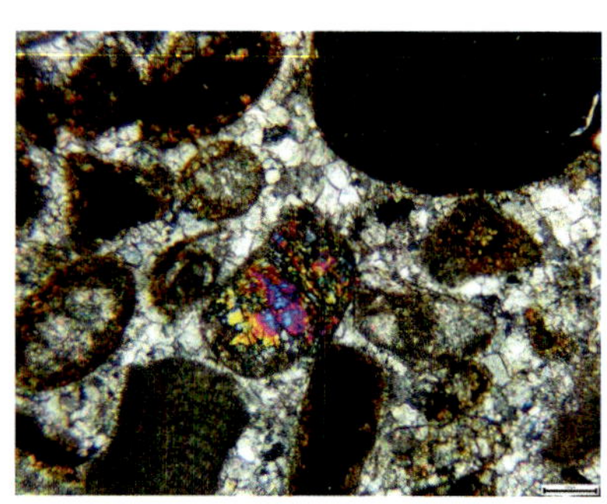

Figure 9. (S I.5) – stone, microscope (HGI, M. Belak)

6. Stele (S I.6)
*Location*: AMZ, inv. no. AMZ KS-108.
*Stone*: Salla marble (W. Prochaska 2017).

Figure 10. (S I.6) – stele (AMZ, I. Krajcar)

Figure 11. (S I.7) – stele (GMS, B. Suntešić)

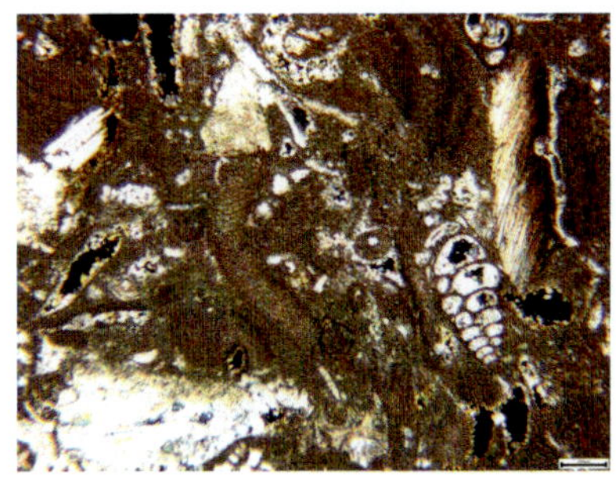

Figure 12. (S I.7) – stone, microscope (HGI, M. Belak)

*Preservation*: Fragmentarily preserved. The majority of the base and a smaller part of the tenon survive, with the surface moderately weathered.
*Description*: *H* 49 cm, *L* 54 cm, T 10 cm. The rectangular, simply framed relief panel features a centaur in a staggering position, depicted with the frontal torso and the face and horse legs in profile to the right. His left hand remains hidden behind a willowing drapery, while he is holding a tree branch in his uplifted right hand and is brandishing it towards an enemy on the right-hand side, of whose figure only an indistinct fragment of a silhouette remains. A thin stick-like object between the centaur and the silhouette has been interpreted as a spear piercing the former's chest, but such interpretation remains questionable. The workmanship is somewhat crude, but still with sense of proportion, movement and natural shapes.
*Date*: 2nd century or first half of 3rd century.
*Publication*: Brunšmid 1905: 55–56, no. 108; Djurić 2013: 14, no. 61.

7. Stele (S I.7)
*Provenance*: A chance find at an unspecified time in the 1950s or 1990s, in the area of the SE cemetery.
*Location*: GMS, inv. no. GMS-3123 AZA.
*Stone*: Badenian biocalcarenite. A possible origin: the mountains of N Croatia (M. Belak 2018).
*Preservation*: Fragmentarily preserved. The upper part of the portrait niche, whose left border is missing, and the lower part of the blind (?) inscription panel survive. The surface is heavily pitted and weathered all over, with the portrait traits much obliterated.
*Description*: *H* 70 cm, *W* 70 cm, T 30 cm. The preserved part of the monument is composed of a relief niche flanked by pillars or a smooth and thick band-like border, of which only the right one partially survives. The relief contains half-figures of two small persons, probably a boy on the left and a girl on the right side, standing in front of their parents. The child on the left side, somewhat smaller than the one on the right, has a roundish, slightly triangular face, and the hair combed from the forehead towards the crown, leaving his ears visible. He is wearing a tunic with a rounded neckline and a cloak draped in V-shaped folds over the chest, fastened with a round brooch (?) on the right shoulder, folded over the left shoulder and hanging down the left arm, which is bent at the elbow. Another possibility, but somewhat less likely, is that the folds on the boy's left shoulder represent his mother's or his father's hand in the gesture of embracing. The child on the right-hand side has a roundish face with indiscernible hairdo, and is wearing a mantel (*palla*?) hanging in rounded folds across her chest. The workmanship seems to be mediocre, but this is hard to assess properly due to a heavy wear of the surface throughout.
*Date*: 3rd century (?)
*Publication*: Škrgulja 2018: 245, no. 267. Internet databases: *lupa* 22453.

8. Stele (S I. 8)
*Provenance*: Infrastructural works in 1787 in the area of the SE cemetery.

Figure 13. (S I.8) – front of the stele (AMZ, I. Krajcar)

Figure 14. (S I.8) – right short side of the stele (O. Harl, *lupa* 8817)

*Location*: AMZ, inv. no. AMZ KS-931.
*Stone*: Gummern marble (H. Müller, 2013).
*Preservation*: Fragmentarily preserved. The topmost part of the pediment is cut off and the lower part of the inscription panel is missing. There is some chipping on the frames, lion-bodies and columns, with the faces of the portrayed people seriously abraded, and the inscription panel cracked in the middle.
*Description*: *H 134.5* cm, W 87.5 cm, T 22 cm; type 5a (?). The preserved part of the stele contains the following segments: a lion pediment with a central Medusa head between two acanthus leaves; a rectangular niche flanked by columns with Corinthian capitals, Tuscan bases and shafts spirally fluted in the upper part and scaled in the lower, holding half-bodies of the deceased set against a background of a ribbed scallop shell; an inscription panel flanked by (mostly destroyed) columns with Corinthian capitals, divided from the portrait niche by a smooth narrow band. Half-figures of five persons are depicted in the portrait niche: a young man in the left background, a woman slightly in front of him, another (older) man to the right of the woman, and two girls in the foreground. The man on the far left has an oval, elongated, slightly triangular face with down-turned lips and a short-cropped hairstyle with an M-shaped fringe. He is wearing a long- and wide-sleeved tunic and a *sagum* fastened with a T-shaped brooch on the right shoulder, and is holding a scroll in his left hand while pointing at it with his right. The woman has a block-like face, a low and tight wrapped headdress, large roundish locks on the forehead and short geometricized strands of hair at the sides, leaving the ears uncovered. She is dressed in a long- and wide-sleeved tunic and a cloak hanging on her shoulders, and is wearing a beaded necklace with apparently a matching bracelet on the right hand, in which she is holding a small roundish object, probably a fruit. The

man on her right has an elongated face and oversized ears, and a short-cropped hairstyle with an M-shaped fringe. He is dressed in a long- and wide-sleeved tunic and a fringed *sagum* fastened with a T-shaped brooch on the right shoulder, and is holding a scroll in his left hand. In the foreground two girls are depicted, a smaller one in front of the woman and a bigger one embraced by both the woman and the man on the far right. Both are dressed in long- and wide-sleeved tunics, wearing the same headgears and hairdos as their mother; the bigger girl is wearing also the same necklace and is holding a diptych and a stylus, and the smaller one is holding a wreath. The short sides are decorated with a vine tendril with grapes. The topmost part of the pediment is cut off so that only the lions' hind legs are missing, but a rectangular dowel hole at either end of the upper surface renders the monument's complete restoration uncertain. Given the measurements' ratio, the stele could have had a base beneath the inscription panel, but lacking a broader decorative band between the portrait niche and the inscription panel, and thus probably featuring a reduced variant of type 5a. The workmanship is crude, but still fairly accurate in the treatment of the drapery and human figures, apart from some disproportions in rendering the hands and ears of the man on the far right.

*Inscription*: *Domu(!) (a)etern(a)e et perpe/[t]u(a)e securitati. C(a)enius / [---]ianus v(ir) e(gregius) v(ivus) f(ecit) Fl(avio) Tibe/[rino?/riano?] ṿ(iro)? [e(gregio)?] q(ui) v(ixit) an(nos) XXX /*[5] *[---]ni(a)e Urs(a)e q(uae) v(ixit) an(nos) / [---]n(a)e q(uae) v(ixit) an(nos) IIII / [---]q(uae) v(ixit) an(nos) / [---*

Ligatures: 1: TE, NE, ET. Lines 3 and 4: reading of Branka Migotti (forthcoming); *Tibe/[rian]i[co?]*: earlier editors; *Fl(avio) Tibe/[riano ---]*: *lupa*.

Translation: To the eternal home and lasting security. Caenius [...]ianus, *vir egregius* (of the equestrian order) erected (this monument) in his lifetime for Flavius Tiberianus (?) *vir egregius* (?), who lived 30 years (or more), for ... Ursa, who lived ... years, for ..., who lived 4 years, .... who lived ... years ...

Commentary: For the abbreviated (?) or wrong form *domu(i)* see an inscription from Ostia (*CIL* VI 3651 = *EDCS*-05700228). *Domus aeterna* and *securitas* are also mentioned on a tombstone from Aquileia (*Inscr. Aquil.* 2980 = *EDCS*-08000212). C(a)enius is a rare *gentilicium*, a few times attested in Italy (three times in Tarquinia and once in Aquileia), twice in Gallia Narbonensis, once in Dalmatia and Sicilia, and three times in Pannonia (Solin and Salomies 1994: 40 [Caenius]; 52 [Cenius]; *OPEL* II: 18, s.v. Caenius; in Aquileia: *Inscr. Aquil.* 938). Tiberinus or Tiberianus (not particularly rare: *OPEL* IV: 121), were cognomina derived from *praenomina* (Kajanto 1965: 175). Ursus (Ursa) was a popular name everywhere throughout the Roman period, also among Christians, and may have been a 'translation' name in the Celtic-speaking regions (*OPEL* IV: 187–188; Lochner von Hüttenbach 1989: 172; Minkova 2000: 169).

*Date*: Second half of 3rd century.

*Publication*: *CIL* III 3985; Gregl and Migotti 2000; *AE* 1999: 1245; Mander 2013: 297, no. 654. Internet databases: *lupa* 8817; *HD*048572; *EDCS*-26600434.

*II. Sarcophagi*

1. Sarcophagus (S II.1)

*Provenance*: Building works at the beginning of the 19th century at no. 9 Lađarska Street, in the area of the SW cemetery.

*Location*: Built into the house at no. 9 Lađarska Street.

*Stone*: Badenian biocalcarenite. A possible origin: the mountains of N Croatia (M. Belak 2018).

*Preservation*: The chest fragmentarily preserved, with the back and short sides missing. The surface of the front is badly weathered and porous throughout, with staining at places.

*Description*: L 203.5 cm, H 71 cm, W?; type 1. The chest is rectangular, with plain lateral fields flanking a moulded inscription panel.

*Inscription: D(is) M(anibus) v(iva) f(ecit) / Aelia [P ?]ublii filia Matro/na sibi et suis.*

Figure 15. (S II.1) – sarcophagus (M. Jambrović)

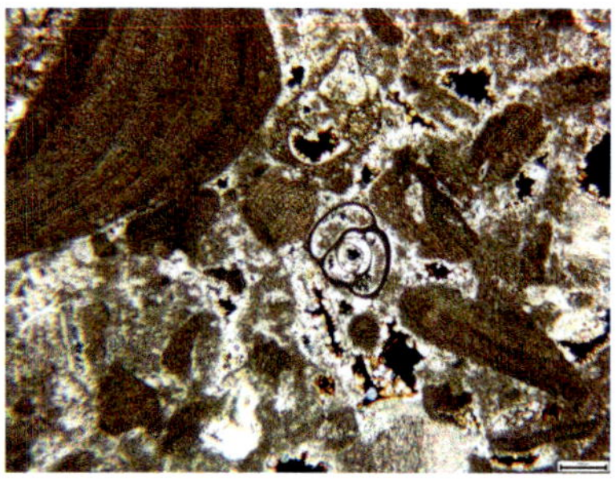

Figure 16. (S II.1) – stone, microscope (HGI, M. Belak)

Line 2: *Aelia Pia matrona*: Proházka 2013: 293 fig. 3; *AE* 2013: 1205; *EDCS*; this reading does not correspond to the existing letters.
Translation: To the spirits of the departed. Aelia Matrona, daughter of Publius, made (this monument) in her lifetime for herself and her family.
Commentary: According to Mócsy 1959: 57/6 (p. 211), a native woman. Several inhabitants of peregrine status were given Roman citizenship under Hadrian.
*Date*: Second half of 2nd century.
*Publication*: *CIL* III 3981; *AIJ*: p. 238; Vuković 1994: 81, no. 59; Burkowsky 1996: 70, n. 13; Buzov 2002: 177–178, fig. 4; Gračanin 2004: 26 no. 13; Migotti 2013b: 195–196, type II.5.1; Prohászka 2013: 293 fig. 3; *AE* 2013: 1205. Internet databases: *lupa* 22319; *HD*074384; *EDCS*-28800744.

2. Sarcophagus (S II.2)
*Provenance*: A chance find at the beginning of the 19th century at no. 9 Lađarska Street in the area of the SW cemetery.
*Location*: Built into house at no. 9 Lađarska Street, Sisak.
*Stone*: Badenian biocalcarenite. A possible origin: the mountains of N Croatia (M. Belak 2018).
*Preservation*: The chest fragmentarily preserved, with the back side missing. The broken off left spandrel of the left niche has been restored without tendrils, while the pilaster's capital has been restored as featuring two superimposed bands decorated (apparently arbitrary) with relief rectangles, therefore mismatching the right pilaster's capital. The surface is stained on the front and short sides at places, the heads of all figures are damaged with weathering and abrasions, and there is natural porosity especially in the left niche and all over the inscription field.
*Description*: L 205 cm, H 71.5 cm, W 68 cm; type 2. The front of a rectangular chest is holding a moulded inscription panel and two lateral arched niches with concave backgrounds, each featuring a small, nude, winged Eros of the 'baby' type (rounded face, wavy hair topped by a knot and framing the face in rich, swirling curls, head bowed, legs crossed) leaning on a reversed torch pulled through a small wreath, and holding the torch with one hand and the wreath with the other. The architectural setting of the front comprises two arched triple-band niches resting on pilasters decorated with two wide and deep vertical flutes in the upper half and several superimposed rows of overlapping scale leafs in the lower. The pilasters are topped by Tuscan capitals composed of two superimposed bands in resemblance to their bases, while the spandrels are each taken by an elongated triangular receptacle sprouting three tendrils. The short sides of the sarcophagus, each provided with a clamp hole, are walled in a continuous line to the left and right of its front, and are each taken by a moulded rectangular niche featuring a Dioscurus. The Dioscuri are depicted in the statuary pose with the legs slightly apart, wearing a cloak fastened with a round brooch on the right shoulder and a *pileus* on the head. Both are leading a horse by the reins with one hand and lifting up a standard in the other. The craftsmanship is difficult to assess due to the wear and damage, but it seems to be of moderate quality. Both the Erotes and the Dioscuri are rendered with a natural look about them and with some sense of movement, proportion and volume, but with a lot of anatomical inaccuracies (oversized heads, stumpy and ill-articulated limbs, especially arms and horses' front legs).
*Inscription*: *D(is) M(anibus). / Pontio Lupo, Aug(ustali) col(oniae) / Sisc(iae), scribae munic(ipii) Faus(tiniani) / Pontia Victorina soror /*[5] *et Domit(ius) Crescens A() D() E()* ++.Translation: To the spirits of the departed. To Pontius Lupus, *Augustalis* of *colonia Siscia* and in charge of public records of *municipium Faustinianum*, Pontia Victorina, his sister, and Domitius Crescens ... (unknown abbreviations).
Line 2: The beginning letters *P* and *L* in *Pontio Lupo* are *longae*. Line 3: *col(oniae) Sisc(ianorum) scribae munic(ipii) Faust(a)*: *lupa* and *EDCS*, less plausible. Line 5: *ET* in ligature. Remains of two more letters seem to be visible after *E*, hence the meaning of the supposedly abbreviated letters is quite uncertain.

Figure 17. (S II.2) – sarcophagus (M. Jambrović)

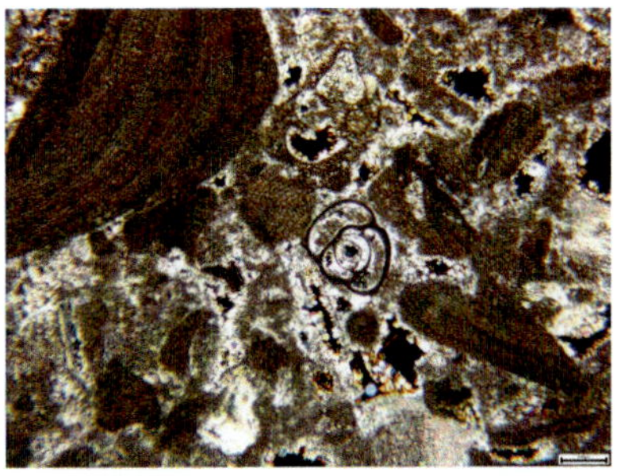

Figure 18. (S II.2) – stone, microscope (HGI, M. Belak)

Commentary: *Colon(ia) Siscia* is mentioned in *CIL* VI 32680 (= *EDR*107064 = *EDCS*-22500003), hence the supplement seems better than *Sisc(ianorum)*. *Pede Faustiniano* is mentioned in a funerary inscription from Rome, erected to Iulius Nero, a soldier from the Third praetorian cohort, whose *origo* is noted as *oriundus in Pannonia superiore pede Faustiniano* (*CIL* VI 2494a = *EDR*104357 = *EDCS*-18800430). An *eques singularis*, Candidius Valentinus, attested in another epitaph from Rome, was *natione Pannonius, cives Faustianus* (*CIL* VI 3241 = *EDR*152488 = *EDCS*-18900495).

The *gentilicium* Pontius is well documented in Italy, and it is very likely that the patron of Lupus and his sister, or his family came from Italy to settle at Siscia. The Pontii are not particularly frequent but are attested in most of the western provinces (Solin and Salomies 1994: 147; Mócsy 1959: 158; Alföldy 1969: 112; *OPEL* III: 153). At Siscia, they must have been a significant family since the name also appears on the *tesserae* (Radman-Livaja 2014: 246). Lupus and his sister probably belonged to a local family. The cognomen Lupus was very widespread in Hispania and Dalmatia, but is not rare elsewhere (*OPEL* III: 39), while Victorinus/a (Kajanto 1965: 71, *passim*) was particularly characteristic of Pannonia after the Marcomannic Wars (*OPEL* IV: 168–169; Barkóczi 1964: 328). Neither occurs on the *tesserae*. Mócsy noted that *scriba* could have been the most important representative of Roman authority in small Pannonian towns (1974: 222).

*Date*: Later 2nd century or first half of 3rd century. The earlier chronology is based mostly on the iconography of the Erotes and Dioscuri, while the later date remains equally acceptable on various grounds (see Chapter II.3.1).

*Publication*: *CIL* III 3974; *AIJ*: p. 238; Šašel 1974: 715, no. 9 (= Šašel 1992: 607); Vuković 1994: 80–81, no. 59; Buzov 2002: 176–177, fig. 2; Gračanin 2004: 25 no. 11; Migotti 2005a; Migotti 2013b: 183–187, type II.1.1; Migotti 2016b; Šačić Beća 2017: 110–111, 153–154. Internet databases: *lupa* 4310; *HD*074367; *EDCS*-28701050.

3. Sarcophagus (S II.3)

*Provenance*: Building works at the beginning of the 19th century at no. 9 Lađarska Street, in the area of the SW cemetery.

*Location*: Built into house at no. 9 Lađarska Street, Sisak.

*Stone*: Badenian biocalcarenite. A possible origin: the mountains of N Croatia (M. Belak 2018).

*Preservation*: The chest fragmentarily preserved, with the back and short sides missing. The front is stained and weathered with abrasion and natural porosity all over, and cracked vertically and diagonally along the entire middle height, with the join filled with greyish cement.

*Description*: L 196 cm, H 71.5 cm, *W*?; type 4. The chest is rectangular, with the majority of the front taken by a moulded inscription panel framed by a wide flat border.

*Inscription*: *D(is) M(anibus). / M(arco) Mulvio Narcisso [an](norum) LX et M(arco) / Mulvio Narcissiano an(norum) XXVI, Aug(ustali) / col(oniae) Sept(imiae) Sisc(iae). /[5] Mulvia Furnia marit[o] et fil(io) carissim(is) v(iva) f(ecit).*

Line 5: Reading the name as Rufina (thus *CIL* and Gračanin) is not plausible, since, as is clear from the good photograph, only the first letter of the second name of the woman is uncertain. In any case, her second name surely ended in *-ia*, although a cognomen is expected, which usually does not have the ending typical of *gentilicia*.

Translation: To the spirits of the departed. To Marcus Mulvius Narcissus, 60 years old, and to Marcus Mulvius Narcissianus, 26 years old, *Augustalis* of the *colonia Septimia Siscia*. Mulvia Furnia had (the tombstone) made in her lifetime for her dearest husband and son.

Commentary: The patron of the Mulvii from this inscription, or patron's family, was almost certainly

Figure 19. (S II.3) – sarcophagus (M. Jambrović)

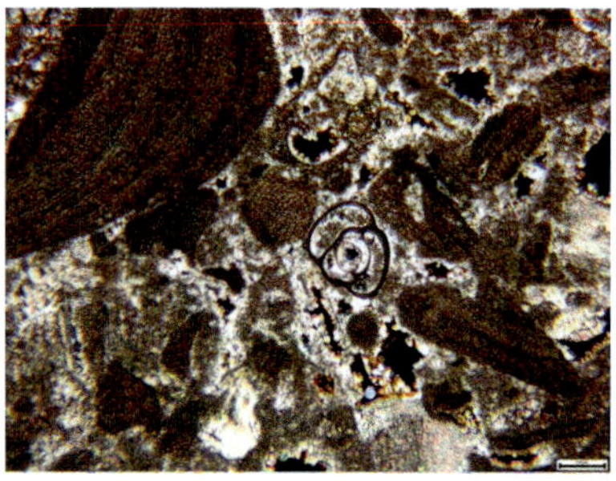

Figure 20. (S II.3) – stone, microscope (HGI, M. Belak)

from northern Italy (according to *OPEL* III: 90, it occurs 23 times in northern Italy and four times in Pannonia). The names Mulvius and Narcissus do not appear on the *tesserae*; the Greek cognomen Narcissus, often attested among freedmen (Lochner von Hüttenbach 1989: 112), does not seem to have been popular before the 3rd century AD. A cognomen *Furnius* is not listed in Solin and Salomies 1994, but a personal name Furnia does occur once each in Pannonia, Moesia Superior, and Inferior (*OPEL* II: 156).

*Date*: 193–235, on account of the title *colonia Septimia*, bestowed on Siscia by Septimius Severus.

*Publication*: *CIL* III 3973; *AIJ*: p. 238; Vuković 1994: 80–81, no. 59; Buzov 2002: 176–177, fig. 2; Gračanin 2004: 24–25 no. 10; Migotti 2013b: 187–189, type II.2.1; cf. Prohászka 2013: 293–294 fig. 5; *AE* 2013: 1205. Internet databases: *lupa* 4313; *HD*074366; *EDCS*-28701049.

*Illustrations*: Photos: 1 – sarcophagus (M. Jambrović); 2 – stone, microscope (HGI, M. Belak).

4. Sarcophagus* (S II.4)

*Provenance*: Rescue excavations in the 1950s in the SE cemetery.

*Location*: GMS, inv. no. GMS-54 AZA.

*Stone*: Badenian sandy biocalcarenite. A probable origin: Mts Ivanščica or Kalnik, NW Croatia (M. Belak 2018).

*Preservation*: The chest completely preserved; stained, weathered, damaged, and cracked at places, especially along the edges.

*Description*: L 225 cm, H 90 cm, W 125 cm, TB 48–55 cm (raising gradually from left to right); type 3a. The chest is rectangular, with the edges cut to receive a lid, and with both the empty inscription field and two curved-backed lateral niches moulded. The niches, topped by Norico-Pannonian volutes, each feature a winged mourning 'baby'-type Eros (plump-body, round chubby-cheeked face, elaborately curled hair, only one wing depicted, legs crossed) in a frontal position standing on a low base, leaning on a reversed torch and resting his head on his raised hand. The workmanship is unskilled, with pronounced anatomical inaccuracies and bad proportions.

*Date*: Second half of 2nd century or early 3rd century.

*Publication*: Burkowsky 2000: 78, 85, no. 30; Buzov 2002: 75; Migotti 2013b: 192, type II.3.4. Internet databases: *lupa* 22448.

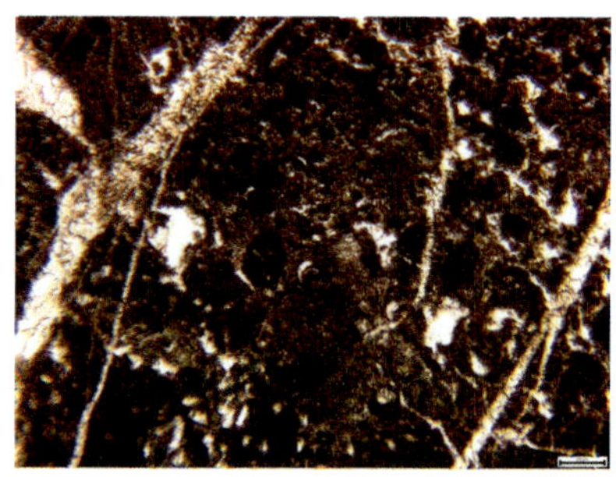

Figure 22. (S II.4) – stone, microscope (HGI, M. Belak)

Figure 21. (S II.4) – sarcophagus (GMS, B. Suntešić)

Figure 23. (S II.5) - sarcophagus (T. Leleković)

5. Sarcophagus* (S II.5)
*Provenance*: Unknown; probably originating from the SW cemetery.
*Location*: Until several years ago in the courtyard of a private house on V. Janića Cape Avenue in Sisak, but gone missing in the meantime.
*Stone*: Probably lithothamnion limestone (macroscopic characterization through a photograph, M. Belak, 2017).

Figure 24. (S II.5) - detail (T. Leleković)

*Preservation*: The chest was completely preserved, with some damage, cracks and staining all over, especially along the edges. A large chip was missing from a top corner of one of the short sides, and a groove was cut through the edge of the opposite short side for reuse. The piece was placed awkwardly behind a tree, which prevented a secure identification of the type. Nevertheless, traces of an architectural moulding at both ends on the front were discernible, suggesting the classification as given below.
*Description*: L 180 cm, H 63 cm, W 85 cm, TB 29 cm; type 4. The roughly worked chest was rectangular with flat edges and a moulded inscription panel flanked by 12-cm-wide flat lateral bands.
*Date*: Second half of 2nd century or first half of 3rd century.
*Publication*: Migotti 2013b: 188–189, type II.2.3.

6. Sarcophagus* (S II.6)
*Provenance*: Building works in 1866 in the area of the SW cemetery.
*Location*: Courtyard of a private house at no. 4 Gundulićeva Street, Sisak.
*Stone*: Badenian biocalcarenite. A probable origin: the Banovina, S of Sisak (M. Belak 2018).
*Preservation*: The chest fragmentarily preserved. Only the bottom with the lowermost part of the inscription panel survives, featuring letters *DRO* of the next to last line of the epitaph. The moulding is well preserved, but the remainder of the surface is heavily pitted and stained.

Figure 25. (S II.6) – sarcophagus (T. Leleković)

*Description*: L 175 cm, *H 42* cm, W 87 cm, TB 42 cm; type 1. The roughly worked rectangular chest has flat edges and a moulded inscription panel flanked by 32-cm-wide plain lateral panels.
*Inscription*: The text, as recorded in *CIL*: *Aurel(iae) Veneriae ob hon(orem) / et flore(m) iuventu[tis] subit(a) / morte occupatae mi(hi ?) pient(issimae) / virgini. P(ublius) Ael(ius) Iulianu[s] /*[5] *ex num(ero) colleg(ii) dendro(phorum) / [v(ivus) f(ecit)].*
Translation: (This monument) was erected in his lifetime by Publius Aelius Iulianus, member of one of the units of the association of timber workers (worshippers of Cybele), to Aurelia Veneria because of her dignity and blossom of her youth, to my most devoted girl, snatched by a sudden death.
Commentary: It is not clear who P. Aelius Iulianus was, certainly not the girl's father, perhaps her (future?) husband. *Ex numero collegii* also occurs in an inscription from Verona (*fabrum*: *CIL* V 3387 = *EDCS*-04202432) and Aquae Sextiae (Aix-en-Provence, *centonariorum*: *CIL* XII 526 [p. 814] = *ILN* III 37 = Liu 2009: 336 no. 21 = *EDCS*-08401106).
The imperial *nomina* reveal that the family of Iulianus, or he himself was awarded citizenship under Hadrian; his reign signified an important phase of urbanization in both Pannonias (Fitz 2003). Iulianus is a frequent cognomen in the western part of the Empire, particularly in the Celtic-speaking provinces (*OPEL* II: 199–200; Lochner von Hüttenbach 1989: 87–88), but also in Dalmatia, where it is not characteristic of the indigenous population (Alföldy 1969: 223–224). The family of Veneria may have acquired citizenship under

Figure 26. (S II.6) – stone, microscope (HGI, M. Belak)

Marcus Aurelius or as late as the *constitutio Antoniniana*. Her cognomen was popular everywhere, particularly in Italy and in the Late Roman period. As a feminine (theophoric) name (*OPEL* IV: 153–154; Alföldy 1969: 323), it was also characteristic of slaves (Lochner von Hüttenbach 1989: 176; Solin 1996: 27–28).
*Date*: Later 2nd century or first half of 3rd century.
*Publication*: *CIL* III 10858; *AIJ*: p. 239; Selem 1980: 201; Gračanin 2004: 36–37 no. 29; Vukelić 2011: 113, 177; Burkowsky 1996: 70, fn. 13; Buzov 2002: 177–178; Migotti 2013b: 196–197, type II.5.2. Internet databases: *lupa* 4318 (without photo); *HD*074532; *EDCS*-29900787.

7. Sarcophagus (S II.7)
*Provenance*: Building works at the beginning of the 19th century at no. 9 Lađarska Street, in the area of the SW cemetery.
*Location*: Built into the house at no. 9 Lađarska Street, Sisak.

Figure 27. (S II.7) - sarcophagus (O. Harl, *lupa* 4311)

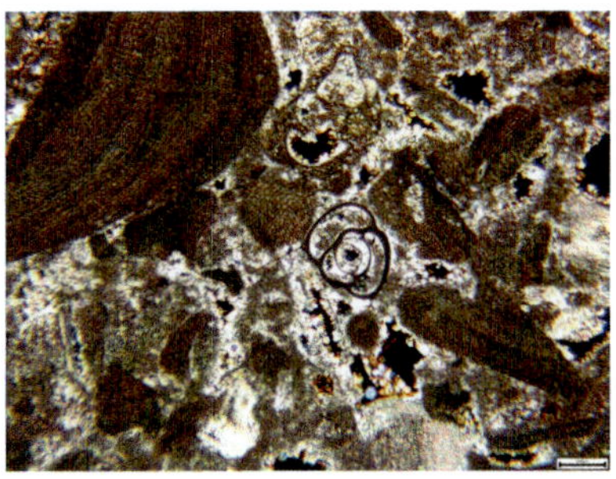

Figure 28. (S II.7) - stone, microscope (HGI, M. Belak)

*Stone*: Badenian biocalcarenite. A possible origin: the mountains of N Croatia (M. Belak 2018).
*Preservation*: The chest fragmentarily preserved. The back and short sides are missing, while the front is heavily stained and weathered with abrasion, and spoiled by natural porosity throughout. Due to the undulating ground surface, the stone is unevenly walled, so that at the left side only a small lowest part is missing, while at the right side missing are Eros' lower legs.
*Description*: L 200 cm, *H* 57 cm (probably H 62 cm); *T?*; type 3a. The rectangular chest holds a moulded inscription panel and equally moulded lateral niches topped by Norico-Pannonian volutes and flanked by wide flat bands remaining from the front surface. Each of the niches feature a mourning winged Eros shown in the guise of a small boy leaning on a reversed torch with one hand and resting his cheek on the other, and with the legs crossed. Their heads are bowed and their plump faces framed with wavy hair rendered in parallel stiff ridges and stopping short of the neck. The workmanship is crude, with pronounced anatomical inaccuracies and stiffness.
*Inscription*: *D(is) M(anibus). / Septumiae Marcell(a) e / qu(a)e vixit ann(os) XX. Cassia / Candida fili(a)e karissi(mae).*
Letters are regular but worn; the lower half of the inscription field was left empty; punctuations marks between the words.

Translation: To the spirits of the departed. To Septumia Marcella, who lived 20 years. Cassia Candida to her dearest daughter.
Commentary: The Cassii are frequent in the Roman Empire, another Cassia is attested on a *tessera*: Cassia Ignastia, Ignastia so far unknown (Radman-Livaja 2014: 182–183). The name Candidus/Candida was particularly well attested among the Sisciani: nine times on stone monuments, as well as on seven *tesserae* (Radman-Livaja 2014: 179). The name Marcellus/Marcella is frequent everywhere, but particularly in northern Italy, Hispania, Celtic-speaking provinces, and Dalmatia (*OPEL* III: 54–55), where it was also borne by Christians (Alföldy 1969: 237). It does not appear on the tags, but Marcellinus/ Marcellina does (Radman-Livaja 2014: 226).
*Date*: End of 2nd century or first half of 3rd century.
*Publication*: *CIL* III 3992; *AIJ*: p. 238; Cermanović 1965: 100 no. 18; Vuković 1994: 81, no. 59; Buzov 2002: 176–177; Gračanin 2004: 31 no. 22; Migotti 2013b: 190–191, type II.3.2. Internet databases: *lupa* 4311; *HD*074392; *EDCS*-28800752.

8. Sarcophagus (S II.8)
*Provenance*: A chance find prior to 1926 probably in Sisak, and possibly in the area of the N cemetery.
*Location*: AMZ, inv. no. AMZ KS-969.
*Stone*: Upper-Jurassic/Lower-Cretacious micritic limestone. A probable origin: the Kordun, SW of Sisak (M. Belak 2018).
*Preservation*: The chest and lid completely preserved. The lid was placed onto the rectangular chest to fit the front line, while its back wall remained protruding by 18 cm. The left half of the lid's back, including the respective *acroterion*, seems to have been cut to fit the line of the chest, but the rest remained intact. The surface is badly weathered with cracks and staining throughout, and with cuts and chips at places, mostly around the edges and on the blind inscription panel.
*Description*: Chest: L 225 cm, H 89 cm, W 104; chest type 3a; lid: L 232 cm, H 37 cm, W 116 cm; lid type 1b. A low-gabled lid with corner *acroteria* and a 17-cm-wide

Figure 29. (S II.8) – sarcophagus (AMZ, I. Krajcar)

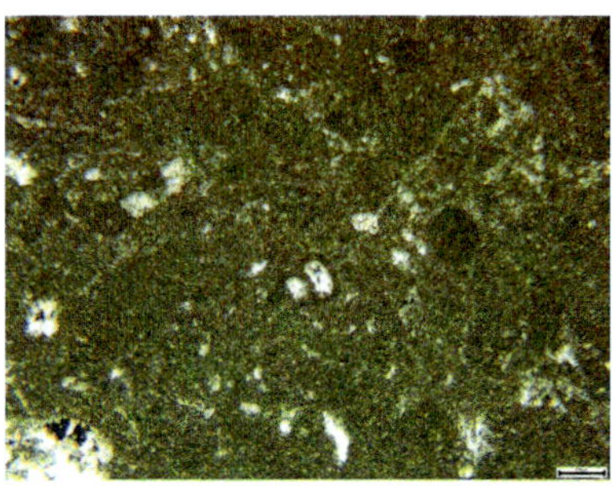

Figure 30. (S II.8) – stone, microscope (HGI, M. Belak)

levelling along the entire length of the ridge is sitting on a rectangular chest whose upper edges are cut for holding the lid. The chest and lid probably do not match, as the latter was not cut to fit the cutting on the chest's upper edges. The back of the chest is worked more roughly than the sides, while the front is taken by a blind moulded inscription panel and moulded, curved lateral niches topped by Norico-Pannonian volutes. The niches each hold a winged Eros in the guise of an older boy with only one wing depicted, with a plump face and a fairly short wavy hair, standing on a base in a frontal position and striding towards the Eros on the opposite side. Both of them are supporting a basket on one shoulder and are holding a roundish object in the opposite, lowered hand. The objects are not discernible, but from analogies can be recognized as a bunch of grapes, revealing the figures as season Erotes (*genii*), i.e. personifications of autumn with a basket of grapes on one shoulder and a single grape in the opposite hand. The workmanship is fairly primitive and stiff, without plasticity and proportion.
*Date*: First half of 3rd century.

*Publication*: Migotti 2013b: 191–192, type II.3.3. Internet databases: *lupa* 22341.
*Note*: The arrival in the AMZ of a sarcophagus from Sisak in 1926 is recorded in the Sisak archive in the AMZ, but with no description (Vukelić 2011: 85, 225, fns 568–570). On the other hand, the sarcophagus in question is the only one in the AMZ whose findspot and time of arrival have so far been considered as unknown; therefore, this set of data makes a plausible match. Judging from the archival data, this monument can probably be identified with the one mentioned as having been found on the way towards the River Odra (Vukelić), which points to the area of the northern cemetery of Siscia.

9. Sarcophagus (S II.9)
*Provenance*: Working of a vineyard in the mid 19th century in the area of the SW cemetery.
*Location*: Missing.
*Stone*: No data; possibly marble, presumed on the basis of a sophisticated iconographic programme.
*Preservation*: The sarcophagus has gone missing in the meantime, but the drawings of its front and short sides, made by the museologist Mijat Sabljar in the 19th century, have been recently found in the Archive of the AMZ.
*Description*: Measurements not recorded; type 3a. The upper edges were equipped with a cutting to fit a lid, and there were clamp holes in the short sides. The middle part of the front was taken by a moulded inscription panel flanked by simply framed arched niches, each featuring the same scene, but with small differences in details, or, rather, in execution. The scenes comprised a winged Eros standing on a sort of (natural?) pedestal in a 'ballet' pose associative of the typical Satyr and Nymph iconography. He holds his folded right leg with

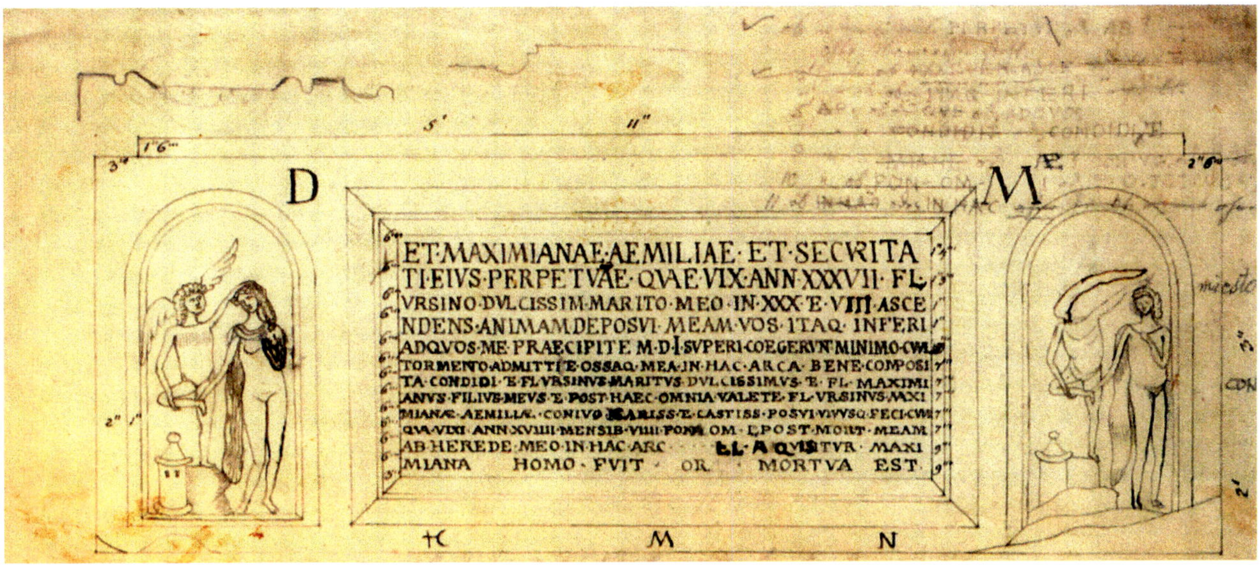

Figure 31. (S II.9) – scan of the drawing – front side (after Vukelić 2009)

his right arm and is lifting his left arm to reach the head of a woman at his side, who is wearing a mantle that leaves her naked body exposed. This scene was interpreted as the standard motif of Eros and Psyche. Such interpretation, however, is not likely given a considerable deviation from the standard picture of Eros and Psyche as two figures of the same size, hugging and kissing each other (Koch and Sichtermann 1982: 213–217; Hermary *et al.* 1986: 884, nos 411–416; *lupa* 4122, 4305, etc.). A more likely interpretation is therefore either Eros and Venus (cf. Schmidt 1997: 223–224, nos 340-355; *lupa* 3848, 15861, etc.) or Eros and Helena (cf. Blanc and Gury 1986: 1024–1027, nos 589–605; *lupa* 2991, 5004, etc.). The scenes on the short sides were placed within simply framed rectangular niches. The left one featured a long-haired nude aiming a bow and arrow at a frontal, long- and curly-haired female figure with exposed upper body and an indistinct depiction of the lower section, and with the uplifted arms turning into branches. The scene was interpreted as Perseus killing Medusa, but a more likely explanation seems to be Apollo and Daphne, the latter turning into a laurel tree (cf. Palagia 1986: 345–346, nos 8–29). However, the latter interpretation implies some controversial details, such as a most unusual gesture of Apollo, a striking resemblance of Daphne to the Minoan Snake Goddess, and the general absence of this scene from Norico-Pannonian funerary iconography. The depiction in the right short-sided niche contained a fairly small snake-like monster in the middle, flanked on the left by a long-haired male nude with a cloak thrown over his lowered left arm and a large conical object in his uplifted right hand. At his right side is a frontal, nude, long- and curly-haired standing female figure with her arms folded behind her back and her lower legs crossed. The scene was interpreted as Perseus and Andromeda, but Hercules and Hesione seems to be a more plausible interpretation because of the club in the god's hand and because of the fact that, according to the comparative artistic evidence, the otherwise rare motif of Hercules and Hesione (Weicker 1912) seems to be fairly familiar in Norico-Pannonian art (Pochmarski 2001: 132, Abb. 11; *lupa* 569, 1485, 3958, etc.). In view of the originality and sophistication of the scenes, it is more probable that the crudeness of the drawings reflects their author's artistic inadequacy rather than the quality of the original stone carvings.

Inscription: *D(is) M(anibus) // et Maximianae Aemiliae et securita/ti eius perpetuae quae vix(it) ann(os) XXXVII Fl(avio) / Ursino dulcissim(o) marito meo in XXX et VIII asce/[5]ndens animam deposui meam vos itaq(ue) Inferi / ad quos me praecipitem Di Superi coegerunt minimo cum / tormento admittite ossaq(ue) mea in hac arca bene composi/ta condidit Fl(avius) Ursinus maritus dulcissimus et Fl(avius) Maximi/[10]anus filius meus et post haec omnia valete. Fl(avius) Ursinus Maxi/mianae Aemiliae coniug(i) kariss(imae) et castiss(imae) posui vivusq(ue) feci cum / qua vixi ann(is) XVIIII mensib(us) VIIII poniq(ue) me post mort(em) meam / ab herede meo in hac arc[a iussi] /[15] A[---]I qui et Iul(ia) [Ma]xi/miana homo fuit mortua est / h(oc) m(onumentum) h(eredem) n(on) s(equetur).*

Line 2: *{et}*: EDH.

Translation: To the spirits of the departed, and to Maximiana Aemilia, who lived 37 years, and to her eternal security. While I have not yet reached the 38th year of age, I laid down my soul to my beloved husband Flavius Ursinus. Therefore, the gods of the underworld, to whom I was sent by the gods who dwell in heaven, abruptly and without suffering, admit my bones lying in this well-constructed chest, provided for me by my dearest husband Flavius Ursinus and my son Flavius Maximianus. And after everything was accomplished, I bid you farewell! I, Flavius Ursinus, commissioned this monument in my lifetime for my dearest and most

Figure 32. (S II.9) – scan of the drawing – short sides (after Vukelić 2009)

chaste wife Maximiana Aemilia, with whom I lived 19 years and 9 months, and I have ordered that when I am dead, my heir should place me in the same chest. Maximiana was a human being, and now she is dead. This monument should not be inherited.
Commentary: Unusual is the inverted word order in the name of Maximiana Aemilia, instead of the expected Aemilia Maximiana. What makes this inscription extremely singular is its composition in the form of alternating passages by both the deceased Aemilia Maximiana and her husband Flavius Ursinus.
The *gentilicium* Aemilius is frequent everywhere, in Italy and also in the provinces, particularly in Hispania, Gallia Narbonensis, Dalmatia, and Pannonia (*OPEL* I: 29–30; *Epigraphik-Datenbank* Clauss/Slaby); in Dalmatia it occurs more often in the Late Roman period (Alföldy 1969: 55), which is in accordance with the cognomina of the deceased; none appear on the *tesserae*. Flavii are well attested in Pannonia, which corresponded to the first phase of the urbanization of the province (Mrozewicz 2010; 2015); Flavii are also well documented at Siscia, 11 times, including members of the Roman army serving abroad (Šašel Kos 2017b), which is not at all surprising for a colonia Flavia.
*Date*: First half to mid 3rd century.
*Publication*: *CIL* III 3989; Gračanin 2004: 29–30 no. 20; Vukelić 2009 (= *AE* 2009: 1041); *CLEPann* 32; Migotti 2017b: 504–506. Internet databases: *HD*072126 = *lupa* 4316 (without illustration) = *EDCS*-28800750.

10. Sarcophagus (S II.10)
*Provenance*: Building works in 1882 in the area of the SE cemetery.
*Location*: AMZ, inv. no. AMZ KS-349.
*Stone*: Pohorje marble (H. Müller, 2013).
*Preservation*: The chest completely preserved. The topmost part of the inscription panel and the niches is missing, but the inscription and the figures survive entirely, as well as slight traces of a cutting along the edges for holding a lid. The edges are mostly broken off and the bottom is chipped at places. Weathering, staining and cracks are visible all over the chest, the largest one running the whole length of its front, and a wide vein of brownish colour appears at the bottom of the left short side and along nearly the whole length of the back. *Description*: L 203 cm, H 70 cm, W 101 cm, TB 18 cm; type 3a. The rectangular chest with a moulded inscription panel is flanked by niches with curved backs, topped by Norico-Pannonian volutes and framed with outer bands and inner fillets, each holding a frontal standing figure. In the left niche is a young woman of a fairly robust figure in the statuary pose, with her head turned slightly to her left. She is wearing a toga and is holding a huge scroll in her left hand and a roundish object, probably a fruit, in her right. She has an elongated, slightly triangular and block-like face, almond-shaped, heavy-lidded eyes topped by heavy brows, a regularly shaped nose with the tip broken off, heart-shaped lips, the hair arranged in thick wavy strands drawn from a central parting to the sides, leaving the ears uncovered and descending to the neck in a thick roll which is then folded upward into a flat chignon on top of the head. The youth in the right niche, also in the statuary pose and with the head turned slightly to his right, is wearing a belted tunic and short boots, and is holding a casserole or skillet (*trulla*) in his highly raised right hand, while grasping with his left hand at a towel thrown over his left shoulder. He has a sub-triangular face, almond-shaped, heavy-lidded eyes, full lips, and the hair rendered as parallel ridges

Figure 33. (S II.10) – sarcophagus (AMZ, I. Krajcar)

running from the rounded crown towards the forehead to form a V-shaped fringe. The workmanship is average and the modelling fairly crude, resulting in anatomical inaccuracies, but still with some sense of proportion, movement, and drapery. The short sides, initially intended for reliefs, were left as outlined in shallow grooves and only roughly chiselled, and with the clamp cuttings only drafted in the form of grooved rectangles in the middle of the upper edges.

*Inscription*: *[D(is) M(anibus) ?] / Romania N(a)evia clari/ssima femina viva / fecit sibi memoriam / ⁵pr(a)ecurrentibus Cletio /⁵ Romuliano et Aur(elio) / Calemero.*

Translation: To the spirits of the departed (?). Romania N(a)evia, *clarissima femina* (most distinguished woman), had a memorial made for herself in her lifetime, after Cletius Romulianus and Aurelius Calemerus had predeceased her.

Letters are regular, with punctuation marks between the words. Many letters are in ligature: line 1: *NE* in *Nevia*; line 2: *MA* in *clarissima*; line 3: *ME* and *AM* in *memoriam*; line 4: *VR*, *NT* in *precurrentibus*; line 5: *ET*; line 6: *ME* in *Calemero*.

Commentary: The identity of the two men mentioned in the inscription is not clear, although it seems most likely to interpret them as the two husbands of Romania Nevia, who had in turn died before her. A possible but less plausible explanation would be to see the two men as her relatives, involved with her through some legal matters. Equally problematic would be to interpret the deceased as two sons from her two marriages. Most probably Nevia's father belonged to the highest order, since neither of her supposed husbands seems to have been a senator (Barbieri 1952: 384 no. 2198). She would have been called *clarissima femina* even if she were married to men of non-senatorial, probably equestrian rank (cf. Alföldy 2001: 237).

The family name Cletius does not seem to be attested to date in the Roman Empire (Solin and Salomies 1994: 57; *OPEL* II: 64). The *gentilicium* Romanius is documented in Italy, particularly in Cisalpine Gaul, as well as in Gallia Narbonensis (*OPEL* IV: 31; Alföldy 1969: 115), elsewhere it is rare; in Pannonia it is only known in this case and possibly from a Siscian *tessera* (Radman-Livaja 2014: 254). Naevius, Romulianus, and Calemerus are rarely attested names (*OPEL* s. vv.). Calemerus is a Greek name, only a few times documented in the western part of the Empire (Barkóczi 1964: 308; *OPEL* II: 22); he or his family may have come from one of the eastern provinces.

*Date*: Mid, or second half of 3rd century.

*Publication*: *CIL* III 10852; Brunšmid 1909: 156, no. 349; *AIJ* 569; Migotti 2001 = AE 2001: 1631; Gračanin 2004: 35-36 no. 27; Migotti 2007b; Djurić 2013: 3, no. 9; Migotti 2013b: 189–190, type II.3.1. Internet databases: *lupa* 3803 = *HD*047135 = *EDCS*-29000487

11. Sarcophagus (S II.11)

*Provenance*: A chance find prior to 1908, most probably in Sisak.

*Location*: AMZ, inv. no. AMZ KS-477.

*Stone*: Transdanubian travertine (B. Djurić, 2015).

*Preservation*: The chest and lid completely preserved. The lid has been reconstructed from two pieces, with a missing part at its back and the middle *acroterion* restored. The surface is stained all over and pitted at places, with some damage along the chest's lower edge.

*Description*: Chest: L 128 cm, H 51 cm, W 65 cm, TB 9 cm; chest type 5; lid: L 134 cm, H 27 cm, W 67 cm; lid type 3. The gabled and tiled lid contains three flat protuberances, one in the centre and two at the ends, all ending in *acroteria*; the corner *acroteria* are half-rounded and the middle ones three-quarter-rounded. The front of the rectangular chest (reported to contain a 9-cm-thick stone bolster at the right bottom, now inaccessible for inspection) is taken by a simply framed blind inscription field flanked by rectangular panels. Each of the panels is holding a realistically depicted *pelta*, with only the tips of the horns and their central trefoil projections carved in relief, and the remainder

Figure 34. (S II.11) – sarcophagus (AMZ, I. Krajcar)

executed in linear grooving. The workmanship is neat and precise.
*Date:* Probably 3rd century.
*Publication*: Brunšmid 1909: 215, no. 477; Migotti 2013b: 194–195, type II.4.1. Internet databases: *lupa* 5736.
*Note*: The sarcophagus was first published by J. Brušmid (1909: 215, no. 477) with the remark 'presumably from Sisak'. On the other hand, the archival data mentions that, together with the sarcophagus of Severilla (no. II. 22.), another, a small one, found at the parish courtyard, is to be sent to the predecessor of the AMZ (Vukelić 2011: 219, fn. 225). This might well be the sarcophagus discussed here, as it is the only small one in the AMZ, while the acquisition of both ash-chests from Siscia, which could have been mistaken for a small sarcophagus (nos III.1. and III.2.), are duly recorded in the archives, with different findspots adduced.

12. Sarcophagus (S II.12)
*Provenance*: Infrastructural works in 1965 in the area of the SW cemetery.
*Location*: GMS, inv. no. GMS-3156.
*Stone*: Badenian biocalcarenite. A probable origin: the Banovina, S of Sisak (M. Belak 2018).
*Preservation*: The chest completely, and lid fragmentarily preserved. The right peripheral part of the lid is missing, and there is damage, chipping and staining all over, especially around the edges.
*Description*: Chest: L 153 cm, H 74 cm, W 90 cm, TB 37 cm; chest type 1; lid: *L 133* cm, H 39 cm, W 68 cm; lid typologically unclassifiable. The lid is gabled, with some 20-cm-wide flattening along the ridge, and is not hollowed inside. The rectangular flat-edged chest has a stone bolster on the left-hand bottom, and the upper parts of the short sides are provided with rather

Figure 35. (S II.12) – sarcophagus (O. Harl, *lupa* 22451)

Figure 36. (S II.12) – stone, microscope (HGI, M. Belak)

long, narrow, vertical clamp holes with some traces of iron on the inside. The inscription field is moulded and flanked by fairly wide lateral panels, with the inscription taking the whole surface and remaining unintelligible.
*Inscription*: *[---] / [---] / [---] tia / [---] / [---].*
Illegible remains of letters of five lines, containing *c.* 12 to 14 letters each.
*Date*: Probably early 3rd century.
*Publication*: Burkowsky 1996: 76, 78, fig. 7; Burkowsky 2000: 77, no. 27; Buzov 2002: 184–185: fig. 13; Migotti 2013b: 197, type II.5.3. Internet databases: *lupa* 22451.
*Note*: The sarcophagus was initially found as lidless and sitting on an unusually shaped pillow-like pedestal, much larger than the chest itself, but gone missing in the meantime. There is a slight possibility that the lid has been reshaped from the now missing base, for two reasons: i.e. the chest and lid do not match; and the lid is not hollowed on the inside, as is typical of sarcophagi covers.

13. Sarcophagus (S II.13)
*Provenance*: Rescue excavations in the 1950s in the SE cemetery.
*Location*: GMS, inv. no. GMS-49.
*Stone*: Badenian siliciclastic sandy biocalcarenite. A probable origin: the Banovina, S of Sisak (M. Belak 2018).
*Preservation*: The chest completely and lid fragmentarily preserved. Approximately one third of the lid is missing and a part of one of the upper corners of the chest is chipped off. There is some wearing around the edges and staining all over.
*Description*: Chest: L 200 cm, H 53 cm, W 75 cm, TB 14 cm; chest type 7; lid: *L 130* cm, H 30 cm, W 80 cm; lid type 1a. The roughly worked, rather elongated rectangular chest is plain, with flat edges. The *acroteria* of the gabled lid possibly hold a volute-like motif, which, however, is not clearly discernible and can also be the result of rough workmanship or damage to the stone.
*Date*: Second half of 2nd – 4th century.
*Publication*: Burkowsky 2000: 77, no. 28; Migotti 2013b: 200, type II.7.1.

Figure 38. (S II.13) – stone, microscope (HGI, M. Belak)

Figure 37. (S II.13) – sarcophagus (GMS, B. Suntešić)

Figure 39. (S II.14) – sarcophagus (GMS, B. Suntešić)

Figure 40. (S II.14) – stone, microscope (HGI, M. Belak)

14. Sarcophagus (S II.14)
*Provenance*: Rescue excavations in the 1950s in the SE cemetery.
*Location*: GMS, inv. no. GMS-50.
*Stone*: Badenian siliciclastic sandy biocalcarenite. A probable origin: the Banovina, S of Sisak (M. Belak 2018).
*Preservation*: The chest completely preserved with slight damage around the edges and on the surface, and staining all over.
*Description*: L 200 cm, H 53 cm, W 80 cm, TB 14 cm; type 7. The roughly worked, rather elongated rectangular chest is plain, with flat edges.
*Date:* Second half of 2nd – 4th century.
*Publication*: Burkowsky 2000: 78, no. 29; Migotti 2013b: 200, type II.7.2.

15. Sarcophagus (S II.15)
*Provenance*: Rescue excavations in the 1950s in the SE cemetery.
*Location*: GMS, inv. no. GMS-51.
*Stone*: Badenian siliciclastic sandy biocalcarenite. A probable origin: the Banovina, S of Sisak (M. Belak 2018).
*Preservation*: The chest completely preserved, with parts of long sides and one entire short side missing, and with staining all over and some damage around the edges.
*Description*: L 220 cm, H 59 cm, W 83 cm, TB 14 cm; type 7. The roughly worked, rather elongated rectangular chest is plain, with flat edges.
*Date*: Second half of 2nd – 4th century.
*Publication*: Burkowsky 2000: 78, no. 31; Migotti, 2013b: 200, type II.7.4.

Figure 41. (S II.15) – sarcophagus (GMS, B. Suntešić)

Figure 42. (S II.15) – stone, microscope (HGI, M. Belak)

Figure 43 (S II.16) - sarcophagus (GMS, B. Suntešić)

16. Sarcophagus (S II.16)
*Provenance*: Rescue excavations in the 1950s in the SE cemetery.
*Location*: GMS, inv. no. GMS-3157.
*Stone*: Badenian sandy biocalcarenite. A probable origin: the Banovina, S of Sisak (M. Belak 2018).
*Preservation*: The chest fragmentarily preserved, with the bottom and insignificant parts of the front, back and sides surviving.

Figure 44. (S II.16) - stone, microscope (HGI, M. Belak)

*Description*: L 200 cm, *H 40* cm, W 100 cm, TB 14 cm; type 7. The roughly worked, probably elongated rectangular chest is plain, with flat edges.
*Date*: Second half of 2nd - 4th century.
*Publication*: Migotti 2013, 200, type II.7.5.

17. Sarcophagus (S II.17)
*Provenance*: Rescue excavations in the 1950s in the SE cemetery.
*Location*: GMS, inv. no. GMS-53 AZA.
*Stone*: Badenian siliciclastic sandy biocalcarenite. A probable origin: the Banovina, S of Sisak (M. Belak 2018).
*Preservation*: The chest completely preserved, with the majority of one long and one short sides missing. There are heavy cracks on the preserved surface and weathering around the edges, as well as a secondary hole through the long side and one of the short sides, respectively.
*Description*: L 200 cm, H 58 cm, W 80 cm, TB 14 cm; type 7. The roughly worked, rather elongated rectangular chest is plain, with flat edges.
*Date*: Second half of 2nd - 4th century.
*Publication*: Burkowsky 2000: 78, no. 32; Migotti 2013b: 200, type II.7.3.

Figure 45. (S II.17) - sarcophagus (GMS, B. Suntešić)

Figure 46. (S II.17) - stone, microscope (HGI, M. Belak)

Figure 47. (S II.18) – sarcophagus (GMS, B. Suntešić)

18. Sarcophagus* (S II.18)
*Provenance*: Found at an unspecified time as reused in the area of the SE cemetery.
*Location*: GMS, inv. no. GMS-3122.
*Stone*: Mesozoic/Paleocene micritic limestone. A possible origin: the Outer Dinarides; a probable origin: the Kordun, SW of Sisak (M. Belak 2018).
*Preservation*: The chest completely preserved, with cracks, damage and staining all over, especially around the edges.
*Description*: L 156 cm, H 66 cm, W 90 cm, TB 34 cm; type 7. The roughly worked rectangular chest is plain, with the edges cut to hold a lid.
*Date*: Second half of 2nd – 4th century.
*Publication*: Burkowsky 2000: 79, no. 36; Migotti 2013: 201, type II.7.6.

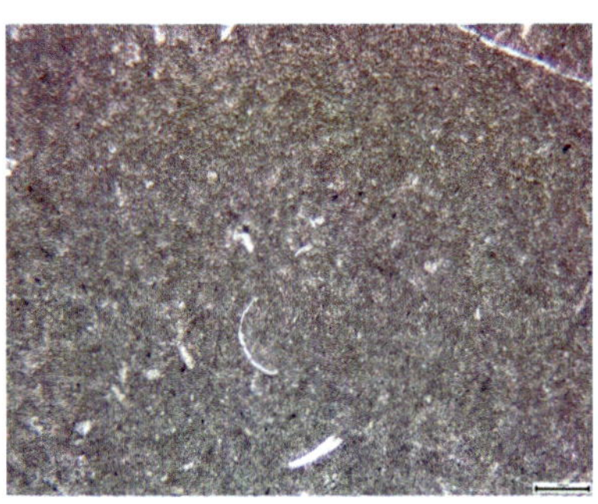

Figure 48. (S II.18) – stone, microscope (HGI, M. Belak)

Figure 49. (S II.19) – sarcophagus (GMS, B. Suntešić)

19. Sarcophagus* (S II.19)
*Provenance*: Possibly stemming from the area of the E cemetery (the site of *Povijesni arhiv*).
*Location*: GMS, inv. no. GMS-27740.
*Stone*: Sandy biocalcarenite. A possible origin: The Banovina, S of Sisak (M. Belak 2018).
*Preservation*: The chest completely preserved, but chipped and battered all around.

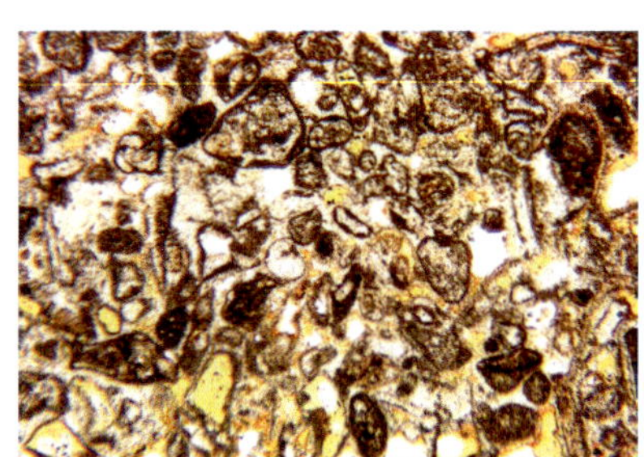

Figure 50. (S II.19) – stone, microscope (HGI, M. Belak)

Figure 51. (S II.20) – sarcophagus lid (GMS, B. Suntešić)

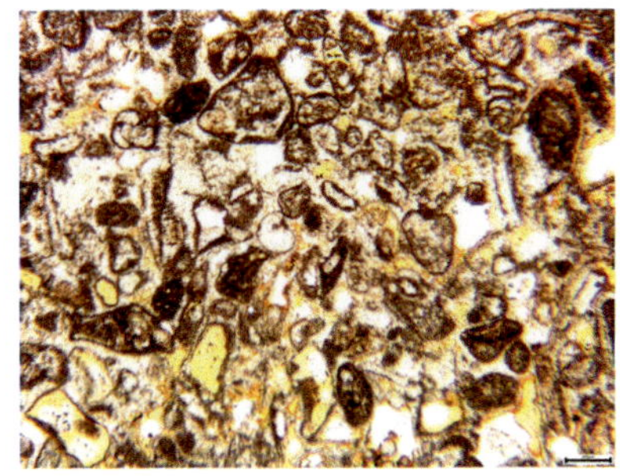

Figure 52. (S II.20) – stone, microscope (HGI, M. Belak)

*Description*: L 162 cm, W 56 cm, H 41 cm, TB 25 cm; a variant of type 7. The roughly worked chest with flat edges is rectangular on the outside and with slightly rounded corners on the inside, reminding of tub-shaped sarcophagi.
*Date*: Second half of 2nd – 4th century.
*Publication*: Unpublished.

20. Sarcophagus lid (S II.20)
*Provenance*: A chance find at an unspecified time in the presumed area of the SW cemetery.
*Location*: GMS, inv. no. GMS-3121 AZA.
*Stone*: Upper-Jurassic/Lower-Cretacious micritic limestone. A possible origin: the Outer Dinarides; a probable origin: the Kordun, SW of Sisak (M. Belak 2018).
*Preservation*: The lid fragmentarily preserved, with roughly two thirds including two *acroteria* missing. Of the remaining two *acroteria* one is nearly completely battered, while the other's tip is broken off, and there are cracks and staining all over.*Description*: *L* 75 cm, H 50 cm, W 112 cm; type 1.1. The plain, roughly worked (unfinished?) gabled lid is provided with corner sub-triangular *acroteria* and a 5-cm-wide levelling along the ridge.
*Date*: Second half of 2nd – 4th century.
*Publication*: Burkowsky 2000: 79, no. 37; Migotti 2013: 202, type II.8.1.

21. Sarcophagus lid (S II.21)
*Provenance*: Found in 2004 during building works in advance of the archaeological excavation at the site of Sveti Kvirin (Saint Quirinus) in the area of the N cemetery; it was found as walled in the enclosure of the nursery school, built there in the 1960s.
*Location*: The archaeological park of Sveti Kvirin; GMS, inv. no. GMS-29162.
*Stone*: Upper-Jurassic/Lower-Cretaceous micritic limestone. A possible origin: the Outer Dinarides; a probable origin: the Kordun and the southern, SW of Sisak (M. Belak 2018).
*Preservation*: Fragmentarily preserved, with approximately one half of the one side missing. There are cracks and staining in black spots all over. A lump of a modern mortared-brick structure, remaining from the previous walling, is attached at one end.

Figure 53. (S II.21) – sarcophagus (GMS, B. Suntešić)

Figure 54. (S II.21) - stone, microscope (HGI, M. Belak)

*Description*: L 235 cm, H 143 cm, W 123 cm; type 2. The roughly worked gabled lid with a 4.5–5-cm-wide levelling along the whole length of the ridge is provided with four corner *acroteria* of a slightly irregular, quarter-round form. There are also nine elongated vertical rows of tiles, topped by eight rows of *imbrices*, to each side of the roof.

*Date*: 2nd - 4th century.

*Publication*: Migotti 2013b: 202, II.8.2.

22. Sarcophagus (S II.22)

*Provenance*: A chance find (probably in building works) prior to 1551 in the area of the SE cemetery.

*Location*: AMZ, inv. no. AMZ KS-351.

*Stone*: Upper-Jurasic/Lower-Cretacious micritic limestone. A probable origin: The Kordun, SW of Sisak (M. Belak 2018).

*Preservation*: The chest and lid completely preserved. There are major cracks, chipping and wear all over the chest, especially around the edges, and also some on the lid, and there is a hole through the back side of the chest.

*Description*: Chest: L 227 cm, H 120 cm, W 146 cm; chest type 6; lid: L 239 cm, H 55 cm, W 145 cm; lid type 2. A gabled and tiled lid (five oblong tiles to a side, joined by four rows of *imbrices*) with a 7-cm-wide flattening along the entire length of the ridge and huge corner *acroteria*, is sitting on a rectangular chest with a string-framed inscription panel in the middle of the front. The inscription panel is in the form of a *tabula ansata* with two very small trapezoidal *ansae*, each holding a roughly carved Chi-Rho. The panels to the left and right of the inscription table each bear a motif of a vase sprouting vines, grapes and wheat-ears, while vine tendrils, birds and animals, carved in low relief, fill the rest of the free space. The workmanship is fairly crude, with the motifs rendered schematically in a markedly ornamental manner.

*Inscription*: *Huic arcae inest Seve/rilla famula XPI(Christi), quae / vixiṭ cum viro novem / continuis annis, cuius /*[5] *post obitum Marcellianus se/dem hanc videtur conlocasse mari/ tus.*

Letters in the first four lines are more regular, with larger space between each, than in the last three lines. Punctuation marks between the words. Letters *L* have a specific shape, with a small *s* added to the horizontal

Figure 55. (S II.22) - sarcophagus (AMZ, I. Krajcar)

Figure 56 (S II.22) – sarcophagus detail (AMZ, I. Krajcar)

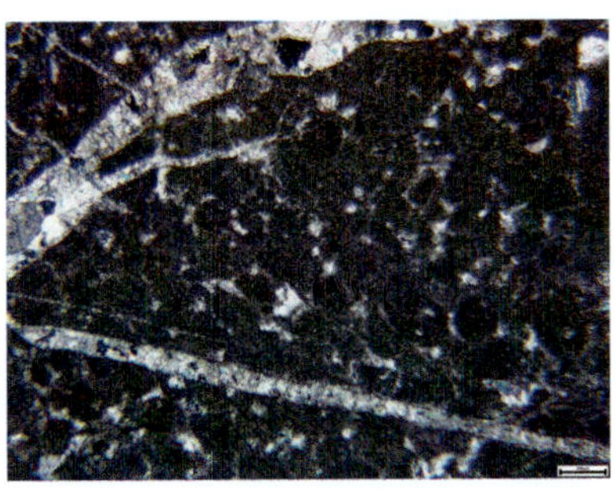

Figure 57. (S II.22) – stone, microscope (HGI, M. Belak)

*hasta*. Line 2: there is a horizontal line above the abbreviation for Christi, *XPI*. *QVAE*: the last three letters seem to be in ligature. Line 5: the last two letters are obliquely squeezed at the end of the line, giving the impression of being cursive. Line 6 and 7: because of the lack of space, line 6 begins much more to the left than the rest of lines, but nonetheless the last three letters of the epitaph had to be carved tightly beneath the 6th line, on the right.

Translation: This is the chest of Severilla, a servant of Christ, who had lived with her husband for nine continuous years, and for whom after her death her husband Marcellianus provided this resting place.

Commentary: The name Severilla and Marcellianus are typical of the Late Roman anthroponymy. Severilla has only been attested in the feminine form and is relatively rare, except in northern Italy and Gallia Narbonensis (*OPEL* IV: 75; *Epigraphik-Datenbank* Clauss/ Slaby; cf. Barkóczi 1964: 324). Marcellianus occurs sporadically, except in Pannonia, where it is attested five times according to the *OPEL* (III: 53; cf. Barkóczi 1964: 317).

*Date*: Early 4th century (after 313 AD). R. Koch (1982: 329) brought up the possibility that the inscription was later than the relief decoration, but this has no justification in either the composition of the motif or in the different levels of the relief and inscription panels. A similar opinion was probably shared by J. Kastelic (1998: 259), who dated the sarcophagus to the end of the 3rd century. It should, however, be dated to shortly after 313, on account of bucolic motifs, typical in the 3rd century, and inconspicuous Chi-Rhos within the *ansae*, which seem to be deliberately adjusted to a sub-cryptic conception from the start. The date proposed in the *EDH*: between the early 4th century and AD 500.

*Publication*: *CIL* III 3996; Brunšmid 1909: 159, no. 351; *AIJ* 581; *ILCV* 1449; Demo 1994: 82 no. 39; Migotti 1997: 39–42; Dresken-Weiland 1998: 109-110, no. 311; Pl. 100. 5; Gračanin 2004: 32 no. 24; Vukelić 2006 = *AE* 2006: 1033; Vukelić 2011: 230–239; Migotti 2013b: 197-199, type II.6.1. Internet databases: *lupa* 3809 = *HD*057288 = *EDCS*-26600437.

*III. Ash-chests*

1. Ash-chest (S III.1)

*Provenance*: A chance find around 1800 in the area of the SE cemetery.

*Location*: AMZ, inv. no. AMZ KS-350.

*Stone*: Badenian sandy biocalcarenite. A probable origin: the mountains of N Croatia (M. Belak 2018).

*Preservation*: The chest completely preserved. The front and back top left corners are partly broken off and the inscription is badly weathered. Despite heavy damage around the edges, slight remains of an internal cutting to hold a lid survive, as well as rectangular clamp holes at the centre top on the short sides.

Figure 58. (S III.1) – ash-chest (AMZ, I. Krajcar)

*Description*: L 95 cm, H 51 cm, W 65 cm, TB 27 cm; type 4. The rectangular chest has a roughly smoothed front holding a moulded inscription panel framed by flat bands on all sides, with the lateral ones slightly wider than the lower and upper ones. The workmanship is rough, especially on the back and short sides.

*Inscription*: *G(aio) Sempro(nio) Severo, cor/nicul(ario) leg(ionis) XIIII G(eminae), (annorum) XXXXI. / Iulia Florentina con/iugi carissimo viva /[5] fec(it).*

Letters are irregular and much worn.

Translation: To Gaius Sempronius Severus, *cornicularius* (adjutant) of the Legion *XIV Gemina* (composed of two), aged 41 years. Iulia Florentina had (this monument) made in her lifetime to her dearest husband.

Commentary: Since the Legion *XIV Gemina* was stationed at Carnuntum, Severus might have been either sent to Siscia on a special mission, or, which seems more likely, he was buried there because he originated from the town and had his family there (see also Radman-Livaja and Vukelić 2015: 402). It may be noteworthy that his and his wife's names are also documented on the *tesserae* (Radman-Livaja 2014: s. vv).

The *gentilicium* Sempronius is very frequent almost everywhere and also in the eastern provinces, but particularly in Italy and Hispania (*OPEL* IV: 64; Alföldy 1969: 119; *Epigraphik-Datenbank* Clauss/Slaby). Severus, too, is well attested almost everywhere, notably also in Pannonia (*OPEL* IV: 76–78), and must have been very popular in Siscia, since it is attested on 12 *tesserae* (Radman-Livaja 2014: 263). Florentinus/a is a name, very well documented in Pannonia (*OPEL* II: 148), to where it may have spread from northern Italy (Mócsy 1959: 174).

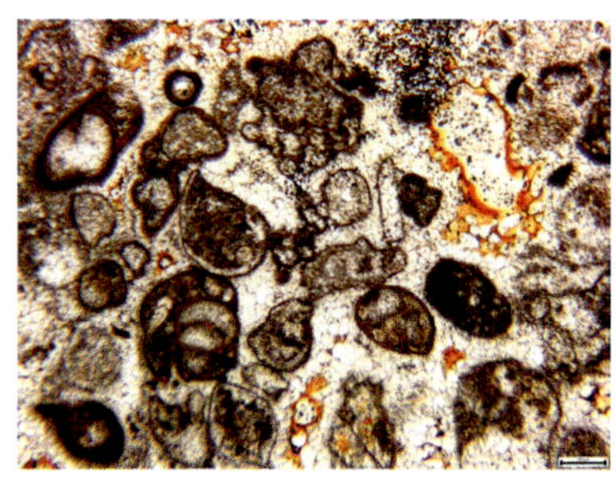

Figure 59. (S III.1) – stone, microscope (HGI, M. Belak)

*Date*: Later 2nd – mid 3rd century. The front is nearly identical in shape with an inscription to the emperor Marcus Antonius Gordianus Pius (238–244) from Daruvar (*AIJ* 586).

*Publication*: *CIL* III 3972 = 10855; Brunšmid 1909: 158, no. 350; *AIJ* 568; Migotti 2007a: 11–12, fig. 5, and Migotti 2013b: 188, Fig. 7a (a wrong picture adduced!). Internet databases: *lupa* 3802 (without photo) = *HD*072122 = *EDCS*-26600431.

2. Ash-chest (S III.2)

*Provenance*: Gardening in 1804 in the area of the SW cemetery.

*Location*: AMZ, inv. no. AMZ KS-343.

*Stone*: Chest – Upper-Jurassic/Lower-Cretacious micritic limestone. A possible origin: the Outer Dinarides; a probable origin: the Kordun, SW of Sisak (M. Belak 2018). Lid – Paleogene coral limestone. A probable origin: the mountains of NW Croatia, notably Žumberak, SW of Zagreb (M. Belak 2018).

Figure 60. (S III.2) – ash-chest, front side (AMZ, I. Krajcar)

*Preservation*: The chest and lid completely preserved. The lid is chipped at the centre bottom, and the chest is damaged around the edges, with a substantial chip at the back lower edge and a major crack between the niche and the top edge in the left upper part. A secondary hole is cut through the centre of the left short side.

*Description*: Chest: L 143 cm, H 58 cm, W 77 cm; chest type 3a; lid: L 1.68 m, H 0.31 m, W 0.90 m; lid type 1a. The gabled lid with corner *acroteria* and flat edges mismatches the rectangular chest in both the material and measurements. The chest's short sides are provided with clamp holes, and the edges hold cuttings to fit a lid, but not the one currently in place. The inscription panel is moulded, while the unframed lateral niches with curved backs, topped by Norico-Pannonian volutes, each feature a frontal standing figure of one of the deceased couple. The left niche holds a standing figure of Septimia Matrona dressed in a peculiar combination of a tunic with long and fairly wide sleeves, a mantle (*palla*) and a narrow folded scarf of *contabulatio* type, holding a scroll in her left hand and pointing at it with her right. She has an oval face, chubby cheeks, almond-shaped and heavy-lidded eyes, a small mouth, the hair arranged in three superimposed layers tapering towards the top, with the rest hanging by the neck in sharp-cut angular strands that leave the ears uncovered. In the right niche M. Aurelius Glabrio is depicted as dressed in a tunic belted with a round buckle and a cloak fastened with a round brooch on the right shoulder, holding a scroll in his left hand and pointing at it with his right. He has a broad and block-like face

Figure 61. (S III.2) – ash-chest, left short side (AMZ, I. Krajcar)

Figure 62. (S III.2) – ash-chest, right short side (AMZ, I. Krajcar)

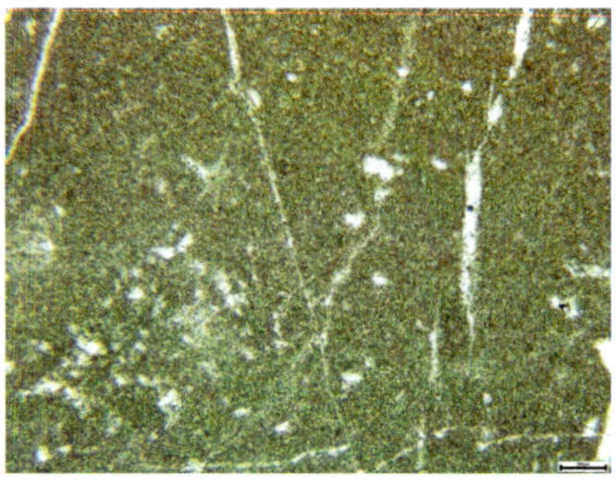

Figure 63. (S III.2) – chest stone, microscope (HGI, M. Belak)

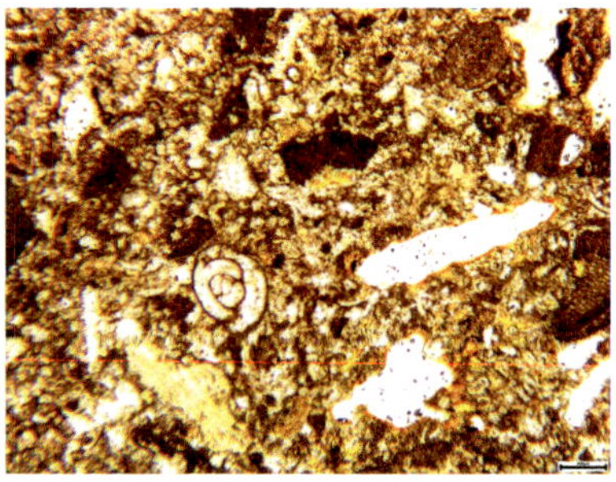

Figure 64. (S III.2) – lid stone, microscope (HGI, M. Belak)

with a trapezoidal upper contour, almond-shaped and heavy-lidded eyes, a straight nose, a short-cropped hair straight on the forehead and sharply receded at the temples, a short moustache, and a curly beard covering half of the cheeks, with the whiskers barely discernible. The Erotes-scenes on the short sides are depicted in simply framed flat panels. The left one holds two Erotes presented as small wingless boys striding towards each other and holding a huge shell between them, through which a secondary hole is pierced. They have round faces, large almond-shaped eyes and tightly closed small lips, and rather short, wavy and curly hair rendered in stiff ridges and ornamental curls. The Erotes on the right short side, striding in opposite sides and supporting a plain garland, are depicted in the manner and style corresponding to those on the left side. All of the reliefs are rendered rather inaccurately and crudely, with some sense of movement and plasticity, especially in the portraits, but with a pronounced lack of proportion and natural shapes.

*Inscription*: *M(arcus) Aurel(ius) Glabrio, / vet(eranus) leg(ionis) XIV G(eminae) ex b(ene)f(iciario) / co(n)s(ularis), vivus sibi et / Septimiae Matronae /[5] coniugi fecit*.

Line 4: *IM* and *AE* in the name *Septimiae* in ligature.

Translation: Marcus Aurelius Glabrio, veteran of the Legion *XIV Gemina* (composed of two), a former *beneficiarius consularis*, made (this monument) in his lifetime for himself and his wife Septimia Matrona.

Commentary: Glabrio may have been stationed at Siscia as a *beneficiarius consularis* (see also Radman-Livaja and Vukelić 2015: 399; Radman-Livaja and Vukelić 2016: 211). Since the urn is from the 3rd century AD, it is not surprising that these names are not attested on the *tesserae*. According to *OPEL*, the cognomen Glabrio was very rare, occurring once in northern Italy, Hispania, Noricum, and Pannonia (only this case noted: II: 167). Matrona does occur another time in Siscia: one Aelia P. f. Matrona is known from a sarcophagus, which she had erected in her lifetime for herself and her family (*CIL* III 3981 = *lupa* 22319; Buzov 2002: 177–178 fig. 4; Migotti 2013b: 195–196, type II.5.1). Supposedly she was an indigenous inhabitant of Siscia (Mócsy 1959: 57/6 [p. 211]), and in case that she could indeed be regarded as such, probably also Septimia Matrona had been a native-born (cf. Barkóczi 1964: 318).

*Date*: 3rd century. The onomastic formula speaks for the first half of the 3rd century, while the portraits are ambiguous. Glabrio's hairdo can be compared to that of Caracalla, but also to various emperors from the mid to the end of the 3rd century. Although the hairstyle of Septimia Matrona is difficult to visualize because of the cracks in stone running over her forehead and diagonally across the front of the scull, the closest parallel seems to be a crudely executed hair-do of the empresses from the late 3rd century.

*Publication*: *CIL* III 3970; Brunšmid 1909: 151, no. 343; *AIJ* 566; *CBI* 313; Gračanin 2004: 22 no. 6; Migotti 2005b; Migotti 2013b: 192–193, type II.3.5. Internet databases: *lupa* 3801 = *HD*072027 = *EDCS*-26600429.

## IV. Tituli

1. *Titulus* (S IV.1)

*Provenance*: Ploughing in 1901 in the area of the SW cemetery.

*Location*: AMZ, inv. no. AMZ KS-344.

*Stone:* Upper-Cretacious/Paleogene sparitized micritic limestone. A possible/probable origin: the mountains of NW Croatia (M. Belak 2018).

*Preservation*: Fragmentarily preserved. Parts on the top and right-hand side are missing, with patches of crust on most of the otherwise polished surface, and with the letters well preserved.

*Description*: H 57 cm, L 55 cm, T 4.5 cm; a rectangular unframed inscription panel.

*Inscription: Dom(us) aet[erna]. / Gaudenti[o] / filio piissim[o] / qui vix(it) an(nos) VI, /[5] m(enses) VI, d(ies) X, h(oras) n(octis) / VIII. Fati m(unus) c(omplevit)*.

Letters are fairly regular; punctuation marks are of specific shape, resembling tiny oblique crosses, except in line 2, where the mark looks like a vertical tiny cross.

Line 2: *Gaudenti[us]*: *AIJ*, *lupa*, and *EDCS*: not plausible; *Gaudenti[o]*, *ILCV*, *EDH*. Line 3: specific shape of letters *F* and *L*.

Translation: Eternal dwelling. To Gaudentius, his most devoted son, who lived 6 years, 6 months, 10 days, 8 hours of the night. He fulfilled the duty imposed upon him by fate.

Figure 65. (S IV.1) – *titulus*
(AMZ, I. Krajcar)

Figure 66. (S IV.1) – stone, microscope
(HGI, M. Belak)

Commentary: The name Gaudentius was much more popular in the Late Roman period and among the Christians than earlier (Kajanto 1965: 260; Alföldy 1969: 209; Barkóczi 1964: 313; Migotti 1997: 52, with references). There is hardly any doubt that the name Gaudentius refers to the son, since the name of the deceased should expectedly be mentioned in an epitaph (cf. *ILJug* I, corrigenda ad *AIJ*, p. 6*). The epitaph may be Christian, but the evidence is not conclusive (see at *Note*).

*Date*: 4th century.

*Publication*: *CIL* III 15181,2; Brunšmid 1909: 153, no. 344; *AIJ* 572; *ILCV* 3659a; Migotti 1997: 52–53, III.c.9; cf. Šašel Kos 2006: 196–197. Internet databases: *lupa* 3805 = *HD*072124 = *EDCS*-30200462.

*Note*: Possibly a Christian inscription, on several grounds. First is the frequency of the name Gaudentius in Christian nomenclature, as well as of the formula *domus aeterna*, which originated in the classical funerary phraseology but was also widely accepted among Christians. The small cross cut at the beginning of the second line is also indicative.

2. *Titulus* (S IV.2)

*Provenance*: Unknown, first recorded in 1866 as kept in the parochial house in Sisak before being transferred to the AMZ.

*Location*: AMZ, inv. no. AMZ KS-348.

*Stone*: Pohorje marble (W. Prochaska 2017).

*Preservation*: Fragmentarily preserved. The stone is broken in two pieces and reassembled, with a fragment on the left side missing; it is chipped around the edges, but the letters are well preserved.

*Description*: H 56 cm, W 32 cm, T 3 cm: The rectangular panel is unframed, with the inscription partly preserved.

*Inscription*: After the enlarged photograph: *[O]pt(imae?) / [memoriae?] Paulini Lucerinis / [---] ab omnibu[s] // [---]R operav[---] / [5] [---]re benedict[---] / [---]T X et tu [---] / [---] a̲ris sac[---] / [--- gr]a̲tias ag[---] / [--- ?.*

The first three lines are carved on the upper fragment; some text between the two may be missing. Nothing of the inscription seems to be missing before the first preserved line, but it is damaged on all other sides. Some

Figure 67. (S IV.2) – *titulus*
(AMZ, I. Krajcar)

letters resemble cursive script. Punctuation marks are present between the words. Reading is uncertain. Line 1: space between the two letters; line 2: *Luceri/nis* written in smaller letters after *Paulini*; probably the end of the line. *AIJ* and *lupa*: *[O]pt(imae?) / [memoriae?] Paulini*, but the supplement is most uncertain (omitted in *EDH*). Line 7 and 8: the beginning *A* is no longer seen.
Translation: ... of Paulinus Lucerinis (genitive?) ... from everybody ...
Commentary: *ILjug* I (Indices AIJ): p. 21* regarded the epitaph as Christian; The name Paul(l)inus is very well documented in Italy and Celtic-speaking provinces, as well as in Dalmatia and Pannonia (Alföldy 1969: 261; *OPEL* III: 129). In Siscia, it occurs twice on the *tesserae*, borne by two men with Roman citizenship (Flavius Paulinus and Domitius Paulinus, Radman-Livaja 2014: 242), but the name was very popular also among the Christians (Alföldy 1969: 261; Minkova 2000: 226). Attested is a Latin *gentilicium* Lucerinius (Solin and Salomies 1994: 107) and a cognomen Lucerinus (Solin and Salomies 1994: 353). However, it is not clear whether *Lucerinis* should be regarded as a nominative or a genitive case (*ILjug* I [Indices *AIJ*]: p. 20*: Lucerinis, *casus*?).
*Date*: Late Roman.
*Publication*: *CIL* III 3991; Brunšmid 1909: 156, no. 348; *AIJ* 574; Demo 1994: 82 no. 40; Migotti 1997: 49 (with references). Internet databases: *lupa* 22585 = *HD*064403 = *EDCS*-26600436.

3. *Titulus* (S IV.3)
*Provenance*: Building works in 1877 in the centre of Siscia (probably a reused find).
*Location*: AMZ, no. AMZ KS-345.
*Stone*: Pohorje marble (W. Prochaska 2017).
*Preservation*: Completely preserved and restored from two pieces. The stone is chipped around the edges and its surface pitted and weathered, with the first line and several other details of the inscription illegible.
*Description*: W 68 cm, H 36 cm, T 5.5 cm. The rectangular inscription panel is unframed, with the inscription mostly preserved but not entirely legible. There is an incised line along the left border, possibly indicating an anticipated but never executed frame.
*Inscription*: Metric inscription: *[---]EC [u]xori / posuit carissim(a)e suae / Maximill(a)e nom[i]na tu(a)e / supEros laudesq(ue) manebun(t) /*[5] *adque (!) utinam su[p]erstes amic(ae?) / reciperes tua mer[i]ta magna / coniugio ereptam / firma requiris (!) ibidem.*
Line 3: *nomina tu<a>*, *EDH*. Earlier editors (also *lupa*): line 5: *am[i]c[o]*; line 8: *Firma requiris*: Firma regarded as a name. Reading is most uncertain.
Translation: ... had (this monument) erected to his dearest wife. The name of his (?) Maximilla will live to see the gods and praise. Hopefully you, my companion (*superstes*?), and dear (wife) will get as a reward for your great merits, you who had been taken from your marriage. There, you will get them confirmed.
Commentary: The translation of the epitaph is difficult and differs from the earlier interpretations mainly in not regarding Firma as a name, but as an adjective (thus also Gračanin 2004: 37). The allusion to the verse 1. 609 from Virgil' *Aeneis* is cited by Hoffiller: '*semper honos nomenque tuum laudesque manebunt ...*' (your honour, name, and praise will always live ...).
Firmus (strong, firm), if it is here indeed a name, which seems less likely, was one of the most popular names in the western part of the Roman Empire (Kajanto 1965: 258; *OPEL* II: 142–143), not least in Noricum, Dalmatia, and Pannonia (Lochner von Hüttenbach 1989: 79; Alföldy 1969: 204), also appreciated by the indigenous population (Mócsy 1959: 174). In Siscia, it occurs on five *tesserae* (Radman-Livaja 2014: 207). Maximilla is mainly

Figure 68. (S IV.3) – *titulus* (AMZ, I. Krajcar)

attested in the feminine form, it occurs everywhere, not least in Dalmatia and Pannonia (*OPEL* III: 69; Alföldy 1969: 242).
*Date*: Probably Late Roman.
*Publication*: *CIL* III 10864; Brunšmid 1909: 153–154, no. 345; *AIJ* 573; *CLE* 598; *CLEPann* 31; Gračanin 2004: 37–38 no. 30. Internet databases: *lupa* 3806; *HD*072125; *EDCS*-29000474.

4. *Titulus* (S IV.4)
*Provenance*: Working of a vineyard in 1823 in the area of the SW cemetery.
*Location*: The Magyar Nemzeti Múzeum, inv. no. RD 195.
*Stone*: Marble; characterization of the stone failed due to a too small sample (W. Prochaska 2017).
*Preservation*: Fragmentarily preserved. All four edges are broken off, and there are chips, abrasions, and staining all over.
*Description*: *W* 58 cm, *H* 50 cm, T 7 cm.
The *titulus* is a rectangular unframed panel, with the majority of the inscription preserved.
*Inscription*: *D(is) M(anibus). / Positus est hic Leburna / magister mimariorum / [q]ui vicxit* (!) *annos plus / [m]inus centum. / [Al]iquoties mortuus / [sum] set* (!) *sic nunquam. / [Opto v]os ad supEros bene / [va]ler{a}e.*
Letters are regular, some letters *T* have oblique upper hasta. *Hederae* between the words.
Translation: To the divine spirits of the departed. Here lies Leburna, master of the actors in mimes, who lived more or less a hundred years. I died several times but never in this manner. I hope you are well in the upper world.
Commentary: The text may be metric, but not necessarily (*CLEPann* 64). A similarly witty funerary text, written in Greek, has come to light in Aquileia, erected for a mime actress (*mima*) Bassilla by one of her co-actors (*biologos*) Heraclides (*Inscr. Aquil.* 710; Zaccaria 1994: 86 no. 1: ... πολλάκις ἐν θυμέλαις, ἀλλ'οὐχ οὕτω δὲ θανούσῃ). Actors and actresses in comic performances are epigraphically well documented, particularly in Rome and other Italian cities (for Cisalpina, see Zaccaria 1994).
The name Leburna is a *hapax*, obviously derived from the Liburni, who lived along the northern coastal regions of the eastern Adriatic and on the islands. The name Liburnus occurs once in Dalmatia (*OPEL* III: 25).
*Date*: Second half of 3rd or first half of 4th century.
*Publication*: *CIL* III 3980; *ILS* 5228; *AIJ* 570 (+ corr. *ILJug* p. 6*); *AE* 2006: 34; *CLEPann* 64; Šašel Kos 2006: 195–197; Migotti 2017a: 31–34. Internet databases: *lupa* 3804; *HD*072123; *EDCS*-26600433.

Figure 69. (S IV.4) – *titulus*
(The Magyar Nemzeti Múzeum)

*VII. Funerary buildings (aedicualae/mausoleums, underground chambers)*

1. *Aedicula* (S VII.1)
*Provenance*: Archaeological excavation of a *villa suburbana* in 1952, situated in the presumed area of the N cemetery.
*Location*: GMS, inv. no. GMS-9.
*Stone*: Pohorje marble (W. Prochaska 2017).
*Preservation*: A pediment fragmentarily preserved. The left corner is broken off, and the separately made bottom part of the cornice is missing. The workmanship is neat and precise, in spite of a textured surface and the lack of final smoothing.
*Description*: *W* 136 cm, *H* 33.5 cm, T 17.5 cm. The pediment is moulded and its whole surface taken by a motif of a ribbon-tied, probably laurel-leaf wreath with a central gem. Two rectangular (3.5 cm deep) dowel holes for taking the roof are visible on the sides of the pediment close to its peak. Furthermore, two rounded holes (3 cm deep) for receiving the columns are cut on the roughly worked underside, the left one 38 cm and the right one 50 cm away from their respective corners. These measurements suggest that the broken-off left corner was 14 cm long, the original total length of the pediment amounting to *c.* 150 cm.
*Date*: Second half of 2nd century or 3rd century.
*Publication*: Burkowsky 2000: 76, no. 23; Migotti 2014: 117; Škrgulja 2018: 244, no. 265. Internet databases: *lupa* 22449.

Figure 70. (S VII.1) – *aedicula* (O. Harl, *lupa* 22449)

2. *Aedicula* (?) (S VII.2)
*Provenance*: Found in 1936 in unknown circumstances, possibly in the area of the E cemetery.
*Location*: GMS, inv. no. GMS-18 AZA.
*Stone*: Pohorje marble (W. Prochaska 2017).
*Preservation*: A fragment of a roof with a console and square panels. One panel is preserved entirely, while only a third of another survives, and there are other slight damages in places.
*Description*: *L* 52 cm, *W* 50 cm, H 22 cm. The cornice is decorated with a frieze of alternating acanthus and shear-shaped leaves. The preserved panel has a relief mask-like head, while in the opposite panel a mythological (Dionysian?) figure with curly hair seems to be depicted. The workmanship of the reliefs is schematic, and the surface is roughly worked.
*Date*: Later 2nd or early 3rd century.
*Publication*: Burkowsky 1999: 68. Internet databases: *lupa* 22455.
*Note*: The determination as a funerary building remains questionable because the findspot cannot be securely established.

Figure 71. (S VII.2) – *aedicula* (O. Harl, *lupa* 22455)

3. *Aedicula* (?) (S VII.3)
*Location*: GMS, inv. no. GMS-27701.
*Stone*: Most probably Pohorje marble (W. Prochaska 2017; M. Belak 2018).
*Preservation*: A well-preserved and slightly stained corner fragment of a cornice and ceiling.
*Description*: *L* 52 cm, *W* 32 cm, H 27 cm. The cornice is decorated with a frieze of alternating acanthus and shear-shaped leaves and a stylized Lesbian (?) cyma, divided by a fillet. The roughly worked surface of the ceiling is provided with a square dowel hole for joining the column.
*Date*: 2nd or 3rd century.
*Publication*: Unpublished. Internet databases: *lupa* 22456.
*Note*: Because of the lack of any clue to the provenance, the identification of this piece as a funerary building remains even more questionable then the one in the previous entry.

Figure 72. (S VII.3) – *aedicula* (O. Harl, *lupa* 22456)

*VIII. Unclassifiable*

1. Inscription (stele?, *titulus*?, altar?) (S VIII.1)
*Provenance*: A chance find before 1866 in the area of the N cemetery.

*Location*: Missing.
*Inscripion*: *Ti(berio) Claudio / Pontio / (centurioni) leg(ionis) XV Apo(llinaris) / an(norum) XXV. / H[e]res / [...?].*
Translation: To Tiberius Claudius Pontius, centurion of the Legion *XV Apollinaris*, 25 years old. Heir ...
Commentary: Tiberius Claudius Pontius was very young and obviously did not start his career as a simple soldier but as *centurio*, possibly his father (who must have been given citizenship as early as under Tiberius, or, alternatively, under Claudius) had also served in the Roman army. Pontius' family, which must have belonged to the indigenous upper class, perhaps originated from Siscia (suggested by Mócsy 1959: 211, no. 57/1). He was either sent to Siscia on a special mission, or his legion may have been stationed in the town, when he died. This would explain Pontius' high rank in the legion, but may be in contradiction with the fact that his tombstone was commissioned for him by his heir (cf. Radman-Livaja and Vukelić 2015: 401). The Legion *XV Apollinaris* may have been temporarily stationed at Siscia during the revolt of the three Pannonian legions in AD 14. Before its arrival in Carnuntum early under Claudius, the legion was briefly stationed at Vindobona (Šašel Kos 1995: 237; Mosser 2002; Mosser 2003: 141; 261 [Siscia]; 144–145 [Vindobona]; Radman-Livaja 2012: 162–165; 169–170, with earlier citations; Šašel Kos 2014: 80; Radman-Livaja and Vukelić 2015: 401).
*Date*: Mid, or second half of 1st century.
*Publication*: *CIL* III 10853; Mosser 2003: 261, no. 188; Vuković 1994: 83, no. 65; Radman-Livaja and Vukelić 2015: 401–402. Internet databases: *HD073473*; *EDCS-29000488*.

2. Inscription (altar?) (S VIII.2)
*Provenance*: Gardening in 1876 in the area of the SE cemetery.
*Location*: AMZ, inv. no. AMZ KS-352.
*Stone*: Crystalline (marmorized) limestone. A possible origin: Mt Medvednica above Zagreb (M. Belak 2018).
*Preservation*: Fragmentarily preserved. A small fragment survives, with insignificant remains of an indistinct relief on the broken-off side.
Description: *H* 29 cm, *W* 14 cm, *T* 7 cm. An inscribed stone plate, the full thickness of which remains unknown, shows indistinct remains of a relief within a moulded frame on the preserved side.
*Inscription*: *------] / C[---] / M[---] / VS+[---] /[5] pos[(uit?* or: *-uerunt?] / [------*
Hoffiller published the fragment as broken off on all sides; however, the photo in *lupa* shows that the left edge has been preserved, rendering the reading in the *lupa* as the only correct one. *AIJ* and *EDH* copy the text as if damaged also on the left. Line 4: *[---]VSI(?)[---]*: *EDH*; line 5: *[---]PO S[---]*, *EDH*.
*Date*: 2nd century (?).
*Literature*: Brunšmid 1909: 161, no. 352; *AIJ* 577. Internet databases: *lupa* 22737; *HD*072128.

Figure 73. (S VIII.2) – front side (AMZ, I. Krajcar)

Figure 74. (S VIII.2) – left short side (AMZ, I. Krajcar)

Figure 75. (S VIII.2) – stone, microscope (HGI, M. Belak)

3. Inscription (stele?, *titulus*?, altar?) (S VIII.3)
*Provenance*: A chance find before 1726 in the area of the SE cemetery.
*Location*: Missing.
*Preservation*: Fragmentarily preserved. Most of the inscription panel survived, but is broken off around all the edges.
*Description*: The rectangular inscription panel was higher than wide, with a framing (?) on the left side, sketched as a thin line on the drawing by Marsigli.
*Inscription*: *V(ivus) f(ecit) / M(arcus) Minic[i]/us Saturni[nus], / [ve]t(eranus) ex op[t]ion[e cl(assis)] /[5] [prae] toriae Rave[n/nat(is)], annor(um) [L]XXX, [sibi] / et Crispiae Ce[l]e[rinae] / coniugi caris[simae] / [et] Minici(i)s Cele[ri], /[10] Cert[o], Grat[o f(ilis)] / et Minici(i)s Ca[lvo et ] / [Pris]cae pare[ntibus] / b(ene)m(erentibus) p(osuit).*
Line 4: *ex op[t]ion(e) [cl(assis)], EDH*. Line 13: *[t(itulum)] m(emoriae) p(osuerunt), lupa*; *b(ene)(?), EDH*.
Translation: Marcus Minicius Saturninus, veteran and former *optio* of the praetorian navy from Ravenna, 80 years old, had (this tombstone) erected in his lifetime for himself and Crispia Celerina, his dearest wife, and to Minicii Celer, Certus, and Gratus, his sons, and to Minicii Calvus and Prisca, his most deserving parents.
Commentry: If Saturninus' tombstone could be dated to the late second half of the 1st century, which seems plausible, he could be regarded as one of the first colonists in Siscia. His case could be linked to the military diploma (*CIL* XVI 14 = *EDCS*-12300218), issued under Vespasian on 5 April AD 71 and found at Salonae, in which it is stated that the veterans who had served in the navy of Ravenna under Sex. Iulius Bassus were settled in Pannonia (... *veteranis qui militaverunt in classe Ravennate sub Sex. Iulio Basso ... sunt deducti in Pannoniam* ... cf. Mrozewicz 2015: 152).
His parents, who had the same *gentilicium*, must have been of freedmen origin; it has been suggested that they originated from Dalmatia, but this is not certain (Mócsy 1959: 211, no. 57/2). The *gentilicium* Minicius, which is particularly frequent in Italy and Hispania (*OPEL* III: 82; *Epigraphik-Datenbank* Clauss/Slaby), does not occur elsewhere in Siscia and neither on the *tesserae*, but the name Saturninus (a very popular cognomen in Italy and all western provinces) is attested on seven lead tags (*OPEL* IV: 51–53; Radman-Livaja 2014: 259), Celer on two, Gratus as well, Calvus on one, and Prisca on five, while Crispius, Celerina and Certus are not documented (Radman-Livaja 2014: 184 [Celer]; 213 [Gratus]; 178 [Calvus]; 248 [Prisca]). Crispia Celerina was most probably a native of Siscia (Mócsy 1959: 211, no. 57/2), since the *gentilicium*, although rare otherwise, has been attested three times at Siscia, also occurring several times in Gallia Narbonensis (*OPEL* II: 85; Crispus and Crispinus are very well documented on the *tesserae*, Radman-Livaja 2014: 192).

Figure 76. (S VIII.3) – scan of the drawing (after Marsigli 1726)

*Date*: Later 1st or early 2nd century.
*Publication*: Marsigli 1726: 113, Tab. 44/4; *CIL* III 3971; Vuković 1994: 88, no. 79; Gračanin 2004: 23 no. 7; Radman-Livaja and Vukelić 2015: 399, 404. Internet databases: *lupa* 4312; *HD*074365; *EDCS*-26600430.
*Note*: The stone was drawn by L. F. Marsigli, but with no clue as to the kind of monument.

4. Inscription (S VIII.4)
*Location*: Missing.
*Inscription*: *D(is) M(anibus). / L(ucio) Marcio / [pa]tri(?) ca/ [ri]ssim[o(?).]P[.] / [---]?.*
Mommsen did not himself see the inscription, but he saw the original copy (*Katancsich geogr. 1, 443 ex schedis Barichevich, quas ipsas vidi inter Katancsichiana*). He noted only two letters missing in the 3rd line, thus *[pa]tri* would have been the only possible supplement, not *[fra]tri*. There is no space for the cognomen of Lucius Marcius. If the copy is reliable, nothing is missing after the last line.

Translation: To the spirits of the departed. To Lucius Marcius, dearest father (?) ...

Commentary: According to Mócsy (1959: 212, no. 57/14), the family would have come from northern Italy. Another Marcius is documented at Siscia: a name of a centurion, Marcius P[-], is inscribed on a mattock (Radman-Livaja 2013). The *gentilicium* Marcius is very frequent, particularly well attested in Italy, Hispania, Gallia Narbonensis, Dalmatia, and especially also in Pannonia, in some cases it could have been an assonance name (*OPEL* III: 56–57; *Epigraphik-Datenbank* Clauss/Slaby), and it also appears on one lead tag from Siscia (Radman-Livaja 2014: 226–227).

*Date*: Perhaps the 2nd century because of the lack of a cognomen, but not earlier than the end of the 1st century, due to the invocation to the *D(is) M(anibus)*.

*Publication*: *CIL* III 3988. Internet databases: *HD*074390; *EDCS*-28800749.

5. Inscription (*titulus*, stele, *aedicula*?) (S VIII.5)

*Location*: GMS no. GMS-27713.

*Stone*: Gummern, Pohorje or Salla marble (W. Prochaska 2017).

*Preservation*: The surviving part of the inscription panel is well preserved, with the frame battered.

*Description*: *W 40* cm, *H 37* cm, T 18 cm. The inscription panel was framed with a border of unknown shape.

*Inscription*: *---]? / M(arco) Iu[---] / con[iugi ---] / aeḍ[ili(?) ---] / [---.*

Regular letters, larger in the first preserved line; left margin of the inscription field is preserved; possibly no line is missing above.

Translation: ...? To Marcus Iu[], (her) husband, aedile? ...

Commentary: *Gentilicia* starting in *Iu-* are many, possibly *Iulius* or *Iunius*. *Ael[ia?]* has been suggested as the first reading, but the remains of the last visible letter rather favour the reading *aed[ili]*, proposed in *lupa* as an alternative.

*Date*: 2nd century (?).

*Publication*: Radman-Livaja 2014: 250–251. Internet databases: *lupa* 22445; *EDCS*-62500015.

Figure 77. (S VIII.5) – inscription (O. Harl, *lupa* 22445)

6. Inscription (S VIII.6)

*Location*: Missing.

*Inscription*: *---] / Crisp/iae [---] / Triṃ[---] / P[..] con/[iug] i I/[---] / [---].*

Slightly different reading in *EDH* (*[------] / Crisp/iae [--] / TRIM[-] / P[--]CON / [--]II / [------]*), which does not correspond entirely to the transmitted copy.

Translation: ... To Crispia ... (his) wife (?) ...

Commentary: Crispia in this epitaph may well have been a native of Siscia, as was, according to Andreas Mócsy (1959: 57/2 [p. 211]), the above-mentioned Crispia Celerina, wife of Marcus Minicius Saturninus, veteran and former *optio* of the praetorian navy from Ravenna and perhaps one of the first colonists in Siscia (*CIL* III 3971 = *lupa* 4312 = *EDCS*-26600430; Radman-Livaja, Vukelić 2015: 399; 404). This does seem plausible, because the *gentilicium*, although rare otherwise (except in Gallia Narbonensis: *OPEL* II 85), has been attested altogether three times at Siscia. One Quintus Crispius Gallicanus is known from a dedication to Silvanus Domesticus (*CIL* III 15181 = *AIJ* 546 = *HD*071840).

*Date*: Probably 2nd or 3rd century, although only based on the use of a *gentilicium*.

*Publication*: *CIL* III 3984; Mócsy 1959: 211, no. 57/2; Gračanin 2004: 27-28 no. 16. Internet databases: *HD*074387; *EDCS*-28800747.

7. Inscription (S VIII.7)

*Location*: Missing.

*Inscription*: *D(is) <M=D>(anibus) / C(ai) Urbici Firmi <d=T>[..?]e<c=X>(urionis) / col(oniae) Sept(imiae) Sisc[i] anor(um). / Urb(icia?) Cond(olla?) vixit [---] /*[5] *[---.*

Probably badly transcribed; emendations of the first three lines were proposed by Mommsen.

Translation: To the spirits of the departed of Gaius Urbicius Firmus, councillor (*decurio*?) of the colonia Septimia Siscia. Urb(icia) Cond(olla?) who lived ...

Commentary: Urbicius is a relatively rare *gentilicium*, twice attested in Pannonia, including this tombstone (*OPEL* IV: 185). Mommsen's suggestion to supplement the name in the last line as *Urb(ica)*, accepted by *EDCS*, but *Urb(icia)* seems much more likely (thus also *EDH*). *Cond(uctor)* is supplemented in the *EDCS*, but does not seem convincing. Several *cognomina* starting in *Cond-*, are attested, all of them are only documented once, except the Celtic name Condollus (Delamarre 2007: 72), which occurs three times, once in Noricum (*OPEL* II: 72; Kakoschke 2012: 359).

*Date:* End of 2nd or in the 3rd century.

*Publication*: *CIL* III 3976; Gračanin 2004: 26 no. 12. Internet databases: *HD*074369; *EDCS*-28701052.

Figure 78. (S VIII.8) – fragment (GMS, B. Suntešić)

8. Stele (?), sarcophagus (?) (S VIII.8)
*Provenance*: Unknown, possibly from the area of the E cemetery.
*Location*: GMS, inv. no. GMS-27730.
*Stone*: Upper-Jurassic/Low Cretacious micritic limestone. The origin: the mountains of NW Croatia, most probably the Banovina, S of Sisak (M. Belak 2018).
*Preservation*: A fragmentary preserved panel with one raised border.
*Description*: *L 60* cm, *W 33* cm, *T 23-34* cm. The flat niche (?) has a wide, band-like frame and no traces of an inscription or a relief, and is heavily chipped and pitted all over.
*Date*: 2nd – 4th century.
*Publication*: Unpublished.

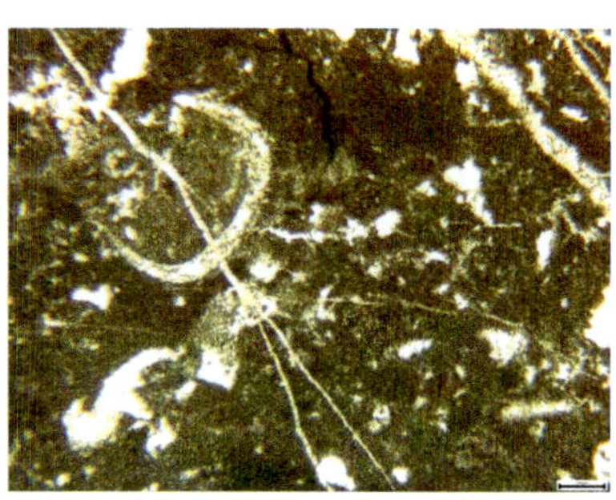

Figure 79. (S VIII.8) – stone, microscope (HGI, M. Belak)

9. Inscription (S VIII.9)
*Provenance*: A chance find in the 18th century at an uncertain place, possibly in the SE cemetery.
*Location*: Missing.
*Stone*: Marble (according to the *CIL*).
*Preservation*: According to the *CIL*, it was a slab restored from fragments.
*Inscription*: *In memoriam Aurel(iae) / Rufinn(a)e coniugi obse/quentissimae ad[q(ue)] pu/dicissimae et innocent/*[5]*tissimae A[u]r[el(ius)] Nican/[d]er [---]VNLCV[---] / [------] / d(ies) XV ob [memoriam] / [p]osuit.*

Line 2: *Rufinae*, *EDH*, wrongly. *N* and *E* are in ligature, possibly also *NAE*, although this is not noted in the *CIL*. However, all other feminine datives in the text are correct. Lines 4 and 5: *et innocent/tissimae*: not noted in the *EDCS*.
Translation: In memory of Aurelia Rufinna, most indulgent, chaste, and blameless wife. Aurelius Nicander ... 15 days, had (this monument) erected in her memory (?).
Commentary: The Greek name Nicander occurs very rarely in the western part of the Empire, except in Italy, often also among freedmen (Solin 1996: 217; *Epigraphik-Datenbank* Clauss/Slaby); according to the *OPEL* once in northern Italy, Hispania, Gallia Narbonensis, and Noricum, as well as twice in Pannonia (including this epitaph) and in Moesia Inferior (III: 100). Rufinus/a is a very frequent Latin name, also in Pannonia (*OPEL* IV: 33–34); in Siscia, the female name Rufina occurs several times (once as Rufinna), while Rufinus twice appears on the *tesserae* (Radman-Livaja 2014: 254): the name is very well attested in the town.
*Date*: 3rd or 4th century (?)
*Publication*: *CIL* III 3983; Vuković 1994: 85, no. 72; Gračanin 2004: 27 no. 15. Internet databases: *HD*074386; *EDCS*-28800746.
*Note*: The finding place is designated in the *CIL* as *Sisciae in arce*, which was explained by V. Vukelić (2011: 257, fn. 637) as possibly suggesting the 16th-century military fort 2 km SE of the town centre as a secondary findspot, indicating at the same time the SE cemetery as a possible original findspot.

10. Inscription (S VIII.10)
*Provenance*: A chance find in 1876 at an unknown place.
*Location*: AMZ, inv. no. AMZ KS-355 (misplaced).
*Stone*: Marble (after Brunšmid 1909).
*Preservation*: A small fragment survived.
*Description*: *H 12* cm, *W 11* cm, T 2.5 cm; an inscribed stone slab.

Figure 80. (S VIII.10) – scan of the drawing (after the *AIJ*)

*Inscription*: *------] / [---]C[---] / [---]V(?)DO[---] / [------*
According to the drawing from Brunšmid and in the *AIJ*, the first visible letter in the second inscribed line is most probably a *V*.
*Date*: probably Late Roman.
*Publication*: Brunšmid 1909: 161, no. 355; *AIJ* 578. Internet databases: *HD*072129; *EDCS*-11301129.

11. Inscription (S VIII.11)
*Provenance*: Found in the area of the SW cemetery at an unknown date before 1873.
*Location*: Missing.
*Stone*: Not mentioned in the literature.
*Inscription*: *Lucernio qui / vix<i=E>t annos / VII et Gaudenti/ anus frater ipsiu/⁵s germanus qui vi/xit ann<o=V>s IIII. Cre/ cian(a)e ux[ori ..].*
According to Mommsen's transcription, the *E* in the first line was longer than other letters. Line 7: possibly *<G>r(a)ecian(a)e*, thus Barkóczi, *OPEL*, *EDH*, but not necessarily: Mommsen has *Creciana* (so also *Epigraphik-Datenbank* Clauss/Slaby), and did not comment on the name; emended name does not appear in the *CIL* indexes among the *cognomina*, but neither does Creciana.
Translation: Lucernio who lived 7 years and Gaudentianus, his full brother, who lived 4 years. To Greciana, (his) wife ...
Commentary: Only one case of the name Lucernio (from Hispania) is noted in *OPEL* (III: 34), some more are listed in the *Epigraphik-Datenbank* Clauss/Slaby. The names Creciana and Gr(a)eciana have not been attested to date elsewhere (Barkóczi 1964: 330/29 no. 29; *OPEL* II 169; not listed in Solin and Salomies 1994), while the name Graecinus occurs several times (Solin and Salomies 1994: 339; *OPEL* II 169). Gaudentianus, as much as Gaudentius, is a typical Late Roman name (Alföldy 1969: 209).
*Date*: Late Roman.
*Publication*: *CIL* III 3987; Gračanin 2004: 29 no. 19. Internet databases: *HD*074389; *EDCS*-28800748.

12. Inscription (S VIII.12)
*Location*: Missing.
*Stone*: Not mentioned in the *CIL*.
*Inscription*: *Memoriae aeternae. Aur(elius) / Reginus Alexsius (!) / Cara coniux posuit et / Amantia.*
Mommsen remarked that the transcription might have been emended ('Interpolatam credo'). Line 1-4: Gračanin (2004: 27) regarded *cara* as an adjective and Amantia as a name, which seems less likely. *EDH* decided for the reading: *Aur(elius) / Reginus Alexsi<a>(!) / cara coniux posuit et / amantia*. However, this explanation can hardly be correct, since it is arbitrary to change Alexsius to Alexsia and *amantia* is not a Latin word. *Cara* and *amantia* are also not regarded as names in the *Epigraphik-Datenbank* Clauss/Slaby. In the indexes of *CIL*, Amantia from this tombstone is listed among *cognomina*.
Translation: To the eternal memory. Aurelius Reginus Alexsius. Cara (his) wife had (the monument) erected and Amantia.
Commentary: According to the *OPEL*, the cognomen Reginus is attested ten times in Gallia Belgica and both Germaniae and several times in Aquitania and Gallia Lugdunensis, while it is rare elsewhere (IV: 25). Reginus/ Riginus also appears twice on the Siscian *tesserae* and may have been regarded as an assonant name in the Celtic-speaking regions (Radman-Livaja 2014: 252; Delamarre 2007: 152). Alexsius is a Latinized and slightly corrupted form of the Greek name Alexios; the name was particularly popular in the later Byzantine Empire, while earlier it occurs sporadically (*OPEL* I: 42). Carus (also Karus) is a frequent name, particularly in Gallia Narbonensis, Hispania, and Pannonia (*OPEL* II: 39), and was obviously a name with affinities in the Celtic-speaking provinces, a 'translation' name (Meid 2005: 190); it also occurs on two Siscian lead tags (Radman-Livaja 2014: 182). The name Amantius/a is attested three times in northern Italy and twice in Dalmatia, otherwise it occurs sporadically (*OPEL* I: 46).
*Date*: Probably Late Roman.
*Publication*: *CIL* III 3982; Gračanin 2004: 26-27 no. 14. Internet databases: *HD*074385; *EDCS*-28800745.

13. Inscription (S VIII.13)
*Provenance*: Ploughing in 1903 in the area of the SW cemetery.
*Location*: AMZ, inv. no. AMZ KS-353 (misplaced).
*Stone*: Marble (after Brunšmid 1909).
*Preservation*: A tiny fragment of a slab survives.
*Description*: *H* 5 cm, *W* 7 cm, T 1.6 cm. The slab was inscribed on both sides, with only a letter *A* discernible on one side.
*Date*: Probably Late Roman.
*Publication*: Brunšmid 1909: 161, no. 353.

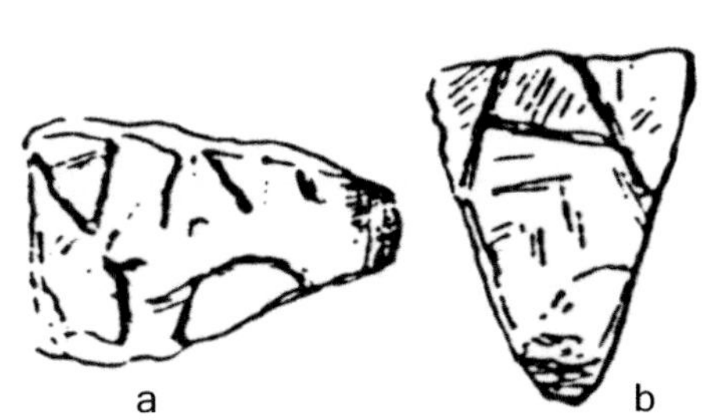


Figure 81. (S VIII.13) – scan of the drawing (after the *AIJ*)

14. Inscription (S VIII.14)
*Location*: AMZ, inv. no. KS-354 (misplaced).
*Stone*: Marble (after Brunšmid 1909).
*Preservation*: A tiny fragment of a slab survives.
*Description*: *H 14* cm, *W 14* cm, T 1.5 cm. The slab was inscribed on both sides.
*Inscription*: Side A: *---]sia[e] Cn[---] / [--- anni]s plus m[inus [---] / [--- v]ixit civ(?)[---*
Side B: *--- in]noce[nt---] / [---] ridu[m?---*
Letters are irregular, typical of the Late Roman period, the lower *hasta* of the letter *L* is not horizontal but turned obliquely towards the bottom. Line 1 (A): . .? . *An]sia Cn[....*, proposed in *AIJ*.
Translation: Side A: ... to ...sia ... who lived more or less ... years ...
*Date*: Probably Late Roman.
*Publication*: Brunšmid 1909: 161, no. 354; *AIJ* 576. Internet databases: *HD*049786; *EDCS*-11301127.

*a* *b*

Figure 82. (S VIII.14) – scan of the drawing (after the *AIJ*)

15. Inscription (*titulus*?, sarcophagus?) (S VIII.15)
*Provenance*: Gardening prior to 1830 in the area of the Late Roman section of the SE cemetery within the town walls.
*Location*: Missing.
*Preservation*: An inscription presumably preserved in its entirety.
*Description*: A horizontal rectangular panel was taken by a five-lined inscription topped by two unusually composed Christian pictograms. On the left was a saltire cross with the Greek letters *alpha*, omeag, pi, and rho in the interstices, and on the left a saltire type of monogrammatic cross accompanied by an alpha and omega motif.
*Inscription*: *Felicissimae et sacmeneuno* (sic!) *coniugi. / Dominus Victorinus maritus sepulcrum / eorum collocavit. Felicissima qu(a)e vixit / cum eo annis XIII anno urbis DXVII sarunam /*[5] *[---]VIISI[---]C[---]DX[---]PHY[---] deo.*
Mommsen did not comment on the inscription. Gračanin suggested that the inscription may have been counterfeit. Line 4: *sar[.]unam*: Gračanin.
Translation: To Felicissima (to the most happy?) and ? consort. Master Victorinus, her husband, erected their tomb. Felicissima who lived with him for 13 years, in the 517th year of the town foundation ... to the god.
Commentary: Mommsen did not regard *Sacmeneunus* as a name, since it is not noted by him in the indexes. The name Felicissimus/a is frequent especially among freedmen, it is well attested particularly in Italy, but also in Dalmatia after the mid 2nd century AD, less so in Pannonia (Barkóczi 1964: 312; Alföldy 1969: 201–202; *OPEL* II: 137). Victorinus/a, derived from Victor and very popular in Pannonia, occurs much more frequently after the Marcomannic Wars (Kajanto 1965: 68; Barkóczi 1964: 328; Lochner von Hüttenbach 1989, 182; *OPEL* IV: 168–169).
*Date*: Late Roman.
*Publication*: Artner 1830: 24; *CIL* III 3996a; *ILCV* 1686; Migotti 1997: 41–42; Gračanin 2004: 33 no. 25. Internet databases: *HD*074868; *EDCS*-29200145.
*Note*: The inscription, recorded as a funerary stone and itself mentioning a *sepulchrum*, could have been a *titulus*, although the layout of the text rather speaks for a sarcophagus.

16. Funerary sculpture (*aedicula*?, grave enclosure?) (S VIII.16)
Provenance: Found as reused in a fill in the 2007 archaeological excavations, probably in the area of the N cemetery.
*Location*: GMS, inv. no. GMS-22187.

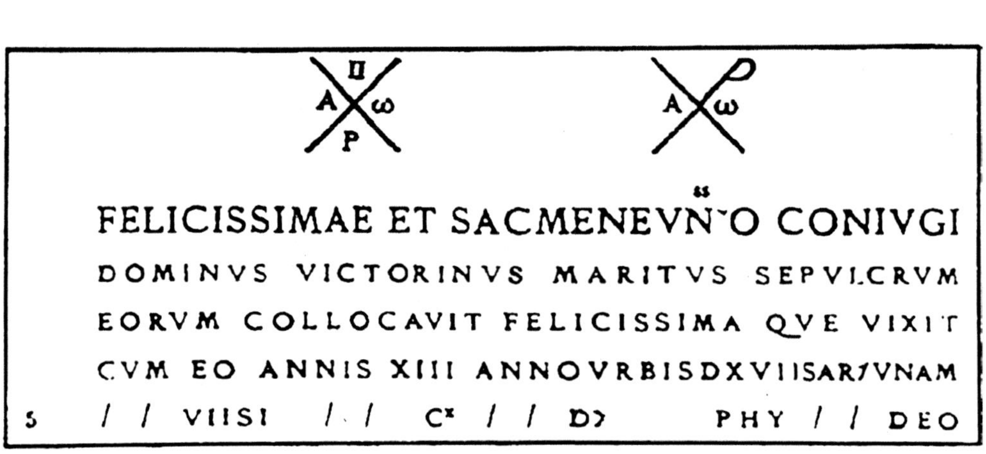


Figure 83. (S VIII.15) – scan of the drawing (after the *CIL*)

Figure 84. (S VIII.16) – sculpture (GMS, B. Suntešić)

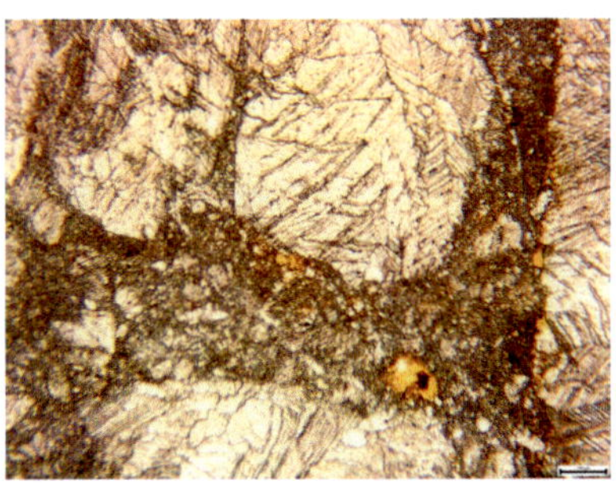

Figure 85. (S VIII.16) – stone, microscope (HGI, M. Belak)

*Stone*: Badenian biocalcarenite. A probable origin: the mountains of N Croatia (M. Belak 2018).
*Preservation*: A lion sculpture fragmentarily preserved, with the head and the right side of the body chipped off.
*Description*: *L* 19 cm, H 13 cm, *T* 6.5 cm. The statuette of a lion is attached onto a slightly undulating, irregularly carved plate-like base. The lion is squatting on the hind legs and is holding a supine bull's head in his front paws, with its tail curving along the left side of its body. The lion's mane is rendered in parallel, vertical, hatched ridges, and the workmanship is crude, stiff and schematic.
*Date*: Second half of 2nd or 3rd century.
*Publication*: Burkowsky 2007: 11–13; Nikoloska and Burmaz 2008: 214–218.
*Note*: The statuette was either one of a pair of lions watching the entrance to a grave, or, but less likely, it was a fragmented lion top of a stele. It was found inside the town but close to the town walls, in a Roman fill dated to the time span between the 2nd and 4th centuries. Therefore, it comes from a secondary findspot, and was recovered together with several artefacts and stone monuments pointing to a religious context, leaving open two possibilities. Either it dates from the period before the erection of the town walls in the Severan period, when its findspot could have been in the area of the cemetery, or it was brought there in a later period from the cemetery outside the walls.

17. Column on an inscribed pedestal (S VIII.17)
*Location*: AMZ, inv. no. AMZ KS-97.
*Stone*: Pyroxene marble, probably from Pohorje (M. Belak 2018).
*Preservation*: Fragmentarily preserved. A complete pedestal and the lower part of a column survive with the pedestal slightly chipped around the edges and the column crusted at places.
*Description*: *H* 37 cm, W 17 cm, T 27 cm. The pedestal, which holds the inscription on its front, is composed of upper and lower oblong parts and a concave section between them, and is supporting a spirally fluted column made in one piece with the base. The column itself sits on a simple rounded base and holds a relief of a straight and stiff animal (most probably a sphinx') leg along its whole height, tapering towards the bottom, with the paw resting on the base and showing five strong claws.
*Inscription*: *Ianuaria // zaesis*.
Translation: Ianuaria (you will) live.
Commentary: The form should be *zeses*, transcribed from Greek, which often occurs in Late Roman (early Christian) inscriptions. The word could also be abbreviated as *z(eses)*. Zaesis, however, should not be explained as a name (thus in *ILJug* p. 6*). The name Ianuarius/Ianuaria appears on the curse tablet from Siscia (Curbera and Jordan 1996) and also on six of the

Figure 86. (S VIII.17) – funerary column (AMZ, I. Krajcar)

lead tags (Radman-Livaja 2014: 215–216). Due to the great popularity of the name in Pannonia (*OPEL* II 189–190) and elsewhere in the Celtic-speaking provinces because of similarity with certain Celtic names (a 'Deckname', Wedenig 2014: 342–343; 347), this is not at all surprising. The name Ianuarius, however, became one of the most frequent names both in the Late Roman pagan and in Christian inscriptions. In Rome this was the second most popular name among the Christian population, after Leo, Lea (Solin 2003a: 27, 30, cf. 19).

*Date*: Probably 4th century.

*Publication*: *CIL* III 3986; Brunšmid 1905: 50, no. 97; Kovács 2007: 70; *AIJ* 571 (add. *ILJug* p. 6*); Gračanin 2004: 28–29 no. 18. Internet databases: *HD*035605; *EDCS*-26600435.

*Note*: judging from the inscription, the column most probably stems from a funerary context – a building or enclosure.

***The presumed territory of Siscia* (S[t])**

*I. Stelai*

1. Stele (S[t] I.1)

*Provenance*: Building works in 1964 in Topusko, 38 km SW of Sisak.

*Location*: GMK, inv. no. GMK-397a; temporarily kept in the open in Topusko.

*Stone*: Upper-Miocene sandstone appearing locally on any geomorphological elevation underneath the Quaternary layers. The origin: local (M. Belak 2018).

*Preservation*: Fragmentarily preserved. The upper part is broken off, but the inscription panel survives, albeit chipped on the top left and around the edges, and weathered and stained all over; still, the letters are fairly well preserved.

*Description*: *H* 184 cm, W 90 cm, T 33 cm; typology uncertain (see at *Note*). The inscription panel is moulded on all sides but the bottom one, leaving free only narrow strips of smooth stone at the lateral sides and a broader one at the bottom.

*Inscription: D(is) M(anibus). / L(ucio) Val(erio) L(uci) f(ilio) Verec/undo, domo Sisc(ia) / vet(erano) leg(ionis) XV Apol(linaris) an(norum) /[5] LXV, quo minus in / funus expensum / erat quod colle/gius(!) veteranorum / contulit ad titu/[10]lum. / H(ic) s(itus) est.*

Letters are fairly regular, larger in the last line, there are several *T longae*.

Translation: To the spirits of the departed. To Lucius Valerius Verecundus, son of Lucius, from Siscia, veteran of the Legion *XV Apollinaris*, 65 years old. His funeral was less expensive due to the college of

Figure 87. (S[t] I.1) – stele (M. Jambrović)

Figure 88. (S[t] I.1) – stone, microscope (HGI, M. Belak)

veterans who contributed towards the monument. He lies here.

Commentary: Verecundus may have taken part in the First Jewish War under Vespasian (AD 66–73); it is highly unlikely to date the tombstone between AD 31 and 70 (Acrudoae 2013: 388–389). He belonged to an association of veterans, which obviously existed at Siscia, and which contributed towards the expenses of his tombstone. *Collegia veteranorum* have only rarely been attested to date: at Aquileia, Ateste, Carnuntum, and Ancyra (Bennett 2006: 93–99 (Ancyra); Aquileia: *CIL* V 784 = *EDR*116867; Ateste: *CIL* V 2475 = *EDR*130463; Carnuntum: *CIL* III 11189 = *HD*073138; *CIL* III 11097 = *HD*000911). Since Siscia is noted as Verecundus' *origo*, this may mean that Topusko did not belong to the *ager* of the town.

The Valerii were one of the most illustrious upper-class families at Siscia (Šašel Kos 2017b: 179). The name Verecundus (having the sense of honour, modest) is frequent everywhere in the western part of the Empire (*OPEL* IV: 157) and may have been an assonance name (Mócsy 1959: 195–196; Lochner von Hüttenbach 1989: 179; Delamarre 2007: 196); at Siscia, it also occurs on one *tessera* (Radman-Livaja 2014: 281).

*Date*: 108–116 (after Mosser: having been recruited at Siscia, Vrecundus had come to Carnuntum and returned to Siscia when discharged).

*Publication*: *ILJug* 3117; Šegvić 1988 (*AE* 1989: 617); Mosser 2003: 102, 258, no. 182; Radman-Livaja and Vukelić 2015: 404. Internet databases: *lupa* 5740; *HD*035205; *EDCS*-10102113.

*Note*: The upper part of the stele could have been of any of the types recorded for the soldiers of the 15th legion *Apollinaris* in the given period: triangular (free pediment), rectangular or pentagonal (with or without a pediment), and rounded (Mosser 2003: 20–24, Taf. 1; Weber-Hiden 2014: 73, Abb. 2, Phase III). Given the preserved dimensions, the stele must have had an upper part probably containing a pediment within a rectangular panel. There is no knowing how high was the lower part, as some of it remains buried in the ground. However, the presence of a considerable base would not accord with the general typology of the 15th legion's funerary monuments as produced by Mosser (2003) and Weber-Hiden (2014).

2. Stele (S[t] I.2)

*Provenance*: A chance find in the 2000s during building works in Topusko, 38 km SW of Sisak.

*Location*: GMK, inv. no. GMG-1113a; temporarily kept in the entrance hall of the hotel Toplica in Topusko.

*Stone*: Upper-Miocene sandstone appearing locally on any geomorphological elevation underneath the Quaternary layers. The origin: local (M. Belak 2018).

*Preservation*: Completely preserved. The column-shafts of the portrait niche and inscription panel are severely damaged, pitted and stained, with cuts all over, and the portraits are extremely weathered and obliterated. The blind, unsmoothed inscription panel has suffered

Figure 89. (S[t] I.2) – stele (B. Djurić)

Figure 90. (S[t] I.2) – detail (B. Djurić)

Figure 91. (S[t] I.2) – stone, microscope (HGI, M. Belak)

a major cut and a pit in its left middle. In addition to multiple damages, roughly worked reliefs and their backgrounds most probably indicate an unfinished state.

*Description*: H 240 cm (268 cm with the tenon), W 80 cm, T 33 cm; a reduced type 5a, with a blind base. The stele is topped by a lion pediment featuring two short-haired, wingless floating Erotes with legs crossed, supporting with both hands a wide-mouthed receptacle. The triangle of the pediment is slightly narrower than the architrave, with the free space on both ends taken by the rams' heads in lions' paws. Below is a rectangular portrait niche flanked by vertically fluted columns with flutes wider in the upper part and narrower in the lower third of the shafts. The columns have composite capitals and Attic bases standing on high parallelepiped shafts and supporting an articulated architrave. In the niche, half-figures of a deceased couple in the pose of *dextrarum iunctio* are depicted against the background of a very roughly worked or unfinished shell, reminding of a double-arched niche. They are both wearing clothes of indistinct shape, with long sleeves and oval necklines, the latter slightly V-shaped in the man. The woman has an oval, slightly block-like face, large almond-shaped eyes, high arched brows, a straight-edged nose, a two-tiered (?) hairdo with a central parting on the forehead and a wrapped headwear, and with the ears covered. The man has a block-like face, large almond-shaped eyes, high arched brows, a straight-edged nose, a thick cap-like hair rendered in wide, parallel ridges running from the crown towards the forehead and ending in a concave fringe, and a short beard and moustache; he seems to be holding something in his left hand, probably a scroll. The portrait niche is divided from the inscription field by a wide decorative band featuring two heraldic dolphins with intertwined tails, flanked by two additional dolphins, each of them devouring a polyp. A blind and roughly worked inscription panel is flanked by apparently smooth columns with composite capitals and Attic bases. Below is a plain base lacking a frame or moulding, supported by an equally roughly worked rectangular tenon. The workmanship appears to be a combination of rough and stiff execution of reliefs (especially the human figures, and less so the animals) and an unfinished state.

*Date*: AD 117-161 (the time of Hadrian or Antoninus Pius).

*Publication*: Šegvić 2006: 271 (no mention of the stele in the text, with only its portrait niche illustrated).

3. Stele (S[t] I.3)

*Provenance*: From the 19th century walled in a building in the village of Viduševac, 27 km SW of Sisak and 10 km NE of Topusko, but presumed to be from Topusko; in 1990 relocated to Topusko.

*Location*: GMK, inv. no. GMK-433a; temporarily kept in the entrance hall of the hotel Toplica in Topusko.

*Stone*: Upper-Miocene sandstone appearing locally on any geomorphological elevation underneath the Quaternary layers. The origin: local (M. Belak 2018).

*Preservation*: Fragmentarily preserved, with the upper half surviving. The monument is extremely weathered, stained and pitted all over, and probably unfinished. The reliefs are heavily obliterated, the columns mostly chipped off, the faces battered flat for easier walling, and there are traces of lime from the walling at places, especially on the portraits.

*Description*: *H 102* cm, W 80 cm, T 30 cm; probably variant of type 5a. The preserved fragment of the stele contains a lion pediment, a rectangular portrait niche, and a smaller uppermost section of the inscription panel, possibly flanked by columns supporting an architrave. The pediment is low-pitched, with a central cubical *acroterion* featuring a relief rosette flanked by two parrots, and with a robust lion lying along each side, apparently holding a pray in its front paws. The portrait niche is flanked by columns supporting an architrave and holding busts of a deceased couple. The woman on the left-hand side has an oval-triangular face, a slim neck, and a huge headwear. A tiny concave fringe of hair on the forehead is barely discernible, while short geometricized strands of hair behind and below the ears are clearly visible. She is wearing a tunic or an over-tunic with a rounded neckline and with two

Figure 92. (S[t] I.3) – stele (B. Djurić)

Figure 93. (S[t] I.3) – stone, microscope (HGI, M. Belak)

large brooches (apparently of a Norican wing type) on the shoulders, and a cloak hanging straight from the shoulders. Her left hand seems to be resting on the left shoulder of the man on her right, presumingly her husband. He has an oval-triangular face and a slim neck, a relatively thick hair combed in narrow strands forward from the crown, and is wearing a tunic with a thick, rounded neckline, while the type of the cloak remains indiscernible.
*Date*: 2nd, or first half of 3rd century.
*Publication*: Unpublished.
*Note*: The stele is placed awkwardly behind another one, with its lowermost preserved part remaining inaccessible.

4. Stele (S[t] I.4)
*Provenance*: A chance find in 1982 in Topusko, 38 km SW of Sisak.
*Location*: AMZ, inv. no. AMZ KS-341.
*Stone*: Upper-Miocene sandstone appearing locally on any geomorphological elevation underneath the Quaternary layers. The origin: local (M. Belak 2018).
*Preservation*: Fragmentarily preserved, with the upper part missing. There are chips and weathering on the upper and bottom left-hand corners and the right-hand column shaft. The surface of the inscription panel is smoothed but with superficial pits and abrasions, and the base is roughly worked.
*Description*: *H* 175 cm, W 77.5 cm, T 25 cm; probably type 3a or 3b. The preserved part of the stele is composed of an inscription panel flanked by smooth columns with Tuscan bases and stylized Corinthian capitals, which support an articulated architrave, possibly topped by a frieze and a pediment. The plain base with a tenon was set into a roughly worked parallelepiped pedestal measuring 103 x 85 x 35 cm, provided by a central socket measuring 45 x 29 x 25 cm.
*Inscription*: *D(is) M(anibus). / Valentiniae / Festae o(bitae) an(norum) L / post(eris)q(ue) eius. /*[5] *Serena soror / et Ingenua et / Serenus fili(i).*

Figure 94. (S[t] I.4) – stele (AMZ, I. Krajcar)

Figure 95. (S[t] I.4) – stone, microscope (HGI, M. Belak)

Letters are regular, there are punctuation marks between some of the words. Line 1: Brunšmid and Hoffiller (*AIJ* 522) noted remains of erroneously carved letters at the beginning of the line, which were erased, but these remains may also be interpreted as a later damage of the stone. Line 3: *O* has a dot inside: it may have signified *theta nigrum* (thus *EDH*), unless this is a chance dot; similar meaningless dots are several in the text; *(obiit)*, *AIJ* and *lupa*. Line 4: *I longa*.

Translation: To the spirits of the departed. To Valentinia Festa, who died at the age of 50 years, and to her descendants. Her sister Serena and her children Ingenua and Serenus.

Commentary: Valentinius is a rare *gentilicium*, the *OPEL* twice attested in northern Italy, Gallia Narbonensis, and Pannonia (once in this case), and a few times in Gallia Belgica with Germaniae and in Noricum (IV: 140); it is derived from the name Valentinus, which is frequent everywhere, particularly in Italy, Dalmatia, Noricum, and Pannonia (*OPEL* IV: 140; Lochner von Hüttenbach 1989: 173; *Epigraphik-Datenbank* Clauss/ Slaby), and after the Marcomannic Wars (Alföldy 1969: 320). At Siscia, it appears on four *tesserae* (Radman-Livaja 2014: 278). Festus and Festa were most popular names in Siscia (as also elsewhere, *OPEL* II: 139–140), both on stone monuments and on the lead tags, where they appear 19 times and where all bearers of the name seem to have been of peregrine status (Radman-Livaja 2014: 205–206). The name Serenus/ais relatively frequent, also in Pannonia and the Celtic-speaking provinces (*OPEL* IV 71), where it could be regarded as an assonance name (Lochner von Hüttenbach 1989: 149). At Siscia, it occurs on three *tesserae* (Radman-Livaja 2014: 262). The name Ingenuus/a (a child of free-born parents) appears on 14 lead tags, which made it one of the most popular names at Siscia (Radman-Livaja 2014: 216–217); the name is generally well attested in Pannonia and particularly Noricum, as well as in the Celtic-speaking provinces (*OPEL* II: 194), where it could have been an assonance name (Lochner von Hüttenbach 1989: 83–84; cf. Šašel 1955: 136–139 [= Šašel 1992: 51–53]).

*Date*: 2nd century.

*Publication*: Brunšmid 1907: 184, no. 341; *AIJ* 522; Mócsy 1959: 211, no. 56: 2. Internet databases: *lupa* 3800; *HD*055673; *EDCS*-11301119.

*Note*: What looks like scant remains of a frieze above the architrave, can also be ascribed to cuts made in the modern period by honing tools. Admittedly, these two explanations are not mutually exclusive.

5. Stele (S[t] I.5)

*Provenance*: A chance find in 1981 in Topusko, 38 km SW of Sisak.

*Location*: GMK, inv. no. GMK-435a; temporarily kept in a showcase in the hotel Toplica in Topusko.

*Stone*: Upper-Miocene sandstone appearing locally on any geomorphological elevation underneath the Quaternary layers. The origin: local (M. Belak 2018).

*Preservation*: The upper part of a stele fragmentarily preserved. The stone is weathered, chipped and abraded all over.

Figure 96. (S[t] I.5) – stele (B. Djurić)

Figure 97. (S[t] I.5) – stone, microscope (HGI, M. Belak)

*Description*: *H* 35 cm, *W* 58 cm, T 29 cm. The rectangular upper part contains an apparently shell-like stylized arch of a niche encircled by a band border and flanked on both sides by a floating winged Eros (only the left one survives) supporting the niche with both hands. The workmanship is extremely coarse, possibly due to an unfinished state.
Date: Second half of 2nd or early 3rd century.
*Publication*: Unpublished.

6. Stele (S[t] I.6)
*Provenance*: Unknown; until the beginning of the 20th century built into a private house in the village of Odra Sisačka, 3 km NW of Sisak.
Location: AMZ, inv. no. AMZ KS-510.
*Stone*: Badenian sandy biocalcarenite. A probable origin: the mountains of N Croatia (M. Belak 2018).
*Preservation*: Fragmentarily preserved, with the upper and lower segments missing. The stone is cracked diagonally across the lower left part, with larger chips around the lower edges and smaller on the upper edges. The whole surface is heavily weathered and damaged, with the reliefs much obliterated and portrait features mostly unrecognizable.
*Description*: *H* 83 cm, W 88 cm, T 33 cm. An elongated, rectangular, simply framed portrait niche holds half-figures of the deceased, two men and a woman in between, and a motif of a diptych above and a bunch of scrolls below in the far left side. The men are portrayed in exactly the same way, wearing a *sagum* fastened with (probably) a round brooch on the right shoulder and a tunic belted with a round buckle, and holding a scroll in their left hand. They differ in only one detail: the man on the right-hand side has a bird-head sword (or possibly a knobbed centurion's stick) tucked under his left arm. Both have oval, elongated faces and long necks, while their hairdos remains obliterated. The woman between them has an oval, elongated face, and a hair arranged in a chignon on the crown, with geometricized strands of hair (?) at the sides of the neck, possibly representing a veiled coiffure. She is wearing a long-and wide-sleeved tunic (possibly cuffed, see below) with an oval neckline and a cloak hanging down from her shoulders in wide, stiff, parallel folds, and is holding a scroll in her left hand and a small round object in her right. The bands round her wrists (more clearly visible on the right hand) can be interpreted as bangles instead of cuffs. A simply framed relief panel below the portrait niche holds, from left to right: an oval shield and a spear leaning against it, a horseman galloping towards the right with cloak billowing behind him, and in front of him a dog chasing an indistinct animal. On the left short side a female figure is carved, wearing a long robe and holding a fruit in her right hand and a scroll in her left. On the right short side a servant is depicted dressed in a belted tunic and holding a jug in his uplifted right hand. The workmanship is hard to assess due to heavy damage and wear, but some sense of natural movement and proportion is evident in all of the reliefs.
*Date*: Mid or second half of 3rd century.
*Publication*: Brunšmid 1909: 219–220, no. 510; *AIJ* 579; Internet databases: *lupa* 3808.
*Note*: The stele presumably contained an inscription panel below the portrait niche, and a top of unknown shape (lion top?), indicated by three dowel holes with lead channels on the upper surface. Alternatively, it was interpreted as a funerary building of a pillar (*Pfeiler*) type (Scholz 2012: 209–211, Abb. 167).

Figure 98. (S[t] I.6) – stele, front side (AMZ, I. Krajcar)

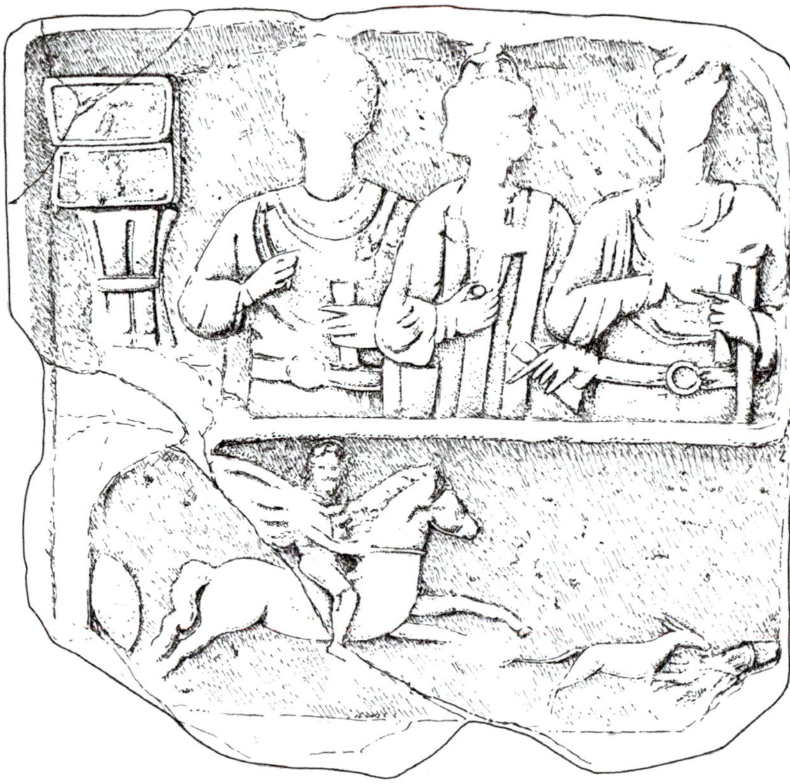

Figure 99. (S[t] I.6) – drawing (M. Galić)

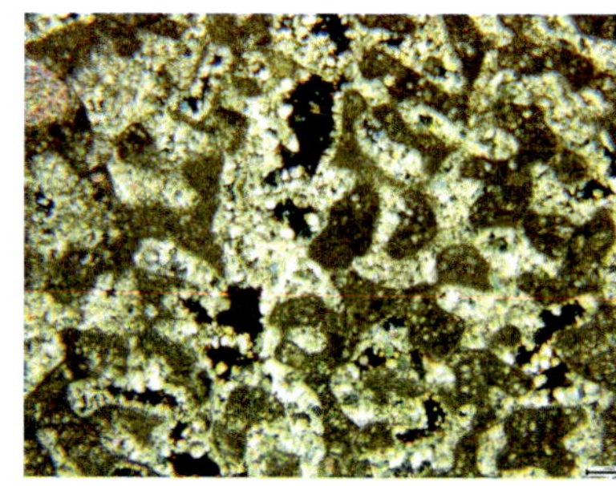

Figure 101. (S[t] I.6) – stone, microscope (HGI, M. Belak)

Figure 100 (S[t] I.6) – stele, left short side (AMZ, I. Krajcar)

Figure 102. (S[t] I.6) – stele, right short side (AMZ, I. Krajcar)

7. Stele (S[t] I.7)
*Provenance*: A chance find during the building of a new church in 1760 in the village of Sela, 7 km SW of Sisak.
*Location*: Missing.
*Inscription*: *Dom(ui) a(eternae) c<o=u>iugi / Messillae factum / qu(a)e vixit / cum conpatre suo /[5] Romano annis XXXVIII. / Hoc est sepulcrum / com(m)une ma<r=t>ito cum.*
According to the copy in the *CIL*, letters in the first line were much larger. Line 1: explained by Mommsen as *dom(u) a(eternae)*, while *EDH* erroneously cited *Dom(ui)(?) (a)etern(a)e(?)*; *Doma Cuiugi* in the *Epigraphik-Datenbank* Clauss/Slaby. Line 4: it is less plausible to restore the word as *conpatri(ota)?*, although the word *conpatriota* does appear on a tombstone from Etruria, Centumcellae (*CIL* XI 3541 = *EDR*127119; cf. also *CIL* VI 7658 from Rome). Line 7: Mommsen emended: *commune cum marito*.
Translation: To the eternal abode. Made for the wife Messilla, who lived 38 years with her godfather Romanus. This grave is shared by her husband.
Commentary: The word *conpater* does not appear in the *Oxford Latin Dictionary* and has not been attested to date in Roman inscriptions. However, see Charles Du Fresne Du Cange *et al.*, *Glossarium mediae et infimae latinitatis*, Niort 1883–1887, s.v. *compater*: *Compater*; Joanni de Janua: *Quasi simul ejusdem filii pater, qui filium alicujus baptizat, vel Christianum facit, vel ad baptismum, vel ad Confirmationem tenet*. (Citation kindly provided by David Movrin). Messilla is a relatively rare name, according to *OPEL*, only attested four times in Dalmatia as a feminine name and twice in northern Italy (III: 78). It may have been an indigenous name (Alföldy 1969: 247).
*Date*: Late Roman; *EDH*: 3rd or 4th century.
*Publication*: *CIL* III 3990; Gračanin 2004: 31 no. 21. Internet databases: *HD*074391; *EDCS*-28800751.

*II. Sarcophagi*

1. Sarcophagus (S[t] II.1)
*Provenance*: A chance find before 1726 in the village of Odra Sisačka 3 km NW of Sisak; found in the riverbed of the Odra, close to its bank.
*Location*: Missing.
*Preservation*: The chest was fragmentarily preserved, with around one third of the upper part missing.
*Description*: L ca. 250 cm (calculated from the drawing in Marsigli). The chest was rectangular, with a moulded (?) inscription panel flanked by simply framed blank lateral niches; probably type 3a.
*Inscription*: *------] / SISCE+A [........] / C(aius) Vitalius Vital[....] / eiusdem et Vital[....] /[5] parentibus [k]arissim[is].*
The inscription field is preserved below and on both lower sides, therefore it can be estimated that in the first visible line *c*. 8 or 9 letters are missing, and in the next two lines three or four. Line 2: *SISCE[--]A[---]*, *EDH*. *EDH* proposed *[--- k]arissim[is ---]* in the last line, but nothing is missing at the end of the epitaph. *[C]arissimis* / *[k]arissimis* would be the most plausible supplement, perhaps also *[r]arissimis*.

Figure 103. (S[t] II.1) – scan of the drawing (after Marsigli)

Translation: ... Gaius Vitalius Vitalis? ... his ... and Vital... to the dearest parents.
Commentary: How to explain *Sisce+a* is not clear, the letter *E* was perhaps inexactly copied. The feminine cognomen Siscia is known from a tombstone from Teutoburgium (Dalj: *CIL* III 10257 = *HD*073719 = *lupa* 3948: Ulpia Siscia). The *gentilicium* Siscius has once been attested (Solin and Salomies 1994: 172; *CIL* III 1464 = *AE* 1980: 758 = *HD*005839 from Sarmizegetusa): Siscius Valerius, centurion of *XIII Gemina Philippiana*. Vitalius is a rare *gentilicium* (*OPEL* IV: 177; derived from the very frequent name Vitalis: *OPEL* IV: 176–177), attested once also in Noricum (Lochner von Hüttenbach 1989: 184). The cognomen of Vitalius was probably Vitalis, although there are some other possibilities (Vitalianus and Vitalio, Vitalus, Vitalicus, Vitalinianus, Vitalinus, and Vitalissimus). At Siscia, Vitalianus and Vitalio occur on one *tessera* each, while Vitalis, either masculine or feminine name, appears on eight tags (Radman-Livaja 2014: 284); it was a popular name in the town.
*Date*: 2nd or 3rd century.
*Publication*: Marsigli 1726: 113, Tab. 44/2; *CIL* III 3975; Vuković 1994: 86, no. 77. Internet databases: *HD*074368; *EDCS*-28701051.

*III. Ash-chests*

1. Ash-chest (S[t] III.1)
*Provenance*: Archaeological excavations in 1990 in Topusko, 38 km SW of Sisak.
*Location*: GMK, inv. no. GMK-431a (chest), GMK-432a (lid); temporarily kept in the entrance hall of the hotel Toplica in Topusko.
*Stone*: Upper-Miocene sandstone appearing locally on any geomorphological elevation underneath the Quaternary layers. The origin: local (M. Belak 2018).
*Preservation*: The chest and lid completely preserved. There are large cuts and chips on the upper surface of the lid, and three of the four *acroteria* are a mostly chipped off.

Figure 104. (S[t] III.1) – ash-chest (M. Jambrović)

Figure 105. (S[t] III.1) – stone, microscope (HGI, M. Belak)

*Description*: Chest: L 88 cm, H 51 cm, W 61 cm; chest type 7; lid: L 107 cm, H 23 cm, W 64 cm; lid type 1b. The extremely roughly worked ash-chest is made of a rectangular chest and a low gabled lid is with four corner *acroteria*, both with flat edges.
*Date*: Probably 2nd or 3rd century.
*Publication*: Unpublished.

2. Ash-chest (S[t] III.2)
*Provenance*: Building works in 1982 in Topusko, 38 km SW of Sisak.
*Location*: GMK, inv. no. GMK-430a; temporarily kept in the entrance hall of the hotel Toplica in Topusko.
*Stone*: Upper-Miocene sandstone appearing locally on any geomorphological elevation underneath the Quaternary layers. The origin: local (M. Belak 2018).
*Preservation*: The lid completely preserved. Three of the four *acroteria* are chipped off, and there are large cuts

Figure 106. (S[t] III.2) – ash-chest lid (M. Jambrović)

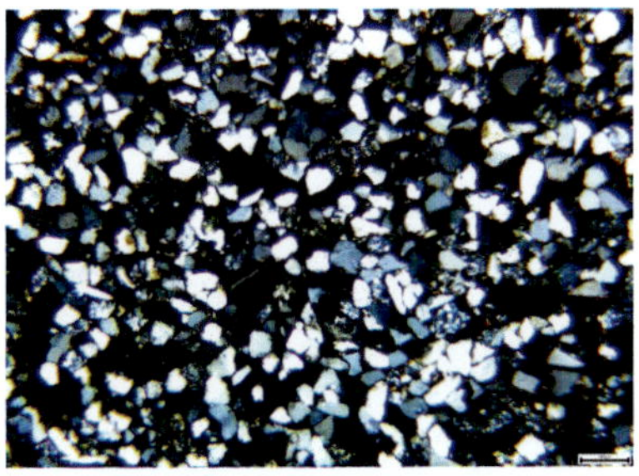

Figure 107. (S[t] III.2) – stone, microscope (HGI, M. Belak)

across the ridge and smaller cuts and pits at various places.
*Description*: L 105 cm, H 30 cm, W 82 cm; type 1b. The lid is low-gabled, with flat edges, four corner *acroteria*, a 3-cm-wide flattening along the ridge, while the borders of the long sides are slanted. The workmanship is fairly rough.
*Date*: Probably 2nd or 3rd century.
*Publication*: Unpublished.

*V. Altars*

1. Altar (S[t] V.1)
*Provenance*: A chance find before 1726 in the village of Odra Sisačka, 3 km NW of Sisak.
*Location*: Missing.
Preservation: The central block (shaft) seems to have been completely preserved, with no knowing whether the altar initially contained a detached base and a top. The upper edge of the front was broken off.

Figure 108. (S[t] V.1) – scan of the drawing (after Marsigli)

*Description*: Measurements (tentatively calculated from the drawing in Marsigli): H 104 cm, W 74 cm, T 60 cm. The altar was in the form of a parallelepiped with the front and the left side (the right one is not recorded) simply framed. The front holds the epitaph and the side panel features a statue representing a full figure of Hercules depicted as a bearded nude leaning on a tree trunk, with a snake at his feet.
*Inscription*: *Q(uintus) Idimiu[s .] / LIFEG VL IAR[...] / avo AN++XIII F[...] / IPLM Syr+?a+NIT[.] /[5] Q(uinto?) Idimi[o?] Q[.]+I[.]+I..] / Patero[.] an[.]T[..] / Q(uinto) Idim[io Q.?] fil(io) [.?] / Viator[i] a[n(norum)] III, / liber(tis) liber<t=F>ab(us) q(ue) /[10] posterisq(ue) suis. / H(oc) m(onumentum) l(ibertos) n(on) s(equetur).*
According to the drawing, the inscription field was framed, the frame is only damaged above, but no line is missing of the epitaph. The inscription is damaged on the surface and on the right, but very few letters are missing on the right. The transcription in the *EDH* and *EDCS* would suggest that more letters are missing. No punctuation mark is visible on the drawing in the 1st line, in some lines they may be damaged. *Impaginatio* could be seen in line 1, in which letters are larger, as well as in line 8, in which much empty space is left before Viator, but it is not centred as line 10, because there is no empty space on the right. The text is most uncertain. Line 2: *v(ir) l(audabilis)?*, most editors. Line 3: *an(norum) LXIII*, most editors, possibly *ann(orum) XXIII*. Supplements of line 7 and 8: Šašel Kos; *EDH* does not regard *viator* as a name: *viator [v]ale(?)*, not plausible.
*EDH*: *Q(uintus) Idimiu[s ---] / LIFEG VL IAR[---] / AVO[-] an(norum) X(?)XIII F[---] / IPLM SYR[-]A[-]NIT[---] / Q(uinto?) Idimi[o? ---] Q[---]I[-]II[---] / PATERO[--] an[---] T[---] / Q(uinto?) Idim[i---] fil[--] [---] / viator [v]ale(?) / liber(tis) liber<t>ab(us)q(ue) / posterisq(ue) suis / h(oc) m(onumentum) <h>(eredem)(?) n(on) s(equetur)*
*EDCS*: *Q(uintus) Idimiu[3] / LIFEG v(ir?) l(audabilis?) Iar[3] / avo[3] an(norum) LXIII f[3] / IPLM Syr[3]a[3]nit[3] / Q(uinto?) Idimi[3] Q[3]i[3]II[3] / Patero[3] an[3]T[3] / Q(uinto?) Idim[3] fil(io?) [3] / viator[3]A[3]III / liber(tis) liber<t=F>ab(us)q(ue) / posterisq(ue) suis / h(oc) m(onumentum) l(ibertos) n(on) s(equetur)*
Translation: Quintus Idimius (?) ... years old ... to Quintus Idimius, son of Quintus (?), Viator, 3 years old, and to his freedmen and freedwomen, and his descendants. This monument should not be inherited by freedmen.
Commentary: *Vir laudabilis* occurs late, not much earlier than the 4th century AD (cf. *Diz. ep.*) and is therefore not compatible with the formula *libertis libertabusque*, mentioned on the tombstone, and neither with the formula that the tombstone should not be inherited.
Among the *gentilicia* attested in Siscia on monuments other than *tesserae*, Idimius is the only non-Latin family name. It was probably an indigenous name, from which the *gentilicium* was derived in the manner of Latin *gentilicia*. It could perhaps be explained as Celtic, since the names in *idi-* commonly appear in various Celtic areas, but the etymology is so far unknown (Alexander

Falileyev, personal communication). Solin and Salomies (1994: 95) cite this inscription for a possible *gentilicium* Idimius. Viator is a frequent name, particularly also in Pannonia and Noricum (*OPEL* IV: 164–165), perhaps an assonance name among the indigenous inhabitants (Mócsy 1959: 196; Barkóczi 1964: 327).
*Date*: 3rd century, on account of the Farnese type of Hercules ('weary Hercules'), a copy of the Classical Greek original from the later 2nd or early 3rd century (cf. Marvin 1983: 355; Kleiner 1992: 338–339, fig. 305).
*Publication*: Marsigli 1726: 113, Tab. 44/3; *CIL* III 3978; Vuković 1994: 86-87, no. 78. Internet databases: *HD*074383; *EDCS*-28800742.

2. Altar (S[t] V.2)
*Provenance*: Building works in 1854 in the village of Odra Sisačka, 3 km NW of Sisak
*Location*: GMS, inv. no. GMS-15 AZA.
*Stone*: Cataclastic breccia marble. A possible origin: the central eastern Alps (M: Belak, 2018).
*Preservation*: A pyramidal altar top fragmentarily preserved. The topmost part and the left lion are broken off, and there are abrasions and chippings all over.
*Description*: *H 110* cm, W 70 cm, T 55 cm. A pyramidal upper part is sitting on a cubical base, featuring various motifs that spread over the front and, from the top to the bottom: a dolphin, two snakes with down-turned heads flanking a trident, two lions by the sides of the trident, depicted with rich manes and the mouths wide open. They are lying perpendicularly on the base, with the bodies in relief and the heads partially in the round. The lion on the left has its front paws positioned right over a ram's head and the one on the right rests its front paws on a horse's head, with both prey animals executed on the cubical base in deep grooved contours; there is a stylized acanthus motif in the free spaces on the sides above each lion. The workmanship is fairly competent, with a sense of natural shapes and proportions.
*Date*: 2nd or 3rd century.
*Publication*: Burkowsky 1999: 71, 124, no. 106; Burkowsky 2000: 81, no. 48; Škrgulja 2018: 244, no. 264. Internet databases: *lupa* 22465.

Figure 110. (S[t] V.2) – right short side (O. Harl, *lupa* 22465)

Figure 109. (S[t] V.2) – front side (O. Harl, *lupa* 22465)

Figure 111. (S[t] V.2) – stone, microscope (HGI, M. Belak)

Figure 112. (S[t] VII.1) – *aedicula* pediment (B. Djurić)

*VII. Funerary buildings*

1. *Aedicula* (?) (S[t] VII.1)
*Provenance*: A chance find in 1981 in building works in Topusko, 38 km SW of Sisak.
*Location*: A private courtyard in Topusko.
*Stone*: Impossible to establish by macroscopy due to the moss-grown surface.
*Preservation*: A pediment nearly completely preserved, with a tiny part of the left corner broken off. The piece is stained and moss-grown all over, with slight pitting and abrasions at places.
*Description*: *L* 176 cm, H 58 cm, W 29 cm. The central part of a pentagonal pediment is taken by a blind inscription panel, topped by an engraved motif of a six-petal rosette in a circle. The pediment, the inscription panel and the trapezoidal corner panels are all framed in the same manner, by a *cyma inversa* moulding. There are two dowel holes on the underside; the front is smoothed, with the back and sides more roughly worked.
*Date*: 2nd – 4th century.
*Publication*: Čučković 1986, 14 (mentioned without illustration); Čučković 2007: 65 (only illustration).
*Note*: The pediment was presumably found in the area of the Roman town and not in the cemetery, but it should be kept in mind that the urban topography of Roman Topusko is insufficiently known. An interpretation as the pediment of a miniature temple is less likely.

*VIII. Unclassifiable*

1. Relief block (*aedicula*?, enclosure wall?) (S[t] VIII.1)
*Provenance*: Unknown; spotted in Petrinja in 1995, and photographed.
*Location*: Missing.
*Stone*: Limestone or sandstone.
*Preservation*: Fragmentarily preserved. The block was broken on the bottom with no clue about the size of the missing part; it was chipped along the right-hand edge, worn-out, and pitted on the surface throughout.
*Description*: Judging from a poor-quality old photograph, the block was a prismatoid, featuring a portrait niche on the front. It was framed by a cable moulding on the top and left short side, and was topped by a band decorated with indistinct floral motif (vine or ivy tendrils?). The portrait niche was taken by half-length figures of a married couple in the gesture of *dextrarum iunctio*. Both were executed in the same way, with elongated faces flattened on the crown and tapered towards the chin, fairly thick necks and angular upper-bodies, and with no clear indication of hairstyles or clothes. There was a diagonal line across the chest of the right-hand side figure, whose slightly more robust shape may be

Figure 113. (S[t] VIII.1) – front side (Z. Gregl)

Figure 114. (S[t] VIII.1) – drawing (M. Galić)

Figure 115. (S[t] VIII.1) – back side (Z. Gregl)

suggesting a man. The workmanship was extremely crude, with flat and schematized rudimentary bodies completely out of natural forms and proportions.*Date*: Possibly 3rd or 4th century.
*Publication*: Unpublished.
*Note*: The photograph of the back and top shows that the stone was not a rectangular slab, but a prismatoidal block with slanted sides, possibly indicating original walling in a funerary structure, or it results from secondary usage.

***The possible territory of Siscia in the south-west* (S[ptSW])**

*I. Stelae*

1. Stele (S[ptSW] I.1)
*Provenance*: Unknown, probably the village of Vukova Gorica, 69 km SW of Sisak.
*Location*: Walled in the sacristy of the church of St. George in Vukova Gorica.
*Stone*: Impossible to establish by macroscopy due to the whitewashed surface.

Figure 116. (S[ptSW] I.1) – stele (V. Jukić)

*Preservation*: Fragmentarily preserved. Only the portrait niche survives, extremely weathered and thickly whitewashed, rendering the reliefs very much blurred.
*Description*: W 93 cm, *H 61* cm, T?; probably type 5 a or 5 b. A rectangular portrait niche, whose framing remains obscured, holds half-figures of a man on the right-hand side, a woman on the left-hand side, and an elongated head of a child of indeterminable sex in front of the woman. The man has an elongated face, rather thick hair combed from the crown towards the forehead in parallel strands, and a hardly discernible beard. He is wearing a *sagum* fastened with a round brooch on the right shoulder, and is holding a volume in his left hand, pointing at it with his right. The woman has an elongated face, a thick and high coiffure parted at the centre and leaving the ears free, and is wearing a tunic and *palla* draped in rounded folds over her chest.
*Inscription*: Only the letter *M* of the invocation *D(is) M(anibus)* is visible below the portrait niche on the right-hand side.
*Date*: Later 2nd or 3rd century.
*Publication*: Unpublished.

2. Stele (S[ptSW] I.2)
*Provenance*: Spotted at an unknown date as walled in a ruinous church at the site of Kneževo brdo in the village of Točak, 68 km SW of Sisak.
*Location*: GMK, inv. no. GMK-10432.
*Stone*: Upper-Jurassic/Lower-Cretaceous micritic limestone. A possible origin: the Outer Dinarides; a probable origin: the Kordun, SW of Sisak (M. Belak 2018).
*Preservation*: Completely preserved, but cracked, chipped and weathered all over.
*Description*: H 120 cm, W 50 cm, T 20 cm; type 8. The stele is composed of completely geometricized components executed in linear, deep-grooved carving

Figure 117. (S[ptSW] I.2) – stele (B. Djurić)

Figure 118. (S[ptSW] I.2) – stone, microscope (HGI, M. Belak)

on a rectangular panel: a pediment with the oversized *acroteria*, a smooth architrave, a rectangular panel with the lower line cut in the middle and with a circular moulded inscription panel holding illegible traces of the inscription, with a stump of the tenon beneath. The workmanship is rude and the structural features rudimentary.
*Date*: Second half of 3rd century.
*Publication*: Šarić 1992: 113–116; Perkić 2012: 205, 238, no. 5.7.2.20., T 8: 6, T. 55: 1.

3. Stele (S[ptSW] I.3)
*Provenance*: Unkown (see at *Note*).
*Location*: Built into the façade of the parish church of St. Peter in the village of Sveti Petar Mrežnički, 69 km SW of Sisak.
*Stone*: Local limestone (after the *AIJ*); micritic limestone (M. Belak 2017, by macroscopy).
*Preservation*: Fragmentarily preserved. A portrait niche merged with the pediment survives, while the inscription panel and possibly a base are missing. The surface is extremely weathered, cracked and whitewash-stained at places, with the portrait features fairly obliterated.
*Description*: *H 84* cm, W 89 cm, *T?* A rectangular, unframed gabled portrait niche is provided with large triangular corner *acroteria* inscribed with the letters *D* and *M*, respectively. It is divided from the (missing) lower section of the stele by a wide strip of single-strand guilloche motif with a central pellet, and is holding half-figures of two men and a woman between and slightly in front of them. The men have similar facial features – oval, slightly triangular faces, fairly big ears, down-turned thin lips, and a short-cropped hair receded on the temples, but differently shaped on the forehead; in the left-hand man the fringe is convex, and in the one on the right it is M-shaped. Their facial hair is not visible due to the wear of the stone, but a stubby beard seems to be discernible in the man on the right, probably the older one of the two, as is further indicated by his nasolabial folds. They are both wearing a *sagum* fastened with a round brooch on the right shoulder. The man on the right is holding a scroll in his left hand, while the one on the left seems to be holding a writing tablet in the left hand and a writing implement in his right. The woman is wearing a tunic with an oval

Figure 119. (S[ptSW] I.3) – stele (B. Djurić)

Figure 120. (S[ptSW] I.3) – drawing (M. Galić)

neckline, and probably a *palla* draped diagonally across the chest, and is holding an oval object in her left hand. She has fairly thick hair (probably) braided at the back, then raised to produce a flat chignon reaching towards the forehead. The workmanship is fairly stiff and crude.

*Inscription*: *D(is) M(anibus) / [------*

Translation: To the spirits of departed. ...

*Date*: Second half of 3rd century.

*Publication*: *AIJ* 498; Migotti 2013a: 309, 317, S. 32. Internet databases: *HD*071787; *EDCS*-11301111.

*Note*: In spite of the fact that the area of the village of Sveti Petar Mrežnički produced a lot of finds indicating an important Roman settlement (possibly Romula Quadrata; see at Ch. II.2), the findspot of this piece cannot be taken for granted. This is because a fragmented sarcophagus lid (S[ptSW] II.1.) in the courtyard of St. Peter's church comes from another site (Perkić 2012: 202, fn. 659).

4. Stele (S[ptSW] I.4)

*Provenance*: A thicket clearing in 1894 in the village of Svojić, 78 km SW of Sisak.

*Location*: AMZ, inv. no. AMZ KS-339.

*Stone*: Upper-Jurassic/Lower-Cretacious micritic limestone. A possible origin: the Outer Dinarides; a probable origin: the Kordun, SW of Sisak (M. Belak 2018).

*Preservation*: Completely preserved. The stele is chipped around the lower edge and on the architrave, and is fairly weathered, especially on the inscription panel.

*Description*: H 170 cm, W 65 cm, T 20 cm; type 9. The stele is composed of the following components: an elongated, arched portrait niche carved into a rectangular panel and merged into a concave-sided pediment, thus producing a tent-like pentagonal upper part with concave ('chipped') corners; a smooth architrave; an inscription panel flanked by smooth columns with triple bases and fan-like stylized capitals; a base originally plain but receiving the last line of the epitaph. The pediment is

Figure 121. (S[ptSW] I.4) – stele (AMZ, I. Krajcar)

Figure 122. (S[ptSW] I.4) – stone, microscope (HGI, M. Belak)

decorated with an elegantly rendered crescent shape while the left concave corner of the pentagonal upper part is framed with a stripe of parallel grooves, possibly engraved at some later point in time. The portrait is rendered as a crudely modelled, slightly triangular head resting on a sturdy neck, with a high rounded crown, smooth hair reaching almost to the line of the eyebrows and peaking towards the nose root, irregularly shaped elongated eyes, a long triangular nose, and small rounded lips. The stele is executed in a combination of rough carving and deep linear engraving, and the craftsmanship is thoroughly crude and unskilled.

*Inscription*: *D(is) M(anibus). / F(---) T(---) et I(---) C(---) / titulum po/sueru<nt=m> fi/*[5]*lio suo Vale/rio Saturni/no mil(iti) l(egionis) M D / CONARIVS / annos vi<x=s>(it) // XXX.*

Letters are irregular, *L* and *M* have specific, cursive-like forms.

Translation: To the spirits of the departed. (Parents) F. T. and I. C. erected the monument (*titulus*) to their son Valerius Saturninus, a soldier of the Legion ..., who lived 30 years.

Commentary: The parents of Saturninus were probably of peregrine status, since otherwise it would be expected that his father's family name should likewise be Valerius; however, the abbreviation *F* in the second line is clearly seen on the stone. The Valerii were one of the most distinguished upper-class families in Siscia and one of the best attested *gentes* in the town generally. Lucius Valerius Valerianus from Siscia was a *flamen divi Claudi* and a high priest of Upper Pannonia in Savaria, the seat of the provincial council (*RIU* 20 = *HD*009523 = *lupa* 7957; Fishwick 2002: 273; Szabó 2006: 221 P 70; 244 P 70). The Valerii were frequent all over the Empire (*OPEL* IV: 143–146). They had probably come from northern Italy, since many are attested in Aquileia (Indexes of *Inscr. Aquil.*; cf. Mócsy 1959: 160; Šegvić 1988: 60). Several other Valerii are known from Siscia, such as one Valeria Saturnina, who dedicated an altar to Liber Pater together with Gaius Volcenius Lupercus (*CIL* III 3956 = 10834 = *AIJ* 540 = *HD*071824 = *lupa* 11578). Some Valerii were members of the Roman army: the above-mentioned veteran of the Legion *XV Apollinaris*, Lucius Valerius Verecundus and Gaius Valerius Spectatus, a soldier of the 8th Praetorian Cohort, known from a tombstone in Rome (*CIL* VI 2689 = Dobó 1975, no. 63 = *EDR*102983; Ricci 1993: 186 no. P. 69). Saturninus was a very popular cognomen in Italy and all provinces of the western part of the Empire; in Siscia, it is also attested on seven lead tags (Mócsy 1959: 189; Kajanto 1965: 54 and *passim*; *OPEL* IV: 51–53; Radman-Livaja 2014: 259).

*Date*: Second half of 3rd or first half of 4th century.

*Publication*: *CIL* III 14333,1; Brunšmid 1907: 183, no. 339; Migotti 2010a; Perkić 2012: 205, *passim*, no. 5.7.2.21 (transcription partly inexact). Internet databases: *lupa* 22325; *HD*032859.

Figure 123. (S[ptSW] II.1) – sarcophagus lid (B. Djurić)

*II. Sarcophagi*

1. Sarcophagus (S[ptSW] II.1)

*Provenance*: Archaeological research in 1997–2001 at the spring Donji vrutak in the village of Barilovečki Cerovac, 71 km SW of Sisak.

*Location*: One half (a) is in the archaeological collection in the medieval Barilović castle, 6 km NE from Barilovečki Cerovac, while another (b) is in the courtyard of the church of Saint Peter in the village of Sveti Petar Mrežnički.

*Stone*: Local limestone (after Perkić).

*Preservation*: A lid completely preserved, split in two halves; one *acroterion* is broken off in fragment *a*.

*Description*: Fragment *a*: *L 110* cm, H 50 cm, W 127 cm; fragment *b*: *L 104* cm, H 50 cm, W 127; total dimensions: L 214 cm, H 50 cm, W 127 cm; type 1a. The roughly worked lid is gabled, and with corner *acroteria*.

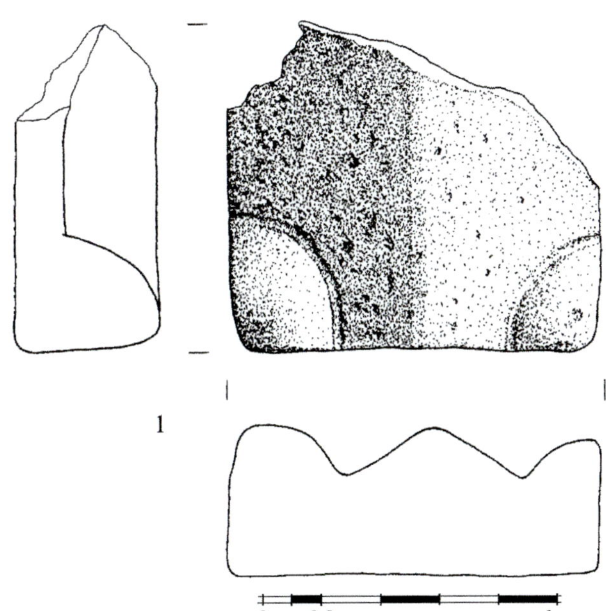


Figure 124. (S[ptSW] II.1) – scan of the drawing (after Perkić 2012)

*Date*: Second half of 2nd or 3rd century.
*Publication*: Perkić 2012: 175, 202, no. 5.7.2.7., T. 49: 1.

2. Sarcophagus (S[ptSW] II.2)
*Provenance*: Archaeological research in 1997–2001 at the spring Donji vrutak in the River Korana, at the village of Barilovečki Cerovac, 71 km SW of Sisak.
*Location*: At the findspot.
*Stone*: Local limestone (after Perkić).
*Preservation*: A lid fragmentarily preserved.
*Description*: L 258 cm, *H 25* cm, W 142 cm; type 1a. The roughly worked lid is gabled, with flat edges and corner *acroteria*.
*Date*: Second half of 2nd or 3rd century.
*Publication*: Zupčić 2007: 218; Perkić 2012: 202, no. 5.7.2.8.

3. Sarcophagus (S[ptSW] II.3)
*Provenance*: Archaeological research in 1997–2001 in the village of Hrnetić, 66 km SW of Sisak.
*Location*: At the findspot in the Kupa River.
*Stone*: Local limestone (after Perkić).
*Preservation*: A chest or lid, allegedly complete but subsequently crushed during army manoeuvres at an unspecified time in the 20th century.
*Date*: Second half of 2nd or 3rd century.
*Publication*: Perkić 2012: 181, fn. 611, 202, no. 5.7.2.9.

4. Sarcophagus (S[ptSW] II.4)
*Provenance*: Archaeological research in 1997–2001 at the site of Rastovac/Milinkovići in the village of Gornja Perjasica, 71 km SW of Sisak.
*Location*: At the findspot.
*Stone*: Local limestone (after Perkić).
*Preservation*: A lid fragmentarily preserved, with a fragment of a gabled side with one complete and one fragmented *acroterion*.
*Description*: *L 108* cm, H 52 cm, W 150 cm; type 1a. The roughly worked lid is gabled, with flat edges and corner *acroteria*.
*Date*: Second half of 2nd or 3rd century.
*Literature*: Perkić 2012: 203, no. 5.7.2.11., T. 50: 1.

5. Sarcophagus (S[ptSW] II.5)
*Provenance*: Archaeological research in 1997–2001 at the site of Bišica greda in the village of Ponorac, *c.* 68 km SW of Sisak.
*Location*: Buried in the ground on the findspot; inaccessible.
*Stone*: Local limestone (after Perkić).
*Preservation*: a lid completely preserved.
*Description*: Dimensions unknown; type 1a. The roughly worked lid is gabled, with flat edges and corner *acroteria*.
*Date*: Second half of 2nd or 3rd century.
*Publication*: Perkić 2012: 204, no. 5.7.2.17.

6. Sarcophagus (S[ptSW] II.6)
*Provenance*: Archaeological research in 1997–2001 at the site of Mravunčev mlin in the village of Blagaj, 72 km SW of Sisak.
*Location*: Walled in a house at the findspot.
*Stone*: Local limestone (after Perkić).

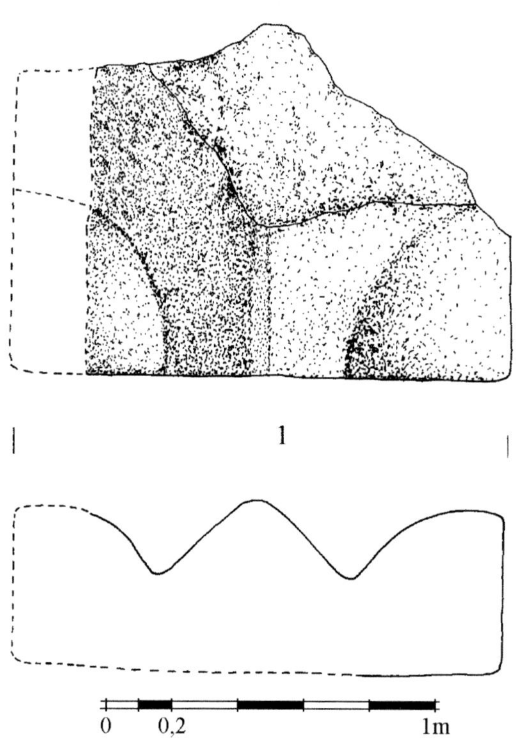

Figure 125. (S[ptSW] II.4) – scan of the drawing (after Perkić 2012)

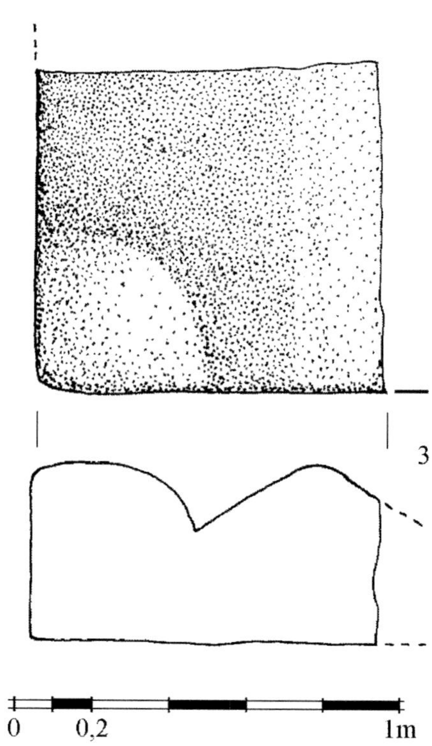

Figure 126. (S[ptSW] II.6) – scan of the drawing (after Perkić 2012)

*Preservation*: A lid fragmentarily preserved, with less than one third surviving, including one *acroterion*.
*Description*: *L 85* cm, H 47 cm, *W 90* cm; type 1a. The roughly worked gabled lid has flat edges and corner *acroteria*.
*Date*: Second half of 2nd or 3rd century.
*Publication*: Perkić 2012: 206, no. 5.7.2.25., T. 57: 3.

7. Sarcophagus (S[ptSW] II.7)
*Provenance*: Archaeological research in 1997–2001 at the site of St George the Martyr church in the village of Gorinci, 82 km SW of Sisak (allegedly brought from the nearby village of Skukani).
*Location*: At the findspot.
*Stone*: Local limestone (after Perkić).
*Preservation*: A lid completely preserved, but restored from many fragments by cementing.
*Description*: Dimensions unknown; type 1a. The roughly worked gabled lid has flat edges and corner *acroteria.*
*Date*: Second half of 2nd or 3rd century.
*Publication*: Perkić 2012: 214–215, 242, no. 5.7.2.64.

8. Sarcophagus (S[ptSW] II.8)
*Provenance*: Archaeological research in 1997–2001 at the site of Bulender in the village of Blagaj, 72 km SW of Sisak.
*Location*: At the findspot.
*Stone*: Local limestone (after Perkić).
*Preservation*: A lid fragmentarily preserved.
*Description*: No specific data recorded. It was probably a roughly worked gabled lid with flat edges and corner *acroteria.*
*Date*: Second half of 2nd or 3rd century.
*Publication*: Perkić 2012: 215, no. 5.7.2.66.

*III. Ash-chests*

1. Ash-chest (S[ptSW] III.1)
*Provenance*: Archaeological research in 1997–2001 in the village of Mateško Selo, 76 km SW of Sisak.
*Location*: At the findspot.
*Stone*: Local limestone (after Perkić).
*Preservation*: A chest fragmentarily preserved. The uppermost part is missing and one corner is chipped off.
*Description*: L 85 cm, *H 43* cm, W 70 cm; type 7. The roughly worked rectangular chest (presumably) had flat edges.
*Date*: 2nd or 3rd century.
*Publication*: Perkić 2012: 201, no. 5.7.2.1., T. 49: 3.

2. Ash-chest (S[ptSW] III.2)
*Provenance*: Archaeological research in 1997–2001 at the site of the Žutuljak spring in the village of Mateško Selo, 76 km SW of Sisak.
*Location*: At the findspot.
*Stone*: Local limestone (after Perkić).

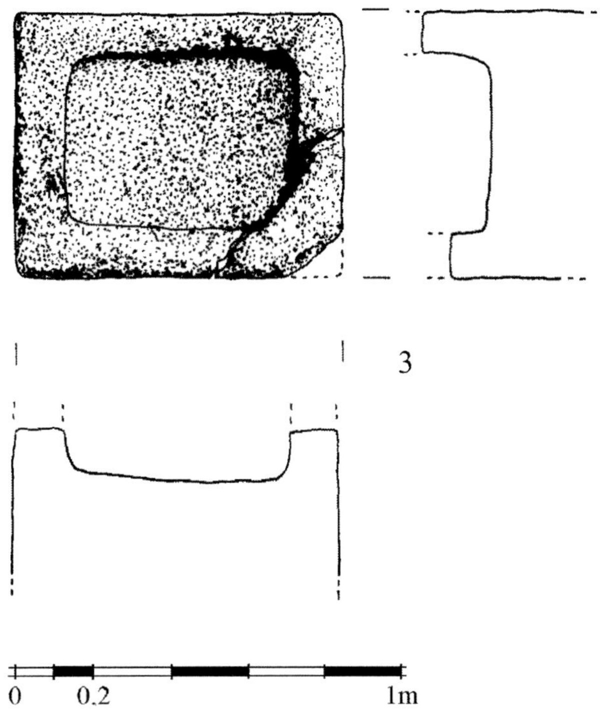

Figure 127. (S[ptSW] III.1) – scan of the drawing (after Perkić 2012)

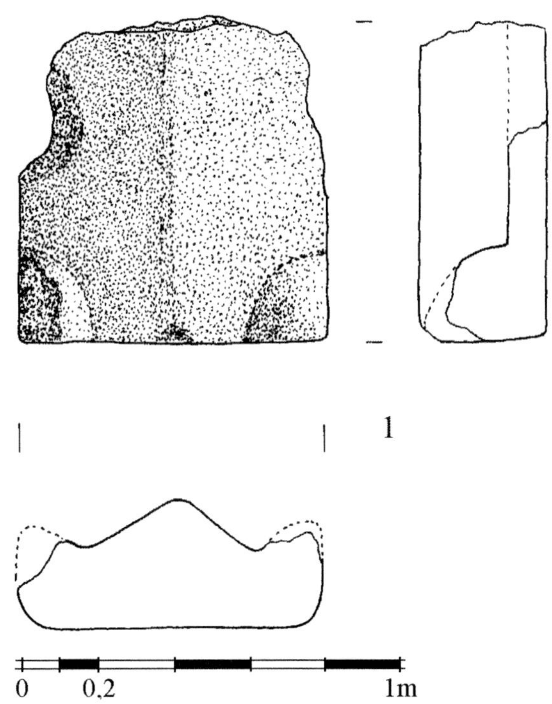

Figure 128. (S[ptSW] III.2) – scan of the drawing (after Perkić 2012)

*Preservation*: A lid fragmentarily preserved, with one side and two pertaining *acroteria* missing.
*Description*: *L 83* cm, H 34 cm, W 81 cm; type 1a. The roughly worked gabled lid has flat edges and corner *acroteria.*
*Date*: 2nd or 3rd century. *Publication*: Perkić 2012: 201, no. 5.7.2.2., T. 48: 1.

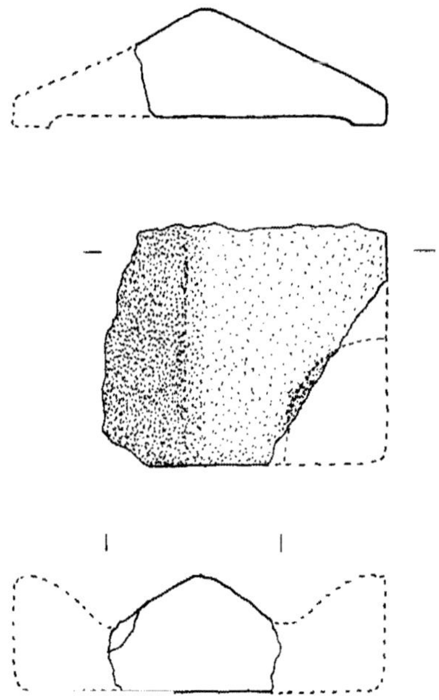

Figure 129. (S[ptSW] III.3) – scan of the drawing (after Perkić 2012)

3. Ash-chest (S[ptSW] III.3)
*Provenance*: Archaeological research in 1997–2001 in the courtyard of M. Mateša in the village of Mateško Selo, 76 km SW of Sisak.
*Location*: At the findspot.
*Stone*: Local limestone (after Perkić).
*Preservation*: A lid fragmentarily preserved. The central part survives, with all four *acroteria* missing.
*Description*: *L* 56 cm, H 28 cm, *W* 64 cm; type 1a. The roughly worked gabled lid has flat edges and (presumably) corner *acroteria*.
*Date*: 2nd or 3rd century. *Publication*: Perkić 2012: 201, no. 5.7.2.3., T. 48: 2.

4. Ash-chest (S[ptSW] III.4)
*Provenance*: Archaeological research in 1997–2001 in the courtyard of M. Mateša in the village of Mateško Selo, 76 km SW of Sisak.
*Location*: At the findspot.
*Stone*: Local limestone (after Perkić).

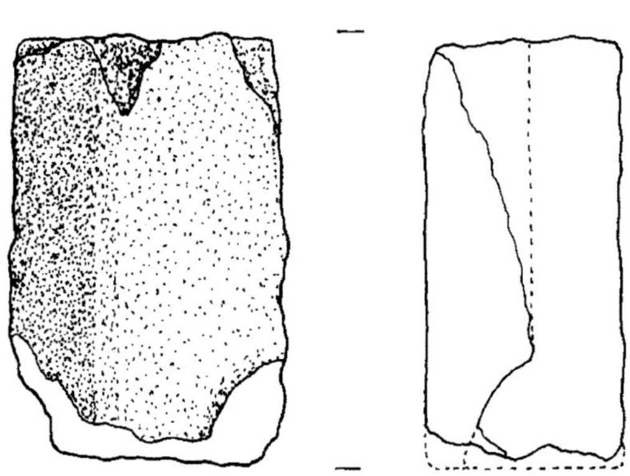

Figure 130. (S[ptSW] III.4) – scan of the drawing (after Perkić 2012)

*Preservation*: A lid fragmentarily preserved. A larger part of a roughly worked lid survives, but for the edges with all four *acroteria*.
*Description*: *L 80* cm, H 40 cm, *W 55* cm; type 1a. The roughly worked gabled lid presumably had flat edges and corner *acroteria*.
*Date*: 2nd or 3rd century.*Publication*: Perkić 2012: 201, no. 5.7.2.4., T. 48: 3.

5. Ash-chest (S[ptSW] III.5)
*Provenance*: Archaeological research in 1997–2001 at the house of J. Banjavčić in the village of Jankovo Selište, 75 km SW of Sisak.
*Location*: Built into a house at the findspot.
*Stone*: Local limestone (after Perkić).
*Preservation*: A lid completely preserved.
*Description*: L 116, H 38 cm, W 80 cm; type 1a. The roughly worked gabled lid has flat edges and corner *acroteria*.
*Date*: 2nd or 3rd century. *Publication*: Perkić 2012: 175, 203, no. 5.7.2.10., T. 49: 2.

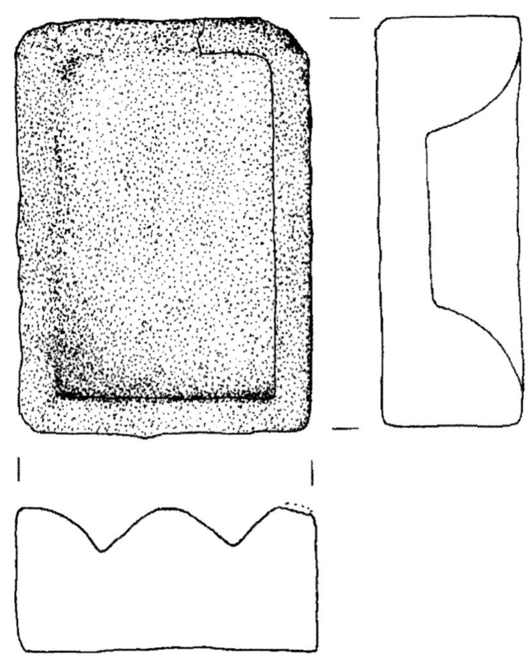

Figure 131. (S[ptSW] III.5) – scan of the drawing (after Perkić 2012)

6. Ash-chest (S[ptSW] III.6)
*Provenance*: Archaeological research in 1997–2001 at the site of Vujaškovići in the village of Ponorac, 68 km SW of Sisak.
*Location*: Built into a house at the findspot.
*Stone*: Local limestone (after Perkić).
*Preservation*: A lid completely preserved, apart from the tip of one *acroterion*.
*Description*: L 95 cm, H 30 cm, W 60 cm; type 1a. The roughly worked gabled lid has flat edges and corner *acroteria*.
*Date*: 2nd or 3rd century. *Publication*: Perkić 2012: 175, 203, no. 5.7.2.12., T. 53: 1.

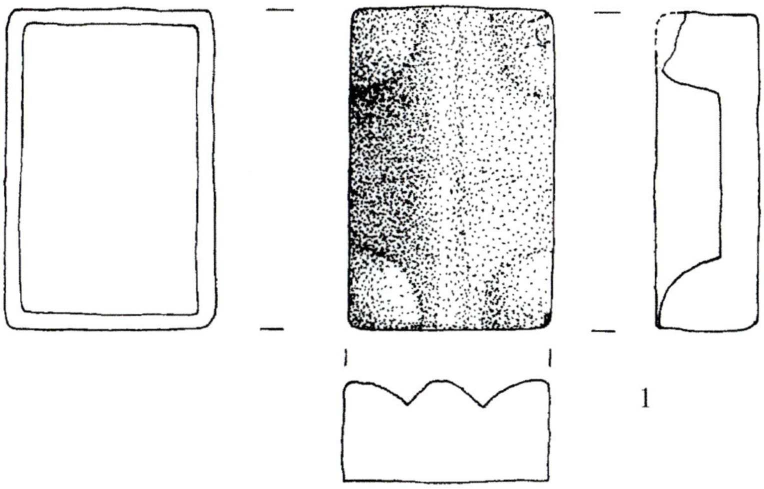

Figure 132. (S[ptSW] III.6) – scan of the drawing (after Perkić 2012)

*Location*: Built into a house at the findspot.
*Stone*: Local limestone (after Perkić).
*Preservation*: A chest fragmentarily preserved, with the bottom and stumps of the sides surviving.
*Description*: L 80 cm, *H 37* cm, W 60 cm; type 7. The roughly worked rectangular chest (presumably) had flat edges.
*Date*: 2nd or 3rd century. *Publication*: Perkić 2012: 175, 203, no. 5.7.2.14., T. 53: 4.

7. Ash-chest (S[ptSW] III.7)
*Provenance*: Archaeological research in 1997–2001 at the site of Vujaškovići in the village of Ponorac, 68 km SW of Sisak.
*Location*: Built into a house at the findspot.
*Stone*: Local limestone (after Perkić).
*Preservation*: A chest fragmentarily preserved, with the bottom and stumps of the sides surviving.
*Description*: L 95 cm, *H 35* cm, W 55 cm; type 7. The roughly worked rectangular chest (presumably) had flat edges.
*Date*: 2nd or 3rd century.*Publication*: Perkić 2012: 175, 203, no. 5.7.2.13., T. 53: 3.

8. Ash-chest (S[ptSW] III.8)
*Provenance*: Archaeological research in 1997–2001 at the site of Vujaškovići in the village of Ponorac, 68 km SW of Sisak.

9. Ash-chest (S[ptSW] III.9)
*Provenance*: Archaeological research in 1997–2001 at the site of Vujaškovići in the village of Ponorac, 68 km SW of Sisak.
*Location*: GMK, inv. no. GMK-10434.
*Stone*: Upper Jurassic-Lower Cretacious micritic limestone. A possible origin: the Outer Dinarides; a probable origin: the Kordun, SW of Sisak (M. Belak 2018).
*Preservation*: A chest completely preserved.
*Description*: L 84 cm, H 60 cm, W 57 cm; type 8. The roughly worked rectangular chest has flat edges and the epitaph in a moulded circular panel in the middle of the front, carved as a deep, V-sectioned groove.
*Inscription*: *D(is) M(anibus). / Nepoti. / Vix(it) an(nis) / X.*
Letters are irregular, impagination clumsy.
Translation: To the spirits of the departed. To Nepos, who lived 10 years.
Commentary: Nepos is a relatively frequent personal name or cognomen (*OPEL* III 98), particularly in Italy and Dalmatia (Alföldy 1969: 251), where it is also well attested among the indigenous inhabitants, notably

Figure 133. (S[ptSW] III.9) – ash-chest (B. Djurić)

Figure 134. (S[ptSW] III.9) – stone, microscope (HGI, M. Belak)

among the Delmatae. Nepos is less well documented in Pannonia (Mócsy 1959: 182; Barkóczi 1964: 319) and indeed the area of Kordun is reaching towards the border region of the province of Dalmatia.
*Date:* 2nd or 3rd century (Šarić: second half of 3rd century).
*Publication*: Šarić 1992: 112–113; Šegvić 1996: 132 no. 1 (*AE* 1993: 1275); Perkić 2012: 175, 204, no. 5.7.2.15., T 8: 4, T. 55: 2. Internet databases: *HD*040144; *EDCS*-03700674.

10. Ash-chest (S[ptSW] III.10)
*Provenance*: Archaeological research in 1997–2001 at the site of Vujaškovići in the village of Ponorac, 68 km SW of Sisak.
*Location*: Built into a house at the findspot.
*Stone*: Local limestone (after Perkić).
*Preservation*: A lid fragmentarily preserved, with a larger corner section and a respective *acroterion* surviving.
*Description*: *L 90* cm, H 37 cm, *W 75* cm; type 1a. The roughly worked gabled lid has flat edges and corner *acroteria*.
*Date*: 2nd or 3rd century.
*Publication*: Perkić 2012: 175, 204, no. 5.7.2.16., T. 53: 2.

11. Ash-chest (S[ptSW] III.11)
*Provenance*: Archaeological research in 1997–2001 at the site of Kneževo brdo in the village of Točak, 68 km SW of Sisak.
*Location*: On military premises at the site of Dugi dol, not far from the findspot.
*Stone*: Local limestone (after Perkić).
*Preservation*: A lid completely preserved apart from the tip of one *acroterion*.
*Description*: L 120 cm, H 25 cm, W 70 cm; type 1a. The roughly worked gabled lid has flat edges and corner *acroteria*.
*Date*: 2nd or 3rd century.
*Publication*: Perkić 2012: 204, no. 5.7.2.18., T. 56: 3.

12. Ash-chest (S[ptSW] III.12)
*Provenance*: Archaeological research in 1997–2001 at the site of Kneževo brdo in the village of Točak, 68 km SW of Sisak.
*Location*: GMK, inv. no. GMK-10433.
*Stone*: Upper-Cretaceous/Paleogene micritic limestone. A possible origin: the mountains of N Croatia, the Outer Dinarides; a probable origin: the Kordun, SW of Sisak (M. Belak 2018).
*Preservation*: A chest fragmentarily preserved, with approximately one-third of the upper part missing.
*Description*: L 97 cm, *H 52* cm, W 70 cm; type 8. The rectangular chest, with (presumably) flat edges, shows feeble traces of parallel grooves (tool marks) on the inner side of the bottom. There are remains of an unintelligible inscription in a fragmentarily preserved,

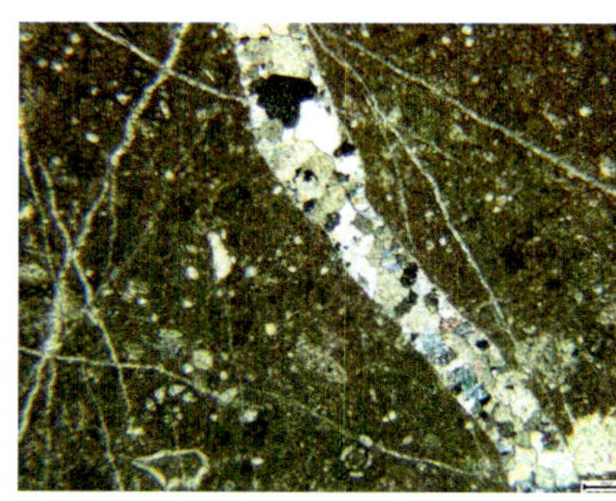

Figure 136. (S[ptSW] III.12) – stone, microscope (HGI, M. Belak)

Figure 135. (S[ptSW] III.12) – ash-chest (B. Djurić)

Figure 137. (S[ptSW] III.16) – ash-chest (B. Djurić)

moulded circular panel in the middle of the front, carved as a deep, V-sectioned groove. The workmanship is rough.
*Date*: 2nd or 3rd century.
*Publication*: Perkić 2012: 205, no. 5.7.2.19., T 8: 3, T. 56: 1.

13. Ash-chest (S[ptSW] III.13)
*Provenance*: Archaeological research in 1997–2001 in the village of Mala Crkvina, 63 km SW of Sisak.
*Location*: Built into a house at the findspot.
*Stone*: Local limestone (after Perkić).
*Preservation*: A chest fragmentarily preserved, with more than half surviving and with the edges broken off.
*Description*: *L* 65 cm, *H* 25 cm, W 80 cm; type 8. The roughly worked rectangular chest (presumably) had flat edges.
*Date*: 2nd or 3rd century.
*Publication*: Perkić 2012: 206, 240, no. 5.7.2.22., T. 57: 4.

14. Ash-chest (S[ptSW] III.14)
*Provenance*: Archaeological research in 1997–2001 in the village of Mala Crkvina, 63 km SW of Sisak.
*Location*: Built into a house at the findspot.
*Stone*: Local limestone (after Perkić).
*Preservation*: A chest fragmentarily preserved, with the back, parts of the short sides, and the edges missing.
*Description*: L 88 cm, *H* 40 cm, W 44 cm; type 8. The roughly worked rectangular chest has flat edges and a slightly sunken, blind circular inscription panel on the front.
*Date*: 2nd or 3rd century.
*Publication*: Perkić 2012: 206, 240, no. 5.7.2.23., T 8: 2, T. 57: 1.

15. Ash-chest (S[ptSW] III.15)
*Provenance*: Archaeological research in 1997–2001 in the village of Mala Crkvina, 63 km SW of Sisak.
*Location*: Built into a house at the findspot.
*Stone*: Local limestone (after Perkić).
*Preservation*: A chest fragmentarily preserved, with more than half surviving.
*Description*: *L* 95 cm, H 90 cm; W 75 cm; type 8. The roughly worked rectangular chest has flat edges.
*Date*: 2nd or 3rd century.
*Publication*: Perkić 2012: 206, 240, no. 5.7.2.24., T. 57: 2.

16. Ash-chest (S[ptSW] III.16)
*Provenance*: Archaeological research in 1997–2001 at the site of the memorial to Croatian soldiers in the village of Generalski Stol (80 km SW of Sisak), possibly brought there from the site of Crikvišće in the area of the villages of Vali/Lauši.
*Location*: At the findspot.
*Stone*: Local limestone (after Perkić).
*Preservation*: A chest fragmentarily preserved, with the bottom surviving and most of the upper part missing.
*Description*: L 78 cm, *H* 37 cm, W 58 cm; type 7. The roughly worked rectangular chest (presumably) had flat edges.
*Date*: 2nd or the 3rd century.
*Publication*: Perkić 2012: 172–173, 206, no. 5.7.2.26., fig. 16a.

17. Ash-chest (S[ptSW] III.17)
*Provenance*: Archaeological research in 1997–2001 at the site of the memorial to Croatian soldiers in the village of Generalski Stol (80 km SW of Sisak), possibly brought from the site of Crikvišće in the area of the villages of Vali/Lauši.
*Location*: At the findspot.
*Stone*: Local limestone (after Perkić).
*Preservation*: A chest fragmentarily preserved, broken in two parts, with most of the upper section missing.
*Description*: L 79 cm, *H* 32 cm, W 57 cm; type 7. The roughly worked rectangular chest (presumably) had flat edges.
*Date*: 2nd or 3rd century.
*Publication*: Perkić 2012: 172–173, 207, no. 5.7.2.27., fig. 16b.

18. Ash-chest (S[ptSW] III.18)
*Provenance*: Archaeological research in 1997–2001 at the site of St George's church in the village of Mateško Selo, 76 km SW of Sisak.
*Location*: Built into the perimeter wall of St George's church.
*Stone*: Local limestone (after Perkić).
*Preservation*: A lid completely preserved, cracked along the length of one side, with three *acroteria* damaged.
*Description*: L 114 cm, H 35 cm, W 85 cm; type 1a. The roughly worked gabled lid has flat edges and corner *acroteria*.
*Date*: 2nd or 3rd century.
*Publication*: Perkić 2012: 207, no. 5.7.2.28., T. 51: 1.

19. Ash-chest (S[ptSW] III.19)
*Provenance*: Archaeological research in 1997–2001 at the site of St George's church in the village of Mateško Selo, 76 km SW of Sisak.
*Location*: Built into the perimeter wall of St George's church.
*Stone*: Local limestone (after Perkić).
*Preservation*: A lid fragmentarily preserved, with the top partially surviving.
*Description*: *L 73* cm, H 36 cm, *W 58 cm*; type 1a. The roughly worked gabled lid (presumably) had flat edges and corner *acroteria*.
*Date*: 2nd or 3rd century.
*Publication*: Perkić 2012: 207, no. 5.7.2.29., T. 51: 2.

20. Ash-chest (S[ptSW] III.20)
*Provenance*: Archaeological research in 1997–2001 at the site of St George's church in the village of Mateško Selo, 76 km SW of Sisak.
*Location*: Built into a house at the findspot.
*Stone*: Local limestone (after Perkić).
*Preservation*: A lid fragmentarily preserved; a smaller mid-upper part survives, with tool marks on the inner side.
*Description*: *L 40* cm, H 28 cm, W 80 cm; type 1a. The roughly worked gabled lid (presumably) had flat edges and corner *acroteria*.
*Date*: 2nd or 3rd century.
*Publication*: Perkić 2012: 207, no. 5.7.2.30., T. 50: 3.

21. Ash-chest (S[ptSW] III.21)
*Provenance*: Archaeological research in 1997–2001 at the site of St George's church in the village of Mateško Selo, 76 km SW of Sisak.
*Location*: Built into a house at the findspot.
*Stone*: Local limestone (after Perkić).
*Preservation*: A lid fragmentarily preserved, with a small corner fragment with the respective *acroterion* surviving.
*Description*: *L 18* cm, *H 28* cm, *W 32* cm; type 1a. The roughly worked gabled lid had flat edges and corner *acroteria*.
*Date*: 2nd or 3rd century.
*Publication*: Perkić 2012: 207–208, no. 5.7.2.31., T. 50: 4.

22. Ash-chest (S[ptSW] III.22)
*Provenance*: Archaeological research in 1997–2001 at the site of St George's church in the village of Mateško Selo, 76 km SW of Sisak.
*Location*: At the findspot.
*Stone*: Local limestone (after Perkić).
*Preservation*: A lid fragmentarily preserved, with the central part, broken off along all four sides, surviving.
*Description*: *L 108* cm, H 30 cm, *W 80* cm; type 1a. The roughly worked gabled lid had flat edges and corner *acroteria*.
*Date*: 2nd or 3rd century.
*Publication*: Perkić 2012: 208, no. 5.7.2.32., T. 50: 5.

23. Ash-chest (S[ptSW] III.23)
*Provenance*: Archaeological research in 1997–2001 at the site of St George's church in the village of Mateško Selo, 76 km SW of Sisak.
*Location*: Built into the baptistery of St George's church.
*Stone*: Local limestone (after Perkić).
*Preservation*: A chest completely preserved, apart from one fragment at a top inner corner.
*Description*: L 97 cm, H 46 cm, W 62 cm; type 7. The roughly worked rectangular chest has flat edges.
*Date*: 2nd or 3rd century.
*Publication*: Perkić 2012: 208, no. 5.7.2.33., T. 52: 1.

24. Ash-chest (S[ptSW] III.24)
*Provenance*: Archaeological research in 1997–2001 at the site of St George's church in the village of Mateško Selo, 76 km SW of Sisak.
*Location*: Built into the altar of St George's church.
*Stone*: Local limestone (after Perkić).
*Preservation*: A lid completely preserved, apart from some damage to two of the four *acroteria*.
Description: L 120 cm, H 33 cm, W 80 cm; type 1a. The roughly worked gabled lid had flat edges and corner *acroteria*; type.
*Date*: 2nd or 3rd century.
*Publication*: Perkić 2012: 208, no. 5.7.2.34., T. 51: 3.

25. Ash-chest (S[ptSW] III.25)
*Provenance*: Archaeological research in 1997–2001 at the site of St George's church in the village of Mateško Selo, 76 km SW of Sisak.
*Location*: Built into the altar of St George's church.
*Stone*: Local limestone (after Perkić).
*Preservation*: A lid fragmentarily preserved, with a fragment of a gabled side with the respective *acroterion* surviving.
*Description*: *L 36* cm, H 24 cm, *W 74* cm; type 1a. The roughly worked gabled lid had flat edges and corner *acroteria*.
*Date*: 2nd or 3rd century.
*Publication*: Perkić 2012: 208, no. 5.7.2.35., T. 52: 8.

26. Ash-chest (S[ptSW] III.26)
*Provenance*: Archaeological research in 1997–2001 at the site of St George's church in the village of Mateško Selo, 76 km SW of Sisak.
*Location*: Built into the altar of St George's church.
*Stone*: Local limestone (after Perkić).
*Date*: 2nd or 3rd century.
*Preservation*: A chest fragmentarily preserved, with the the upper part missing and a diagonal crack across the bottom.
*Description*: L 85 cm, *H 32* cm, W 57 cm; type 7. The roughly worked rectangular chest (presumably) had flat edges.
*Publication*: Perkić 2012: 208–209, no. 5.7.2.36., T. 52: 4.

27. Ash-chest (S[ptSW] III.27)
*Provenance*: Archaeological research in 1997–2001 at the site of St George's church in the village of Mateško Selo, 76 km SW of Sisak.
*Location*: Built into the altar of St George's church.
*Stone*: Local limestone (after Perkić).
*Preservation*: A chest fragmentarily preserved, with approximately two-thirds of the lower part surviving.
*Description*: L 86 cm, *H 38* cm, W 56 cm; type 8. The roughly worked rectangular chest (presumably) had flat edges. There are feeble traces of tool marks on the inner side of the bottom, and a blind, slightly sunken circular inscription field on the front, whose upper part is broken off.
*Date*: 2nd or 3rd century. *Publication*: Perkić 2012: 209, no. 5.7.2.37., T 8: 5, T. 52: 7.

28. Ash-chest (S[ptSW] III.28)
*Provenance*: Archaeological research in 1997–2001 at the site of St George's church in the village of Mateško Selo, 76 km SW of Sisak.
*Location*: Built into the altar of St George's church.
*Stone*: Local limestone (after Perkić).
*Preservation*: A lid fragmentarily preserved as broken in five fragments that are individually built into the altar of St George's church.
*Description*: Dimensions are not recorded; type 1a. The roughly worked gabled lid had flat edges and corner *acroteria*.
*Date*: 2nd or 3rd century. *Publication*: Perkić 2012: 209, no. 5.7.2.38., T. 52: 2, 3, 5, 6, 9.

29. Ash-chest (S[ptSW] III.29)
*Provenance*: Archaeological research in 1997–2001 at the site of St George's church in the village of Mateško Selo, 76 km SW of Sisak.
*Location*: Built into the foundations of the south church wall.
*Stone*: Local limestone (after Perkić).
*Preservation*: A lid fragmentarily preserved, with the topmost part missing.
*Description*: L 120 cm, *H 30* cm, W 83 cm; type 1a. The roughly worked gabled lid had flat edges and corner *acroteria*.
*Date*: 2nd or 3rd century.
*Publication*: Perkić 2012: 209, no. 5.7.2.39., T. 54.

30. Ash-chest (S[ptSW] III.30)
*Provenance*: Archaeological research in 1997–2001 at the site of St George's church in the village of Mateško Selo, 76 km SW of Sisak.
*Location*: Built into the foundations of the south church wall.
*Stone*: Local limestone (after Perkić).
*Preservation*: A lid fragmentarily preserved, with the topmost part missing.
*Description*: L 130 cm, *H 35* cm, W 88 cm; type 1a. The roughly worked gabled lid had flat edges and corner *acroteria*.
*Date*: 2nd or 3rd century.
*Publication*: Perkić 2012: 209, no. 5.7.2.40., T. 54.

31. Ash-chest (S[ptSW] III.31)
*Provenance*: Archaeological research in 1997–2001 at the site of St George's church in the village of Mateško Selo, 76 km SW of Sisak.
*Location*: Built into the foundations of the south church wall.
*Stone*: Local limestone (after Perkić).
*Preservation*: A lid fragmentarily preserved, with the topmost part missing.
*Description*: L 116 cm, *H 33* cm, W 83 cm; type 1a. The roughly worked gabled lid had flat edges and corner *acroteria*.
*Date*: 2nd or 3rd century. *Publication*: Perkić 2012: 209-210, no. 5.7.2.41., T. 54.

32. Ash-chest (S[ptSW] III.32)
*Provenance*: Archaeological research in 1997–2001 at the site of St George's church in the village of Mateško Selo, 76 km SW of Sisak.
*Location*: Built into the foundations of the south church wall.
*Stone*: Local limestone (after Perkić).
*Preservation*: A lid fragmentarily preserved, with the topmost part missing.
Description: L 117 cm, *H 32* cm, W 90 cm; type 1a. The roughly worked gabled lid had flat edges and corner *acroteria*.
*Date*: 2nd or 3rd century.
*Publication*: Perkić 2012: 210, no. 5.7.2.42., T. 54.

33. Ash-chest (S[ptSW] III.33)
*Provenance*: Archaeological research in 1997–2001 at the site of St George's church in the village of Mateško Selo, 76 km SW of Sisak.
*Location*: Built into the foundations of the south church wall.
*Stone*: Local limestone (after Perkić).
*Preservation*: A lid fragmentarily preserved, with the topmost part missing.

*Description*: L 116 cm, *H 28* cm, W 93 cm; type 1.1. The roughly worked gabled lid had flat edges and corner *acroteria*.
*Date*: 2nd or 3rd century.
*Publication*: Perkić 2012: 210, no. 5.7.2.43., T. 54.

34. Ash-chest (S[ptSW] III.34)
*Provenance*: Archaeological research in 1997–2001 at the site of St George's church in the village of Mateško Selo, 76 km SW of Sisak.
*Location*: Built into the foundations of the south church wall.
*Stone*: Local limestone (after Perkić).
*Preservation*: A lid completely preserved.
*Description*: L 156 cm, H 40 cm, W 93 cm; type 1a. The roughly worked gabled lid had flat edges and corner *acroteria*.
*Date*: 2nd or 3rd century.
*Publication*: Perkić 2012: 210, no. 5.7.2.44., T. 54.

35. Ash-chest (S[ptSW] III.35)
*Provenance*: Archaeological research in 1997–2001 at the site of St George's church in the village of Mateško Selo, 76 km SW of Sisak.
*Location*: Built into the foundations of the south church wall.
*Stone*: Local limestone (after Perkić).
*Preservation*: A chest fragmentarily preserved, with approximately one half missing.
*Description*: *L 60* cm, H 55 cm, W 90 cm; type 8. The roughly worked rectangular chest with flat edges has a slightly sunken, blind circular inscription panel on the front.
*Date*: 2nd or 3rd century.
*Publication*: Perkić 2012: 210, no. 5.7.2.45., T. 54.

36. Ash-chest (S[ptSW] III.36)
*Provenance*: Archaeological research in 1997–2001 at the site of St George's church in the village of Mateško Selo, 76 km SW of Sisak.
*Location*: Built into the foundations of the north church wall.
*Stone*: Local limestone (after Perkić).
*Preservation*: A lid fragmentarily preserved, with the topmost part missing.
*Description*: L 130 cm, *H 38* cm, W 88 cm; type 1a. The roughly worked gabled lid had flat edges and corner *acroteria*.
*Date*: 2nd or 3rd century. *Publication*: Perkić 2012: 211, no. 5.7.2.46., T. 54.

37. Ash-chest (S[ptSW] III.37)
*Provenance*: Archaeological research in 1997–2001 at the site of St George's church in the village of Mateško Selo, 76 km SW of Sisak.
*Location*: Built into the foundations of the north church wall.
*Stone*: Local limestone (after Perkić).
*Preservation*: A lid fragmentarily preserved.
*Description*: *L?*; H 65 cm, W 93 cm; type 1a. The roughly worked gabled lid has flat edges and corner *acroteria*.
*Date*: 2nd or 3rd century. *Publication*: Perkić 2012: 211, no. 5.7.2.47., T. 54.

38. Ash-chest (S[ptSW] III.38)
*Provenance*: Archaeological research in 1997–2001 at the site of St George's church in the village of Mateško Selo, 76 km SW of Sisak.
*Location*: Built into the foundations of the north church wall.
*Stone*: Local limestone (after Perkić).
*Preservation*: A lid fragmentarily preserved.
*Description*: L 107 cm, *H 32* cm, *W?*; type 1a. The roughly worked gabled lid has flat edges and corner *acroteria*.
*Date*: 2nd or 3rd century.
*Publication*: Perkić 2012: 211, no. 5.7.2.48., T. 54.

39. Ash-chest (S[ptSW] III.39)
*Provenance*: Archaeological research in 1997–2001 at the site of St George's church in the village of Mateško Selo, 76 km SW of Sisak.
*Location*: Built into the foundations of the north church wall.
*Stone*: Local limestone (after Perkić).
*Preservation*: A lid fragmentarily preserved.
*Description*: L 111 cm, *H 35* cm, *W?*; type 1a. The roughly worked gabled lid has flat edges and corner *acroteria*.
*Date*: 2nd or 3rd century.
*Publication*: Perkić 2012: 211, no. 5.7.2.49., T. 54.

40. Ash-chest (S[ptSW] III.40)
*Provenance*: Archaeological research in 1997–2001 at the site of St George's church in the village of Mateško Selo, 76 km SW of Sisak.
*Location*: Built into the foundations of the north church wall.
*Stone*: Local limestone (after Perkić).
*Preservation*: A lid fragmentarily preserved.
*Description*: L 99 cm, *H 42* cm, *W?*; type 1a. The roughly worked gabled lid has flat edges and corner *acroteria*.
*Date*: 2nd or 3rd century.
*Publication*: Perkić 2012: 211, no. 5.7.2.50., T. 54.

41. Ash-chest (S[ptSW] III.41)
*Provenance*: Archaeological research in 1997–2001 at the site of St George's church in the village of Mateško Selo, 76 km SW of Sisak.
*Location*: Built into the foundations of the north church wall.
*Stone*: Local limestone (after Perkić).
*Preservation*: A lid fragmentarily preserved.
*Description*: L 112 cm, *H 30* cm, *W?*; type 1a. The roughly worked gabled lid has flat edges and corner *acroteria*.
*Date*: 2nd or 3rd century.
*Publication*: Perkić 2012: 212, no. 5.7.2.51., T. 54.

42. Ash-chest (S[ptSW] III.42)
*Provenance*: Archaeological research in 1997–2001 at the site of St George's church in the village of Mateško Selo, 76 km SW of Sisak.
*Location*: Built into the foundations of the north church wall.
*Stone*: Local limestone (after Perkić).
*Preservation*: A lid fragmentarily preserved.
*Description*: L 122 cm, *H 34* cm, *W?*; type 1a. The roughly worked gabled lid has flat edges and corner *acroteria*.
*Date*: 2nd or 3rd century.
*Publication*: Perkić 2012: 212, no. 5.7.2.52., T. 54.

43. Ash-chest (S[ptSW] III.43)
*Provenance*: Archaeological research in 1997–2001 at the site of St George's church in the village of Mateško Selo, 76 km SW of Sisak.
*Location*: Built into the foundations of the north church wall.
*Stone*: Local limestone (after Perkić).
*Preservation*: A lid fragmentarily preserved.
*Description*: L 136 cm, H 37 cm, W 99 cm; type 1a. The roughly worked gabled lid has flat edges and corner *acroteria*.
*Date*: 2nd or 3rd century. *Publication*: Perkić 2012: 212, no. 5.7.2.53., T. 54.

44. Ash-chest (S[ptSW] III.44)
*Provenance*: Archaeological research in 1997–2001 at the site of St George's church in the village of Mateško Selo, 76 km SW of Sisak.
*Location*: Built into the foundations of the north church wall.
*Stone*: Local limestone (after Perkić).
*Preservation*: A chest fragmentarily preserved.
Description: *L?*, H 31 cm, W 90 cm; type 7. The roughly worked rectangular chest has flat edges.
*Date*: 2nd or 3rd century.
*Publication*: Perkić 2012: 212, no. 5.7.2.54., T. 54.

45. Ash-chest (S[ptSW] III.45)
*Provenance*: Archaeological research in 1997–2001 at the site of St George's church in the village of Mateško Selo, 76 km SW of Sisak.
*Location*: Built into the foundations of the north church wall.
*Stone*: Local limestone (after Perkić).
*Preservation*: A chest fragmentarily preserved.
*Description*: *L 80* cm, *H 35* cm, *W 66* cm; type 7. The roughly worked rectangular chest (presumably) had flat edges.
*Date*: 2nd or 3rd century.
*Publication*: Perkić 2012: 212–213, no. 5.7.2.55., T. 54.

46. Ash-chest (S[ptSW] III.46)
*Provenance*: Archaeological research in 1997–2001 at the site of St George's church in the village of Mateško Selo, 76 km SW of Sisak.
*Location*: Built into the foundations of the north church wall.
*Stone*: Local limestone (after Perkić).
*Preservation*: A lid fragmentarily preserved.
*Description*: L 100 cm, *H 26* cm, *W 62* cm; type 1a. The roughly worked gabled lid has flat edges and corner *acroteria*.
*Date*: 2nd or 3rd century.
*Publication*: Perkić 2012: 213, no. 5.7.2.56., T. 54.

47. Ash-chest (S[ptSW] III.47)
*Provenance*: Archaeological research in 1997–2001 at the site of St George's church in the village of Mateško Selo, 76 km SW of Sisak.
*Location*: Built into the foundations of the north church wall.
*Stone*: Local limestone (after Perkić).
*Preservation*: A chest fragmentarily preserved, with the topmost part missing.
*Description*: L 86 cm, *H 62* cm, W 64 cm; type 7. The roughly worked rectangular chest (presumably) had flat edges.
*Date*: 2nd or 3rd century.
*Publication*: Perkić 2012: 213–214, no. 5.7.2.60.

48. Ash-chest (S[ptSW] III.48)
*Provenance*: Archaeological research in 1997–2001 at the site of St Michael's church in the village of Erdelj, 80 km SW of Sisak, probably brought there from the site of Crikvišće in the nearby village of Vali.
*Location*: Built into the church at the findspot.
*Stone*: Local limestone (after Perkić).
*Preservation*: A fragmentarily preserved chest front, broken off and chipped around the edges. There is a diagonal stylolithic seam in the upper third, and the stone is weathered and pitted all over.
*Description*: *L 59* cm, *H 36* cm, *T?*; type 1. The roughly worked front of the chest is taken by a moulded inscription panel, flanked on the left lateral side by a grooved motif of a peltate *ansa*. An identical motif, which should have been executed on the opposite lateral side, is not discernible.
*Inscription*: *+++DONI / Ascani / ann(os) vix(it) XXX [---] / [---(?)*
Line 1: possibly *L*, *A*, and *I*, but also *DON* may not be absolutely certain.
Translation: (The memorial) of ... Ascanius ... Lived 30 years.
Commentary: Except in Italy, the cognomen Ascanius is only sporadically attested: a freedman C. Farax Ascanius from Scarbantia, Pannonia Superior is known (*CIL* III 4257 = *RIU* 1: 186 = *EDCS*-26600175), as well as a freedman L. Cusius Ascanius from Tilurium in Dalmatia (*AE* 1999: 1231 = *EDCS*-14800172).
*Date*: 2nd or 3rd century.
*Publication*: Perkić 2012: 172–174, 214, no. 5.7.2.62., fig.18.

Figure 138. (S[ptSW] III.48) – ash-chest (B. Djurić)

Figure 139. (S[ptSW] III.50) – ash-chest (Lj. Perinić)

49. Ash-chest (?) (S[ptSW] III.49)
*Provenance*: Archaeological research in 1997–2000 in the village of Srednji Poloj, 74 km SW of Sisak.
*Location*: Built into a house at the findspot.
*Stone*: Local limestone (after Perkić).
*Preservation*: A fragmentarily preserved inscription.
*Description*: L 90 cm, *H 25* cm, W 60 cm; presumably type 8. The fragment most probably belonged to the front side of a roughly worked rectangular ash chest with flat edges and with a (presumed) circular inscription panel engraved.
*Date*: 2nd or 3rd century.
*Publication*: Perkić 2012: 214, no. 5.7.2.63.

50. Ash-chest (?) (S[ptSW] III.50)
*Provenance*: Archaeological research in 1997–2001 at the site of Vali (Lauši), near the settlement of Generalski Stol, 80 km SW of Sisak.
*Location*: Built into a house in the village of Vali (Lauši).
*Stone*: Local limestone (after Perkić).
*Preservation*: Fragmentarily preserved.
*Description*: Dimensions are not recorded (*c.* 40 x 30 cm). A small fragment of an inscription possibly belonged to an ash-chest's front side.
*Inscription*: *[---]RO PRỌ[---] / [---]I S O RI+[---] / [---*
No line seems to be missing above, on all other sides, however, the inscription is fragmentary.
*Date*: 2nd or 3rd century. *Publication*: Perkić 2012: 157, fig. 19, 214, no. 5.7.2.61.

*V. Altars*

1. Altar shaft (?) (S[ptSW] V.1)
*Provenance*: A chance find in the village of Mrežnički Varoš, 68 km SW of Sisak; first recorded in 1707.

*Location*: Recorded in 1938 as built into a mill house at the findspot, now missing.
*Stone*: local limestone (after the *AIJ*).
*Preservation*: An inscribed block fragmentarily preserved, broken in two pieces built-in separately (*a* and *b*). The right-hand side border was battered and the surface much weathered. The recorded inscription seems to have been much longer than visible on the photograph in the *AIJ*.
*Description*: Fragment *a*: *H* 50 cm, *W* 55 *cm*, *T?*; fragment *b*: *H* 39 cm, *W* 30 *cm*, *T?* The block was a parallelepiped with a moulded inscription panel on the front. The moulded panel on one of the sides featured a motif of two snakes (or, rather, dolphins) with down-turned heads and intertwined tails.
*Inscription*: *[D(is) M(anibus)]. / [C(aio) Numisi]o Ursin(o) / [et Pompo]niae Iuven/[ti]llae /⁵ [et Num]isiae Rufi/[nae]. / [C(aius) Numisius R]ufus fil(ius) / [et Numisia C(ai) f(ilia) Mar]/[cellina sor]or /¹⁰ [inceptum pere]gerunt.*
According to the *CIL*, the inscription (still entirely preserved at the time) was copied and sent to Kukuljević from his friend, who, however, did not pay attention to the division of lines. These were reconstructed by Hoffiller (*AIJ*), on the basis of two preserved fragments. Ligatures: line 2 (at the end): *IN*; line 10 (at the end): *NT*.
Translation: To the spirits of the departed. To Gaius Numisius Ursinus and Pomponia Iuventilla and Numisia Rufina. Gaius Numisius Rufus, son, and Numisia Marcellina, daughter of Gaius, sister, provided for the monument to be finished.
Commentary: The Numisii had very likely come from Italy or perhaps from Hispania or Gallia Narbonensis to settle in Siscia, since the *gentilicium* is elsewhere rare (Radman-Livaja 2014: 239, 274). The *gentilicium* Pomponius is also well documented in Italy and Hispania, as well as in Dalmatia, while it is rare in Pannonia, where it only occurs three times (*OPEL* III: 151–152). The cognomen Ursinus is well attested particularly in Dalmatia (mainly in the Late Roman period: Alföldy 1969: 317) and Noricum (Lochner von Hüttenbach 1989: 172), and less so in Pannonia (possibly from Dalmatia: Mócsy 1959: 194; Barkóczi 1964: 326), elsewhere it is not frequent (*OPEL* IV: 186–187). According to the *OPEL*, the name Iuventilla (masculine form is not attested) only occurs once in Gallia Narbonensis, Lugdunensis, and Pannonia (in this case), as well as four times in Dalmatia (II: 211; Mócsy 1959: 177). Its erroneous classification as an Illyrian name was corrected by Géza Alföldy (1969: 225, with citations). The name Rufinus/a (also Rufinna) is very well attested in Siscia, twice also on the *tesserae* (Radman-Livaja 2014: 254; cf. *OPEL* III: 33–34). It is derived from Rufus (red-haired), which is frequent everywhere, but particularly in Italy, Hispania, Gallia Narbonensis, Dalmatia, and Pannonia (*OPEL* IV: 35–36); at Siscia, it also appears on three *tesserae* (Radman-Livaja 2014: 255; cf. Mócsy 1959: 188). The name

Figure 140. (S[ptSW] V.1) – altar shaft, scan of the figure (after the *AIJ*)

Marcellinus/Marcellina, which is frequent everywhere (it may have spread from Italy to Dalmatia and Pannonia, cf. Mócsy 1959: 180), also occurs on three lead tags at Siscia (Radman-Livaja 2014: 226; cf. *OPEL* III: 53–54; Alföldy 1969: 236–237).
*Date*: 2nd century.
*Publication*: *CIL* III 3935 (+ p. 2328, 114); *AIJ* 496. Internet databases: *HD*071784; *EDCS*-26600414.
*Note*: According to archival data brought in the *AIJ*, one of the sides of this tombstone was taken by a motif of snakes, suggesting that the monument was a funerary altar. A figural relief on the opposite side was not recorded, which means that most probably it had been damaged or completely obliterated, as often happens with these kinds of funerary stones (cf. Handy and Pochmarski 2014: 84). It should also be noted that the motif of snakes by a trident is very rare in Norico-Pannonian funerary art, while dolphins are ubiquitous.

*VIII. Unclassifiable*

1. Inscription (altar?) (S[ptSW] VIII.1)
*Provenance*: A chance find in the village of Sveti Petar Mrežnički, 69 km SW of Sisak.

Figure 141. (S[ptSW] VIII.1) – the front side (AMZ, I. Krajcar)

Figure 142. (S[ptSW] VIII.1) – the right short side (AMZ, I. Krajcar)

*Location*: AMZ, inv. no. KS-965.
*Stone*: Lower-Cretaceous micritic limestone. A possible origin: the Outer Dinarides and the southern Pokuplje; a probable origin: the Kordun (M. Belak 2018).
*Preservation*: Fragmentarily preserved. A smaller part of the right side of a presumably block-like stone survives, with insignificant remains of an inscription and a side relief. The border of the inscription panel is battered, the surface stained, and the short side much damaged, but the letters are well preserved.
*Description*: *H* 72 cm, *W* 41 cm, *T* 23 cm. The framed inscription panel and both short sides of the block were probably decorated with the motif of a scroll within a moulded frame (the right-side one survives).
*Inscription*: *------?] / [---]elia[e] / [---]Ceni / [--- uxo?]ri / [---? posteris]que suis /*[5] *[v(ivus)] f(ecit).*
Line 1 and 2: *? Aure]lia[e / ....ceni, AIJ; [---]elia[e] / [---]C(?)eni, EDH.* Remains of an *E* are visible in line 1, but it is not entirely clear if the empty space before the letter *C* in line 2 is large enough to regard the letter as the initial of a name. Line 3: *uxori* is not certain. Line 5: *F longa.*
Translation: ... To ...elia ... his wife (?) ... and to his descendants (had the monument erected) in his lifetime.

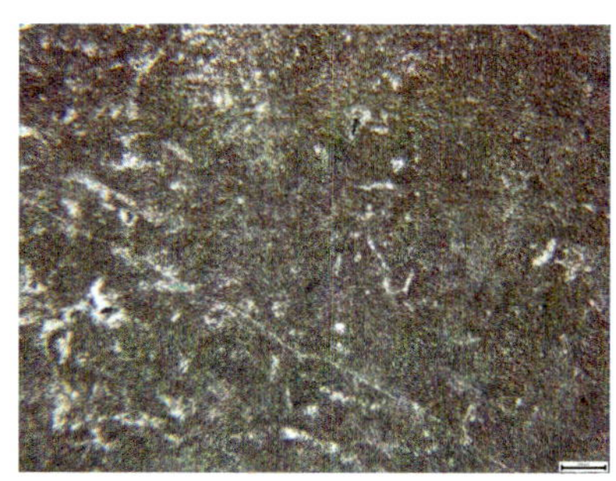

Figure 143. (S[ptSW] VIII.1) – stone, microscope (HGI, M. Belak)

Commentary: The *gentilicium* could have been Aelia or Aurelia, both very frequent in Pannonia, as also elsewhere (*OPEL* I: s.vv.), but there are very many possibilities to restore a non-imperial *gentilicium*. There is no Latin cognomen starting in *Ceni-*, but the *gentilicium* Cenius is known from Siscia (*CIL* III 3985 = *HD*048572, see above; Solin and Salomies 1994: 52). The cognomen Genis is once attested in Moesia Superior (*OPEL* II: 165).
*Date*: 2nd century (?).

*Publication*: *AIJ* 495; *ILSl* 1, 165 (erroneously regarded as from Gradac near Metlika, cf. *AIJ* p. 219). Internet databases: *HD*068514; *EDCS*-11301719.

2. Relief (S[ptSW] VIII.2)
*Provenance*: A chance find in the village of Sveti Petar Mrežnički, 69 km SW of Sisak.
*Location*: Built into the church of Saint Peter in the village of Sveti Petar Mrežnički.
*Stone*: Yellowish limestone (after the *AIJ*); impossible to establish by macroscopy due to a layer of whitewash on the surface.
*Preservation*: A partially preserved and fairly weathered relief panel.
*Description*: *H 68* cm, *W 51* cm, *T*? A rectangular panel is taken by an unframed, irregularly hollowed niche, which holds a relief of a female nude striding towards left, with her face turned towards right, and with her right foot firmly on the ground. She has an oval-triangular face framed by hair or a wrapped headdress reaching to the level of the ears, with an indeterminable triangular shape sticking out on the left side. She is holding a round object (a *patera*? a mirror? a mistakenly depicted tambourine with a superfluous handle?) in her right hand, while grasping with her left at a dress or a cloak thrown over her left shoulder. The workmanship is extremely crude, with no sense of natural features and anatomical proportions.
*Date*: 3rd century (?).
*Publication*: *AIJ*: p. 219, no. 499.

Figure 144. (S[ptSW] VIII.2) – relief (B. Djurić)

*Note*: Given the measurements, the piece could have been a lateral relief panel of the front of a sarcophagus or a shaft relief. The figure is hard to establish with certainty: maenad? Venus? a syncretised maenad and Venus?

3. Stone block (S[ptSW] VIII.3)
*Provenance*: Unknown; first recorded as walled into a private house in the village of Erdelj, 80 km SW of Sisak, probably brought there from the site of Crikvišće in the nearby village of Vali.
*Location*: Hrvatski povijesni muzej/The Croatian History Museum, Zagreb, inv. no. 25312.
*Stone*: Lower-Cretacious micritic limestone. A possible origin: the Outer Dinarides; a probable origin: The Kordun, SW of Sisak (M. Belak 2018).
*Preservation*: Fragmentarily preserved. The left side is missing and the surface is weathered all over and cracked at places.
*Description*: *W 70* cm, H 58 cm, T 37 cm. The front side of a rectangular block is taken by a relief of a horseman riding towards left, depicted in a two-lobed niche shaped to closely accommodate the figure. The workmanship is extremely course, with no sense of natural movement and anatomical proportions, especially as concerns the human figure.

Figure 145. (S[ptSW] VIII.3) – monument (Hrvatski povijesni muzej)

Figure 146. (S[ptSW] VIII.3) – stone, microscope (HGI, M. Belak)

*Date*: 3rd century (?).
*Publication*: Valentić and Prister 2002: 58, no. 96; Perkić 2012: 173–174.
*Note*: When first published, the stone was dated to the 15th century, but it is more likely a Roman monument.

***The possible territory of Siscia in the south-east*** **(S[ptSE])**

*I. Stelai*

1. Stele (S[ptSE] I.1)
*Provenance*: A chance find in the village of Japra, 51 km SE of Sisak.
*Location*: ZMBiH, inv. no. 1420.
*Stone*: Upper-Cretacious bioclastic limestone. Possible origin: the Outer Dinarides (M. Belak 2018).
*Preservation*: Fragmentarily preserved. The upper-left part survives with segments of the pediment and inscription panel. The stele is damaged along the side edges and on the capital by the inscription panel, and is chipped on the upper edge, with the uppermost part possibly missing.
*Description*: *H 50* cm, *W 28 cm*, T 25 cm. The upper part was probably rectangular, with a moulded pediment sitting on an articulated architrave and a narrow frieze featuring a shear-shaped cyma. The spandrel is decorated with a palmette and scrolled leaves, and

Figure 147. (S[ptSE] I.1) – stele (ZMBiH, E. Bujak)

Figure 148. (S[ptSE] I.1) – stone, microscope (HGI, M. Belak)

the pediment corner with a bud-like floral motif. The inscription panel is moulded and flanked by a spirally fluted column, and is topped not by a capital, but a construction resembling a double base composed of three pairs of *tori* and two high *scotiae* in between. The workmanship is fairly competent.
*Inscription*: *M(arcus) M[---] / C(ai) f(ilius) T[---] / XXVI [---] / [-----*
Translation: Marcus M[-], son of Gaius, T[-] ...
Only the upper-left corner of the inscription field is preserved. The man was at least 26 years old, but was not older than 29 years.
*Date*: 1st or early 2nd century.
*Publication*: Paškvalin 2012: 95–96, no. 31.

2. Stele (?) (S[ptSE] I.2)
*Provenance*: A chance find during iron slag exploitation in the 1930s in the village of Blagaj, 51 km SE of Sisak.
*Location*: Missing.
*Stone*: Sandstone (according to D. Sergejevski) or limestone (according to V. Paškvalin).
*Preservation*: Fragmentarily preserved, with the inscription panel and stumps of corner *acroteria* surviving. The stone was battered and chipped around the edges, especially on the left side and on the bottom, where a part may have been missing.
*Description*: *H 118* cm, W 85 cm, T 20 cm. A rectangular slab provided by corner *acroteria* has an inscription panel within a moulding composed of a wider band of water leaves framed with a string of a bead-and-reel ornament, and by an outer ornamental band featuring rosettes in acanthus scrolls. The workmanship is fairly competent.
*Inscription*: *D(is) M(anibus). / Ulp() Nonn[a?]/tioni an(norum) XL. / Camp(anius?) Marcia/*[5]*nus v(ivus) s(ibi) et / Iul(ia) Marcian[a] / paren(tes) pii[..] / I IVET[...] / [------?.*
Letters are regular. On the basis of a photograph, a missing final *A* is questionable. Line 2: *Ulp()* cannot be supplemented because of uncertain gender. Line 3: *O* is smaller; line 4: *MP* in ligature, *I* is carved inside *C*, as also in line 6. Lines 7-8: probably *pii[ss]/i[m]i*. One line may be missing at the end of the epitaph, but this is not certain.
Translation: To the spirits of the departed. To Ulp() Nonnatio, 40 years old. Campanius Marcianus [erected]

Figure 149. (S[ptSE] I.2) – scan of the figure (after Sergejevski 1940)

in his lifetime for himself and Iulia Marciana, pious parents ... .

Commentary: The names Nonnatio or Nonntio are not listed in *OPEL*; either would be a *hapax*. *Nonntio* is included in the catalogue of names in Alföldy 1969: 254 and regarded as Celtic. It is not clear whether the deceased was male or female; Sergejevski and Alföldy decided for the latter, Paškvalin, *EDH*, *Epigraphik-Datenbank* Clauss/Slaby, and Mesihović (2011: 361–362) for the former. Campanius is the most probable supplement, other two possibilities for Latin *gentilicia* beginning with *Camp()* would have been Campasius and Campatius (Solin and Salomies 1994: 44). C(aius) Amp(eius?) has been suggested by Sergejevski as another possibility for *CAMP()*; *gentilicia* starting in *Amp()* are several. Marcianus/a is everywhere a frequent cognomen and a personal name (*OPEL* III: 55–56), in Pannonia particularly towards the end of the 2nd century; in the Celtic-speaking provinces this may have been an assonance name (Radman-Livaja 2014: 227).

*Date*: Probably 2nd century.

*Publication*: Sergejevski 1940: 12 no. 3, fig. 3; *ILJug* 1479A; Mesihović 2011: 361–362; Paškvalin 2012: 39, no. 18. Internet databases: *HD*033757; *EDCS*-10100501.

*Note*: The monument was classified as a *titulus*-stele on account of the *acroteria*. It seems to be of amalgamated form, but closer to a stele for two reasons: its thickness and the possibility that there was a missing part below the inscription panel.

3. Stele (S[ptSE] I.3)

*Provenance*: Found in 1955 as reused in a medieval grave in the village of Cikote near Prijedor, 57 km SE of Sisak.

*Location*: MK.

*Stone*: Badenian sandy biocalcarenite. A possible origin: mountains in N Croatia and in the valleys of the Sana and Una Rivers, NW Bosnia and Herzegovina; a probable origin: Mts Kozara or Prosara, NW Bosnia and Herzegovina (M. Belak 2018).

*Preservation*: Fragmentarily preserved. The upper part is missing, being broken at the shoulder-level of the female figure in the portrait niche, and the bottom-right edge of the base is chipped. The stone is slightly worn-out and pitted all over, but the pits on the back of the niches and the inscription panel around the letters seem to be chisel marks rather than damage. The relief scene on the base is quite weathered.

*Description*: *H* 179 cm, W 79 cm, T 20 cm; probably type 5a or 5b. The preserved part of the stele is composed of a portrait niche, an inscription panel, and a base. The niche is flanked by smooth columns sitting on the bases composed of a plinth and three superimposed *tori*, and is divided from the inscription panel by a frieze of ivy

Figure 150. (S[ptSE] I.3) – stele (M. Jambrović)

Figure 151. (S[ptSE] I.3) – stone, microscope (HGI, M. Belak)

tendrils springing from the central calyx. It holds a half-figure of a woman wearing a wide, bulbous, pleated over-dress fastened by large brooches of uncertain type (possibly a knee or a sharply profiled [*kräftig profilierte*] form), of which only the elongated triangular heads are visible on the shoulders. A large hoop is suspended from each of the brooches, falling short of the breasts. Underneath the over-dress the woman is wearing a long- and tight-sleeved tunic, with two bangles (rather than double-banded cuffs) on the wrists. She seems to be holding a swaddled baby in her arms, supporting its head with her left hand, while the area of her belly is mostly hidden behind a blind rimmed roundel, whose function remains obscure. The inscription panel is flanked by columns topped by stylized composite capitals and sitting on the bases identical to those in the portrait niche. A simply framed elongated base features a mythological episode involving three characters. The central one is shown frontally in a long stride between two warriors turned towards each other; she is spreading her arms and touching them both as in the gesture of preventing them from clashing (a Sabine woman separating the Romans and Sabines?; Athena, Achilles and Hector in a scene from the Iliad?). The goddess is dressed in a *peplos* with an overfold belted under the breasts, while the warriors are wearing short tunics and crested helmets. They are holding a buckled shield in their left hand, and weapons in their right; the warrior on the left probably has a dagger and the one on the right side a spear. The workmanship of the relief in the portrait niche is fairly crude (arms), contrary to the much better and livelier scene executed on the base. The latter's workmanship is, however, hard to estimate accurately because of the heavy weathering.
*Inscription*: *D(is) M(anibus). / Ulp(ia) Mate/ra an(norum) XXXV. / Ael(ius) Quartu(s) /*[5] *uxori fe/cit.*
Translation: To the spirits of the departed. Ulpia Matera, 35 years old. Aelius Quartus (had the monument) made for (his) wife.
Commentary: Aelii and Ulpii are well attested in Siscia, some were undoubtedly native inhabitants. Matera, possibly Celtic or an assonance name, is only attested a few times, but is twice documented in Pannonia (*OPEL* III 64; *Epigraphik-Datenbank* Clauss/Slaby), once also on a *tessera* from Siscia (Radman-Livaja 2014: 230). She may have belonged to the indigenous population, as also her husband, since the colourless name Quartus was often also a translation name.
*Date*: 2nd or 3rd century.
*Publication*: Sergejevski 1957: 118–120 no. 9, fig. 3 (drawing); *ILJug* 164. Internet databases: *HD*033136.
*Note*: A discrepancy between both the subject-matter and the workmanship of the reliefs in the portrait niche and those on the base suggests that the former was possibly made to order after the purchase of the monument.

4. Stele (S[ptSE] I.4)
Provenance: Found in 1955 as reused in a medieval grave in the village of Cikote near Prijedor, 57 km SE of Sisak.
*Location*: MK.
*Stone*: Badenian sandy biocalcarenite. A possible origin: mountains in N Croatia and in the valleys of the Sana and Una Rivers (BA) NW Bosnia and Herzegovina; a probable origin: Mts Kozara or Prosara (northern BA) (M. Belak 2018).
*Preservation*: Fragmentarily preserved. The stele has been broken into four parts, of which three survive today: two reassembled as one whole (1), and the third (2) exhibited separately. Missing are the bottom-right segment of the inscription panel and the entire upper section of the stele, including the topmost part of the portrait niche. The surface is slightly stained and pitted all over, but the reliefs and letters are fairly well preserved.
*Description*: Measurements taken as a whole, comprising all three fragments: *H 187* cm, W 91 cm, T 7.5 cm. The preserved part of the stele comprises a portrait niche flanked by spirally fluted columns sitting on Attic bases, and a moulded inscription panel surrounded by wide smooth bands of the remainder of the stele's face; the first two lines of the inscription take the space between the portrait niche and the inscription panel. In the portrait niche thigh-length figures of two men and a woman between them are depicted. The men are wearing long- and tight-sleeved tunics and *saga* fastened with a round brooch on the right shoulder. The one on the left is resting his right hand on the edge of his cloak and is holding a scroll in his left, while the one on the right is clutching his scroll with both hands. The woman is wearing a long- and tight-sleeved tunic with a rounded neckline and possibly a beaded bracelet on the left wrist, a stiff cloak with wide lapels (?) falling over her arms and down her body, and is holding a fruit in her right hand and a bulbous toilet bottle in her left. Faint traces of a narrow, elongated tenon are preserved on the bottom, measuring 61 x 4 cm. The workmanship is thoroughly stiff and crude, but still without pronounced anatomical inaccuracies; the drapery is mostly lacking in elaboration, apart from some breast and shoulder folds of the *saga*.
*Inscription*: *D(is) M(anibus) et perpetuae se/curitati. // Tatoniae Procu/lae ann(orum) XXXIII con(iugi) /*[5] *pi{t}*

Figure 152. (S[ptSE] I.4) – upper part of the stele (M. Jambrović)

*ent(issimae) M(arcus) Aur(elius) Surus / vet(eranus) leg(ionis) X g(eminae) ex b(ene)f(iciario) et / MM(arci) Aur<e>li(i) Optatus / et Firminianus ma/tr(i) infelicissim(ae) /*[10] *et libert(i) Aur(elii) Baosus / et Sumpa et Zipandus / et Quintus et Caro/ni libert(o) θ(obito) / hh(eredes) f(aciendum) c(uraverunt).*

Letters are regular and long, the first line is carved on the architrave; letters *D* (at the beginning) and *M* (at the end) are twice as large and extend across two lines. Originally, the inscription surface was slightly damaged and some letters were adapted to this damage. Line 4: *AE* in lig. *O* is smaller; line 5: *MAVR* in ligature; line 6: *ET* twice in lig.; line 7: *MM*: *m(agistri) m(unicipii)*, Sergejevski wrongly. Line 9: *ME* or *MAE* in lig.; line 10: *ET*, *RT*, *AVR* in lig.; line 11: both *ET* in lig., as well as *VMP* in Sumpa and *AN*; Zipanidus: Škegro, perhaps by mistake; line 12: both *ET* and *NT* in lig. Line 13: *RT* in lig.; *n(atione) I(talici) libert(i) q(ui et)*: Sergejevski erroneously; *theta nigrum* often occurs in the epitaphs of the soldiers of the Legion *X gemina* (*ILJug*). *Caro/ni(us?) libert(ae)q(ue)*: *AE*; *et Caroni libert(o)q(ue)*: Alföldy 1961: 140 no. 38.

Translation: To the spirits of the departed and to eternal security. To Tatonia Procula, his most devoted wife, 33 years old, Marcus Aurelius Surus, veteran of the Legion *X gemina* (twin), former beneficiarius (soldier granted special privileges), and Marci Aurelii Optatus and Firminianus, to (their) unfortunate mother. And freedmen Aurelii Baosus, Sumpa, Zipandus, and Quintus. And to the deceased freedman Caro. The heirs had (this monument) made.

Commentary: The Latin *gentilicium* Tatonius/a has twice been attested to date: once in this case and once elsewhere in Pannonia (*OPEL* IV: 109; Solin and Salomies 1994: 182). Proculus/a is a well-documented name everywhere in the Roman Empire (*OPEL* III: 166–167; *Epigraphik-Datenbank* Clauss/Slaby), including Pannonia, and was one of the most popular names in Siscia, where it is also attested on 20 *tesserae* (Radman-Livaja 2014: 249). The frequent name Surus (*OPEL* IV: 102) can be explained in different ways, since the origin of the name may have been Celtic, Illyrian in the broad sense of the word, Thracian (Dana 2014: 339–340) or Semitic, or it may denote a Syrian origin of its bearer (Alföldy 1969: 303; Minkova 2000: 260; Solin 2003b: 668–669; Meid 2005: 289; Radman-Livaja 2014: 269). Of the Latin names of the couple's sons, Optatus is not rare, while Optata is documented once on a lead

Figure 153. (S[ptSE] I.4) – lower part of the stele (detached) (M. Jambrović)

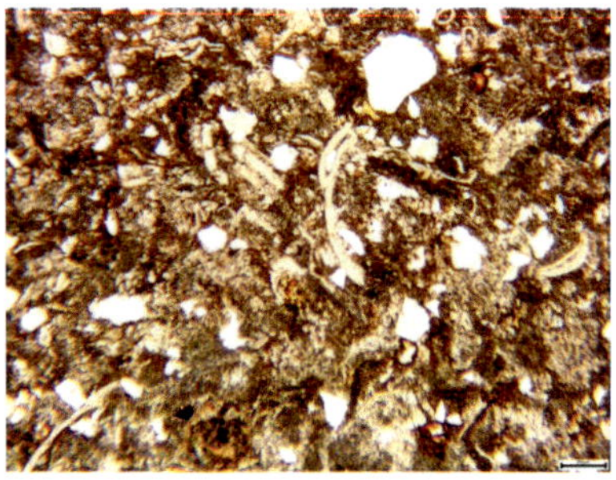

Figure 154. (S[ptSE] I.4) – stone, microscope (HGI, M. Belak)

tag from Siscia (Radman-Livaja 2014: 240). Firminianus, however, is only sporadically attested, except in Pannonia, where it occurs five times (*OPEL* II: 142). The Latin name Caro (Solin and Salomies 1994: 309) is obviously rare, since *OPEL* only cites this inscription (II: 38). Of the names of the freedmen, all three have only been documented in this epitaph (cf. also *Epigraphik-Datenbank* Clauss/Slaby): for Baosus, see *OPEL* I: 111; Sumpa could have been male or female (*OPEL* IV: 99); for Zipandus see *OPEL* IV: 190.

*Date*: 3rd century.

*Publication*: Sergejevski 1957: 116–118 no. 8, Pl. III (*AE* 1958: 66); *ILJug* 163; *CBI* 450; Škegro 1997: 100, no. 106; Bojanovski 1982: 114, Pl. 8. Internet databases: *HD*019714.

Figure 155. (S[ptSE] I.5) – stele (M. Jambrović)

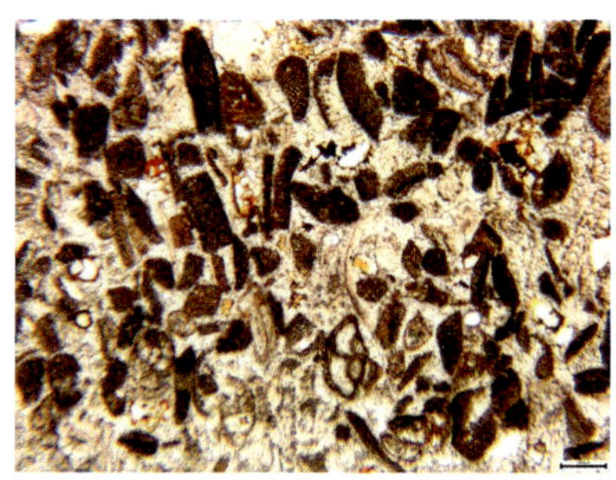

Figure 156. (S[ptSE] I.5) – stone, microscope (HGI, M. Belak)

5. Stele (S[ptSE] I.5)

*Provenance*: Possibly from the village of Cikote near Prijedor, 57 km SE of Sisak.

*Location*: MK.

*Stone*: Badenian bicalcarenite. A possible origin: mountains in N Croatia and in the valleys of the Sana and Una Rivers (BA) NW Bosnia and Herzegovina; a probable origin: Mts Kozara or Prosara (northern BA) (M. Belak 2018).

*Preservation*: Fragmentarily preserved. The surviving parts comprise the portrait niche with a narrow architrave, the dividing decorative band, and a smaller part of the blind inscription panel. There is a vertical crack along the whole height of the right-hand side of the stele; the upper border of the architrave is chipped, the columns and capitals abraded and/or chipped, and the whole surface of the stone extremely weathered, with the facial features of the portrayed people heavily obliterated.

*Description*: *H 112* cm, W 85 cm, T 16–17 cm; probably type 5a or 5b. The niche is flanked by fluted columns topped by indistinct capitals and sitting on high, multi-segmented bases; the flutes seem to be running in opposite sides within the two halves of each column. Within the niche a man on the right-hand side and a woman on the left-hand side are depicted. The woman has a roundish face resting on a stout neck, and a high hairdo with a rounded crown and no inner elaboration, and with small tufts of hair sticking out at the sides of the lower cheeks; it remains unclear whether the ears were exposed or not. She is wearing a tunic with a rounded neckline and a *palla* draped across her chest in V-lines and falling down the shoulders in smooth stiff segments looking like extremely wide sleeves; the possibility that these are the tunic sleeves is less likely. The woman is holding a round object (fruit?) in her right hand, while her left hand is stuck between the side edges of her *palla*. The man has an oval face with a pointed chin and a rounded crown, and a short-cropped hairstyle; a high-set small earlobe seems to be discernible on the left side. He is wearing a *sagum* fastened on the right shoulder by an indistinct shape of brooch, and falling down his left arm in vertical stiff folds, and is clutching a scroll in his left hand. The wide decorative band between the portrait niche and the inscription field has an elongated scroll on

the right-hand side and blurred motifs on the left. The workmanship is hard to assess accurately due to the wear, but the human figures, especially as concerns their clothes, are rendered schematically and crudely.
*Date*: First half of 3rd century.
*Publication*: Unpublished.

6. Stele (S[ptSE] I.6)
*Provenance*: A chance find during iron slag exploitation in the 1930s in the village of Blagaj, 51 km SE of Sisak.
*Location*: Missing (not existing in Prijedor, as presumed by V. Paškvalin).
*Stone*: Sandstone (according to D. Sergejevski) or limestone (according to V. Paškvalin).
*Preservation*: Fragmentarily preserved. When last seen, only the portrait niche survived, and the fragment was chipped around the edges, with the woman's head and the children's heads and faces severely battered. The remainder of the reliefs were fairly well preserved as far as can be discerned from the poor-quality published photograph.
*Description*: *H 115* cm, W 89 cm, T 21 cm. The arched niche, whose front border is decorated with a motif of water leaves, possibly framed by a string of bead-and-reel ornament, rests on spirally fluted columns topped by stylized Corinthian capitals; only stumps of the stele's top above the arch remain, decorated with indistinct floral motifs. A family of five (parents with three children in front of them) is depicted in the niche under the figure of an eagle, whose spread wings evoke a shell-like background. The woman on the left has an oval elongated face, eyes angled downwards-angled eyes, a long slender neck decorated with two beaded necklaces, and hair arranged in superimposed layers on the crown. She is wearing a long- and tight-sleeved tunic (no neckline visible) and a *palla* draped in parallel folds across the chest, and is resting her right hand on the shoulder of the child in front of her, and her left on her husband's left shoulder. The man on the right has an elongated block-like face, eyes angled downwards, a long neck, short-cropped hair with a concave, ornamentally arranged fringe on the forehead, and possibly a short beard. He seems to be wearing a tight- and long-sleeved tunic and a *sagum* draped in V-lines across the chest, fastened by a knee (?) brooch on the right shoulder. His left hand rests on the shoulder of the child on the right-hand side. Of the three children in the foreground, the left and the middle one seem to be girls wearing variously draped *pallae*. In the left-side one the two ends of the *palla* are crossed diagonally on the chest in a typical Late Roman style, while the middle one is wearing it the same way as her mother, but featuring also a rounded neckline. The girl on the left side seems to be resting her left hand on the left shoulder of the girl in the middle. The sex of the child on the right-hand side is not discernible either from the clothes or portrait features. The workmanship is thoroughly rough, with sketchily and geometrically executed drapery and human forms with pronounced anatomical inaccuracies, especially in the execution of the arms and hands.
*Date*: First half of 4th, rather than the 2nd or 3rd century, as previously dated.
*Publication*: Sergejevski 1940: 10–11, no. 1, fig. 2; Paškvalin 2012: 123–124, no. 60.

Figure 157. (S[ptSE] I.6) – scan of the figure (after Sergejevski 1940)

*IV. Tituli*

1. *Titulus* (S[ptSE] IV.1)
*Provenance*: A chance find prior to 1902 in Rakanske Barice, 55 km S of Sisak, in the ruins of a Roman mausoleum housing a sarcophagus.
*Location*: ZMBiH, inv. no. 98.
*Stone*: Pohorje marble (W. Prochaska 2017).
*Preservation*: Fragmentarily preserved. The slab is broken in several pieces and reassembled. It is damaged on the surface and chipped around the edges, with the right and left bottom corners chipped off, and with some (but not much) of the stone possibly missing on the lower edge. Smaller parts of the inscription are missing and/or are illegible.
*Description*: L 83.5 cm, *H 49* cm, T 4 cm. The inscribed rectangular slab lacks a moulding or a framing. The latter was presumed by Paškvalin to have disappeared through wearing or battering, but no such traces are discernible on the surface.

Figure 158. (S[ptSE] IV.1) – *titulus* (ZMBiH, E. Bujak)

*Inscription*: *Militia in[s]igni raptus tri/eteride sexta sedibus his / situs est miserabilis Heli/[odo]rus. / Heliodorus et C(?) I(?)I(?)/[5]tiane parentes miseri fe[cerunt].*

Letters are irregular, typical of the Late Roman period. Letters *A* have small inverted *V* instead of the horizontal *hasta*. Line 2: first and last *I longa*; line 3: first *I longa*; line 5: *NE* and *NT* in lig. Lines 4 and 5: *EDH*: *C(?)I(?)I(?)/tian(a) e*, but the name of the mother should be in nominative, which expectedly has a Greek ending.

Translation: Snatched away from his excellent military career in his sixth period of three years (i.e. when he was 18), the pitiable Heliodorus now lies in this resting-place of the dead. Heliodorus and ...tiane, unfortunate parents, had (the monument) made.

Commentary: According to *OPEL*, the name Heliodorus is only sporadically attested except in Italy, Hispania, and Pannonia (Alföldy 1969: 214; II 175–176), partly also among freedmen and Christians (Alföldy 1969: 214; II 175–176). In Rome, several are socially high-ranking *liberti* (Solin 1996: 205).

*Date*: 3rd or 4th century.

*Publication*: *CIL* III 8367a = 13242 (+ p. 2328, 159); Patsch 1900: 63–64, Fig. 37; *CLE* 535; Paškvalin 2012: 36 no. 7. Internet databases: *HD*058262 (with external photograph of a cast); *EDCS*-31400322.

## ANDAUTONIA

### *The village of Ščitarjevo* (A)

*I. Stelai*

1. Stele (A I.1)

*Provenance*: Vineyard cultivation in 1958 in the area of the S cemetery.

*Location*: Missing.

*Preservation*: Fragmentarily preserved at the time of discovery. The stele was reassembled from two pieces and cracked all over, with the inscription panel and a smaller part of the pediment surviving.

*Description*: Measurements were not recorded. It transpires from the poor-quality and incomplete newspaper illustration (Degmedžić 1958) that the stele was composed of a moulded inscription panel and a triangular pediment with a central rosette, with no knowing whether it also had a base. A low quality photograph does not show clearly the form of the upper part of the stele; as its pediment does not appear to be free, it may have been a lion pediment, or, more probably, it was contained within a rectangular upper part.

*Inscription*: *Sex(tus) Valerius / Sex(ti) f(ilius) Quir(ina) Sec/undus Med(iolano) / vet(eranus) leg(ionis) XIII Gem(inae). /[5] [Va]l(erius) Victori(nus) f(ilius) ei/us h(eres) f(aciendum) c(uravit).*

Line 3: *med(icus)*, Degmedžić, not plausible.

Translation: Sextus Valerius Secundus, son of Sextus, of the voting-district *Quirina*, veteran of the Legion XIII *Gemina*. Valerius Victorinus, his son and heir, provided (for the monument) to be made.

Commentary: Secundus was one of the most widespread Latin cognomina, and was particularly popular in northern Italy, the provenance of the veteran (Mócsy 1959: 189–190; Barkóczi 1964: 323; *OPEL* IV: 59–61; Alföldy 1969: 291–292). Interestingly, in Pannonia, the name may have been an assonance name among the Celtic-speaking indigenous inhabitants, or a translated name, thus also at Siscia, where it occurs on 11 *tesserae* (Radman-Livaja 2014: 261). The Valerii were one of the most widespread non-Imperial *gentilicia* (Alföldy 1969: 131–133; *OPEL* IV: 143–146); they had probably emigrated from northern Italy to settle in the

Figure 159. (A I.1) – scan of the figure (after Degmedžić 1958)

prosperous Pannonian towns (Mócsy 1959: 160; Šegvić 1988: 60; cf. Barkóczi 1964: 327). They were one of the best documented upper class families at Siscia and Poetovio: M. Valerius Maximianus, a victorious general in the Marcomannic Wars, was the first known Roman senator from Pannonia (Šašel 1982: 567–568 [1992: 176–177]; Filippini and Gregori 2014). *Victorinus* is more characteristic for the late Roman period (Kajanto 1965: 71, *passim*; Alföldy 1969: 327–328; Lochner von Hüttenbach 1989: 182; *OPEL* IV: 168–169). Sex. Valerius Secundus was one of the 46 veterans from Mediolanum; his case was analysed and commented with others from Transpadana by Yves Burnand, who defined them as members of municipal middle class (2002: cf. p. 218).
*Date*: 2nd century (?).
*Publication*: Degmedžić 1958; *ILJug* 1133; Gregl 1996: 13. Internet databases: *lupa* 4320; *HD*034618; *EDCS*-10001136.

*II. Sarcophagi*

1. Sarcophagus* (A II.1)
*Provenance*: A chance find of unknown date from the site of Kutelo, in the area of the S cemetery.
*Location*: The Archaeological Collection Ščitarjevo, inv. no. A-19212.
*Stone*: Upper-Miocene sandstone. A probable origin: Mt Medvednica above Zagreb (M. Belak: 2018).
*Preservation*: The chest completely preserved. It is cracked on one of the short sides, chipped around the lower edges, especially on one of the long sides, and is stained and moss-grown in places all over.
*Description*: L 210 cm, H 60 cm, W 82.5 cm, TB 25 cm in the middle; type 7. The rectangular flat-edged chest has the unequally thick bottom, provided with cushion-like bolsters at both ends; their thickness is 31 cm and 42 cm, respectively. The short sides are also of uneven

Figure 160. (A II.1) – sarcophagus (Lj. Perinić)

Figure 161. (A II.1) – stone, microscope (HGI, M. Belak)

thickness (11.5 to 13 cm). The chest is equally roughly worked on the inside and outside, and there is a mass of undressed stone slightly protruding from the outer bottom of one of the short sides.
*Date*: 2nd half of the 2nd century – 4th century.
*Publication*: Unpublished.

### *The presumed territory of Andautonia* (A[t])

*I. Stelai*

1. Stele (A[t] I.1)
*Provenance*: Road construction in 1884 on a hillock above the site of a Roman quarry on the slope of the hill Medvednica above the town of Zagreb, 12 km NW of Ščitarjevo.
*Location*: AMZ, inv. no. AMZ KS-359.
*Stone*: Badenian sandy biocalcarenite. A possible origin: the mountains of NW Croatia; a probable origin: Mt Medvednica above Zagreb (M. Belak 2018).

*Preservation*: Fragmentarily preserved and probably unfinished. The majority of the portrait niche and the upper part are missing, while the remainder is well preserved, apart from a slight damage on the moulding of the inscription panel and some pitting on the base. The surfaces are worked to a different degree; the inscription panel is polished, the majority of the remaining surface lightly dressed, and the centre of the base with the tenon only roughly worked.
*Description*: *H 127* cm, W 54 cm, T 18 cm. The preserved part of the stele is composed of a plain base with a tenon, a moulded inscription panel, and an unframed portrait niche with a rounded lower edge, featuring truncated geometricized busts of the deceased. Only a lightly worked bust supporting a smoothed neck on the right-hand side survives, as well as a trefoil motif between the two busts. The workmanship is difficult to assess due to the fragmentary and unfinished state.
*Inscription*: *C(aius) Iulius Maximi / lib(ertus) Adietumarus / an(norum) C h(ic) s(itus) e(st). / Testamento /*[5] *fieri iussit / sibi et Sporillae / uxori / an(norum) L.*
Letters are regular and carefully carved, punctuations marks are visible between the words. Line 1: *MA* in ligature. Line 6: *sibi*: the second *I* and *B* in ligature; *T* has a wavy upper *hasta*. Note the characteristic form of letters *V*, with the first *hasta* more inclined than the other, in lines 1, 2, and 5, but not in line 7.
Translation: Gaius Iulius Adietumarus, freedman of Maximus, a 100 years old, lies here. In his will he provided for (this tombstone) to be erected to him and his wife Sporilla, aged 50.
Commentary: According to Mócsy (1959: 49/1 p. 210), Adietumarus' master C. Iulius Maximus was most probably of Italian descent. However this is far from certain, since the name was very popular in Dalmatia and Pannonia among the indigenous population (Alföldy 1969: 242–243; Radman-Livaja 2014: 231), and also among the epichoric inhabitants of the Emona area (Šašel Kos 2017a). It is attested on 13 *tesserae* from Siscia; in the Celtic-speaking provinces it may have been a 'translation' name (Radman-Livaja 2014: 231). Adietumarus is a Celtic name (Mócsy 1959: 162; *OPEL* I: 22, where only this tombstone is cited; Meid 2005: 94: with the meaning 'great in zeal'; Delamarre 2007: 12). The *gentilicium* Iulius is one of the most widely attested *nomina gentilia* in the Roman Empire, not counting the Aurelii: (*OPEL* II: 201–207; see also Mócsy 1959: 148; Barkóczi 1964: 298). Sporilla (the corresponding masculine form has not been attested) seems to be only documented in this inscription and once in Noricum (*Speratus / Sironis / et Sporilla / Commodi / lib(erta) v(ivi) f(ecerunt) sibi*, *CIL* III 5441 = *HD*038623, in the region of Flavia Solva; see *OPEL* (IV: 91 and *Epigraphik-Datenbank Clauss / Slaby*). The name may have derived from Sporus (from the Greek name *Sporos*, see Lochner von Hüttenbach 1989: 153), hence, other explanations are perhaps no longer valid (Mócsy [1959: 191] cited A. Alföldi for the opinion that the name would have been Thracian).

Figure 162. (A[t] I.1) – stele (AMZ, I. Krajcar)

Figure 163. (A[t] I.1) – stone, microscope (HGI, M. Belak)

*Date*: 2nd half of the 1st century or the early 2nd century.
*Publication*: *CIL* III 10867; Brunšmid 1909: 164, no. 359; *AIJ* 482; Degmedžić 1957: 109–110; Internet databases: *lupa* 3789; *HD*071782; *EDCS*-29000468.
*Note*: The stele was presumed never to have been used, on account of its unfinished state and the fact that no traces of a Roman cemetery could have been established in the vicinity.

2. Stele (A[t] I.2)
*Provenance*: Until 1878 kept in the premises of the Archbishopric in Zagreb, 12 km NW of Ščitarjevo.
*Location*: AMZ, inv. no. AMZ KS-357.
*Stone*: Pohorje marble (W. Prochaska 2017).
*Preservation*: Fragmentarily preserved, with most of the inscription panel and probably the entire inscription surviving. The left border is mostly battered and the right one completely so, with light abrasions, cracks and staining on the remaining surface.
*Description*: *H* 60 cm, *W* 52 cm, T 15 cm. The inscription panel is flanked by spirally fluted columns.
*Inscription*: *Aconia[e] / C(ai) filiae / Salviae / C(aius) Iulius /[5] Paternus / uxori / rarissimae / [------?.*
Although the upper part of the inscription field is broken off, no line seems to be missing. Its lower part is also damaged, thus it is uncertain whether any other line followed the last preserved one. Letters are regular, while punctuations marks are not visible.
Translation: To Aconia Salvia, daughter of Gaius. Gaius Iulius Paternus (had the monument erected) to his most special wife.
Commentary: The Latin *gentilicium* Aconius (Solin and Salomies 1994: 5) is a rare *nomen gentile*, either as Acconius or Aconius (attested in Rome and [northern] Italy, as well as sporadically in Gallia Belgica and Germaniae, Hispania, Gallia Narbonensis, Aquitania, and Dalmatia, and Pannonia: Alföldy 1969: 54; *OPEL* I: 18 and 20). One of the three Pannonian individuals, documented by *OPEL*, was from Siscia, C. Aconius Maximus, known from a military diploma (*CIL* XVI 18 = *HD*032916). The name is twice attested on the *tesserae* from nearby Siscia (Radman-Livaja 2014: 157; cf. Mócsy 1959: 150). Salvia also occurs on two *tesserae* from Siscia (Radman-Livaja 2014: 257). Aconia Salvia could well have been a native Pannonian; her ancestors could have come from northern Italy, where the *cognomen* Salvius has best been attested (*OPEL* IV: 45). For the Iulii, see above (I/1). Since the name Paternus is well documented in Italy, Galliae, and Hispania, but is rather rare elsewhere, the person may theoretically have come from one of the Celtic speaking provinces; the name also occurs on a *tessera* from Siscia (Radman-Livaja 2014: 242; cf. Mócsy 1959: 184; Barkóczi 1964: 320; *OPEL* III: 127–128).
*Date*: Later 1st century or the 2nd century.
*Publication*: *CIL* III 10866; Brunšmid 1909: 162 no. 357; *AIJ* 480; Degmedžić 1957: 109–110; Mócsy 1959: 211, no. 57: 8; Djurić 2013: 9, no 40; Internet databases: *lupa* 3787; *HD*071781; *EDCS*-29000467.

Figure 164. (A[t] I.2) – stele (AMZ, I. Krajcar)

3. Stele (A[t] I.3)
*Provenance*: Infrastructural works in 1992 in the village of Odra, 11 km SW of Ščitarjevo.
*Location*: AMZ, inv. no. KS-930.
*Stone*: Gummern marble (H. Müller 2013).
*Preservation*: Completely preserved, apart from the finial at the peak of the pediment. The stele was broken diagonally in two large fragments and several small chips and restored. The surface is lightly stained and abraded at places, but is generally well preserved; type 3a.
*Description*: H 187 cm, W 85.5 cm, T 16.5 cm; type 3a. The stele is composed of a lion pediment with a central Medusa head and two birds of prey, an architrave, a frieze with a floral scroll, an inscription panel flanked by scaled columns with Corinthian capitals and Attic bases, and an elaborated base. The base is composed of three *aedicula*-type niches, the central one topped by an elongated scrolled arch with two rosettes in the scrolls, and the lateral ones by triangular pedimental frames. The lateral niches each feature a winged mourning Eros with the legs crossed and the head supported on one hand and on a reversed torch pulled through a wreath, held in the other hand. The central niche holds a fully armed horseman in a parade gait with a spear, a shield, and possibly a helmet and muscle armour. There was also a tenon underneath the base, obliterated in the process of restoration. The workmanship is uneven – better in the decorative motifs and somewhat worse in the composition of low-relief figures revealing a good sense of movement and realistic detail, but also showing a considerable lack of anatomical proportion.
*Inscription*: *T(ito) Fl(avio) Aterigis / fil(io) Ateboduo / an(norum) LXXV, ve[te]r(ano) / emerito coh(o)r(tis) I[I] Var(cianorum) /[5] eq(uitatae), cornuclario* (!) */ praef(ecti), et Crispinae / coniug(i) an(norum) LV / et Fl(avio) Augurino nep(oti) / an(norum) V. /[10] H(eres) f(aciendum) c(uravit).*
Letters are regular, with punctuation marks between the words. *Impaginatio* is careful: lines 1 and 3 are carved for one letter towards the right, lines 7, 9, and 10 are centred.

Figure 165. (A[t] I.3) – stele (AMZ, I. Krajcar)

Line 2: *ve[te]r(ano)*: earlier editors; however, *TE* is visible although damaged; thus also in *EDH*. Line 4: *coh(ortis)*: earlier editors; a letter is clearly visible after *H*, most probably an *R*. *Va[r]cianorum*: A. Rendić-Miočević 2013: 355, by inadvertence; *Va(---)*: internet databases. Actually, *V* and *A* are in ligature, while *R* is visible on the photo. Line 6: *Crispi[n](ae)*: earlier editors; however, letters are clearly visible.

Translation: To Titus Flavius Ateboduus, son of Aterix, 75 years old, veteran who completed his term in the Second mounted cohort of the Varciani, adjutant to the prefect, and to his wife Crispina, 55 years old, and to his grandson Flavius Augurinus, 5 years old. The heir provided (for this monument) to be erected.

Commentary: Ateboduus belonged to the Varciani, the indigenous population settled at Andautonia and its territory; he returned to his native country. It could be inferred from his rank, that he knew Latin and was literate (cf. Radman-Livaja and Ivezić 2012: 141; Radman-Livaja 2014: 126, fn. 678). *Cohors II Varcianorum equitata* was stationed in Lower Germania, at Remagen (Cichorius 1900: 347–348; Saria 1955: 364; Spaul 2000: 329, cf. 315; cf. A. Rendić-Miočević 2013: 357). Note the very rare occurrence of *veteranus emeritus*: one is attested in Rome (*CIL* VI 669 = *EDCS*-17300811).

According to *OPEL*, the name Ateboduus has only been attested three times in Noricum to date (*OPEL* I: 82); it is Celtic (Delamarre 2007: 28) and may have been related to the crow-like war goddess of the Celts (Lochner von Hüttenbach 1989: 19; Kakoschke 2012: 267–268). The Flavii are very well attested in Italy and throughout the western part of the Empire, but particularly in Dalmatia and Pannonia, including nearby Siscia (*OPEL* II: 145–147; cf. Mócsy 1959: 149; Alföldy 1969: 38–41; Radman-Livaja 2014: 207–208); this is to be expected also at Andautonia, particularly if the town was a *municipium Flavium*, which is very probable (Šašel 1970: 71 [= 1992: 580]). One T. Flavius Bonio from Andautonia, a horseman of the *ala Frontoniana*, is known from Aquincum (*CIL* III 3679 = *Tit. Aq.* II 616 = *HD*068590 = *lupa* 2726). The name Aterix was a *hapax* until the discovery of this tombstone, only known from Mursa (*ILJug* 292: *Urbana Aterigis liberta*; *OPEL* I: 84; Delamarre 2007: 29). Augurinus is well attested in Italy, also in northern Italy, elsewhere it occurs sporadically (Mócsy 1959: 165; Alföldy 1969: 159; *OPEL* I: 95); the name Crispinus/a is documented everywhere, particularly also in northern Italy (Mócsy 1959: 171; *OPEL* II: 85), and appears on six *tesserae* from Siscia (Radman-Livaja 2014: 192).

*Date*: Early 2nd century.

*Publication*: Šegvić 1996: 132 no. 2; Djurić 2013: 1, no. 2; A. Rendić-Miočević 2013 (*AE* 2013: 1206). Internet databases: *lupa* 8814; *HD*056602; *EDCS*-32300431.

4. Stele (A[t] I.4)

*Location*: Built in the base of the statue of St. Anna in the village of Hraščina, 39 km NE of Ščitarjevo.

*Stone*: Pohorje marble (W. Prochaska 2017).

*Preservation*: Fragmentarily preserved. The pediment, the portrait niche, the dividing decorative band and the uppermost part of the inscription field survive. The surface is heavily weathered, damaged and stained, with some of the architectural elements and facial features much obliterated. Tiny remains of the mass of stone along the sides of the pediment indicate a pediment contained within a rectangular upper panel, or a pentagonal upper part.

*Description*: *H 107* cm, W 87 cm, *T?* The preserved part of the stele is composed of a pediment featuring a central Medusa head flanked by two birds of prey, a narrow plain architrave, a rectangular flat-backed portrait niche holding busts of a man and a woman, a dividing decorative band holding a frieze of ivy leaves, and a simply framed inscription panel. The niche is flanked

Figure 166. (A[t] I.4) – stele (T. Leleković)

by scaled or spirally fluted columns with capitals and bases of the Tuscan type. The woman on the left has a roundish block-like face, down-turned lips, and oversized ears. She is wearing a round, sub-trapezoidal hat, possibly rimmed, leaving visible a slightly concave horizontal fringe on the forehead, while huge ball-like rolls of hair are sticking out on both sides of her cheeks. She is dressed in a tunic with an oval neckline and a sleeveless over-tunic with two huge brooches of the so-called *kräftig profilierte* type on the shoulders. Two torques around her neck are visible; the smaller one is smooth, possibly with a pendant, and the larger one twisted. The man on the right-hand side has a roundish block-like face, down-turned lips, a beard presumed but not clearly discernible, and the hair arranged in strong, parallel strands combed from the crown towards the forehead and temples in the Trajanic fashion. He is wearing a pleated tunic with a straggly V-neckline. The craftsmanship is obscured by the wear, but seems to be competent and especially accurate in treatment of the drapery.

*Inscription*: *M(arcus) Ulp(ius) Gemin[us] / Savi f(ilius) an(norum)[---] / Volus[ia---] [------*

Letters are regular and elegant.

Translation: Marcus Ulpius Geminus, son of Savus, ?-year-old, .... Volusia ...

Commentary: The Ulpii are attested everywhere (cf. Alföldy 1969: 42–43; *OPEL* IV: 179–181) and also several times at Siscia, including six lead tags (Radman-Livaja 2014: 276). The name Savus has only been documented twice (*OPEL* IV: 53; Delamarre 2007: 162), one *Savi f(ilius)* is also attested in Gallia Narbonensis (Carpentorate: *CIL* XII 1160: *EDCS*-08500841). In the case of Geminus, the name of his father most probably refers to the Savus River. The name Geminus is documented almost everywhere, but is more frequent in Italy and western provinces (*OPEL* II: 164; cf. Mócsy 1959: 175; Alföldy 1969: 210). According to *OPEL*, the name is attested three times in Pannonia, but also occurs on two *tesserae* from Siscia (Radman-Livaja 2014: 212). The *gentilicium* Volusius, derived from the rare name (perhaps *praenomen*) Volusus (Lochner von Hüttenbach 1989: 185; cf. Salomies 1987: 61), occurs almost everywhere, but more frequently in Italy (Alföldy 1969: 139; *OPEL* IV: 183; cf. Mócsy 1959: 212, no. 61; Delamarre 2007: 205).

*Date*: First half of the 2nd century.

*Publication*: *AIJ* 456. Internet databases: *HD*069291; *EDCS*-11301101.

5. Stele (A[t] I.5)

*Provenance*: Until 1873 built into the premises of the Archbishopric in Zagreb, 12 km NW of Ščitarjevo.

*Location*: AMZ, inv. no. AMZ KS-358.

*Stone*: Badenian sandy biocalcarenite. A possible origin: the mountains of NW Croatia; a probable origin: the Banovina, S of Sisak (M. Belak 2018).

*Preservation*: Fragmentarily preserved. The uppermost and the lower parts of the stele are missing. The uppermost part of the pediment was cut off, but a ram's head sitting on its right side testifies to a lion pediment, and a socket above Medusa's head to a central fin; there is no knowing whether the stele had a base. The surface is weathered and stained all over, with damage and chipping particularly heavy along the leftmost side of the stele and on the deceased's face.

*Description*: *H 158* cm, W 86 cm, T 28 cm; type 4. The preserved part of the stele is composed of a lion pediment holding a Medusa head flanked by acanthus leaves, a frieze of lotus leaves, a rectangular portrait niche and an inscription panel divided by a decorative band holding a frieze of scrolled rosettes. The portrait niche and the inscription panel are flanked by columns with Corinthian capitals, double-*torus* bases and shafts decorated with elongated leaves tied with a ribbon, spanning continuously the whole height of the stele. The deceased has an oval face, the nose chipped off, the eyes and lips obliterated, short curly hair and a beard styled in the Hadrianic fashion. He is portrayed as a *togatus* bust set against the background of a wide-ribbed scallop shell. The workmanship is accurate, with anatomical shapes and proportions, and naturally executed decorative motifs and drapery.

*Inscription*: *D(is) M(anibus). / L(ucio) Egnatuleio / L(uci) f(ilio) Florentino /[a]n(norum) XXX pater /5 [f](aciendum) c(uravit)]. H(ic) s(iti) s(unt).*

Letters are regular but very worn, punctuation marks are not well visible.

Translation: To the spirits of the departed. To Lucius Egnatuleius Florentinus, son of Lucius, 30 years old. His father had (this monument) made. They (!) rest here.

Commentary: The *gentilicium* Egnatuleius only occurs sporadically, according to *OPEL* (II: 114), once in

Figure 167. (A[t] I.5) – stele
(AMZ, I. Krajcar)

Figure 168. (A[t] I.5) – stone,
microscope (HGI, M. Belak)

northern Italy, twice in Hispania, and once in Pannonia (on this tombstone: cf. Mócsy 1959: 154). The name further occurs a few times in Africa and several times in the rest of Italy, mainly in Rome (*EDCS*). The name Florentinus is frequent everywhere, but particularly in Pannonia and in the Celtic-speaking provinces (*OPEL* II: 148; Mócsy 1959: 174; Barkóczi 1964: 312; Alföldy 1969: 205). The name also appears on the *tesserae* from Siscia; its early bearers may have come from northern Italy (Radman-Livaja 2014: 208).

*Date*: First third of the 2nd century.

*Publication*: *CIL* III 10860; Brunšmid 1909: 163–164, no. 358; Schober 1923: 241; *AIJ* 481; Degmedžić 1957: 108–109; Gregl and Migotti 2000: 130–131, 164 fig. 9; Cambi 2002: 152–153, fig. 233. Internet databases: *lupa* 3788; *HD*057336; *EDCS*-29000466.

*Note*: There is the possibility that the stone stems from Sisak and not from Ščitarjevo, as stated in the *CIL*.

6. Stele (A[t] I.6)

*Provenance*: Infrastructural works in 1989 in the village of Donji Čehi, 13 km SW of Ščitarjevo, within a tumuli cemetery.

*Location*: AMZ, inv. no. KS-929.

*Stone*: Badenian sandy siliciclastic biocalcarenite. A possible origin: the Banovina S of Sisak; a probable origin: Mt Medvednica above Zagreb (M. Belak 2018).

*Preservation*: Fragmentarily preserved. Missing are the base, the middle *acroterion*, the side sections of the lower frame of the inscription panel, and the lower part of the right pilaster. There are larger superficial chips on the lower part of the left pilaster and the head of the right lion, and minor chipping at various places, including the inscription panel.

*Description*: *H* 186 cm, W 81 cm, T 22 cm; type 5a. The preserved part of the stele comprises a lion pediment holding a Medusa head flanked by acanthus leaves, a narrow spiral frieze, a rectangular flat-backed niche, flanked by smooth columns with Corinthian capitals and Tuscan bases, a decorative rosette band, and an inscription panel flanked by smooth pilasters with Corinthian capitals and a geometrically decorated band on the bottom. Tiny remains of a base below the band of rosettes survived, but were obliterated in the process of conservation and restoration. In the niche half-figures of three adult deceased are depicted (two women and a man on the right-hand side) and a bust of a boy in front of the women. The man has a pear-shaped face, fairly thick hair receded on the temples, executed in pronounced parallel ridges, a thick beard and moustaches, and roundish eyes with accentuated pupils. He is wearing a tunic and a *sagum* fastened on the right shoulder with a round brooch, and is holding a whip in his right hand. To the left of him two women are depicted, dressed and accessorized in the same manner and looking very similar. They have plump cheeks, small lips, roundish eyes with accentuated pupils, a wrapped headdress arranged in three superimposed layers, a plaited fringe on the forehead framing the temples and covering the ears, with the only difference between the two being the two locks of hair on the forehead of the woman in the middle.

Figure 169. (A[t] I.6) – stele (AMZ, I. Krajcar)

Figure 170. (A[t] I.6) – stone, microscope (HGI, M. Belak)

They are wearing pleated tunics with almost straight necklines, tubular necklaces (torques) with lunular pendants, and stiff smooth cloaks with a V-cut on the chest and with long, narrow, cuffed sleeves, fastened (or just decorated) at the beginning of the V-cut with a round dotted brooch. There is a knee brooch on the left woman's right shoulder in a purely decorative role, and she is also wearing a bracelet. Between and in front of the women, and embraced by the left-one's hand, is a small boy dressed in a tunic, with a plump face, the hair combed forward from the crown and rendered in parallel stiff ridges producing a concave fringe. The portraits are detailed and expressive, and the workmanship fairly competent, but for some anatomical inaccuracies, especially in the size of the left woman's hand.

*Inscription*: *D(is) M(anibus). / Valens et / Melania / Caeserni Aviti /*[5] *servi sibi et / Valentinae an(orum) XX / et Donico an(norum) II / filiis karissimi[s].*

Letters are elegant and *impaginatio* is careful: lines 1, 3, 5, and the last one are centred. There are punctuation marks between the words.

Translation: To the spirits of the departed. Valens and Melania, slaves of Caesernius Avitus, (had this monument erected) to themselves and to Valentina, 20 years old, and Donicus, 2 years old, their dearest children.

Commentary: The Caesernii came from Aquileia to settle at an early date at Emona, which can be regarded as their secondary centre. From Emona its members spread to various towns of Noricum, but particularly to Pannonia, as Romanization progressed in these two provinces. A part of the wealth of the Caesernii may have been based on the iron industry. Many of their freedmen were *seviri* and *Augustales*. On the *gens Caesernia* see Šašel 1960 and 1981 (= 1992) and Zaccaria 2006. Avitus is rather rare in northern Italy and Dalmatia, but very frequent in Hispania, Galliae, Noricum, and Pannonia; the name was obviously popular in the Celtic-speaking provinces (*OPEL* I: 97–98; cf. Mócsy 1959: 165), where it may have been an assonance name (Lochner von Hüttenbach 1989: 26–27; Meid 2005: 254–255; Delamarre 2007: 34; Kakoschke 2012: 288–289). The name also occurs on three *tesserae* from Siscia (Radman-Livaja 2014: 171): the patron of the family must have most probably been a local resident. According to *OPEL* (II: 107), the Celtic name Donicus has only been attested twice, once at Pola in Histria (*CIL* V 165 = *EDR*136937 = *EDCS*-04200251) and once in Dalmatia, borne by a soldier from Celeia (*OPEL* II: 107; Alföldy 1969: 191; Delamarre 2007: 88). The name Melanio is only sporadically attested, Melania is not noted in *OPEL* (III: 73; cf. Kajanto 1963: 85). Valens is a common name and most popular also in Siscia, as is clear from the *tesserae* (*OPEL* IV: 139–140; Radman-Livaja 2014: 277), while Valentinus/a has a similar distribution, but occurs more often after the Marcomannic Wars (*OPEL* IV: 140–141; cf. Mócsy 1959: 194; Alföldy 1969: 320).

*Date*: Mid 2nd century.

*Publication*: Gregl and Migotti 2000: 129–130, fig. 3b (*AE* 2002: 1125); Migotti 2008a (with references; *AE* 2008: 1081); Mander 2013: 637. Internet databases: *lupa* 8816; *HD*046270; *EDCS*-30101116.

7. Stele (A[t] I.7)
*Provenance*: Archaeological excavations in 1959 at the site of Držićeva Street, Zagreb, 12 km NW of Ščitarjevo. The stele was found in a frescoed masonry tomb (4 x 3 m) within a Roman cemetery, where it seems it was moved after both monuments had collapsed in a disaster.
*Location*: AMZ, inv. no. AMZ KS-934.
*Stone*: Pohorje marble (H. Müller 2013).
*Preservation*: Fragmentarily preserved. Parts of the inscription panel and the relief panel survive, restored from four fragments. A corner pediment piece in the shape of a lion, reported to have been found as well, cannot be traced in the AMZ, and is documented only by a very low-quality, totally blurred photograph (Gorenc 1960: figure on p. 25).

Figure 171. (A[t] I.7) - stele (AMZ, I. Krajcar)

*Description*: *H 107* cm, *W 27* cm, T 10 cm; probably type 3a or 3b. The preserved fragment of the stele contains three segments: an inscription panel flanked by columns decorated with vines springing from a cantharus, and with only the initial letters in three superimposed lines of the inscription preserved; a dividing decorative band of spiral leaves below; a rectangular relief niche framed on the left side by a band of stylized acanthus leaves, featuring the raised right hand of a maenad, and some drapery above. The workmanship is mediocre, with neatly executed decorative motifs, but with geometricized drapery.
*Inscription*: *---] / G+[---] / E+[---] / Ṃ[---] / [------.*
Line 2: after *G*, a vertical *hasta* is preserved; line 3: after *E*, an oblique *hasta* is preserved, perhaps a *V* or an *X*.
*Date*: 2nd century.
*Publication*: Gorenc 1960; Gregl 1996: 12; Djurić 2013: 11, no. 47.

8. Stele (A[t] I.8)
*Provenance*: A chance find during ploughing in 2012 in the village of Lekneno, 2 km SE of Ščitarjevo.
*Location*: The Archaeological Collection Ščitarjevo, inv. no. A-2284.
*Stone*: Pohorje marble (W. Prochaska 2017).
*Preservation*: Completely preserved. A small fragment from the right top corner is chipped off, while the right-hand columns of the portrait niche and the inscription panel are damaged. The surface is fairly weathered, with the inscription and portrait features much obliterated.
*Description*: Type 5a. The stele is composed of a lion pediment with a central Medusa head flanked by acanthus leaves and a pine-cone finial, a cable-patterned architrave, a rectangular, flat-backed portrait niche flanked by vertically fluted columns (Attic bases, Corinthian capitals), a palmette decorative band, an inscription panel flanked by columns (vertical flutes in the upper part, scale-leaf pattern in the lower, Attic bases, Corinthian capitals), a base flanked by acanthus-decorated panels, featuring a motif of two heraldic dolphins with down-turned heads by a trident, and a tenon with the pedestal. The portrait niche holds half-figures, or, rather, busts of a man on the right-hand side, a woman on the left-hand side, and a bust of a boy between and in front of them. The man has a somewhat elongated oval face and a rounded crown, thick hair with a very slightly concave fringe, and prominent ears; he is wearing a *sagum* fastened on the right shoulder with a brooch of indistinct type. The woman has a roundish face, a pointed chin, round eyes with arched eyebrows, a two-tiered wrapped bonnet with two thick strands of hair sticking out at the sides of her face and covering the ears. She is wearing a tunic with a rounded

Figure 172. (A[t] I.8) – stele
(The Archaeological Collection Ščitarjevo)

thickened neckline, a cloak V-shaped across the chest, and possibly a close-fitting segmented or tubular necklace. The facial features and clothes of the boy are the same as his father's. The workmanship is hard to asses due to the wear, but seems to be somewhat crude, with the lack of natural shapes in the portraits, and with the architectural part executed somewhat better.

*Inscription*: *D(is) M(anibus) / C(aius ?) Marcius [---] / testamento fieri iussit / sibi et con(iugi ?) M/*[5]*odestinae [---] / et Vindici f(ilio) ann(orum) ++ / [---] / [---].*

Reading of Milan Lovenjak; letters are very damaged, several remaining illegible.

Translation: To the spirits of the departed. Gaius Marcius ... In his will he provided for (this tombstone) to be erected to him and his wife (?) Modestina ... and (his) son Vindex, ... years old ...

Commentary: Marcius is a frequent Latin *gentilicium*, particularly in southern Italy and Italy generally, and in Pannonia (Alföldy 1969: 97–98; *OPEL* III: 56–57), also occurring on a tombstone from Siscia in this catalogue (*CIL* III 3988 = *HD*074390). It could have been a *pseudo-gentilicium*, derived from cognomen Marcus (Kakoschke 2012: 130–131). The name Modestinus/a is relatively rare, except in Rome and Italy and Hispanic provinces (*Epigraphik-Datenbank Clauss / Slaby*; *OPEL* III: 84). Vindex is a Latin cognomen, neither frequent, nor rare (particularly not in Pannonia: *OPEL* IV: 171), while in Celtic-speaking provinces it may have been an assonance name, based on the adjective *vindos* ('white, happy'; Lochner von Hüttenbach 1989: 183). The name also occurs on two *tesserae* from Siscia (Radman-Livaja 2014: 283).

*Date*: First half of the 2nd century.

*Publication*: Unpublished (Lovenjak and Pintarić, forthcoming).

*Note*: The description was made on the basis of the insight into the monument, the publication of which is being prepared by Milan Lovenjak and Tatjana Pintarić. Therefore, it was described without the measurements, and with only a small illustration adduced for a preliminary information.

9. Stele (A[t] I.9)

*Provenance*: Grave-digging in the 1950s at the site of the Križevci municipal cemetery, 43 km NE of Ščitarjevo.

*Location*: Gradski muzej Križevci / The Museum of Križevci, inv. no. GMK-26.

*Stone*: Pohorje marble (W. Prochaska 2017).

*Preservation*: A lion top completely preserved. The upper half of the right lion is broken off and put back in place, and the piece is much weathered and damaged all over.

*Description*: W 40 cm, H 110 cm, T 25 cm. Between the lions a head of Serapis is carved with long hair and moustaches, and with its back supported against a truncated prismatic shaft.

*Date*: 2nd or 3rd century.

*Publication*: Demo 1982: 76–78; Migotti 2013a: 310: no. 4, 31: LT 4.

Figure 173. (A[t] I.9) – lion top
(Gradski muzej Križevci, O. Blagec)

Figure 174. (A[t] I.10) – lion top (AMZ, I. Krajcar)

10. Stele (A[t] I.10)
*Provenance*: A chance find of unknown date at the site of Gornji Bukovac (Zagreb), 12 km NW of Ščitarjevo.
*Location*: Mounted on a courtyard wall of a private house at no. 1 Gornji Bukovac.
Stone: Pohorje marble (W. Prochaska 2017).
*Preservation*: A lion top fragmentarily preserved. The lion on the left-hand side, as well as the back of the female head, are broken off, and the stone is fairly weathered and completely stained.
*Description*: H 45 cm, *W 78 cm*, T 34 cm. A female head placed between two lions has an oval face, full cheeks, a wide nose, elongated eyes, and full, closed, slightly smiling lips. A wreath of spirally arranged strands of hair encircles her face, while the preserved part of her high scalp is smooth and conspicuously rounded on the crown.
*Date*: 2nd or 3rd century.
*Publication*: Gregl 1996: 14.

Figure 175. (A[t] I.11) – stele (AMZ, I. Krajcar)

11. Stele (A[t] I.11)
*Provenance*: A chance find at the end of the 19th century at the village of Kerestinec, 24 km SW of Ščitarjevo, in a Roman cemetery.
*Location*: AMZ, inv. no. AMZ KS-356.
*Stone*: Badenian biocalcarenite. The possible origin: the mountains of N Croatia (M. Belak 2018).
*Preservation*: Fragmentarily preserved and inserted into a modern plastered pedestal. The stele is chipped off along the tip, as well as along the lower and right-hand edges. The surface is much damaged, abraded, pitted and weathered, with both the relief and the epitaph obliterated.
*Description*: *H 153* cm, *W 68* cm, T 22 cm. The stele is shaped as an irregular oblong slab with a sub-trapezoidal upper part and with the outlines of its structural components (a pediment, a portrait niche and an inscription panel) incised into the surface. The pediment is contained within a rectangular or a pentagonal panel, with two letters *M* flanking a standing human figure of indiscernible sex, outlined by shallow grooves and extending into the niche below. It has a rounded head, a longish straight nose, a barrel-like body, and tube-shaped upper arms and lower legs. The rectangular, shallow flat-backed portrait niche contains two more figures with slightly triangular

Figure 176. (A[t] I.11 – stone, microscope (HGI, M. Belak)

Figure 177. (A[t] II.1) - sarcophagus (AMZ, I. Krajcar)

faces and wedge-shaped noses, rendered similarly as the previous one, but more in relief than grooving. The inscription panel is smooth. The workmanship is extremely crude and incompetent, with human figures so primitively executed as to render their sex obscured.
*Inscription*: *M(---) M(anibus)? / Pontius / pos(u)it si[bi ] / [e] t co[niu]g[i] /⁵ [e]t M(arco ?) C[---] / A[---].*
The inscription field is damaged and therefore also several letters; these are irregular and clumsily carved, the same letters are of different shape. Line 1: perhaps *MM(anibus) d(is)*: *EDCS*; *M(---) M(---)*: *EDH*. Perhaps *M(anibus) M()* or *M(e)m(oria)*?; the first *M* may indeed be an error for *D(is)*, as was supposed by Hoffiller (*AIJ* 483).
Translation: To Manes (?). Pontius had (this monument) erected for himself, his wife, and Marcus(?) ....
Commentary: According to *OPEL* (III: 153), Pontius as a personal name or a cognomen has only been attested to date three times in northern Italy, twice in Hispania, and once in Dalmatia, as well as twice in Pannonia, once in this case. A centurion of the Legion *XV Apollinaris*, Ti. Claudius Pontius, is known from Siscia (*CIL* III 10853 = *HD*073473; Fitz 1993: 325 no. 222; Mosser 2003: 261 no. 188; Radman-Livaja and Vukelić 2015: 401–402). The name Pontius in the feminine form is attested on two *tesserae*, once as a *gentilicium*, but once perhaps as a personal name (Radman-Livaja 2014: 246). Mócsy supposed that Pontius may have been an indigenous name (Mócsy 1959: 184).
*Date*: Second half of the 2nd or 3rd century.
*Publication*: *CIL* III 15183; Brunšmid 1909: 162, no. 356; *AIJ* 483; Degmedžić 1957: 110–111. Internet databases: *lupa* 3790; *HD*071783; *EDCS*-32700130.

*II. Sarcophagi*

1. Sarcophagus (A[t] II.1)
*Provenance*: Archaeological excavation of a Roman cemetery in 2000 in the village of Samoborski Novaki, 25 km SW of Ščitarjevo.
*Location*: AMZ, inv. no. KS-951.
*Stone*: Upper-Miocene sandstone appearing locally on any geomorphological elevation underneath the Quaternary layers. The origin: local (M. Belak 2018).

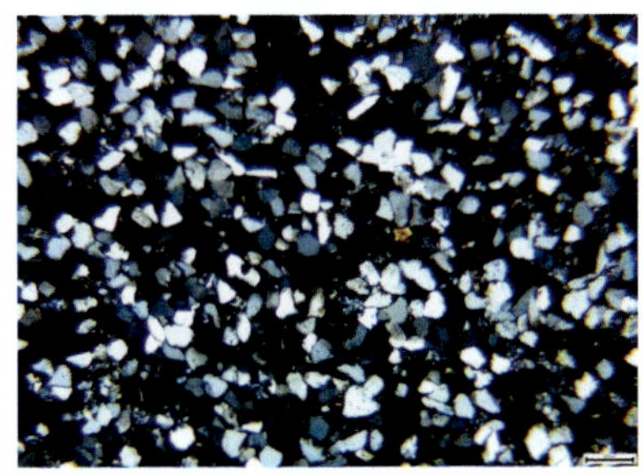

Figure 178. (A[t] II.1 - stone, microscope (HGI, M. Belak)

*Preservation*: A chest completely preserved, but chipped along the upper and lower edges.
*Description*: L 200 cm, H 56cm, W 76 cm; type 7. The very roughly worked rectangular chest has an internal cutting around the edges for taking a lid.
*Date*: Second half of the 2nd century - 4th century.
*Publication*: Galić and Radman-Livaja 2002.

*VI. Medallions*

1. Medallion (A[t] VI.1)
*Provenance*: A chance find in 1868 in the village of Majur, 45 km NE of Ščitarjevo, as protruding with its tip from the surface of the road.
*Location*: GMB, inv. no. GMB-3428.
*Stone*: Most probably Pohorje marble (W. Prochaska 2017).
*Preservation*: Fragmentarily preserved. The lower part of the rim and the tenon front are chipped off to leave visible a socket (standing for a dowel hole) for joining to some kind of a pedestal (altar, wall, etc.). The border and the reliefs are much weathered and abraded, with two major chips out of the left side, and a vein of brownish colour running through the whole width of the piece and sticking out along the upper rim.
*Description*: H 78 cm, W 69 cm, T 20-24 cm. A pyramidal base holds a rounded-back tondo in the role of a portrait niche with a banded rim, which appears to be smooth but still possibly holds traces of the original geometricized decorative pattern. Half-figures of a man on the right-hand side and a male youth on the

Figure 179. (A[t] VI.1) - medallion (GMB, Z. Tokić)

left side are depicted in the niche. The man has an oval elongated face, slightly plump cheeks with sticking-out ears, fairly thick hair with receding temples, and a short curly beard. He is wearing a *sagum* fastened on the right shoulder with a round brooch. The youth has an oval elongated face, prominent ears, and a short-cropped hair with slightly receding temples, and is dressed in a tunic with a rhomboid neckline. The man is embracing the youth with his right arm and is resting his left hand on the latter's upper arm. The workmanship is fairly crude, as reflected in geometricized drapery and an oversized man's left hand.

*Date*: 3rd century.

*Publication*: Medar 2012: 14 (only mentioned, with a photograph).

*Note*: This piece shows a couple of unusual features, one technical and another structural. The first one, figuring as an iron-coloured hoop all around the perimeter is actually a layer of impurity in the stone. The second one, unparalleled in the known comparative material, has been produced by taking off (originally or secondarily) the rim from the upper third downwards, thus simulating a small roof which is otherwise typical of this kind of monument, but rendered in a naturalistic form. There is the possibility that such structural shape had been conceived from the start, but was rejected in the course of carving.

*VIII. Unclassifiable*

1. Inscription (A[t] VIII.1)

*Provenance*: A chance find, first spotted in the present location in the 1970s.

*Location*: Built into the parish house in the village of Čučerje, 14 km NW of Ščitarjevo.

*Stone*: Marble (characterization of the stone failed, as the sample was too small; W. Prochaska 2017).

*Preservation*: Fragmentarily preserved. An incomplete inscription panel survives, with borders on the upper and right sides of the front battered and visible as faint imprint traces.

*Description*: *H 49* cm, *W 51* cm, *T 11* cm. The rectangular inscription panel has a smoothed front, slightly worked side(s), and 11-cm-wide band-like borders on all four, or at least three sides (save the bottom) of the front. The surface is much weathered and the inscription partly illegible.

*Inscription*: *[---]vero / [---] p(rimo)p(ilo) leg(ionis) / [---] leg(ionis) IIII Fl(aviae) / [--- Pr]oculin(a)e /$^{5}$ [---]* Θ(obitae) ann(orum) L / [------.

Lines 1 and 2: missing are *praenomen*, *gentilicium*, and possibly the voting tribe of the centurion, as well as the name of the legion in which he served as *primuspilus*. Line 4: missing is perhaps the legion's title *felix* and the *gentilicium* of Proculina, probably his wife.

Translation: To ... Verus? or Severus?, ... *primuspilus* of the ... legion ... of the Fourth Flavian Legion ... to Proculina ... who died 50 years old ...

Commentary: Verus, if not fragmentary, is one of the most wide-spread personal names and cognomina in the western part of the Empire (*OPEL* IV: 160–161; cf. Mócsy 1959: 196; Barkóczi 1964: 327), also among the indigenous population (Alföldy 1969: 325); interestingly, at Čučerje a dedication to Hercules was discovered, erected by one P. Aelius Verus (*CIL* III 4007 = *HD*057296). Severus was even more popular (*OPEL* IV: 76–78), also occurring on 12 Siscian *tesserae* (Radman-Livaja 2014: 263). The name Proculinus/a is not particularly frequent, except in Italy, Hispania, Dalmatia, and

Figure 180. (A[t] VIII.1) - inscription (AMZ, I. Krajcar)

Pannonia (*EDCS*; *OPEL* III: 166; cf. Mócsy 1959: 186); after the Marcomannic Wars the name is attested among the Syrians along the Danubian *limes* (Barkóczi 1964: 321). Proculus/a, however, was one of the most popular names at Siscia, since it appears on 20 *tesserae* (Radman-Livaja 2014: 249). The unknown centurion had a career in the army, but only his service in the *IV Flavia* is known. The legion was founded by Vespasian in AD 70 and was stationed at Burnum in Dalmatia, but was transferred to Moesia (Singidunum) under Domitian around 86, due to the Dacian invasion.
*Date*: Later 1st or 2nd century.
*Publication*: Bekić and Lovenjak 2002 (*AE* 2002, 1126). Internet databases: *HD*046271; *EDCS*-30101117.
*Note*: The sides are somewhat more roughly worked than the front, while their thickness remains unknown, leaving undecided whether the slab was a stele or a *titulus* (an altar being the least likely possibility).

2. Statue of Icarus (finial of a tombstone or a tumulus burial) (A[t] VIII.2)
*Provenance*: A chance find in 1892 in the village of Vugrovec, 12 km NW of Ščitarjevo.
*Location*: AMZ, inv. no. AMZ KS-33.
*Stone*: Salla marble (W. Prochaska 2017).
Preservation: Fragmentarily preserved. The head and feet, as well as a part of the base and the lowermost part of the right wing are broken, and there are abrasions and staining all over the body.

Figure 181. (A[t] VIII.2) – statue of Icarus (AMZ, I. Krajcar)

*Description*: *H* 55.5 cm, W 42 cm. Icarus is shown as a winged nude wearing a mantle (*paludamentum*, *chlamys*) fastened with a round brooch on the chest and falling in rich wavy folds down the sides of the body and between the legs, reaching down to the ground. The weight of the body is on the right leg, while the left is slightly drawn back and to the side. The arms are lowered and slightly set aside in a firm grip of the wing-handles, shown against the background of short feathers rendered in a decorative scaled pattern, and long ones as parallel vertical ridges and grooves. The workmanship is competent, revealing a well-proportioned natural musculature, sense of movement, and accuracy in the treatment of the drapery.
*Date*: 2nd or 3rd century.
*Publication*: Brunšmid 1904: 229–230, no. 33; Degmedžić 1957: 92–95; Sanader 2012: 133–134. Internet databases: *lupa* 22379.

3. Shaft block (A[t] VIII.3)
*Provenance*: A chance find in 1979 in the village of Repišće, 32.5 km SW of Ščitarjevo, in the area of a tumuli cemetery.
*Location*: Gradski muzej Jastrebarsko / The Museum of Jastrebarsko, inv. no. 2.
*Stone*: Badenian sandy siliciclastic biocalcarenite. A possible origin: the mountains of N Croatia; a probable origin: Mt Medvednica above Zagreb (M. Belak 2018).
*Preservation*: Completely preserved. The stone is chipped along the upper right border of the front, with major chips out of the upper right side and the back. It is extremely weathered, with the relief depiction battered in the upper half.
*Description*: H 87 cm, W 60 cm, W 32 cm. The monument is shaped as a parallelepiped apparently smoothed on all four sides, but roughed with weathering on the front. The front side is framed by a lightly engraved line (clearly visible on the lower border and barely discernible at places on the left and upper borders) possibly featuring a much-weathered motif of a footed vessel (a cantharus or a crater) with a tiny tendril at each side (or snakes?) rising from its stem upwards. The workmanship is rough, but still hard to assess properly, due to the wear.
*Date*: 2nd or 3rd century.
*Publication*: Gregl 1990: 101, fn. 9; Gregl 1997: 21 (both papers only mentioned the stone, without illustration).
*Note*: Given the archaeological context, and the lack of clamp or dowel holes, this piece could have belonged to any building within a tumulus or its vicinity, but can tentatively be interpreted also as an altar shaft with the inscription battered.

Figure 182. (A[t] VIII.3) – shaft (AMZ, I. Krajcar)

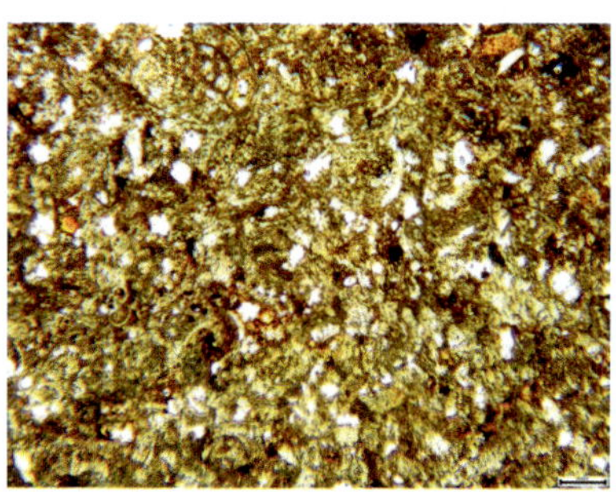

Figure 183. (A[t] VIII.3) – stone, microscope (HGI, M. Belak)

***The possible territory of Andautonia in the north-west* (A[ptNW])**

*I. Stelai*

1. Stele (A[ptNW] I.1)
*Provenance*: Building works in 1857 in the village of Lobor, 42.5 km NE of Ščitarjevo.
*Location*: AMZ, inv. no. 360.
*Stone*: Badenian biocalcarenite. A possible origin: the mountains of N Croatia (M. Belak 2018).
*Preservation*: Fragmentarily preserved. The lower section is missing, and both upper corners chipped of, with the surface cracked at places and extremely weathered, leaving the relief figures heavily obliterated. The restoration of the stele remains uncertain; it possibly contained a top (as suggested by the remains of a dowel hole on the right top-corner surface, while the one on the left side is lost to breakage) and a base.
*Description*: *H* 160 cm, W 99 cm, 20 cm. The preserved part of the stele contains a simply framed portrait niche with the concave background, topped by a Norico-Pannonian volute, accommodating three standing portraits of the deceased: two soldiers standing with legs slightly apart, and a woman in-between in a statuary pose. The men are depicted in a similar way, with an oval face, short hair rendered in straight parallel strands barely raised from the scalp and combed towards the front, ending in a concave fringe; both have a short curly beard, while moustaches are barely discernible due to the wear, especially in the left-hand figure. They are wearing a belted tunic fastened with a square brooch and a *sagum* fastened on the right shoulder with a round brooch, and are equipped with a sword tucked into a scabbard ending in a round chape, suspended on a baldric with a round fastener. Both are holding a scroll in the left hand; the man on the right is pointing at it with his right hand, while the man on the left is holding an indistinguishable object with a roundish top in his right hand. The woman in the middle has an elongated oval face, a band-like wrapped headdress leaving visible the forehead fringe of slightly waved hair, parted in the centre and stopping short of the ears; there is a close-fitting fretted band necklace around her neck. She is dressed in native attire: a long- and wide-sleeved tunic with an oval neckline, a sleeveless belted over-tunic fastened on the shoulders with fairly large brooches of uncertain type (an elongated rectangular head, a bow with wing-like side projections, and a foot ending in a knob) and decorated with a round brooch on the chest. The over-tunic reaches to below the woman's knees, with an ankle-length under-skirt worn underneath and visible below. She is holding a round object, probably a fruit, in her left hand on which she is wearing a fretted bracelet, and is possibly resting her right hand on her son's left shoulder. The short sides of the stele are decorated with a much damaged and obliterated motif of vines and grapes. The workmanship is fairly competent, with some sense of anatomical proportions, postures, and natural shapes in spite of a fairly hard execution.
*Inscription*: *D(is) M(anibus). / M(arco) Coc(ceio) Superiano >(centurioni) leg(ionis) X G(eminae) ann(orum) / XXX mens(ium) VIIII et Val(erio) / Luciliano mil(iti) co(ho) r(tium) praet(orianarum) /*[5] *[an]n(orum) XL me(n)s(ium) III. Sep(timia) Lucilla / [m]ater pientissima et Coc(ceius) / [---]us sig(nifer) leg(ionis) X [---] / [------.*
Letters are irregular and clumsily carved. Ligatures: line 2: a small *O* carved in the first *C* in *Coc(ceio)*; *VP* in *Superiano*. Line 3: *ME* in *mens(ium)*; line 5: the same ligature.
Translation: To the spirits of the departed. To Marcus Cocceius Superianus, centurion of the Tenth Twin Legion, 30 years and 8 months old, and to Valerius

Figure 184. (A[ptNW] I.1) - stele (AMZ, I. Krajcar)

Figure 185. (A[ptNW] I.1) - drawing (K. Rončević)

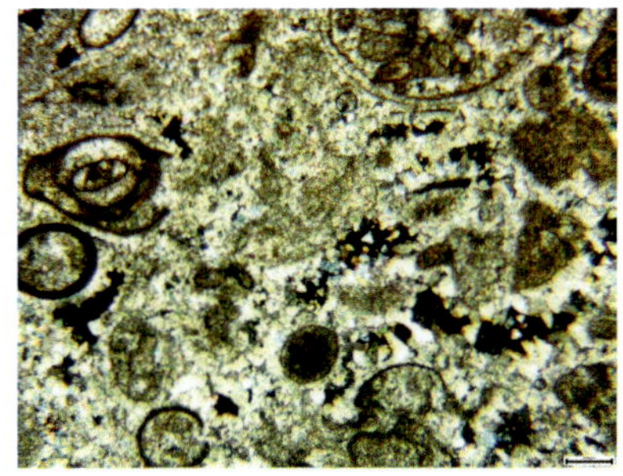

Figure 186. (A[ptNW] I.1) - stone, microscope (HGI, M. Belak)

Lucilianus, a praetorian, 40 years and 3 months old. Septimia Lucilla, their most pious mother, and Cocceius ...us (?), a standard-bearer of the Tenth Legion [had (this monument) erected] ...

Commentary: The Imperial *gentilicium* Cocceius is attested almost everywhere, but is not frequent, except in Pannonia, Noricum, and both Moesiae (Alföldy 1969: 41–42; *OPEL* II: 67); the cognomen Superianus is very rare: according to *OPEL* (IV: 100), it is only attested in this case in Pannonia, as well as three times in Dacia. The Valerii, who are very frequent everywhere (*OPEL* IV: 142–146), were one of the most distinguished upper class families in near-by Siscia (Šašel Kos 2017b). The name Lucillus/a is spread everywhere, but more so in Italy and Celtic-speaking provinces; in Pannonia, it is attested more frequently after the Marcomannic Wars (*OPEL* III: 35; Mócsy 1959: 179; Barkóczi 1964: 316; Alföldy 1969: 233). The name Lucilianus is rather rare (*OPEL* III: 35; Barkóczi 1964: 316). The Septimii are documented everywhere, but are particularly frequent in Pannonia (*OPEL* IV: 69–70; cf. Barkóczi 1964: 299). If Septimia Lucilla was mother of both deceased soldiers, it seems that she had them by two different husbands. Cocceius with the unknown cognomen, the *signifer* of the Tenth Legion, must have been the father of the first soldier, a centurion in the same legion.

*Date*: First half of the 3rd century.

*Publication*: *CIL* III 4114; 10888; Brunšmid 1909: 165–166, no. 360; Schober 1923: 71–72, no. 154; *AIJ* 455; Alföldy 1965a: 139, no. 12; Migotti 2011a (*AE* 2009, 1042). Internet databases: *lupa* 3110; *HD*068798; *EDCS*-26600396.

2. Stele (A[ptNW] I.2)

*Provenance*: Archaeological excavations in 1998–2016 in Lobor, 22.5 km NW of Ščitarjevo.

*Location*: Walled at the findspot in Lobor.

*Stone*: Conglomeratic sandstone. A probable origin: Mts Ivanščica or Kalnik, NW Croatia (M. Belak 2018).

*Preservation*: A pediment fragmentarily preserved. The left half of the pediment and the associated spandrel remain, chipped on all four sides, much weathered and pitted all over, with remains of red paint around the dog's body; a (secondary?) groove runs diagonally across the left-hand upper corner.

Figure 187. (A[ptNW] I.2) – stele (K. Filipec)

Figure 188. (A[ptNW] I.2) – stone, microscope (HGI, M. Belak)

*Description*: *W 59* cm, *H 53* cm, T 24 cm. The simply framed pediment, contained within a rectangular upper part of the monument, holds a figure of a dog in a vivid movement, looking upwards and turning towards the figure in the missing part of the pediment. The spandrel probably contains a snake-like figure of a dolphin with a huge down-turned bulbous head, or a sea monster. The workmanship is fairly competent, with the figures rendered sketchily but with the sense of movement.
*Date*: Later 2nd or 3rd century.
*Publication:* Filipec 2017: 103, no. 7.
*Note*: All finds in the section of *possible territory* (except no I.1.) stemming from the village of Lobor (I.2, VII.1–7) were recovered during the excavations led by Krešimir Filipec in 1998–2016 around the hilltop church of *Majka Božja Gorska* (Our Lady of the Mountain), at the complex archaeological site yielding strata from the Bronze Age to the modern period, including Roman sculpture and church architecture from the early Christian era to the Baroque and into the present day.

*VII.* Aediculae – mausolea – *funerary chambers*

1. *Aedicula* (A[ptNW] VII.1)
*Provenance*: Archaeological excavations in 1998–2016 in Lobor, 42.5 km NW of Ščitarjevo (see at *Note* in no. I.2.).
*Location*: Temporarily at the findspot in Lobor.
*Stone*: Probably Pohorje marble (see at *Note*).
*Preservation*: A fragmentarily preserved soffit slab with entablature. It is chipped off on three sides and damaged on the remaining one, with possible traces of a secondary use in the Romanesque church.
*Description*: *L 105* cm, *W 86* cm, H 30 cm. The fragment is provided with one rectangular hole in the preserved corner on the inner surface, whose side contains a decorative band composed of a stylized Lesbian cyma, topped by a plain band provided with equally spaced rectangular holes (four of them in the undamaged part), resembling a row of 'negative' dentils. The workmanship seems to be either rather crude or the piece remained unfinished.
*Date*: Later 2nd or 3rd century.
*Publication:* Filipec 2017: 105, no. 13.
*Note*: Marble monuments from Lobor were not characterized, but three samples taken from the leftover scraps at the site revealed the origin from Pohorje (W. Prochaska 2017).

Figure 189. (A[ptNW] VII.1) – *aedicula*, side view (K. Filipec)

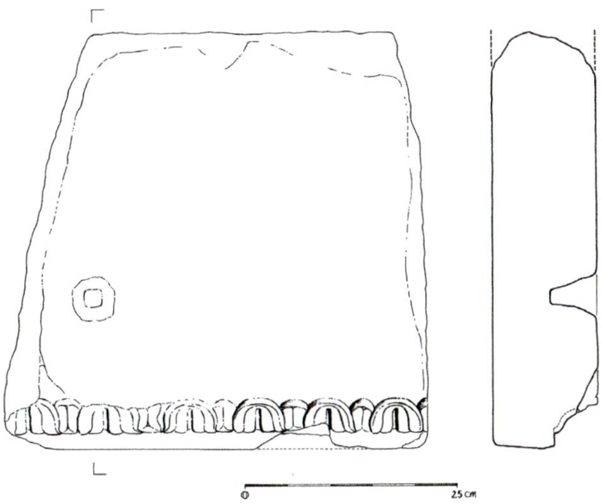

Figure 190. (A[ptNW] VII.1) – *aedicula*, underneath view (after Filipec 2017)

2. *Aedicula* (A[ptNW] VII.2)
Provenance: Archaeological excavations in 1998–2016 in Lobor, 42.5 km NW of Ščitarjevo (see at *Note* in no. I.2.).
*Location*: Temporarily at the findspot in Lobor.
*Stone*: Badenian sandy biocalcarenite. A probable origin: Mts Ivanščica or Kalnik, NW Croatia (M. Belak 2018).
*Preservation*: A fragmentarily preserved corner block. The stone is broken on the bottom and right-hand side, and is heavily weathered and pitted all over, especially on the upper part, including the figure's face and the objects in her hands.
*Description*: *H* 51 cm, *W* 40 cm, *T* 49 cm. The cubical corner block bears reliefs on the front and the left side, while

Figure 191. (A[ptNW] VII.2) – *aedicula* fragment (K. Filipec)

Figure 192. (A[ptNW] VII.2) – stone, microscope (HGI, M. Belak)

its back, left in a rough state, holds a rectangular clamp hole for fixing onto the next member of a funerary structure. On a simply framed panel on the front of a servant girl is carved wearing a wide-sleeved (? – as appears on the right arm, but not on the left) tunic with a rounded neckline, and a belted over-tunic with deep, vertical folds between the legs; the presumed brooches have probably been rendered invisible through damage. She is holding a mirror in her uplifted right hand, while supporting a large box on her left hand. On the left side of the block, a tree with bushy branches is depicted. The workmanship appears to be crude, but this can partly be a false impression due to the poor preservation of the piece.
*Date*: Later 2nd or 3rd century.
*Publication*: Filipec 2017: 102, no. 2.

3. *Aedicula* (A[ptNW] VII.3)
*Provenance*: Archaeological excavations in 1998–2016 in Lobor, 42.5 km NW of Ščitarjevo (see at *Note* in no. I.2.).
*Location*: Temporarily at the findspot in Lobor.
*Stone*: Badenian sandy biocalcarenite. A probable origin: Mts Ivanščica or Kalnik, NW Croatia (M. Belak, 2018).
*Preservation*: A fragmentarily preserved side wall. Roughly one half of the block is missing on the right side; it is chipped around the edges and on the back, and badly weathered and pitted throughout.
*Description*: *H* 117 cm, *W* 55 cm, T 31 cm. The rectangular elongated block is part of a side wall of a funerary structure. In a semi-cylindrical, simply framed niche a maenad is depicted in the right profile, with her knees slightly bent and her feet barely apart, wearing a drapery wound around her thighs and knotted in front. Her hair is falling tidily down her nape reaching the shoulders, while her uplifted arms suggest the playing of cymbals, which, however cannot be discerned in the relief. In the background of her upper body a folded drapery (?) is depicted, possibly part of maenad's clothes; below it on the right, the letters *ais* are shallowly carved. A wide decorative band above the niche is taken by an undulating body of a sea monster turned towards the right. The workmanship is coarse, especially in the execution of anatomic details, but the movement and drapery of the lower body have a natural touch about them.

Figure 193. (A[ptNW] VII.3) – *aedicula* fragment (K. Filipec)

Figure 194. (A[ptNW] VII.2) – stone, microscope (HGI, M. Belak)

*Date*: Later 2nd or 3rd century.
*Publication*: Filipec 2017: 102, no. 4.

4. *Aedicula* (A[ptNW] VII.4)
*Provenance*: Archaeological excavations in 1998–2016 in Lobor, 42.5 km NW of Ščitarjevo (see at *Note* in A[ptNW] I.2.).
*Location*: Temporarily at the findspot in Lobor.
*Stone*: Badenian sandy biocalcarenite. A probable origin: Mts Ivanščica or Kalnik, NW Croatia (M. Belak, 2018).
*Preservation*: A fragmentarily preserved side wall. It was broken in two pieces and cemented together, with roughly one third of the front and some of the upper edge missing. The upper part of the front, including the figures' faces, is heavily damaged, but what remains of the reliefs is fairly well preserved.
*Description*: *H 142* cm, *W 61* cm, T 29 cm. The rectangular elongated block has a central niche with a depiction of a female figure standing on a pedestal in a typical gesture of holding her folded drapery in her uplifted left hand. Because of a full dress and a frontal statuary pose, the figure could theoretically be interpreted as the owner of the tomb (cf. *lupa* 14548, Aquileia), but an analogy from Noricum (*lupa* 400) suggests a maenad or some other mythological figure. The right-hand side panel

Figure 195. (A[ptNW] VII.4) – *aedicula* fragment (K. Filipec)

Figure 196. (A[ptNW] VII.4) – stone, microscope (HGI, M. Belak)

is taken by two superimposed semi-cylindrical, arched, simply framed niches. In the lower one a winged 'baby'-type Eros with a plump face and curly hair braided on the crown is shown in the guise of a season (autumn or summer) personification. Eros is depicted with his legs slightly apart and with only the right-hand side wing shown, and is wearing a cloak fastened on the right shoulder with a round brooch. In his uplifted left hand he is holding a fruit, while supporting with his right more fruits contained in the fold of his cloak at the chest. In the niche above, a naked female figure with the legs pressed tight together and the uplifted left hand is depicted (Venus rather than a maenad), with traces of red paint along her left hip and thigh. The narrow side panel is taken by an ivy tendril springing from a palm-like base. The workmanship is mediocre, but still with some sense of natural shapes, proportions and movement.
*Date*: Later 2nd or early 3rd century.
*Publication*: Filipec 2017: 103, no. 5.

5. *Aedicula* (A[ptNW] VII.5)
*Provenance*: Archaeological excavations in 1998–2016 in Lobor, 42.5 km NW of Ščitarjevo (see at *Note* in no. I.2.).
*Location*: Temporarily at the findspot in Lobor.
*Stone*: Badenian sandy biocalcarenite. A probable origin: Mts Ivanščica or Kalnik, NW Croatia (M. Belak 2018).

Figure 197. (A[ptNW] VII.5) – *aedicula* fragment (K. Filipec)

Figure 198. (A[ptNW] VII.5) – stone, microscope (HGI, M. Belak)

*Preservation*: An architrave beam completely (?) preserved. It is chipped at the upper left side, and is much weathered and pitted all over.
*Description*: L 120 cm, H 30 cm, W 47 cm. The beam is slightly thinned at both ends, with clamp and dowel holes at the thinned sections, and has one short side decorated in relief. In a simply framed niche a genre scene is depicted featuring a woman sitting on a stone or a stool behind a goat and milking it into a large milking pot; the goat has a slender body, a short tail and tiny horns. The workmanship appears to be quite good in spite of the heavy damage, showing a sense of a vivid movement and natural shapes.
*Date*: Later 2nd or 3rd century.
*Publication*: Filipec 2017: 103, no. 6.
*Illustration*: Photos: 1 – *aedicula* fragment (K. Filipec); 2 – stone, microscope (HGI, M. Belak).

6. *Aedicula* (A[ptNW] VII.6)
*Provenance*: Archaeological excavations in 1998–2016 in Lobor, 42.5 km NW of Ščitarjevo (see at *Note* in no. I.2.).
*Location*: Temporarily at the findspot in Lobor.
*Stone*: Badenian sandy biocalcarenite. A probable origin: Mts Ivanščica or Kalnik, NW Croatia (M. Belak 2018).
*Preservation*: A fragmentarily preserved cornice. It is broken off and chipped on all sides, and pitted at places, but what survives of the relief is well preserved.

Figure 199. (A[ptNW] VII.6) – *aedicula* fragment (K. Filipec)

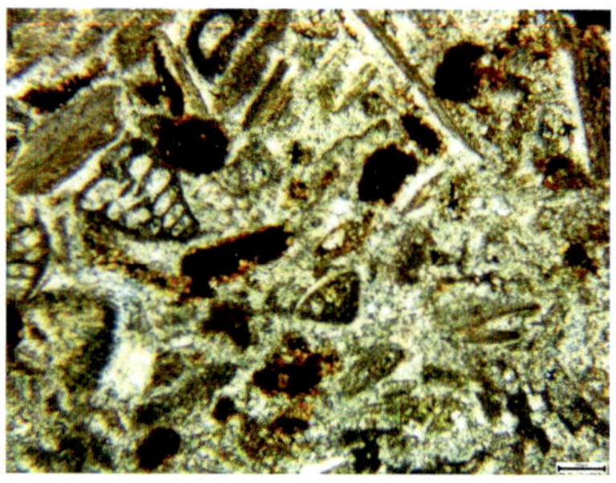

Figure 200. (A[ptNW] VII.6) – stone, microscope (HGI, M. Belak)

*Description*: *L 53* cm, *H 24* cm, *W 43.* A cubical fragment of the cornice features two superimposed friezes, one composed of tongue-like water leaves with a central groove, and the lower one containing a string of acanthus leaves with slight traces of red paint at places. The decorative motifs are stylized, but the workmanship is fairly competent and precise.
*Date*: Later 2nd or 3rd century.
*Publication*: Filipec 2017: 103, no. 8.

7. *Aedicula* (A[ptNW] VII.7)

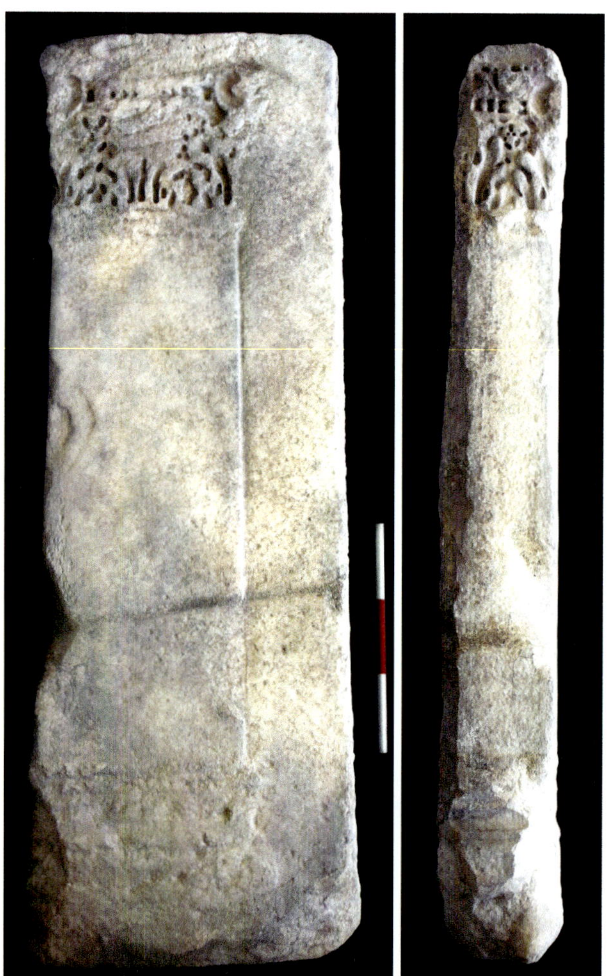

Figure 201. (A[ptNW] VII.7) – *aedicula* fragment, front and short side (K. Filipec)

*Provenance*: Archaeological excavations in 1998–2016 in Lobor, 42.5 km NW of Ščitarjevo (see at *Note* in A[ptNW] I.2).
*Location*: Temporarily at the findspot in Lobor.
*Stone*: Probably Pohorje marble (see *Note* at A[ptNW] VII.1).
*Preservation*: A fragmentarily preserved wall. A segment of the wall survives, possibly cut vertically to reduce the thickness for secondary use. It is chipped around the edges and the reliefs are damaged at places.
*Description*: H 118.1 cm, *W 40* cm, T 14–15 cm. The end of the fragmented wall is decorated with a pilaster sitting on a high, elaborate base and topped by a Corinthian capital. The capital is composed of acanthus leaves executed in the 'lace style' and two superimposed bands of geometric decoration. It is depicted also on the side of the wall, with the pilaster shaft defined in incised lines. The workmanship is hard to assess because of the heavy damage on the reliefs, but a quality work still transpires.
*Date*: Later 2nd or 3rd century.
*Publication*: Filipec 2017: 104, no. 10.

*VIII. Unclassifiable*

1. A reworked funerary monument (?) (A[ptNW] VIII.1)
*Provenance*: Placed by the road at the village of Paka, 43 km NE of Ščitarjevo; reported to have been brought there in the 1930s from an unknown site in the vicinity.
*Location*: Not characterized because of inaccessibility at the time of inspection.
*Preservation*: A water basin holding a Roman relief is partially submerged in the ground, so that its complete shape and the state of preservation remain obscured. The relief figure is much weathered, chipped and stained, especially around the mouth area.
*Description*: L 113 cm, W 72 cm, H 35 cm. The water basin, constructed at a spring site, has the form of a receptacle with rounded short sides, fixed between two slightly

Figure 202. (A[ptNW] VIII.1) – water basin (GMV, Archives)

lower, roughly dressed rectangular slabs. It seems to be reworked (and relocated) from a Roman monument of unknown shape and function, possibly a funerary one. The front wall features a mask-like human (probably male) pear-shaped face, whose lower part is blurred with damage. It has over-sized almond-shaped eyes and hair parted at the centre and sticking out sideways in large hatched scrolls. The workmanship is hard to assess due to the wear, but the iconography and technique of execution reveal an indigenous prehistoric sculptural tradition.
*Date*: 2nd or 3rd century.
*Publication*: Fulir 1969: 389–390.
*Note*: The possibility for ascribing this piece a funerary use comes from the motif of the male head, representing a simplified version of leafy mask-like faces from the circle of Dionysian or water-deity iconographies, often used in funerary contexts (cf. *lupa* 5261, 13328).

2. Funerary slab (stele or a built-in grave relief) (A[ptNW] VIII.2)
*Provenance*: A chance find in unknown circumstances at the village of Mađarevo, 47 km NE of Ščitarjevo (donated to the GMV in 1959).
*Location*: GMV, inv. no. 43825.
*Stone*: Badenian biocalcarenite. A possible origin: the mountains of N Croatia (M. Belak 2018).
*Preservation*: A fragmentarily preserved portrait niche. Around two thirds of the niche survives, pitted and crusted in places, with the right brow and nose of the figure chipped off.
*Description*: *H 41* cm, *W 38* cm, T 15 cm. A rectangular slab contains a niche with an arch to accommodate the deceased's head, with the lower part simply framed or flanked by a geometricized half-column on the right side. The hair is depicted as short parallel perpendicular strokes on the rounded crown-top. The face, resting on a fairly slender neck, is triangular in shape, as are the nose and eyes with deep irregular holes for irises and longish arched cuts for the brows, while the mouth is slit-like. The workmanship is very crude and the human figure rendered in a thoroughly geometrical manner.
*Date*: Late Roman.
*Publication*: Goss 2007: 88, no. 70.

Figure 203. (A[ptNW] VIII.2) – slab (Lj. Perinić)

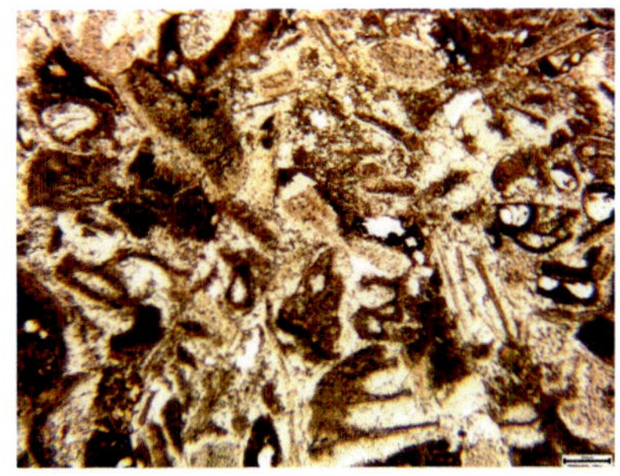

Figure 204. (A[ptNW] VIII.2) – stone, microscope (HGI, M. Belak)

***The possible territory of Andautonia in the north-east* (A[ptNE])**

*I. Stelai*

1. Stele (A[ptNE] I.1)
*Provenance*: A chance find in 2001 in the former riverbed of the Drava near the village of Gabajeva Greda, 81.5 km NE of Ščitarjevo.
*Location*: Muzej grada Koprivnice / The Museum of Koprivnica, inv. no. MGKc-10960.
*Stone*: Salla marble (W. Prochaska 2017).
*Preservation*: A lion top completely preserved. It is unfinished, with various segments unevenly worked; the left-hand-side lion is completed to a higher degree than the remainder of the sculpture. Tool marks are visible all over, and there is some chipping on the lower part of the female face.
*Description*: L 160 cm, H 68 cm, W 44 cm. Between the antithetic lying lions an unfinished round-faced female head with a voluminous lobed coiffure is sculpted, with its back supported against a truncated prismatic shaft.
*Date*: 2nd or 3rd century.
*Publication*: Kulenović and Muštra 2002.
*Note*: The stone was found in the riverbed of the Drava, and therefore probably intended for a town farther downstream, most likely Mursa. Nevertheless, it is still possible that it was lost during the unloading near its findspot, the more so as a similar example stems from Križevci, some 31 km SW of Gabajeva Greda (A[t]I.9).

Figure 205. (A[ptNE] I.1) – lion top (Lj. Perinić)

AQUAE BALISSAE

***The town of Daruvar* (AB)**

*I. Stelai*

1. Stele (AB I.1)
*Provenance*: A chance find in the mid 19th century at the site of Podborje, 2 km SW of the centre of Daruvar.
*Location*: AMZ, inv. no. AMZ KS-955.
*Stone*: Pohorje marble (H. Müller, 2013).
*Preservation*: Fragmentarily preserved. The top of the pediment and the section below the portrait niche are missing, the left-hand side column of the portrait niche is partly chipped off and the right-hand side one completely battered. The pediment's top and the majority of the spandrels are also chipped, and the whole surface extremely weathered, obliterating the reliefs, especially the columns and facial features of the deceased.
*Description*: *H* 75 cm, W 82 cm, T 19.5 cm; probably type 5a or 5b. The rectangular flat-backed portrait niche with half-figures of three deceased people (two men and a woman between them) is flanked with spirally fluted (?) columns. It rests on a decorative band featuring a frieze of animals running in opposite directions, and is surmounted by a rectangular top holding a Medusa head flanked by two birds, while originally the spandrels were each taken by a sea monster, whose only slight remains survive. All of the portrayed deceased have oval elongated faces with blurred facial features. The man on the right has short curly hair, and seems to be bearded. He is wearing a tunic and a *sagum* fastened on the right shoulder with a round brooch, and is holding a sword or a centurion's stick in his left hand. The man on the left-hand side seems to be beardless, with short curly hair. He is also wearing a tunic and a *sagum* fastened on the right shoulder with a round brooch, and is pointing with the fingers of his right hand at an unknown object held in his left, hidden behind the woman's arm. At the same time he seems to be holding in the same hand an object of the shape identical with the one in the other man's hand, which remains unexplainable. The woman has fairly thick hair rounded on the crown and with an apparently concave fringe that extends towards the cheeks, framing the face and covering the ears. A central parting is not clearly visible, but is suggested by a peaked form in the middle of the forehead; a tight-fitting wrapped headdress can be presumed, but this remains inconclusive. The woman is wearing a long- and wide-sleeved pleated tunic, a cloak (*palla*) wrapped around her shoulders and hanging in a large fold across her chest, and a beaded necklace; she is probably holding something in her right hand, folded across her chest. The workmanship is hard to assess due to the wear, but the carving is rather crude and stiff (especially the clothes), with anatomical disproportions (arms and hands).
*Date*: End of 2nd or first half of 3rd century.

Figure 206. (AB I.1) – stele (AMZ, I. Krajcar)

Figure 207. (AB I.2) – lion top (O. Harl, *lupa* 26400)

*Publication*: Schejbal 2004: 104, Map IV/7, 111–112, fig. 16; Migotti 2013a: 308, no. 30. Internet databases: *lupa* 22761.

2. Stele (?) (AB I.2)
*Provenance*: A chance find in building works in 1966 some 500 m to the NW of the centre of the Roman town.
*Location*: ZJD.
*Stone*: Badenian sandy calcarenite. A probable origin: the Slavonian Mountains (M. Belak 2018).
*Preservation*: A lion top completely preserved. It is very roughly worked on the back and extremely weathered, but also possibly unfinished, as indicated by picked sections all over the lions' bodies.
*Description*: H 35 cm, L 89 cm, W 25 cm. The top is composed of two back-to-back lying lions with frontal heads and with an indistinct conical feature (possibly a head) between them. Despite the heavy weathering a glimpse of a coarse workmanship is allowed.
*Date*: 2nd or 3rd century.
*Publication*: Schejbal 2004: 104, Map IV/6, 112, fig. 17. Internet databases: *lupa* 26400.

Figure 208. (AB I.2) – stone, microscope (HGI, M. Belak)

3. Stele (?) (AB I.3)
*Provenance*: A chance find in building works in the 1960s at the site of the Silvanus temple in the western section of the Roman town.
*Location*: ZJD.

Figure 209. (AB I.3) – stele (O. Harl, *lupa* 26399)

*Stone*: Badenian sandy siliciclastic biocalcarenite. A probable origin: Mts Psunj or Papuk in the Slavonian Mountains (M. Belak 2108).
*Preservation*: Fragmentarily preserved, with the right-hand part of a niche mostly surviving, and with its top right corner slightly chipped. The surface is worked quite smooth, including the reverse, but is weathered along the section of a diagonal break and on the left side of the face. The portrait niche probably contained two complementary arched segments.
*Description*: *H 34* cm, *W 28* cm, T 10 cm. The arched unframed niche with a rounded back seems to be unfinished in a halo-like segment above the man's head, and also in a superfluous stone mass by his neck and left shoulder. The niche holds a relief head of an apparently beardless man with a markedly elongated face, large almond eyes, a straight geometricized

Figure 210. (AB I.3) – stele, right short side
(Lj. Perinić)

Figure 211. (AB I.3) – stone, microscope (HGI, M. Belak)

nose, small, full lips, a prominent rounded chin, rather thick and smooth short hair receded at the temples, and elongated high-set ears; he is wearing a *sagum*. The workmanship is rather coarse and primitive.

*Date*: Mid or later 3rd century.

*Publication:* Schejbal 2004: 104, Map IV/4, 112–113, fig. 19. Internet databases: *lupa* 26399.

*Note*: There is a very slight possibility that this is a niche from a funerary building. This is suggested by a well-worked short side(s) and back, and a chip out of the right short side, near to the lower edge, which could have possibly developed from subsequent damage to an original dowel hole.

*II. Sarcophagi*

1. Sarcophagus (AB II.1)

*Provenance*: A chance find during building works in 1921 in the area of the S cemetery.

*Location*: Missing.

*Stone*: Sandstone (after Schejbal).

*Preservation*: The chest was completely preserved.

*Description* (from archival data): L 219 cm, H 45 cm, W 118 cm; type 7. The plain rectangular chest had flat upper edges.

*Date*: 2nd – 4th century.

*Publication*: Schejbal 2004: 104, Map IV/11, 113; The Archives of the AMZ 42, Daruvar (a letter dated March 2nd, 1921).

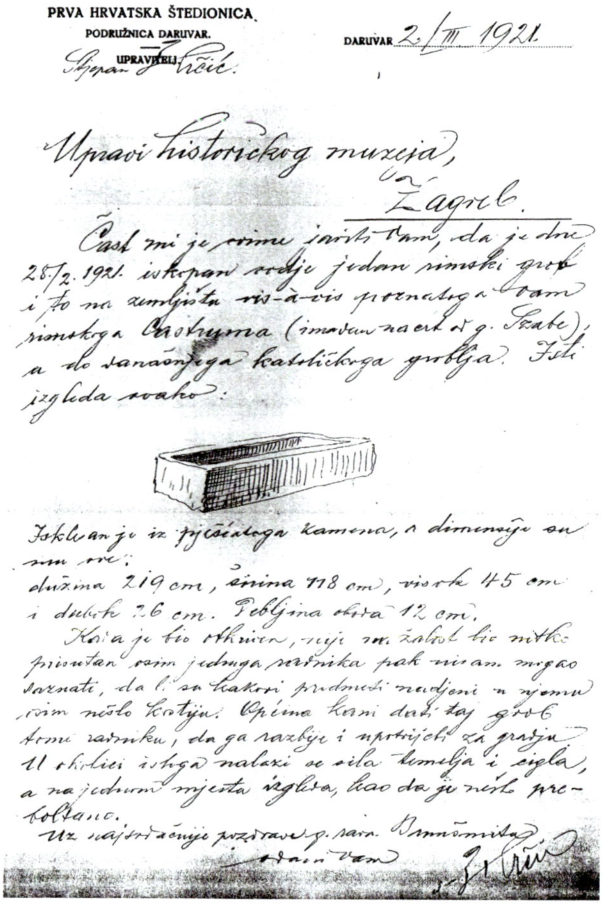

PRVA HRVATSKA ŠTEDIONICA.
PODRUŽNICA DARUVAR.
UPRAVITELJ Stjepan Ž. Ivčić.

DARUVAR 2/III 1921.

Upravi historičkog muzeja,
u Zagreb.

Čast mi je ovime javiti Vam, da je dne 28/2. 1921. iskopan ovdje jedan rimski grob i to na zemljištu vis-à-vis poznatoga Vam rimskoga Castruma (imovina nacrt g. Szabe), a do današnjega katoličkoga groblja. Isti izgleda ovako:

Isklesan je iz pješčanoga kamena, a dimensije su mu ove:
dužina 219 cm, širina 118 cm, visok 45 cm i dubok 26 cm. Debljina okvira 12 cm.

Kad je bio otkriven, nije na žalost bio nitko prisutan osim jednoga radnika pak nisam mogao saznati, da li su kakovi predmeti nadjeni u njemu osim nešto kostiju. Općina kani dati taj grob tome radniku, da ga razbije i upotrijebi za gradju. U okolici istoga nalazi se sila temelja i cigla, a na jednom mjestu izgleda, kao da je nešto prekoltano.

Uz najsrdačnije pozdrave g. ravn. Brunšmida
odani Vam
[signature]

Figure 212. (AB II.1) – scan of the drawing – a page from the Archive

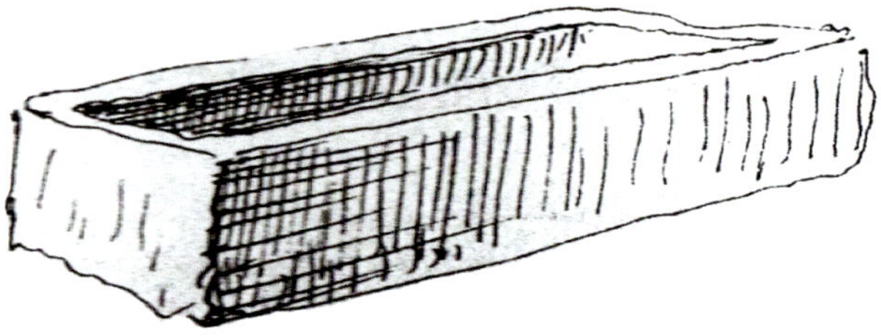

Figure 213. (AB II.1) – scan of the drawing – sarcophagus (AMZ, Archives)

2. Sarcophagus (AB II.2)
*Provenance*: A chance find during the dredging of the River Toplica in 1933 in the western part of the Roman town.
*Location*: Missing.
*Preservation*: Allegedly found complete (lid not mentioned) but subsequently broken and walled into the river bank.
*Description*: Described as 'a large Roman stone sarcophagus'.
*Date*: 2nd – 4th century.
*Publication*: Schejbal 2004: 104, Map IV/10, 113-114 (no illustration).

*V. Altars*

1. Altar (AB V.1)
*Provenance*: A chance find in 1920 at the site of Kućište, *c.* 2.5 km west of the centre of Daruvar.
*Location*: AMZ, inv. no. AMZ KS-953.
*Stone*: Badenian biocalcarenite. A possible origin: the mountains of N Croatia (M. Belak 2018).
*Preservation*: A completely preserved altar shaft. The rear bottom section with the respective short-side lower corners is broken off, but with practically no damage to the reliefs. The chipped-off fragment has been substituted by modern concrete infill, and the stone is extremely weathered, cracked, stained and pitted all over.
*Description*: W 87 cm, H 147-152 cm, T 70-72 cm. The altar shaft has four sides roughly smoothed, with three of them (except the back) moulded. The front is taken by the inscription, and the sides each feature a bearded male figure standing on a fairly high pedestal, with the frontal body and the head and legs in profile. The one on the right short side is bearded, but his counterpart on the left short side is possibly beardless, or his beard was obliterated from the wear of the stone surface. Both men are wearing a tunic and a *sagum* fastened on the right shoulder (brooch not depicted), and a headgear with a forward inclined apex and a neck-guard, and are holding a curved stick in the right hand. These figures have been variously interpreted as *genii* (*ILJug* 1132), or soldiers with a Phrygian cap, a sword and a shield (Szabó), or Attis with a *pedum* (*lupa*). On balance, they seem to represent a mixture of Oriental (Phrygian cap, *pedum*), and northern, possibly Dacian Barbarians (long beard, posture and appearance very similar to those on

Figure 214. (AB V.1) – altar, front side (AMZ, I. Krajcar)

Figure 215. (AB V.1) – altar, left short side (AMZ, I. Krajcar)

Figure 216. (AB V.1) – altar, right short side (AMZ, I. Krajcar)

Figure 217. (AB V.1) – stone, microscope (HGI, M. Belak)

the Tropaeum Traiani), but not in a mourning pose as typical of funerary iconography. There is a small round hole (diam. 3 cm) on one of the top-corner surfaces, but with no counterpart on the opposite side, while possible features of the same kind on the remaining two corners are lost (if ever existing at all) to edge breakage, suggesting that the single existing hole was not a dowel hole. Therefore there is no knowing whether the altar had a detached top, and the same holds true for a base. The workmanship is thoroughly incompetent, with highly schematized and stiff bodies, lacking any naturalness in either anatomy or drapery.

*Inscription*: *D(is) M(anibus). / P(ublio) Ael(io) P(ubli) fil(io) / Aeliano, scri/bae, dec(urioni), IIIIvir(o) /*[5] *m(unicipii) Iasorum, / an(norum) XLV. Ael(ius) / Laelianus / patri piissimo / f(aciendum) c(uravit).*

Several letters are damaged; the punctuation marks are not well visible. Ligatures: line 4: *IR* at the end of the line; line 5: *VM* at the end of the line; line 8: *IM*. Horizontal line above IIII in line 4; above the first *M* in line 5 and above *AN* in line 6.

Translation: To the spirits of the departed. To Publius Aelius Aelianus, son of Publius, secretary in charge of public records, member of the town council, *quattuorvir* of *municipium Iasorum*, 45 years old. Aelius Laelianus had (this monument) erected for his most pious father.

Commentary: The name *Municipium Iasorum* has not been known before the discovery of this funerary altar, indicating that the entire *civitas* must have been included in the municipal territory. Since Aquae Balissae presumably received municipal status during the reign of Hadrian (*IIIIviri* are mainly known in the *municipia* founded by Hadrian, Alföldy 1965b: 108), the presence of Aelii in the town is to be expected. The frequency of the *gentilicium* in Pannonia (*OPEL* I: 33-38; cf. Mócsy 1959: 150; Barkóczi 1964: 299) confirms the great impact of urbanization in the province under Hadrian (cf. Šašel 1989, 62 [1992, 695]; Fitz 2003, 51). The *cognomina* are not particularly rare; Aelianus is more frequently attested than Laelianus, particularly in Pannonia; according to *OPEL*, Aelianus has occurred 11 times (I: 25–26; cf. Mócsy 1959: 162; Barkóczi 1964: 304), Laelianus two times (III: 16; cf. Mócsy 1959: 177).

*Date*: Mid 2nd century.

*Publication*: Szabó 1934: 81; Alföldy 1964: 218–221 fig. 1, drawing (*AE* 1964: 11); Alföldy 1965b: 107–108, no. 6; *ILJug* 1132; Schejbal 2004: 107–108, fig. 5 a-c. Internet databases: *lupa* 22330; *HD*015574; *EDCS*-10001135.

*VII. Funerary buildings*

1. Portrait relief from an *aedicula* (?) (AB VII.1)

*Provenance*: A chance find in building works before World War II at no. 10 Svačićeva Street, in the vicinity of the thermal complex.

*Location*: AMZ, inv. no. AMZ KS-959.

*Stone*: Badenian low-sandy biocalcarenite. A probable origin: Mts Papuk or Psunj in the Slavonian Mountains (M. Belak 2018).

*Preservation*: Fragmentarily preserved, broken in two pieces. The majority of a niche survives, except for a smaller section of its right lateral side and the lower border, as well as the lower body of the tombstone. The surface is much weathered all over and the stone considerably chipped around the edges. There is an oval-rectangular hole of irregular shape (3.5 cm in diameter, 2.5 cm deep) on the far left side of the upper surface, most probably a dowel hole, pointing to a missing upper part.

Figure 218. (AB VII.1) – stele (AMZ, I. Krajcar)

Figure 219. (AB VII.1) – stele, detail of the top surface (AMZ, I. Krajcar)

Figure 220. (AB VII.1) – stone, microscope (HGI, M. Belak)

*Description*: *H 45* cm, *W 34* cm, T 14 cm. The irregularly hollowed portrait niche is framed by a smooth border, whose upper section is slightly curved to accommodate the crown of a head. It contains half-figures of a woman on the left-hand side, a man on the right-hand side, and a bust of a child in front of the woman. The woman has an oval face, chubby cheeks, large roundish eyes with drilled irises, a rectangular nose, and full, small, slightly parted lips. Her hair is wrapped on the crown and is parted at the centre, falling in thick waves to approximately chin level and covering the ears. Her clothes cannot be clearly perceived: she is possibly wearing a long- and wide-sleeved tunic with an oval neckline, but a double tubular neckline remains unclear. Equally blurred is her cloak, whose wide vertical fold seems to be is falling from the right shoulder down the side of her body, with the rest of it thrown over her left arm. The man has a slightly triangular face somewhat smaller than the woman's, a small prominent chin, one eye bigger and more roundish than the other, both with drilled irises, a wedge-shaped nose, and small, tight, down-turned lips. He is beardless, and has short hair depicted in parallel engraved lines running from the crown towards the forehead and slightly receded at the temples. He is wearing a *sagum* fastened on the right shoulder with a round (?) brooch. The folds of the *sagum* are executed as fairly deeply engraved and loosely spaced lines draped diagonally across the chest. Its fabric is depicted as if being transparent, because underneath the cloak and over the stomach section some densely spaced folds are engraved, as if belonging to a lower part of a tunic made of a thinner and differentially folded fabric. The woman is embracing with both hands the child in front of her, most probably a boy with a block-like elongated face and very large round eyes with drilled irises, whose remaining facial features and the upper body are obliterated through abrasion. The workmanship is extremely rough, schematized and primitive.

*Date*: First half or mid 3rd century.

*Publication*: Unpublished.

*Note*: At first sight the family portrayed looks like the parents and a child, but the fact that the grown-up beardless man's head is slightly smaller than the woman's, leaves open the possibility that a mother and her two sons are represented.

*VIII. Unclassifiable*

1. Statue of Icarus (finial of a tombstone or a tumulus burial) (AB VIII.1)

*Provenance*: A chance find in 1952 as reused in the eastern stretch of the Late Roman and mediaeval town wall.

*Location*: ZJD.

*Stone*: Badenian low-sandy biocalcarenite. A probable origin: Mts Papuk or Psunj in the Slavonian Mountains (M. Belak 2018).

Figure 221. (AB VIII.1) - statue (O. Harl, *lupa* 26397)

Figure 222. (AB VIII.1) - stone, microscope (HGI, M. Belak)

*Preservation*: Fragmentarily preserved, restored from two pieces. The topmost part is missing, and there is widespread weathering and chipping around the edges, especially at the top and the right side, but also on the body.
*Description*: Base: L 20 cm, H 6 cm, W 16 cm; figure: *H 44.5* cm, W 23.5 cm; total *H 50.5 cm*. The statue was carved in one piece with the base, and was executed in high relief with the back only roughly worked. Icarus is shown in a statuary posture with his arms lowered and slightly set aside, wearing a mantle (*paludamentum*, *chlamys*) tied in a knot at the chest, and discernible at the sides of the body and between the legs, reaching down to the feet. The workmanship is rather incompetent, as transpires from the wings and drapery executed in stiff ridges, and the body squat and ill proportioned, with anatomical inaccuracies.
*Date*: 2nd or 3rd century.
*Publication*: Schejbal 2004: 104, Map IV/8, 112–113, fig. 18; Sanader 2012. Internet databases: *lupa* 26397.

### *The presumed territory of Daruvar* (AB[t])

*I. Stelai*

1. Stele (AB[t] I.1)
*Provenance*: A chance find prior to 1873 in the village of Orešac, 40 km NE of Daruvar.
*Location*: Missing.
*Preservation*: It seems to have been completely preserved at the time of discovery.
*Description*: Written data and a linear drawing available from the old references allow for a restoration of the stele as composed of a rectangular top with a Medusa head in the centre and an inscription panel below.
*Inscription*: *M(arcus) Nunn[i]dius / Successus / vete(ranus) coh(ortis) XXXII / c(ivium) R(omanorum) ann(orum) XC. Petro/[5]niae C(ai) f(iliae) Proc(u)lae / coniugi ann(orum) LX, Nun/nidia Vitalis ann(orum) XIX, / M(arco) Nunnidio Saturn[i]no / ann(orum) VII, sibi et suis /[10] v(ivus) p(osuit).*
Translation: Marcus Nunnidius Successus, veteran of the 32nd cohort of Roman citizens, 90 years old, had (this monument) erected in his lifetime for himself and his family: to his wife Petronia Procula, daughter of Gaius, 60 years old; Nunnidia Vitalis, 19 years old, to Marcus Nunnidius Saturninus, 7 years old.
Line 6–7: Nunnidia Vitalis – nominative case, probably by inadvertence. She and M. Nunnidius Saturninus were undoubtedly the children of the couple.
Commentary: The Latin *gentilicium* Nunnidius (Solin, Salomies 1994: 129) is known from stamps (*figlinae*, *praedia*) mainly from Rome and Ostia, but it has not been attested to date in northern Italy and neither in the western provinces (*OPEL* III: 107; cf. Mócsy 1959: 157). The Petronii are well-attested everywhere, particularly also in northern Italy, but also in Pannonia and Dalmatia (Alföldy 1969: 108–109; *OPEL* III: 135; cf. Mócsy 1959: 157–158; Barkóczi 1964: 302); the *cognomen* Proc(u)lus and Proc(u)la is frequent in Italy and all western provinces, particularly in Dalmatia and Pannonia (also occurring in the eastern part of the Empire); the name is attested on 20 *tesserae* from Siscia, where it was one of the most popular names (*OPEL* III: 166–167; cf. Mócsy 1959: 186; Barkóczi 1964: 321; Radman-Livaja 2014: 249, with literature). The well-documented names Saturninus, Successus, and Vitalis also occur on several *tesserae* from Siscia, shedding light on their popularity in south-western Pannonia (Radman-Livaja 2014: 259, 268, 284, with literature); the latter may have been either a masculine or a feminine name.

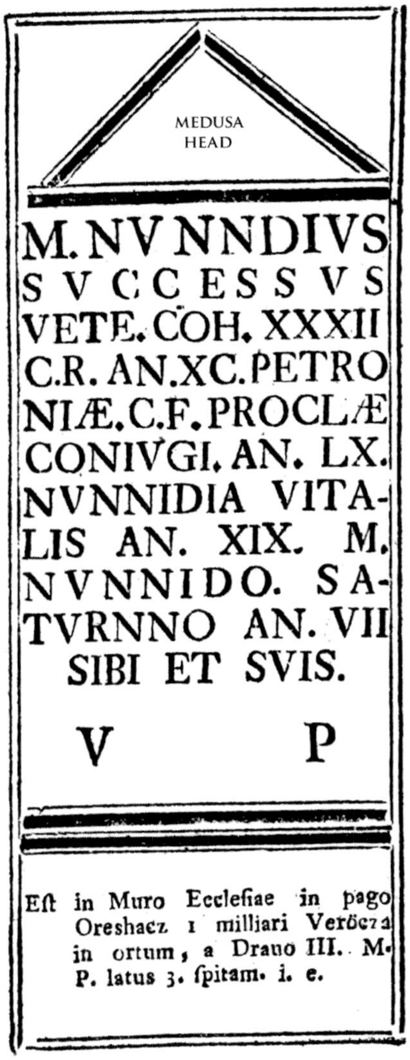


Figure 223. (AB[t] I.1) – scan of the drawing (after Kovachich 1786, readjusted by. B. Migotti)

The cohort *XXXII voluntariorum civium Romanorum* most probably belonged to the earliest auxiliary troops in Pannonia, having been stationed at Siscia (see the tombstone of the physician M. Mucius Hegetor: *CIL* III 10854 = *HD*072121; Šašel 1974: 734 [1992: 616]) until Vespasian, when it would have been transferred to Germania Superior (Lőrinz 2001: 44, with earlier literature). However, decisive evidence is lacking (Radman-Livaja 2004: 18–19); J. Spaul (2000: 47–48) argues for early 3rd century AD for the two inscriptions, but his arguments are not plausible.
*Date*: The Flavian period (70s to 100s). This chronology, established by B. Lőrinz on historical and epigraphic grounds, was challenged by J. Spaul (2000: 47–48) on the basis of the structure of the epitaph; his argumentation is not well founded, and a date in the 3rd century can be further confounded by the monument's structural typology (cf. Djurić 2008, 161–162).
*Publication*: *CIL* III 4006; Kovachich 1786: 908; Lőrinz 2001: 297, no. 480; Migotti 2016a: 175. Internet databases: *lupa* 4288; *HD*074408; *EDCS*-28800763.

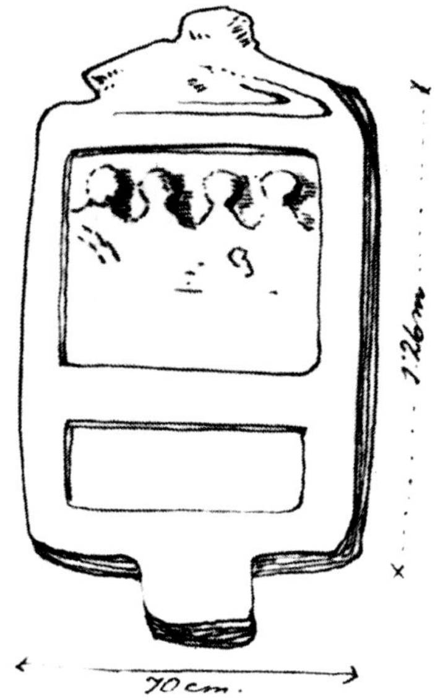


Figure 224. (AB[t] I.2) – scan of the drawing (after Horvat 1962)

2. Stele (AB[t] I.2)
*Provenance*: A chance find prior to 1906 at the medieval hill-fort site of Rudina near the village of Čečavac, 32 km SE of Daruvar.
*Location*: Missing; only a sketchy and imprecise drawing made by Gjuro Szabó in 1906 survives, published in Horvat 1962.
*Preservation*: Completely preserved at the time of discovery, with the left-hand side of the pediment slightly chipped.
*Description*: H 126 cm, W 70 cm, T (not recorded). The stele is composed of a pediment with a top *acroterion*, a rectangular portrait niche featuring four half-figures of the deceased of indiscernible sex, a dividing panel, a low base, and a tenon.
*Date*: 2nd or 3rd century.
*Publication*: Horvat 1962: 13–14.
*Note*: The drawing that provided the details of the description is extremely sketchy and probably imprecise, which leaves the restoration of the type dubious, and the portrait features completely indiscernible. It is hardly likely that the inscription panel was merged with the portrait niche as appears on the drawing, especially not within a panel leaving a very small place for the epitaph. The stele could have also had a lion pediment, given that Szabó produced a drawing of another stele as if taking a simple pediment (Schejbal 2004: 112, fig. 16), while its pediment is actually contained within a rectangular upper part (AB I.1).

3. Stele (AB[t] I.3)
*Provenance*: A chance find before 1902 in Donji Podgradci, 58 km SW of Daruvar.
*Location*: Once in the ZMBiH (inv. no. 206), now missing and not traceable in any of the possible subsequent new locations mentioned in the literature.
*Stone*: Limestone (after the *AIJ*).

Figure 225. (AB[t] I.3) – scan of the figure (after the *AIJ*)

*Preservation*: Fragmentarily preserved at the time of discovery, with a substantial part of the inscription panel surviving. It was chipped off around the edges, the right column was nearly completely battered and the surface weathered and pitted, but the inscription was fairly well preserved.
*Description* (after the *AIJ*): *H 105* cm, W 90 cm, T 18 cm. The inscription panel was flanked by vertically fluted columns.
*Inscription*: *D(is) M(anibus) / Ursioni et / Sirmiae. Po/tamilia Proba /⁵ fratri et sorori / pientissimis / et [?O]giniae matri.*
Punctuation marks between the words are not certain, because the inscribed surface is very worn. Line 2: *ET* in ligature.
Translation: To the spirits of the departed. To Ursio and Sirmia. Potamilia Proba (had this monument erected) for her most pious brother and sister, and to her mother Oginia.
Commentary: Ursio is not very frequent outside Italy and Dalmatia, but according to *OPEL*, it has occurred four times in Pannonia (IV: 187; cf. Mócsy 1959: 194), and is also attested on two *tesserae* (Radman-Livaja 2014: 277). Sirmia is a *hapax* (*OPEL* IV: 85; cf. Mócsy 1959: 191), as is also Potamilia (Solin and Salomies 1994: 148; *OPEL* III: 155; cf. Mócsy 1959: 158; *Epigraphik-Datenbank* Clauss/Slaby); however, the feminine name Potamilla has been attested several times in Italy (*Epigraphik-Datenbank* Clauss/Slaby). Although Probus is a frequent name (cf. Alföldy 1969: 273–274; *OPEL* III: 164–165), it has not been attested on the *tesserae* from Siscia. [O]ginia is Celtic (Meid 2005: 238); 'perhaps the most obvious example of Gaulish names in the region' (Falileyev 2014: 112); it is a rare *gentilicium* (*OPEL* III: 111; cf. Mócsy 1959: 157).

*Date*: Second half of 2nd or 3rd century.
*Publication*: *AIJ* 591; Mesihović 2011: 646–647; Šačić Beća 2017: 128–129, 156–157 . Internet databases: *lupa* 3813; *HD*072136; *EDCS*-11301134.

4. Stele (AB[t] I.4)
*Provenance*: A chance find in 1895 during repairs to the parish church in the town of Bjelovar, 50 km NW of Daruvar. The fragment most probably comes from a nearby site of the Roman cemetery, but was found as walled in the lunette of the main church portal during the building of the church in 1765–1772. Unfortunately, the stele was walled backwards, so that the relief remained hidden, while its back, newly inscribed with a contemporary dedicatory inscription, faced the spectator; it has remained in the same position till the present day.
*Location*: The parish church of Saint Teresa of Ávila in Bjelovar.
*Stone*: Marble (by macroscopy).
*Preservation*: Fragmentarily preserved. A pediment and a relief panel (partly) survive, with a secondary hole for a metal handle pierced through the middle of the pediment.
*Description*: The measurements are approximate, with the thickness reported to be considerable; *H 60* cm, *W 70* cm, *T?* Poor remains along the sides of the pediment do not reveal the stele type with certainty, but they suggest type 5b. The rectangular relief panel is taken by the mythological episode of Iphigenia in Tauris, with Iphigenia boarding a boat featuring a curved rostrum, and with Pylades assisting her; Pylades is depicted against the background of town walls and four soldiers wearing helmets and shields. The scene was most probably flanked by now battered columns which supported the frieze decorated with indistinct motifs. The upper part seems to have been a rectangle holding

Figure 226. (AB[t] I.4) – scan of the drawing (after Medar 1986)

a pediment with a Medusa head flanked by two birds, and with sea monsters in the spandrels. Judging from the drawing, the workmanship was fairly competent, with sense of movement and natural shapes.
*Date*: Late 2nd or early 3rd century.
*Publication*: Lovrenčević 1981: 201; Medar 1986; Cambi 2003.

5. Stele (AB[t] I.5)
*Provenance*: A chance find in 1913 at the site of a smaller Roman settlement in the village of Brusnik, 17 km S of Daruvar.
*Location*: AMZ, inv. no. AMZ KS-954.
*Stone*: Badenian sandy biocalcarenite. A probable origin: Mts Psunj or Papuk in the Slavonian Mountains (M. Belak 2018).
*Preservation*: Completely preserved. Staining, chips and damage predominate on the lower left, in the middle of the inscription panel, and on the back of the niche; there is a large chip out of the lower centre of the top, and serious damage on the left-side lion, with all reliefs heavily weathered.
*Description*: The stele is composed of a shaft (H 155 cm, W 68 cm, T 22 cm) and a separate lion top (H 63 cm, W 95 cm, T 33 cm); total height: (218 cm); type 6. The top is composed of two back-to-back lions lying on a corniced base, with frontal heads and with the front paws resting on a ram's head. Between them are two geometricized superimposed heads with large eyes angled downwards and wedge-shaped noses. The female one above has shoulder-length hair parted in the centre and hanging down in straight strands, and the child below has chubby cheeks and curly hair. The pediment, contained within a rectangular panel, features a pear-shaped Medusa (?) head between acanthus leaves and with *hippocampi* in the spandrels. The rectangular flat-backed portrait niche, flanked by spirally fluted columns, holds half-figures of a man on the left-hand side and a woman on the right-hand side. The man has an oval, block-like face, short hair receded at the temples, a wedge-shaped nose, narrow eyes, and ears set high at temple-level. He is wearing a *sagum* fastened on the right shoulder with a round brooch and a belted tunic with a huge round buckle, and is holding a sword and an under-sized oval-bossed shield. The woman has a rounded face and fairly thick ear-length hair framing her face in a wreath-like manner. She is wearing hoop earrings, a low round cap of the type *pileus Pannonicus*, a long- and wide-sleeved tunic, and a cloak draped across the shoulders and chest; two large brooches are depicted on the shoulders, composed of a knobbed foot and a head provided with two knobs at the sides of a central triangular peak. In her right hand she is holding an indistinguishable oval, loop-like object, possibly a wreath, but presumed to be a bird in the previous publications. There is apparently a huge torque-like hoop around her neck and shoulders, with each end shaped as three knobs, and a plain tubular

Figure 227. (AB[t] I.5) – stele (AMZ, I. Krajcar)

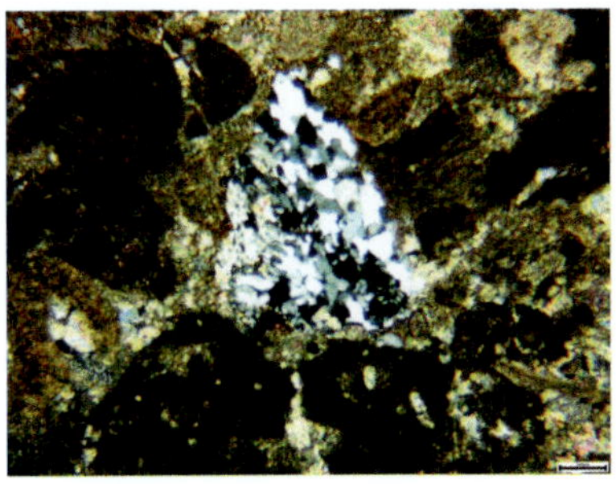

Figure 228. (AB[t] I.5) – stone, microscope (HGI, M. Belak)

torque worn tighter around her neck. A divider between the portrait niche and a simply framed inscription panel is in the form of a narrow band decorated with indistinct floral or geometrical motifs. The workmanship is extremely crude, stiff, and lacking sense of proportion, but efficient in rendering details.
*Inscription*: *D(is) M(anibus) / Aur(elii) Nasonis militis / leg(ionis) IIII Flaviae Antoni/nian(ae) et Priscae Tato/[5]nis matri. Memoriam / pasuerunt* (!) *Proclus et / Proclianus et Provin/cialis Maximiani fi/li(i).*
The inscribed surface is much worn, not all punctuations marks are visible.
Translation: To the spirits of the departed of Aurelius Naso, soldier of the 4th Flavian Antoninian legion and to mother Prisca, daughter of Tato. Proclus, Proclianus, and Provincialis, sons of Maximianus, had the memorial erected.
Commentary: Maximianus, father of the three sons, must have been the (first?) husband of Prisca, since she must have been their mother. Naso and she are depicted as husband and wife, hence she cannot be Naso's mother. Maximianus is attested almost everywhere in western provinces, but was particularly frequent in northern Italy, Noricum, and Pannonia (Alföldy 1969: 241; *OPEL* III: 69; cf. Barkóczi 1964: 318); Naso is a rare name outside Italy, attested a few times in the Iberian peninsula and Gallia Narbonensis, and only once in Pannonia (*OPEL* III: 96; cf. Barkóczi 1964: 319; Delamarre 2007: 139), but it also occurs on a *tessera* in Siscia (Radman-Livaja 2014: 235). The popular name Priscus/Prisca is also documented on five *tesserae* from Siscia (Radman-Livaja 2014: 248, with literature). For well-attested Proclus/Proculus see *supra*, while Proc(u)l(l)ianus is a rare name (*OPEL* III: 166; cf. Barkóczi 1964: 321), less so Provincialis, except in Pannonia (*OPEL* III: 169; cf. Mócsy 1959: 186; Barkóczi 1964: 321; Alföldy 1969: 276). Tato is an indigenous, 'Illyrian'-Pannonian name (Radman-Livaja, Ivezić 2012: 142; cf. Falileyev 2014: 112; cf. Barkóczi 1964: 325; Delamarre 2007: 179).
*Date*: 198–235 (198–222; 211–222 [*EDH, Epigraphik-Datenbank* Clauss/Slaby]; 213–222). These dates are based on different opinions about the period in which the legion *IV Flavia* bore the epithet *Antoniniana* (Mirković and Dušanić 1976: 68–69; Lőrinz 1985: 186–187; Kovács 2017: 262–263).
*Publication*: *AIJ* 590; Pinterović 1975: 126–127; Schejbal 2004: 121; Ubl 2013: XXIII no. 7, Pl. 3.12 (figs 180 and 243). Internet databases: *lupa* 3812; *HD*072135; *EDCS*-11301133.

6. Stele (AB[t] I.6)
*Provenance*: Amateur excavations in 1990 in the village of Kusonje, 15 km S of Daruvar.
*Location*: Muzej grada Pakraca/The Museum of Pakrac, inv. no. MGP-320.
*Stone*: Badenian sandy siliciclastic biocalcarenite. A probable origin: Mts Psunj or Papuk in the Slavonian Mountains (M. Belak 2108).
*Preservation*: Completely preserved shaft and pedestal, but for a top that should have been fixed onto the upper tenon of the stele. The base is chipped and pitted, and the shaft cracked on the upper left side. The front of the shaft is smoothed, and the sides, the background of the relief and the right-hand bottom corner of the inscription field are roughly worked or possibly unfinished. There is some staining and weathering all over.
*Description*: The monument is made of a shaft and a separately carved pedestal, irregularly shaped and very roughly worked. Shaft: H 117 cm, W 65 cm, T 39 cm; top tenon: L 36 cm, H 20 cm, T 8 cm; bottom tenon: L 37 cm, H 23 cm, T 11 cm; total stele H: 160 cm; tenon base: L 84 cm, W 75 cm, H 46 cm, with a rectangular hole measuring 42 x 26 x 12 cm; type 7. The stele is a parallelepiped with a tenon on the upper and bottom surfaces; the former was for holding a separately carved top and the latter for fixing into the pedestal. The majority of the front is taken by the inscription panel, moulded and additionally framed with a band of stylized leafed branches, and with the free space above

Figure 229. (AB[t] I.6) – stele
(O. Harl, *lupa* 10057)

Figure 230. (AB[t] I.6) – stele and the base (O. Harl, *lupa* 10057)

Figure 231. (AB[t] I.6) – stone, microscope (HGI, M. Belak)

the inscription decorated with a motif of an elongated stemmed triangle flanked by sea-griffins. In the last line of the inscription a six-pointed star is crudely carved between the letter *T* and two *C*s. A simply framed field under the inscription panel contains the motif of two short-haired wingless Erotes half striding and half floating in opposite sides and supporting a smooth garland. The workmanship is inaccurate, with rude and stiff Erotes' bodies looking more like they are falling down rather than either floating or striding, and with a flat, geometricized ornamental border.

*Inscription*: *D(is) M(anibus) / Cass(io) Sexto / militi c(o) hort(is) / D Mauror(um) /⁵ Cassi(i) Exsorat/us (!), Surus, Qui/ ntus, fratre/s et her(edes) / tit(ulum) posu(eru)n/t.*

Rustic script, letters are uneven, particularly in the last line. Punctuations marks are carved inconsistently. Line 4: A horizontal stroke across the vertical *hasta* of the letter *D* marks the numeral (Đ), which is unusual. In the last line, after the *T*, a stylized star and a moon (?) are carved.

Translation: To the spirits of the departed. To Cassius Sextus, soldier of the Cohort of the 500 Mauri. Cassii Exsoratus, Surus, and Quintus, brothers and heirs, had the tombstone erected.

Commentary: The *gentilicium* Cassius is frequent in Italy and the provinces, both western and eastern (Alföldy 1969: 73; *OPEL* II: 41); it also occurs on two *tesserae* from Siscia (Radman-Livaja 2014: 182–183, with relevant literature). The name Sextus, a Roman *praenomen*, was also a common cognomen or a personal name; it occurs on 8 *tesserae* from Siscia (Radman-Livaja 2014: 263–264, with earlier literature). Surus/Sura (Sura can be both a masculine or a feminine name) is a popular name (*OPEL* IV: 102), also attested on 28 *tesserae* from Siscia, which makes it one of the most popular names in the town; while it rarely denoted a person from Syria, its origin may variously be explained as Celtic, Illyrian, Thracian, or Semitic (Radman-Livaja 2014: 269, with earlier literature). It is generally well-attested in Pannonia, but also in northern Italy, Noricum, and Dalmatia. The Latin name Exoratus (Solin and Salomies 1994: 325) is often incorrectly carved as Exsoratus and in this case can indeed be regarded as Celtic; it is attested in the Celtic speaking provinces and northern Italy (Delamarre 2007: 100; cf. *OPEL* II: 130–131). The name also occurs on two lead tags from Siscia (Radman-Livaja 2014: 204, with earlier literature). Quintus, which is, like Sextus, a current Latin *praenomen*, is very well documented everywhere also as a *cognomen* or a personal name (*OPEL* IV: 20); it also occurs on four lead tags in Siscia (Radman-Livaja 2014: 251). As could be inferred from the names of the brothers, their family was probably from Pannonia, where Sextus may have been recruited into the cohort.

The cohort was sent, together with the cohort *milliaria Maurorum equitata* to Upper Pannonia in AD 171, which is inferred from the fact that both participated in the first expedition against the Germans in the Marcomannic Wars (*expeditio Germanica prima*: *CIL* VIII 12066 = *EDCS*-24400196; Lőrinz 2001: 272, no. 391). At the end of the wars both cohorts were transferred to Matrica in Lower Pannonia (Lőrinz 2001: 38–39; cf. Spaul 2000: 468–469).

*Date*: Later 2nd or first half of 3rd century (after the year 170), based on the presumption that the *Cohors*

*quingenaria Maurorum* came to Pannonia Superior in 171 and left for Pannonia Inferior between the late 170s and the early 180s (Lőrinz 2001: 39).
*Publication*: Bulat 2001 (*AE* 2001: 1659); Njegovan Stárek 2009. Internet databases: *lupa* 10057; *HD*047147; *EDCS*-24500523.
*Note*: This tombstone has formal and structural characteristics of both a stele (higher than wide, a characteristically decorated base, a tenon-pedestal construction) and an altar (a considerable thickness, a characteristically framed inscription panel). The upper tenon could have supported any kind of a top, either befitting a stele or an altar. Curiously, no separate top was found during the (amateur, but careful and systematic) excavation.

7. Stele (AB[t] I.7)
*Provenance*: A chance find in the mid 19th century, probably in the village of Veliki Bastaji, 10 km NE of Daruvar.
*Location*: AMZ, inv. no. AMZ KS-936.
*Stone*: Gummern marble (H. Müller, 2013).
*Preservation*: Fragmentarily preserved. The upper and lower parts are missing, and the stone is fairly weathered and chipped around the edges.
*Description*: *H 86* cm, W 72 cm, T 20 cm; probably type 3a or 3b. The inscription panel is flanked by columns decorated with overlapping leaf-scales, and the base, framed by a double moulding, holds the *lupa Romana* standing between two plants and turning her head towards the suckling twin brothers. The workmanship is fairly schematic and rigid.
*Inscription*: *Fl(avio) Val[erio] / mil(iti) coh(ortis) [---] / pr(a)etor(iae) [---] / Val(erio) Dig(-) NỊ[---] /*[5] *vẹ(terano ?). Adatili[a] A[n?]/nia mater fil(iis) / pientissimis fec(it).*
Line 1: thus also *EDH*, others: *Val(erio)*. Line 4: reading checked with photo; after *G*, almost certainly a punctuation mark; *Val(erio) Dign[---]*: all earlier editors; of the last visible letter only the vertical hasta has been preserved. Line 5: *vet(erano?)*: earlier editors; possibly *VF*: *v(iva) f(ecit)*, but *fec(it)* at the end of the last line would in this case be pleonastic. The last *A* seems certain. Line 7: *T* and *I* in lig.
Translation: To Flavius Valerius, soldier of ... praetorian cohort, ... to Valerius Dig[-] ..., veteran (?). Mother Adatilia A[n?]nia had (this monument) made for her most pious sons.
Commentary: Flavii are expectedly very well documented in Pannonia (cf. *OPEL* II: 145–147), since urbanization of the province began under Vespasian; the nearby Siscia was *colonia Flavia*. The Valerii, who were one of the most widespread non-Imperial *gentilicia* (cf. Alföldy 1969: 131–133; *OPEL* IV: 143–146), were also one of the best documented upper class families in Siscia (Šašel Kos 2018, forthcoming). Adatilia seems to be a *hapax* (it is neither mentioned in Solin and Salomies 1994 nor in *OPEL* and in *Epigraphik-Datenbank* Clauss/ Slaby). An(n)ius/a, if correctly read and supplemented,

Figure 232. (AB[t] I.7) – stele (AMZ, I. Krajcar)

could be regarded as 'Illyrian' in the broad sense of the word, but could also be Celtic (Alföldy 1969: 150–151; Delamarre 2007: 23).
*Date*: Second half of the 2nd or first half of 3rd century.
*Publication*: *CIL* III 4001 = 10865; *AIJ* 588; Alföldy 1965a; Palágyi 2003: 483, Abb. 4; Schejbal 2004: 104, Map IV/7, 111–112, fig. 15; Djurić 2013: 8, no. 32. Internet databases: *lupa* 3810; *HD*072134; *EDCS*-26600440.

8. Stele (AB[t] I.8)
*Provenance*: Ploughing in 1983 at the site of Treštanovačka gradina in the village of Tekić, 47 km SW of Daruvar.
*Location*: GMP, inv. no. GMP-10.460.
*Stone*: Badenian sandy biocalcarenite. A probable origin: Mts Psunj or Papuk in the Slavonian Mountains (M. Belak 2018).
*Preservation*: Fragmentarily preserved. A part of the top survives, with minor chips, abrasions and staining.
*Description*: *W 38* cm, *H 43* cm, T 12 cm. The top is in the form of a relief slab of unknown full shape, sitting on a smooth cornice and featuring a fragmentarily preserved half-nude male whose head, shoulders, right arm, and left hand are missing. He is reclining on a draped bolster, most likely a cloak with a round brooch attached, with his legs crossed and his left arm resting along the side of his left upper body and left hip. His nudity suggests a mythological character whose precise nature remains obscured: a god, a hero, a genius, satyr, Eros (?). The workmanship is course, revealing a linear style, stiffness, and anatomical inaccuracies.
*Date*: 3rd century (?).
*Publication*: Sokač-Štimac 2005: 5. Internet databases: *lupa* 26276.

Figure 233. (AB[t] I.8) – stele (GMP, B. Knez)

Figure 234. (AB[t] I.8) – stone, microscope (HGI, M. Belak)

*Note*: The iconography of the naked figure indicates an Endymion, but it should be noted that Endymion figures in Norico-Pannonian funerary art are mostly rendered differently from the one here. On the other hand, the figure from Tekić has the closest parallel in a stele-top depiction of Liber and Libera from Pécs (*lupa* 645), with the main difference in the position of the arms, one of which is raised above the head. Arguably, the gesture with a raised arm is typical of all similarly positioned male and female mythological figures.

9. Stele (AB[t] I.9)
*Provenance*: A chance find prior to 1873 amidst Roman ruins in the village of Vetovo, 48 km NE of Daruvar.
*Location*: Missing.
*Preservation*: According to the *CIL*, the inscription was reassembled from two pieces, but not much of it appears to have been missing.
*Description*: The stele contained busts of a man and a woman in the upper niche and an inscription field below.
*Inscription*: *D(is) M(anibus). Gr(a)ecus / DE FA Max() / titu[lum (?) ---] / [---]ÇVOCS /[5] Mumapio / filio.*
The first three lines of the epitaph do not seem to be fragmentary. Line 2: *de Fa(bio?) Max()*, proposed in *EDCS*, not convincing. Lines 2 and 4 cannot be plausibly supplemented; the name of a woman (probably the wife) should be expected.
Translation: To the spirits of the departed. Gr(a)ecus ... to son Mumapius.
Commentary: the cognomen or personal name Graecus occurs more often in Italy (*Epigraphik-Datenbank* Clauss/Slaby), particularly in northern Italy, but is sporadically attested in all western provinces, partly among freedmen (Alföldy 1969: 212; *OPEL* (II: 169–170; Kakoschke 2012: 441). Mumapius seems to be a *hapax*, unless the name has not been correctly copied.
*Date*: Mid 2nd – early 4th century.
*Publication*: *CIL* III 3997; Pinterović 1975: 125; Schejbal 2004: 121. Internet databases: *HD*074397; *EDCS*-28800757.

*II. Sarcophagi*

1. Sarcophagus (AB[t] II.1)
*Provenance*: Amateur excavations in 1842 in the village of Veliki Bastaji, 10 km NE of Daruvar. It was found in an underground funerary chamber with frescoed walls and a mosaic floor, together with several other monuments: a sarcophagus that does not survive, a marble inscription slab, and fragments of several other inscribed slabs; see no. AB(t) VII.1.
*Location*: AMZ, inv. no. AMZ KS-959.
*Stone*: Pohorje marble (H. Müller 2013).
*Preservation*: A lid (allegedly gone missing in the meantime, see *Note*) and a chest with a separately carved base completely preserved. There are slight traces of a cutting on the chest's upper edges to hold a lid, while the short sides are provided with clamp holes. The chest is severely damaged around the upper edges, with part of the upper front missing, and with cracks along the veins and at other places on the front, especially on the right-hand corner. The right-hand lateral niche is partly damaged and the left one even more severely so, with the head of the portrayed person missing, and with the facial features of the surviving head in the right-hand niche completely obliterated by wear and veining.
*Description*: Judging from the preserved photos, at one time the chest was holding a gabled lid of type 1.2 with very low *acroteria*, and with the measurements not recorded, but this was probably not its original place (see *Notes*). The measurements of the rectangular chest are: L 232 cm, W 128 cm, H 78 cm; base: L 261 cm, W 147 cm, H 35 cm; type 3b. The majority of the front is taken by a moulded blank inscription field flanked by moulded lateral niches, each topped by an elaborate type of the Norico-Pannonian volute and holding one of the standing portrait figures of the deceased couple. Both figures are depicted in the guise of *palliati*, and are squeezed within a simply framed, arched niche with a curved back. In the left niche is a frontal female figure in a statuary pose hidden behind heavy drapery, with feet slightly apart. She is resting her left hand on her

Figure 235. (AB[t] II.1) – sarcophagus, front side (AMZ, I. Krajcar)

Figure 236. (AB[t] II.1) – sarcophagus, left short side (AMZ, I. Krajcar)

Figure 237. (AB[t] II.1) – sarcophagus, right short side (AMZ, I. Krajcar)

abdomen, while holding a round object, probably a fruit, in her right hand. Her head has been chipped off, with only a sturdy neck and a strand of hair on its left side visible. Her *palla*, worn over a long- and tight-sleeved tunic, is decorated with embroidery (or a rich lunate necklace) at the area of the collarbones and shoulders, and reaching to the ground; both items are rendered in thick and stiff parallel ridges. The right-hand niche holds a frontal male figure in a statuary pose, holding a scroll in his left hand and pointing at it with his right. He has a rounded, block-like head, chubby cheeks, thick short hair, and is wearing a *pallium* rendered in thick and stiff folds, but still natural enough to reveal movements of the limbs. The simply framed short sides are decorated in low and flat relief giving the effect of woodcarving. The right one features two heraldic panthers, each seated on its hind legs in front of a small tree and flanking a vase filled with fruits, while vines and grapes are springing from it. The left short side is completely covered in arabesque-like vines and grapes, flanked by two small trees. The panthers' fur is rendered in circlets, their heads are bowed, and their lifted outer front legs are touching the vase. The separately carved sarcophagus base is decorated with two superimposed ornamental bands, the upper bearing zigzags and the lower a stylised version of the leaf and dart motif. The

Figure 238. (AB[t] II.1) – sarcophagus, presumed matching lid (AMZ, I. Krajcar)

workmanship is more competent on the short sides, where it is vivaciously decorative, while on the front it shows stiffness and ill proportioning in the rendering of human figures, especially in their hands and feet. The hairdo of the male figure (quite thick hair with strands sticking above the ears) and the outlining of the niches possibly suggest the lack of a final touch.
*Date*: Turn of 3rd and 4th centuries.
*Publication*: Migotti 1996; Migotti 1997: 43–44; Schejbal 2004: 112, Figure 16; 119, Figure 27 a–b; Djurić 2013: 1, no. 3. Internet databases: *lupa* 3811.
*Note*: There is a fragmentary gabled, tiled lid in the Archaeological Museum in Zagreb (inv. no. AMZ KS-960), of which a smaller part with one corner *acroterion* survives (H 43 cm, *W 76* cm, *L 68* cm; type 2). It is considered to be from an unknown site, as no archival data survives for it. However, its iconography (vines, grapes and a small tree or a branch featuring oval-rhomboid leaves) and the style of its execution are astonishingly similar to those of the sarcophagus in this entry. If to this we add that both were carved in Pohorje marble (W. Prochaska 2017), and that their widths correspond (147 cm vs. 76 cm as roughly a half of the whole), the presumption that this lid belonged to the chest in question is very probable. This is further corroborated by an archival photo in which the sarcophagus is seen as covered with an undecorated lid with low *acroteria*, but a relief-decorated lid fragment with a high *acroterion* can be spotted in its vicinity (see *Illustrations*). In addition to the above described motifs, the lid's gable contains traces of a central portrait, while the termination of the only preserved roof *imbrex* is marked by a vertical cut throughout its height.

2. Sarcophagus (AB[t] II.2)
*Provenance*: Archaeological excavations in 1986 at the medieval hill-fort site of Rudina near the village of Čečavac, 32 km SE of Daruvar. The slab was re-used as part of the covering of a Romanesque grave vault in a Benedictine monastery church.
*Location*: GMP, inv. no GMP-10.451.

Figure 240. (AB[t] II.2) – stone, microscope (HGI, M. Belak)

*Stone*: Badenian sandy siliciclastic biocalcarenite. A probable origin: Mts Psunj or Papuk in the Slavonian Mountains (M. Belak 2018).
*Preservation*: Fragmentarily preserved. A fragment of the front wall survives, chipped along the edges.
*Description*: H 72 cm, *W 44.5* cm, T 8-13 cm; a variant of type 6. The preserved fragment of a sub-trapezoidal slab is most probably the right lateral side of a sarcophagus, featuring a lightly engraved and sketchily rendered fish with angular outlines and an impressed annular circle for an eye. The fish is placed next to what appears to be a border section of the *tabula ansata* inscription panel, and below its head a trefoil composed of triangular leaves is engraved. The workmanship is thoroughly coarse and schematic.
*Date*: 3rd or early 4th century.
*Publication*: Migotti 1997: 44–45; Schejbal 2004: 121. Internet databases: *lupa* 26277.

3. Sarcophagus (?) (AB[t] II.3)
*Provenance*: A chance find prior to 1898 in the village of Cage, 34 km S of Daruvar, at the site of a Roman *villa rustica*.
*Location*: Gradski muzej Nova Gradiška/The Museum of Nova Gradiška, inv. no. GMNG-3070.
*Stone*: Badenian biocalcarenite; a possible origin: the mountains of N Croatia (by macroscopy, M. Belak 2018).
*Preservation*: An *acroterion* (?) completely preserved.

Figure 239. (AB[t] II.2) – sarcophagus (GMP, B. Knez)

Figure 241. (AB[t] II.3) – sarcophagus (Lj. Perinić)

*Description*: H 36.5 cm, W 45.5 cm. The hollowed, half-rounded middle *acroterion* of a sarcophagus lid (?) has a flap-like shape at the top centre, holding a high relief of a quadruped, possibly a bear. The animal is turned towards an indistinct object, a plant or an animal, and has its right front leg uplifted. The workmanship is fairly course.
*Date*: 3rd century (?); mediaeval (?).
*Publication*: Brunšmid 1898: 205, fig. 103; Pinterović 1975: 130; M. Mihaljević, in Goss 2007: 74.

4. Sarcophagus (?) (AB[t] II.4)
*Provenance*: A chance find prior to 1898 in the village of Cage, 34 km S of Daruvar, at the site of a Roman *villa rustica*.
*Location*: Missing.
*Stone*: Limestone (after Brunšmid).
*Preservation*: An *acroterion* (?) completely preserved at the time of discovery.
*Description*: H 22 cm, W 36.5 cm. The hollowed, half-rounded middle *acroterion* of a sarcophagus lid had a flap-like shape at the top centre, holding a relief of a winged sea griffin. Judging from the drawing, the workmanship was course.
*Date*: 3rd century (?); mediaeval (?).
*Publication*: Brunšmid 1899: 205, fig. 104; Pinterović 1975: 130.

Figure 242. (AB[t] II.3) – sarcophagus, scan of the drawing (after Brunšmid 1899)

*III. Ash-chests*

1. Ash-chest (AB[t] III.1)
*Location*: Walled in the cemetery church of St. Clare in Novigrad Podravski, 73 km NE of Ščitarjevo.
*Stone*: Marble.
*Preservation*: The chest is completely preserved, apart from the fact that the outer face of its rear side is partly obscured by the church wall. The inner walls are roughly worked in parallel grooves and well preserved, while the front is obliterated by two layers of paint covering it and rendering the relief hardly discernible.

Figure 243. (AB[t] III.1) – ash-chest (V. Jukić)

*Description*: L 75 cm, H 46 cm, W 41 cm, TB 24.5 cm; type 3a. The rectangular flat-edged chest holds a simply framed central inscriptional panel and two lateral gabled niches each featuring a frontal standing figure on a fairly high pedestal. The one on the left seems to be a woman wearing an ankle-length dress and a cloak, with her right arm bent towards the chest and the left one on the abdomen. The figure in the right niche should be a man probably wearing a tunic and a *sagum*, holding an indiscernible object (most probably a scroll) in his left hand and pointing at it with his right.
*Date*: Later 2nd or 3rd century.
*Publication*: Unpublished.

*VI. Medallions*

1. Medallion (AB[t] VI.1)
*Provenance*: Ploughing in 1955 in the village of Biškupci, 35 km SE of Daruvar.
*Location*: GMP, inv. no. GMP-10.456.
*Stone*: Badenian sandy biocalcarenite. A probable origin: Mts Psunj or Papuk in the Slavonian Mountains (M. Belak 2018).
*Preservation*: Fragmentarily preserved. A part of the right side and the lower border are missing, the latter showing a breakage that possibly indicates an original socket for inserting into a pedestal. The piece is weathered, chipped and abraded all over.
*Description*: *W 38* cm, H 47 cm, T 10 cm. The medallion has a form of an irregular tondo with a smooth border; a motif of parallel S-lines taking the whole of the border band can be tentatively discerned, but this can also be an optical illusion coming from the worn out and ruffled surface of the stone. The surface inside the border is taken by half-figures of a woman on the left and a man on the right side. The woman has an oval block-like face, short and thick

Figure 244. (AB[t] VI.1) - medallion (GMP, B. Knez)

Figure 245. (AB[t] VI.1) - stone, microscope (HGI, M. Belak)

hair with a central parting, set high on the crown in a rough 'melon' style, almond-shaped eyes, and high-set ears. She is supposedly wearing a tunic, although apparently not depicted (the oval sagging folds on the breast should belong to an over-dress) and a *palla*, and a beaded necklace, while a thick-rimmed cup is sitting askew on the palm of her right hand. The man, dressed in a long-sleeved tunic and a *sagum* fastened on the right shoulder with a round brooch, has an oval elongated face, thick upswept hair with a blurred fringe, almond-shaped eyes, and high-set ears. He is holding two indistinct objects in his left hand: one oval and another elongated, tapering towards the bottom and ending in an oval finial. The workmanship is very primitive, with crudely stylised figures but still without striking anatomical disproportions.

*Date*: Early 3rd century.

*Publication*: Sokač-Štimac 2004: 26 (described as a sarcophagus fragment). Internet databases: *lupa* 26275.

*VII. Funerary buildings*

1. Funerary slab (AB[t] VII.1)

*Provenance*: Amateur excavations in 1842 in the village of Veliki Bastaji, 10 km NE of Daruvar. It was found in an underground funerary chamber with a frescoed walls and a mosaic floor, together with sarcophagus AB(t) II.1 (see *Note*).

*Location*: ZJD.

*Stone*: Badenian tufitic sandy microsparite limestone. A probable origin: Mts Psunj or Papuk in the Slavonian Mountains (M. Belak 2018).

*Preservation*: Fragmentarily preserved, reassembled from six pieces of various dimensions. A large triangular part has been chipped off from the right-hand top corner; a bottom part holding an additional two lines of the inscription that was recorded previously, is now walled and invisible. The preserved part is mostly undamaged.

Description: *H 100* cm, W 92 cm, T 2.5 cm. The slab has a form of a vertical rectangle with no moulding or framing.

*Inscription*: *Tartareis ruptus fornacibus implicat omnem, / perpetuum vitae quondam datum. Nunc rapit acris / aeternum poenis factum pro crimine flamma. / Hic simili natos meritis pro talibus igni /*[5] *ante obitum genuit perituros tabe parentum; / donec cura Deum miserandi cepit et atro / nunc tulit exitio simulacrum pendere semper / poenas indigne suum et premi nocte maligna. / Nam, Dominum puro velamine semper amictum /*[10] *atque inmortalem, caelique regionibus usum, / et culpa vacuum portantem insignia natum / cuncta patris pressis humano crimine membris/ induit. Et nulli orso sic fas promere verbum / hunc unquam pecasse Deum. Tamen omnia magnus /*[15] *alterius delicta tulit. Sic corpore sumpto, / demissus caelo terras petit, haut* (!) *secus artus / induit humanos, qua, ut tellus ferre valebit /, membrorum meditante Deum de viscere [---] / [---*

The inscription, containing currently visible 18 lines in hexameter verses, was skilfully inscribed in the best tradition of Late Roman rustic capitals (*scriptura actuaria*). Elongated letters are carved close to each other in regular lines, separated by equal distances. There are no punctuation marks between the words and no ligatures. Letters *A*, *E*, *F*, *H*, and *L* contain specific characteristics. The text is composed of regular hexameters, with two exceptions: *caelique* instead of *caeli* in the 10th line and *promere* instead of *promerere* in the 13th line.

Translation: A fire bursting from the furnaces of Tartarus embraces him entirely, him that once had been endowed with eternal life; but now, a severe blaze reaches him as a punishment for his crimes. Before dying, he had generated sons that would perish in a similar manner; until God took mercy on him and let his image expiate forever in the dark and shameful disaster, oppressed with a sinister night. For Lord appeared clad in clean garments, immortal and sinless, with his residence in Heaven and with all the inherent tokens

Figure 246. (AB[t] VII.1) – slab (O. Harl, *lupa* 26165)

of his father, yet oppressed with sins of humanity. It would not be right to say that somebody of such descent and God has ever sinned; yet in his greatness he atoned for others' sins. Sent from Heaven, he assumed a human body and descended to the Earth and with his human body the Earth was able to support ....

Commentary: Imbued with occasionally hardly intelligible allegories and metaphors, this hymn escapes an accurate and incontestable translation, but it evidently addressed the dogma of original sin and redemption.

*Date*: 4th century.

*Publication*: B. Migotti, in: Demo 1994: 122 no. 171; Migotti 1996: 131–135 (*AE* 1996, 1222); Migotti 1997: 47–49; Schejbal 2004: 120–121, fig. 30.

Internet databases: *lupa* 26165; *HD*050387; *EDCS*-03000750.

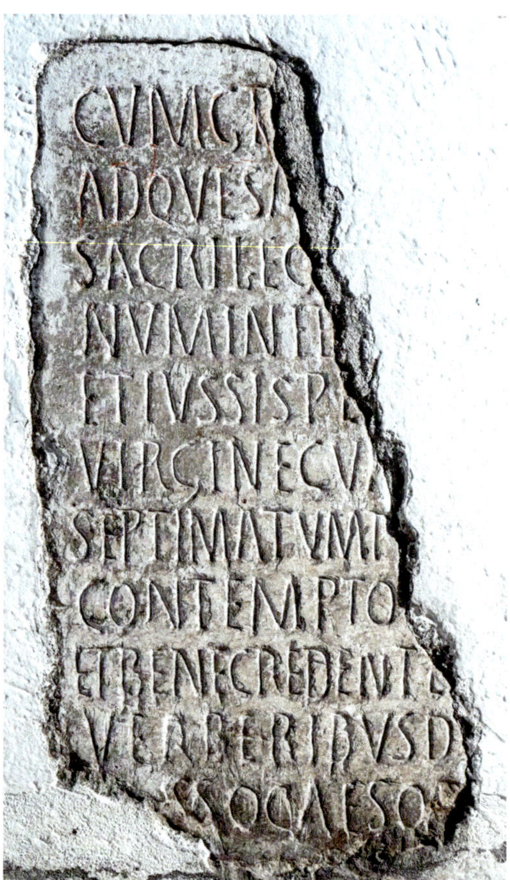

Figure 247. (AB[t] VII.1.1) – slab (O. Harl, *lupa* 26412)

Figure 248. (AB[t] VII.1.2) – slab (O. Harl, *lupa* 26413)

Figure 249. (AB[t] VII.1.3) – slab (O. Harl, *lupa* 26414)

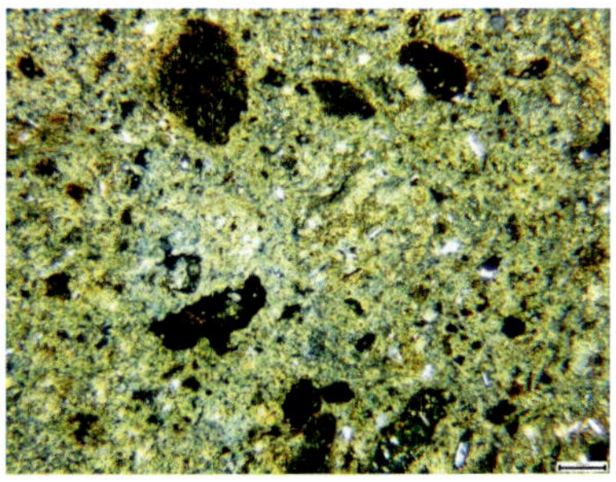

Figure 250. (AB[t] VII.1) – stone, microscope (HGI, M. Belak)

*Note*: The slab comes from the same context as sarcophagus AB(t) II.1. Three smaller marble fragments, made of Pohorje marble (W. Prochaska 2017) evidently containing similar verses, were found at the same time; two of them were mistakenly published in *CIL* III (206* a and b) as *falsae*. In amateur excavations in 1968 another fragment was found, broken into two larger and four smaller pieces. Of all of the mentioned fragments, three survive today: one stored and two walled in the Janković Collection in Daruvar (Schejbal 2004: 120–121, figs 28, 29, and 31). All of them were probably fixed onto inner walls of a burial chamber, which, according to the information from the time of discovery, consisted of several compartments. The walled pieces: 1. *H 61* cm, *L 30* cm (*lupa* 26413); 2. *H 21.5* cm, *L 41* cm (*lupa* 26414); the stored piece: 3. *W 41* cm, *H 22.5* cm, T 4.5 cm (*lupa* 26412).

## VIII. Unclassifiable

1. Inscription (AB[t] VIII.1)
*Provenance*: A chance find in the 19th century near the town of Našice, 69 km SE of Daruvar.
*Location*: Missing.
*Inscription*: *D(is) M(anibus) / [------?] / Claudia domo Inicer(o) / praef(ectus) /alae I Panno(niorum) / [------*
The inscription had been copied by Luka Ilić around the mid 19th century for his book (1852), which remained in manuscript; the transcription seems plausible and can presumably be trusted. However, it remained unknown to Mommsen when he collected material for *CIL* III. Much of the inscription must have either been broken off or illegible.
Translation: To the spirits of the departed. ... of the voting tribe Claudia, from Inicero, commander of the *ala I Pannoniorum*.
Commentary: It is difficult to know which *ala I Pannoniorum* was commanded by the unknown prefect. The unit could have been *ala I Pannoniorum Tampiana* or *ala I Pannoniorum Sabiniana*, but possibly also *ala I Pannoniorum* with no additional title, which served in Moesia and Numidia (Spaul 2000: 167–172; cf. Spaul 1995). Otherwise only *ala Pannoniorum* and *ala II Pannoniorum* have been attested to date in Pannonia. J. Spaul (2000: 170) suggested that the former may have been *prima* although there is no proof for that. *Ala Pannoniorum* had been in Dalmatia under Augustus and was transferred to Pannonia around AD 15; under Vespasian the unit was sent to Moesia Inferior and returned to Pannonia after the Marcomannic Wars (Ferjančić 2015). Tampiana was perhaps recruited in Pannonia as early as under Augustus or Tiberius; both Tampiana and Sabiniana were transferred to Britannia, where Sabiniana had its garrison at Onnum (Halton Chesters), while Tampiana (temporarily in Pannonia towards the end of the 1st century and the beginning of the 2nd) was stationed in Noricum (Lentia, modern Linz) after AD 125 (Lőrinz 2001: 22, with earlier literature; Radman-Livaja 2012: 168–169; Acrudoae 2012).
The prefect's voting tribe was *Claudia*; of autonomous towns in Upper and Lower Pannonia, only inhabitants of Savaria are known to have been inscribed in the *tribus Claudia* (Kubitschek 1889: 270), but *Claudia* must have been the voting tribe of several other Pannonians with Roman citizenship. Incero (Incerum) is a road station known from the *Itinerarium Antonini* along the road Siscia–Mursa (265.8), as well as along the road Siscia–Cibalae (260.5–6; see Mócsy 1968b). It is

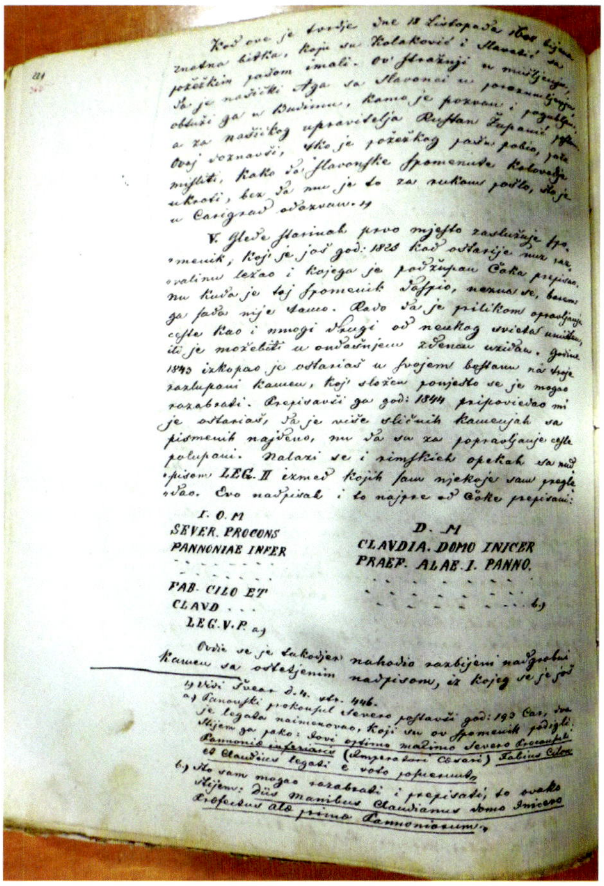
I. O. M
SEVER. PROCONS
PANNONIAE INFER
FAB. CILO ET
CLAVD . . .
LEG. V. P. a)

D. M
CLAVDIA. DOMO INICER
PRAEF. ALAE. I. PANNO.

Figure 251. (AB[t] VIII.1) – scan of a page from Ilić Oriovčanin 1852

interesting that one of the *variae lectiones* for the name of the settlement on the Siscia–Mursa road is Inicero, which seems to be confirmed by this epitaph and may have been more correct. Inicerum has been located at Trešstanovačka gradina near Tekić (Bulat 1977a: 82–83; thus also Barrington Atlas); other locations include Velika (cf. Schejbal 2004: 102–103 and Map 2) and Vetovo (Fodorean 2017: 344–345). The settlement must have been important, since it is described in the *Itinerarium Antonini* as a stopping place for emperors and high officials (*Incero sed mansio Augusti in pretorio est*).
*Date*: 2nd century (?).
*Publication*: Ilić Oriovčanin 1852: 363; Pinterović 1975: 129.
*Note*: The authenticity of this inscription remains slightly questionable.

2. Relief head (AB[t] VIII.2)
*Provenance*: Amateur excavations in 1898 at the site of Treštanovačka gradina in the village of Tekić, 47 km SE of Daruvar.
*Location*: AMZ, inv. AMZ KS-169a.
*Stone*: Biocalcarenite (by macroscopy, M. Belak 2017).
*Preservation*: Fragmentarily preserved. The head is cracked in the upper part, with the nose and smaller segments of the face restored in plaster.
*Description*: *H 22* cm, *T 11* cm (protruding from the background). The head features an elongated oval female face with almond-shaped, heavy-lidded and eyes angled downwards, a wedge-like straight nose and down-turned lips, with the hairdo resembling the so-called 'melon' type. It is parted at the centre, tightly framing the face and covering the ears, and is arranged in unusually wide and stiff strands running from the forehead towards the crown in a cap- or a wig-like form, with the rest of the hair straight. The workmanship is thoroughly rigid and unskilful.
*Date*: Early 3rd century.
*Publication*: Brunšmid 1898: 204–205, fig. 101; Pinterović 1975: 130. Internet databases: *lupa* 22785a.
*Note*: The head is common to any kind of a portrait funerary monument. The unusually cap-like styled hair can either be an imitation of Julia Domna's wig hairstyle, or the hair on the forehead can be an awkward rendering of forelocks often found in female hairdos of various types and in various periods, the Severan empresses included (Nodelman 1983: 113, figs 15 and 16).

3. Relief head (AB[t] VIII.3)
*Provenance*: Amateur excavations in 1898 at the site of Treštanovačka gradina in the village of Tekić, 47 km SE of Daruvar.
*Location*: AMZ, inv. no. AMZ KS169b.
*Stone*: Biocalcarenite (by macroscopy, M. Belak 2017).
*Preservation*: Fragmentarily preserved. The nose is mostly battered, and the whole head much weathered on the surface.
*Description*: *H 18* cm, *T 10* cm (protruding from the background). The head features an oval somewhat block-like female face, with heavy-lidded, almond-shaped eyes that are angled downwards, and straight, slightly down-turned lips, with the hairdo resembling the so-called 'melon' type. It is parted at the centre, tightly framing the face and covering the ears, and is arranged in unusually wide and stiff strands running from the forehead towards the crown in a cap- or a wig-like form, with the rest of the hair straight. The workmanship is thoroughly rigid and unskilful.
*Date*: Early 3rd century.
*Publication*: Brunšmid 1898: 204–205, fig. 102; Pinterović 1975: 130. Internet databases: *lupa* 22785b.
*Note*: As in the previous entry.

Figure 252. (AB[t] VIII.2) – relief head (AMZ, I. Krajcar)

Figure 253. (AB[t] VIII.3) – relief head (AMZ, I. Krajcar)

PERIPHERAL AREAS: A SELECTION OF MONUMENTS BELONGING TO NEIGHBOURING CITIES

*I. Metulum - south of Siscia*

**Josipdol** (territory of the Roman *municipium* Metulum, the province of Dalmatia)

1. Stele or sarcophagus
*Provenance*: A chance find in the 19th century at the site of Čakovac in the vicinity of Josipdol, 91 km SW of Sisak.
*Location*: AMZ, inv. no. AMZ KS-167.
*Stone*: Badenian calcareous sandstone. A possible origin: the mountains of NW Croatia; a probable origin: The Banovina and the southern Pokuplje, SW of Siscia (M. Belak 2018).
*Preservation*: Fragmentarily preserved. A relief niche survives, composed of several pieces cemented together. It is broken off and cracked around the edges, and heavily damaged on both figures' faces and the man's left hand.
*Description*: *H 100* cm, *L 76* cm, T 20 cm. The portrait niche is irregularly hollowed and simply framed, containing frontal standing figures of a married couple in the pose of *dextrarum iunctio.* The man on the right-hand side has an oval face, short cropped hair receded at the temples and forming a peaked fringe on the forehead, over-sized ears, almond-shaped and heavy-lidded eyes with accentuated irises, a triangular nose and pronounced nasolabial folds. He is wearing a tunic with a collar-like neckline and a toga with loosened *umbo*, and is holding a volume in his left hand. The woman has an oval face with almond-shaped and heavy-lidded eyes with accentuated irises (visible in the left eye); her hair is parted at the centre, framing the face and folding backwards below the ears. She is wearing a long-sleeved barrel-shaped dress with a collar-like neckline (or a necklace?) and a bottom hem, girdled above the waist with a ribbon belt decorated with geometrical motifs; there is a beaded bracelet on her right arm. A diagonal groove on the dress is depicted, running from the left thigh to the right lower leg. The same groove seems to be running in the opposite side to produce two crossed lines, but this is most probably a distorted perception resulting from a crack on the stone. Thus, the woman is most probably wearing a dress provided with an over-fold in the lower part. Her left hand rests on her husband's shoulder, while shaking hands with him with her right hand. Both are wearing closed shoes, with a sort of ankle wrapping for the man. The workmanship is generally crude and stiff, but with the sense of natural shapes in drapery, executed differentially depending on the texture (smooth surfaces versus grooved or incised folds).

Figure 254. (Metulum 1) - stele (AMZ, I. Krajcar)

Figure 255. (Metulum 1) - stone, microscope (HGI, M. Belak)

*Date*: Mid 3rd century.
*Publication*: Brunšmid 1905: 90, no. 167; Cambi 1975: 78–79, fig. 7. Internet databases: *lupa* 22690.
*Note*: The motif *dextrarum iunctio* is typical of Roman sarcophagi (though not in Dalmatia), but appears sporadically also on *stelai*. The piece in question has the back side smoothly worked, which does not tip the balance in favour of either a stele or a sarcophagus, but its dimensions point to a stele rather than sarcophagus.

2. *Aedicula*
*Provenance*: A chance find in the 19th century at the village of Čakovac in the vicinity of the settlement Josipdol, 91 km SW of Sisak.

Figure 256. (Metulum 2) – shaft (AMZ, I. Krajcar)

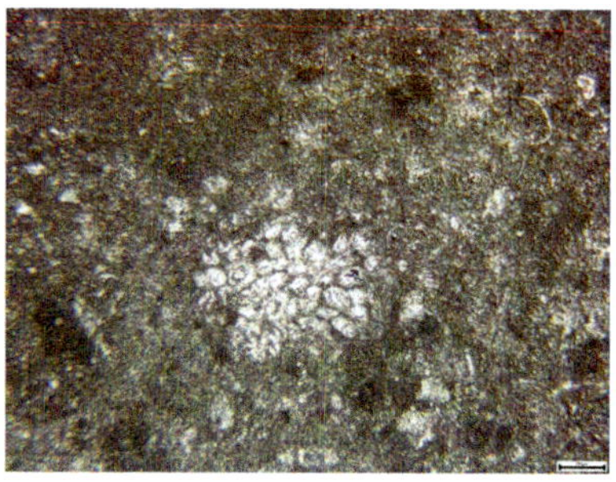

Figure 257. (Metulum 2) – stone, microscope (HGI, M. Belak)

*Location*: AMZ, inv. no. AMZ KS-166.
*Stone*: Lower-Cretaceous micritic limestone. A possible origin: the Outer Dinarides, that is the Kordun and the southern Pokuplje, S-SW of Sisak, as well as the Lika, further south (M. Belak 2018).
*Preservation*: A shaft completely preserved. It is weathered and damaged around the edges, as well as on the face and the right lower leg of the Eros figure. It probably remained unfinished, as is indicated by a lumpy mass of stone on the lowermost part of the relief panel.
*Description*: H 112 cm, L 55 cm, T 55 cm. The block is a shaft of an *aedicula* or a funerary enclosure wall, bearing a wide, shallow, vertical groove of uncertain function on its fairly well worked back side, and three dowel holes on the upper surface. The sides are blank and roughly worked, while the front is taken by a rectangular, moulded, lightly whitewashed relief niche with an irregularly carved back. It holds a mourning-type Eros with the wings barely carved, curly hair, crossed legs, the head bowed and supported on both hands, with the left one resting on the right shoulder, and the right one holding a down-turned torch pulled through a wreath. The workmanship is unskilled and crude, lacking any sense of anatomical proportions.
*Date*: Probably 3rd century.
*Publication*: Brunšmid 1905: 189–190, no. 166; Internet databases: *lupa* 22327.

*II. Poetovio (Pannonia) – NW of Andautonia*

***Petrijanec*** (presumed *mansio Aqua Viva*), 65 km NE of Ščitarjevo

1. Sarcophagus
*Provenance*: Construction works in 1960 at the site of a Roman road.

Figure 258. (Poetovio, Petrijanec 1) – sarcophagus (Lj. Perinić)

Figure 259. (Poetovio, Petrijanec 1) – stone, microscope (HGI, M. Belak)

Figure 261. (Poetovio, Petrijanec 2) – stone, microscope (HGI, M. Belak)

*Location*: GMV, inv. no. GMV-66343.
*Stone*: Badenian sandy biocalcarenite. A probable origin: Mts Kalnik or Ivanščica (M. Belak 2018).
*Preservation*: The chest completely preserved. It is stained at places and slightly chipped along the upper and lower edges.
*Description*: L 212 cm, H 81 cm, W 125 cm; type 7. The rectangular plain chest has two roundish clamp holes on short sides and narrow cuts along the edges for taking a lid. It is roughly worked on the outside and even worse on the inside and the bottom.
*Date*: 2nd – 4th century.
*Publication*: Šarić 1978: 178 (just the site of the findspot of the sarcophagus mentioned); Zorković 1995: 21 (with a photo, but no measurements).

2. Sarcophagus
*Provenance*: Construction works in 1940 at the site of a Roman road.
*Location*: GMV, inv. no. GMV-66344.
*Stone*: Badenian sandy biocalcarenite. A probable origin: Mts Kalnik or Ivanščica (M. Belak, 2018).
*Preservation*: The chest and lid completely preserved. The chest is reassembled from two pieces, with considerable parts of the walls battered off, but still holding together.
*Description*: Chest: L 200 cm, H 40 cm, W 79.5; chest type: 7; lid: L 205 cm, H 22.5 cm, W 92.5 cm; lid type 1b. The low, elongated, plain rectangular chest with flat edges holds a low-gabled lid with a flattened ridge, provided with small corner *acroteria.* Both the chest and lid are roughly worked on the outside with a pointed chisel.
*Date*: 2nd – 4th century.
*Publication*: Šarić 1978: 178 (just the site of the find-spot of the sarcophagus mentioned); Zorković 1995: 19–20 (just mentioned).

***Sračinec***, 64 km NE of Ščitarjevo

1. Sarcophagus or ash-chest
*Provenance*: A chance find in a gravel pit at the site of Turnišče during field-walking in 2010.
*Location*: GMV, inv. no. GMV-66346.
*Stone*: Badenian sandy biocalcarenite. A probable origin: Mts Kalnik or Ivanščica (M. Belak 2018).
*Preservation*: The chest preserved fragmentarily, and the lid completely. Only the lower parts of one long and one short side of the chest survive, as well a lump of the upper corner, while the lid is reassembled from seven pieces.
*Description*: Chest: L 98 cm, *H 46* cm, W 58; chest type 7; lid: L 117 cm, H 25 cm, W 69 cm; lid type: variant of 1b. The plain rectangular chest has flat edges and holds a lid of irregular oblong form, only resembling a gabled type. Both the chest and lid are extremely roughly worked, with pick-tool marks all over (of a carving pick on the outside and a pointed chisel on the bottom).
*Date*: 2nd – 4th century.
*Publication*: Unpublished.

Figure 260. (Poetovio, Petrijanec 2) – sarcophagus (Lj. Perinić)

Figure 262. (Poetovio, Sračinec) - chest (Lj. Perinić)

Figure 263. (Poetovio, Sračinec) - lid (Lj. Perinić)

Figure 264. (Poetovio, Sračinec) - stone, microscope (HGI, M. Belak)

***Ivanec***, 50 km N of Ščitarjevo

1. Shaft-block of a funerary building (?)
*Provenance*: Archaeological excavations in 1998 at the Stari grad site; found as reused in the walls of a Gothic church.
*Location*: The Town Hall, Ivanec.
*Stone*: Miocene (Badenian) biogenic limestone. A probable origin: the northern slopes of Mt Ivanščica (D. Kurtanjek 2011).
*Preservation*: Fragmentarily preserved. The right-hand side and the lower part are missing, and the stone is chipped along all the edges and damaged throughout, including the human figures, especially the one on the short side.
*Description*: *H 49* cm, *L 42* cm, *T 28* cm. The preserved part of the shaft is a parallelepiped whose two sides hold human relief figures. On the front a standing man is depicted in a spherically carved vaulted niche, preserved approximately at thigh level. He has a block-like face resting on a sturdy wrinkled neck, a high rounded crown and a hairdo close to the scalp and combed forward in long, straight, parallel locks to form a slightly rounded fringe receded at the temples; the facial hair is not discernible. He is dressed in a tunic and a *sagum* fastened on the right shoulder with a round brooch, and is most probably holding a volume in his (chipped off) left hand and pointing at it with his right. The side niche seems to be fashioned in the same way as the front one, with a standing human figure whose sex is obliterated by heavy damage, but with a line of a chignon on the crown suggesting a female. She seems to be wearing a conical long- and wide-sleeved dress creased on the left shoulder, with her head turned to the left and her right arm bent at the elbow and raised. With the front figure probably representing the

Figure 265. (Poetovio, Ivanec) shaft-block – front side (AMZ I. Krajcar)

Figure 267. (Poetovio, Ivanec) shaft-block – right short side (AMZ I. Krajcar)

Figure 266. (Poetovio, Ivanec) shaft-block – front and left short side (AMZ I. Krajcar)

owner of the monument, the figure on the side most probably depicts a female servant. The workmanship is hard to assess due to the damage, but the front figure is fairly competently carved, with the good sense of proportion, movement and drapery, contrary to the architecture of the niche, which is either very roughly worked or perhaps not finished.
*Date*: 3rd century.
*Publication*: Belaj and Migotti 2011.
*Note*: It is not possible to assess the kind of monument with complete certainty. Therefore its original shape and description remain debatable, but it was presumably much higher, with the figures carved in full height. Some traces of working on the surface of the short side opposite to the figured one suggest the possibility that this block was fixed onto a surface of another one, as component of a funerary building.

***Varaždinske Toplice***, 54 km NE of Ščitarjevo

1. Portrait niche of a funerary building (?)
*Provenance*: Archaeological excavations in 2011 and 2012 at the site of the Roman baths; found as reused in the enclosure wall of the thermal pool.
*Location*: Zavičajni muzej Varaždinske Toplice / The Museum of Varaždinske Toplice, inv. no. VTKS-30.
*Stone*: Alpine marble (by macroscopy).

Figure 268. (Poetovio, Varaždinske Toplice) – portrait niche of a funerary building (D. Kušan Špalj)

*Preservation*: Completely preserved. The niche has been reassembled from several pieces, with a large chip out of the mid bottom. Both the reliefs and inscription are well preserved.

*Description*: H 54 cm, W 40.5 cm, T 6–7 cm. The stone slab contains a flat inscription panel and above it a simply framed portrait niche, which holds half-length figures of a married couple, the woman on the left-hand side and the man on the right. Both are wearing a stiff and stern expression on their block-like faces, endowed with large almond-shaped eyes, tightly closed down-turned lips, prominent ears and a straight nose with accentuated nostrils. The woman has a wavy hairdo parted in the centre and falling half way down her neck, and turning at the back to form a chignon high on the crown and towards the forehead. She is dressed in a tunic with an oval neckline and a long- and wide-sleeved cloak, apparently decorated with a narrow embroidered band running at some distance from the sleeve's edge, and is wearing earrings with rectangular bezels, a beaded necklace and a matching bracelet on her right hand. Her left hand is resting on her husband's right shoulder, while her right is bent at the elbow, with the middle finger and the forefinger sticking out. The man is wearing a wide- and long-sleeved tunic and a toga whose triangular *umbo* covers his left shoulder. His short cropped hair is combed in long, straight locks from the back towards the front and the temples, producing a slightly M-shaped fringe. He is holding a scroll in his left hand, pointing at it with his middle finger and forefinger. The workmanship is fairly good, with an apparently purposeful stiffness of execution, but with the sense of natural shapes, which is more pronounced in drapery and faces, and is lacking in the execution of the hands.

*Inscription*: *Aur(elius) Aquilinus / dec(urio) DVM(?) et Aurelia / Maximilla.*

*Ligatures*: line 2: *ET*; line 3: *MA*. Nothing is missing of the inscription.

Translation: Aurelius Aquilinus, decurion *dum(?)* and Aurelia Maximilla.

Commentary: It seems that the couple had the tombstone erected in their lifetime, since their age at death has not been noted. The cognomen Aquilinus is frequent in Italy (particularly in northern Italy), Dalmatia, and Noricum (Alföldy 1969: 155; Lochner von Hüttenbach 1989: 18; Kakoschke 2012: 261–262), elsewhere not particularly (*OPEL* I 70). Maximilla is well attested mainly in Italy, Gallia Narbonensis, Dalmatia, and Pannonia (*OPEL* III 69; Barkóczi 1964: 218; Alföldy 1969: 242). Aquilinus was most probably a member of the municipal council of Poetovio, since Aquae Iasae seem to have belonged to the *ager* of Poetovio. This may be corroborated by the altar dedicated to Isis and Serapis by C. Valerius Priscus, *Augustalis* of *colonia Ulpia Traiana Poetovionensis* (Kušan Špalj 2014: 193; *AE* 2014: 1048). The interpretation of the abbreviation *DVM* is uncertain. Possibly *du(u)m(vir)*, particularly because one Metilius Aelianus Dumviranus is known from Šentjanž (Rečica ob Savinji: *CIL* III 5111; Wedenig 1997: 109–111 (C2); Visočnik 2017: 278; *HD*059834), his second cognomen denoting a person born in the year when his father was a *duumvir*.

*Date*: Second half of 3rd century.

*Publication*: Kušan Špalj 2014: 80–81. Internet databases: *HD*075030; *EDCS*-68200028.

*Note*: According to Dora Kušan Špalj, a long-term researcher of Aquae Iasae, this stone cannot be interpreted as funerary because it was found in the fill with many other *spolia*, none of which was funerary. Since this hypothesis cannot be rejected, the true nature of this monument remains unknown, representing an isolated case in Noriko-Pannonian (funerary) art.

Figure 269. (Poetovio, Križovljan) portrait niche of a funerary building (T. Turković and N. Maraković)

***Križovljan***, 65 km NE of Ščitarjevo

1. Portrait niche of a funerary building
*Provenance*: Found at the site of the church of the Holy Cross at an unspecified time.
*Location*: Walled in the façade of the church of the Holy Cross.
*Stone*: Limestone (after Goss 2007).
*Preservation*: Completely preserved. The stone is heavily weathered, with the portraits much obliterated.
*Description*: Approximately 100 cm x 40 cm. There is no knowing whether the stone had an inscription panel immediately below. The portrait niche is trapezoidal with a narrow smooth band for a frame, and with a slight indentation in the middle of both lateral frames. It holds frontal relief heads of seven people: three adults and four children arranged alternately in a straight line. All the heads are pear-shaped, with accentuated cheekbones, huge oval eyes, high-set fairly small rounded ears, triangular noses, tightly closed full lips, and necks strong and trapezoid-shaped in the adults, while nearly completely lacking in the children. A trait of the hairdo shared by all portraits is a concave fringe, while there are two variants of the hair on the crown: one low and tight on the scalp (the leftmost figure and all the children), and the other in two superimposed rows (the other two grown-up persons). Due to the primitive execution and stone wear, the sex of the portrayed adults remains obscure. While the one on the left-hand side seems to be a man, and his counterpart on the right-hand side a woman, the one in the middle is highly uncertain because of a blend of a hairstyle featuring superimposed layers and a beard. The workmanship is extremely crude, but grotesque shapes of human faces leave the impression of a 'style' rather than artistic incapability.
*Date*: Later 3rd or 4th century.
*Publication*: Goss 2007; Migotti 2012a.

*III. Unknown municipality south of* Aquae Balissae

***Mahovljani*** (Bosnia and Herzegovina), 74 km SE of Daruvar

1. Lion top of a stele
*Provenance*: A chance find at the site of Maleševića njive.
*Location*: A private collection in Mahovljani.
*Stone*: Fine-grained marble. A possible origin: anywhere from the central eastern Alps to central-northern Croatia (the Banovina) to northern Bosnia and Herzegovina (Mts Prosara and Vučjak) (M. Belak 2018).
*Preservation*: Completely preserved. The piece has been reassembled from two fragments. It is chipped around the lower edges, and pitied and cracked all over.
*Description*: L 90 cm, H 60 cm, W 30 cm. The top is composed of two back-to-back lions with frontal heads lying on a thin flat base at the sides of an oval-biconical object, most probably a pine cone with the roughly worked surface and no scales depicted. The workmanship is coarse and primitive.
*Date*: 2nd or 3rd century.
*Publication*: Unpublished.

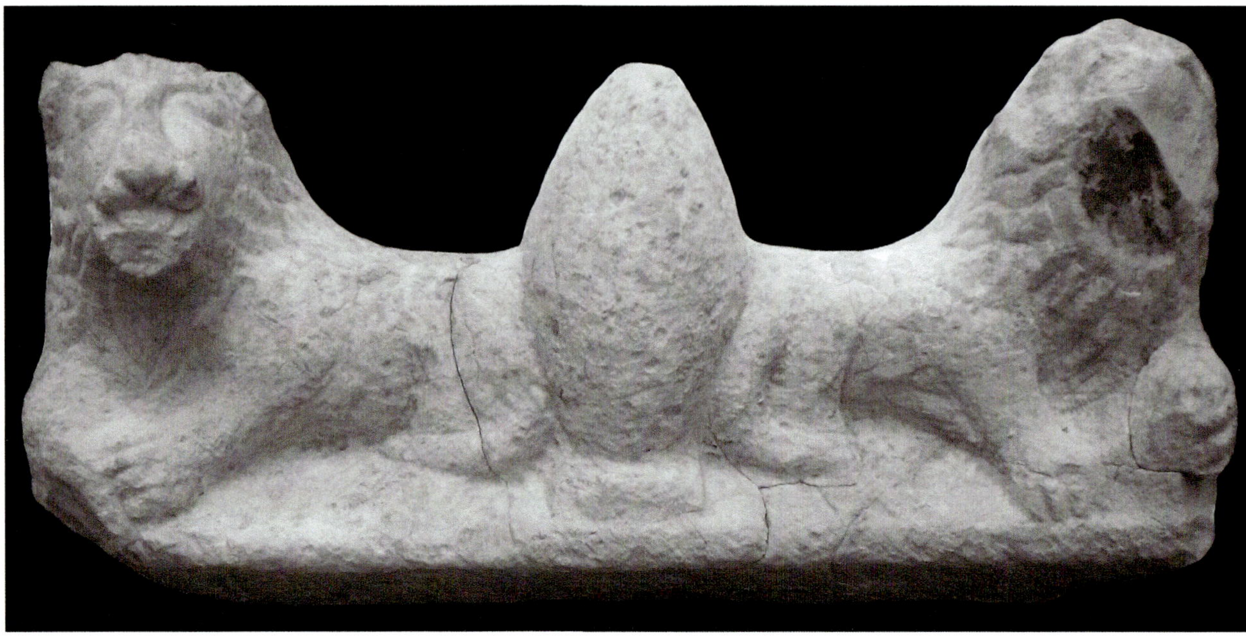

Figure 270. (Poetovio, Mahovljani) lion top
(Turistička organizacija Laktaši /the Tourist Bureau, Laktaši)

Figure 271. (Poetovio, Mahovljani)
- stone, microscope (HGI, M. Belak)

# Chapter II
# Discussion of the evidence

Branka Migotti

## II.1. TERRITORIES OF ANDAUTONIA, AQUAE BALISSAE AND SISCIA

### *Introduction*

Quite a few starting points and tools in tentative fixing of the territories of Roman autonomous towns have been employed in numberless discussions of this issue, such as boundary stones, milestones, literary sources, epigraphic evidence of city officials and soldiers, onomastics, small finds, economic potential, geographical features, the logic of 'spatial geometry', medieval administrative borders, and so on.[1] Nevertheless, the *ager* of the majority of Roman provincial cities remains notoriously conjectural, with many hypothetical boundaries attempted in the literature. The sorest point of the controversy is two sharply contrasted theories, of which one states that all of the land was assigned to the *agri* of respective autonomous towns, while the other claims that a lot of land remained outside such division, resulting in much smaller town territories than customarily presumed.[2] According to the latter theory, the remainder of the land could have pertained to various owners, such as the Emperor or other upper-class private owners, the Army, *praefecturae*, *vici*, *conciliabula*, priestly colleges, mines, and so on.[3] Recently, Martin Lehner criticized this theory, specifically as it had been argued by Jaroslav Šašel.[4] Notwithstanding this criticism, it seems that a very important point for the understanding of the concept of the municipal *ager* indeed lies in the duality between the assigned (cadastrated) land in terms of the laws of land division (*finitima linea)* and the surrounding territories, often marked by rivers or mountains (*arcifinius*), as recorded in written sources. Even if the term *arcifinius* is not taken in the above meaning (as related by Varro in the 1st century BC), the fact still remains that the Roman mentality clearly distinguished between the administered and unadministered land. In this context should be observed Pliny the Elder's *civitates adtributae* (Plin. *NH* 3.138), which represent tribes or cantons situated within or *near* the territories of their closest municipal towns.[5] If now we seek to apply the concept of the difference between the legal and natural boundaries of the municipal *agri* to the present context, a couple of implications emerge. Firstly, the idea here was not to fix the town territories precisely or even more precisely than it had been done before, but to study the monuments of the probable or hypothetical individual town territories in order to check overall relations and potential differences between them.[6] Therefore, the main purpose of dealing with the hypothetical *agri* was practical, that is, focused on the organization of the material, with only a secondary aim to see whether so organized evidence revealed indications of separate territorial identities. For all these reasons, the discussion of the two opposed schools of thought on the town boundaries will not be tackled here in more detail. Presuming that in reality the cadastrated town territories were not as large as claimed by some commentators, the wider areas suggested here as the cities' *possible agri* should actually be perceived as their *arcifinii*, that is, not belonging to the respective towns administratively but standing under and benefiting from their material, social and cultural influences. In the present context this primarily concerns the production of stone monuments in terms of their importation from the town workshops or the production under their influence. So far, the *agri* of all three researched were tentatively defined on the basis of the general assumption for Pannonia, that namely the town territories covered the area measuring roughly 50 or 60 km in diameter, and were accordingly delineated as more or less regular circles.[7] The terms *territory* or *ager* will be used throughout this discussion for practical reasons, implying, however, not only the core areas of the cities as fixed in the previous literature, but also their possible conditional/ extended meanings as explained above. This was done by taking into account the nearest municipalities in the vicinity of all three cities, without exactly counting on previously conceived regular geometric circles of their territories (Figure 1).

[1] Póczi 1980: 239; Laveau 1993; Hudeczek 2008: 9; Lehner 2012; Weber 2012; Ragolič 2014.

[2] That said, it still needs to be kept in mind that the differences in the size of cities across the Empire were enormous. See Laveau 1993: 467; Weber 2012: 223.

[3] Bojanovski 1988: 281; Šašel 1989: 61–73; Weber 2012.

[4] Šašel 1989; Lehner 2012: 45–46.

[5] Cf. Hardy 1912: 130; Whittaker 1994: 19–20; Starac 1999: 30, 117; Campbell 2000: 3–4, 142, *passim*; Campbell 2012: 188; Weber 2012: 222–223.

[6] In seeking to fix the border between Noricum and Upper Pannonia, M. Lehner (2012: 46) realized that practically no elements of the material culture proved specific enough for dependable differentiating between these two provinces.

[7] Fitz 1980b: 149 and Póczi 1980: 239 (taken generally); *AIJ*: 219 and Lolić 2003: 132, fig. 1 (referring to Siscia); Schejbal 2004: 102, Map II (referring to Aquae Balissae).

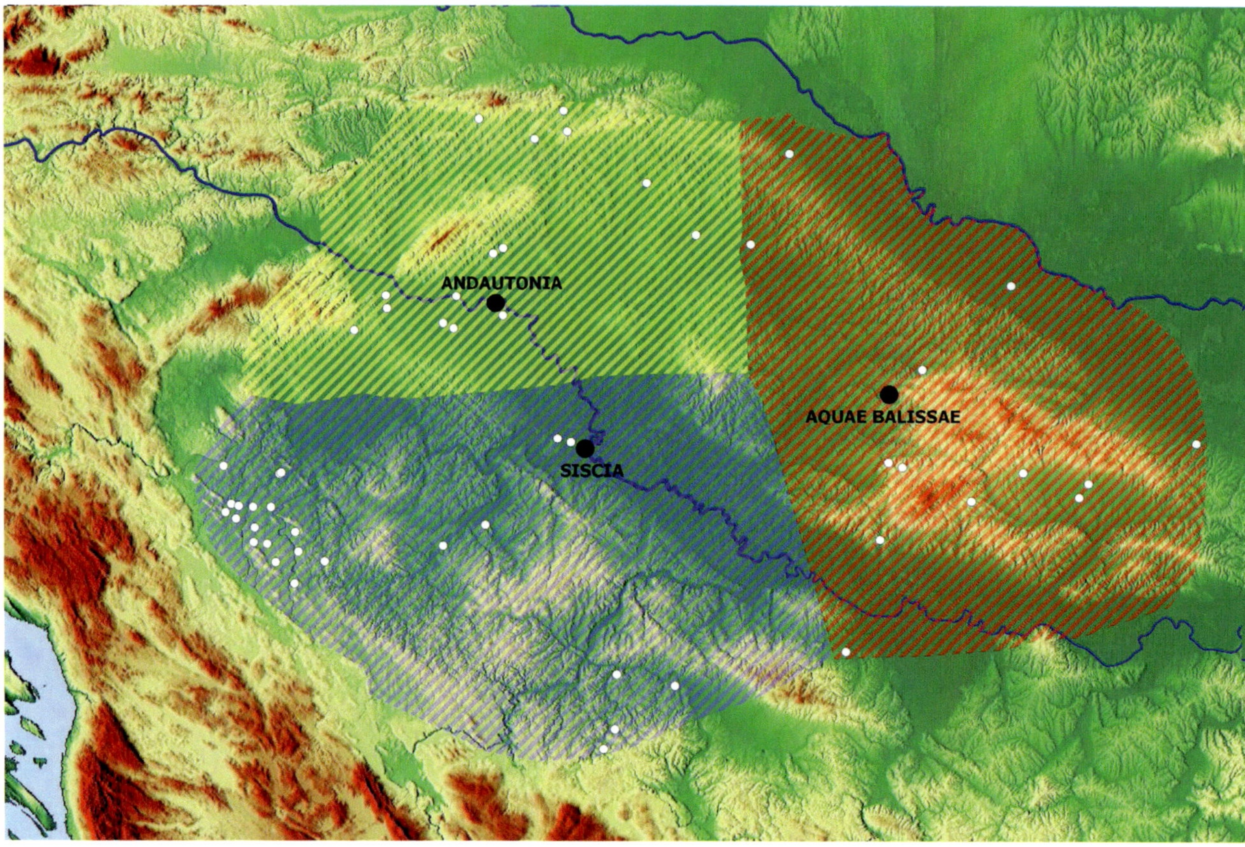


Figure 1. Map of the territories of Siscia, Andautonia and Aquae Balissae
(drawn by: D. Fudurić and T. Leleković)

*Andautonia, Iovia and Aquae Balissae*

The *ager* of Andautonia was presumably fixed as comprising the valleys of the Sava River and its tributaries, and stretching up to the nearest mountains on the western, southern and northern sides, while bordering on the Siscian *ager* some 15 km to the south-east of Ščitarjevo.[8] Given that the nearest autonomous town on the northwest was Poetovio, the territory so fixed of Andautonia presumes that the *ager* of Poetovio reached the northern slopes of Mt Medvednica above the city of Zagreb. Contrary to that, in the Andautonia Catalogue the presumed territory of city stretches to the ridge of Mt Ivanščica, thus comprising in the north-west the sites of Lobor, Mađarevo and Paka, at a distance of 32 km, 47 km and 43 km, respectively, from Ščitarjevo (Figure 2). This was done with the suggestion from Roman land surveyors (*agrimensores, gromatici*) in mind, that the natural borders of assigned and non-administered Roman territories tended to follow natural features, among others mountain peaks.[9] Recently, the *ager* of Poetovio was reconsidered by using the so-called Thiessen polygons as the ultimate tool. On the basis of this methodology alone, the well-known thermal bath of Aquae Iasae (Varaždinske Toplice), 49 km to the southeast of Ptuj, remained outside the territory of Poetovio. On account of that, as well as the distance of only 16 km between Ludbreg (*civitas* Iovia) and Varaždinske Toplice, the inclusion of Aquae Iasae into the *ager* of Iovia's would have been expected. However, this was rejected on the basis of historical-administrative reasons and the importance of Poetovio in the wider surroundings, as arguments for the inclusion of Aquae Iasae in the *ager* of Poetovio.[10] On balance, this issue should still be considered as unresolved, especially in view of the fact that Aquae Iasae should have been relocated to the *ager* of Iovia, at least in Late Antiquity, when Poetovio became a Norican city.[11] The monumental evidence of Aquae Iasae can give a certain insight into this controversy, but cannot help considerably in its clearing. Only one piece was included in the Catalogue (Peripheral areas II. Poetovio, Varaždinske Toplice), but its funerary character is questionable and its iconographic traits insufficiently diagnostic. However, according to a persuasive suggestion of Dora Kušan Špalj, the marble

[8] Nemeth-Ehrlich and Kušan Špalj 2003: 110, maps 1 and 2.
[9] Whittaker 1994: 20; Hudeczek 2008: 9; Weber 2012: 222–223. On the border between Poetovio and Andautonia, see fns 41 and 42 in Ch. II.2.

[10] Ragolič 2014: 336.
[11] Migotti 2012c: 4. This argument should be given due consideration despite its inconclusiveness, because there were cities whose territories (partly) belonged to neighbouring provinces (Leveau 1993: 468).

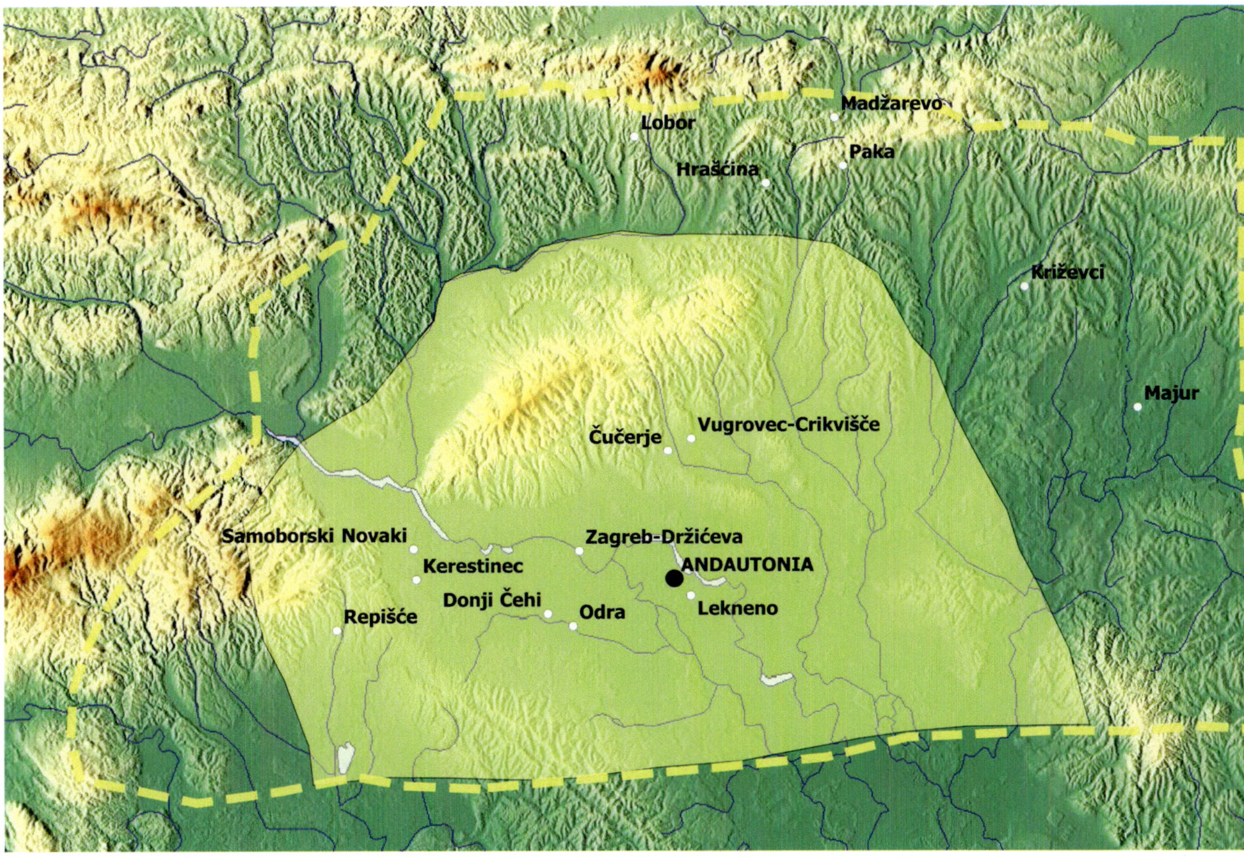


Figure 2. Map of the territory of Andautonia
(drawn by: D. Fudurić and T. Leleković)

relief slabs and blocks holding mythological scenes and heroes, whose original precise findspot in Varaždinske Toplice is unknown, and which were earlier interpreted as parts of a decorative fence around the central thermal fountain, actually belong to a funerary *aediculae*, such as those from Šempeter near Celeia in Noricum.[12] It should be observed that in all the above-mentioned blocks featuring heroes from Aquae Iasae the figures are set within simple rectangular frames, lacking a pedimental structure or Norico-Pannonian volutes so typical of the Celeian production.[13] This could be taken as an indication, although thoroughly inconclusive, towards the prevalence of a Poetovian artistic influence in Aquae Iasae, because non-elaborated frames in funerary buildings seem to have been rare in Celeia and its territory (notably Šempeter), as were those elaborate ones in Poetovio.[14] On the other hand, even if potential artistic influences from Celeia in Aquae Iasae are considered, this cannot be used in the same manner as when discussing the site of Lobor and its territorial affiliation (see at Ch. II.2). This is because Aquae Iasae was more conveniently connected with Poetovio via the Drava and (possibly) Plitvica Rivers. On balance, the monumental evidence cannot help considerably in resolving the issue of the territorial-administrative affiliation of Aquae Iasae, but they do point to its connection with Poetovio. The remainder of the evidence from the presumed Poetovian territory (Catalogue, Peripheral areas II) comprises monuments variegated in kinds and execution: two sarcophagi (Petrijanec), one sarcophagus or ash-chest (Sračinec), a single each shaft block (Ivanec) and portrait niche of a funerary building (Križovljan). All of them but one (Ivanec) are of more or less poor workmanship, and therefore of evidently local scope; all were carved in local stone. The sites of Ivanec and Križovljan only furnished remains of pottery and faint traces of architecture, insufficient for a determination of the kind of Roman settlements there.[15] Contrarily, the village of Petrijanec is known as the location of *mansio* (a post station) Aqua Viva, on the road Poetovio–Siscia, with substantial architecture and other finds recovered from archaeological excavations; more sarcophagi have been reported from this place, one of them marble.[16] Nevertheless, only one monument (Ivanec) of those listed above was recovered in archaeological

[12] Kušan Špalj 2014: 90.

[13] Significantly, all comparative motifs employed by Gorenc in displaying a common artistic background of Norico-Pannonian sculpture, bear the Norico-Pannonian volute (Gorenc 1971: Pls XIII: 1, XIV: 1, XV: 1, XXIII: 2, XXIV: 1, XXV: 1).

[14] Cf. *lupa* 5314, 13270.

[15] On Ivanec, see Belaj and Migotti 2011: 163–167, and on Križovljan, see Migotti 2012a: 387.

[16] Šarić 1978; Zorković 1995: 16–17.

excavations. Overall, the evidence from the peripheral area presumably belonging to Poetovio does not specifically point to any particular production centre, but fits well with the cultural circle comprising south-eastern Noricum and south-western Pannonia.

No matter whether the *ager* of Poetovio started only at the northern slopes of Ivanščica and the Kalničko gorje (the Kalnik mountains), or whether it reached as south as Mt Medvednica, it remains to be seen how the land east of the town of Varaždin (and possibly also Varaždinske Toplice if it belonged to Poetovio) was divided in terms of the administrative affiliation between the town territories of Andautonia and Aquae Balissae, and what the role of Iovia in this division could have been. This is necessary because of several sites whose affiliation remains undecided on account of too large a distance from the two nearest cities: Andautonia (Ščitarjevo) and Aquae Balissae (Daruvar). These are: Novigrad Podravski (73 km from Ščitarjevo, 58 km from Daruvar), Križevci (43 km from Ščitarjevo, 71 km from Daruvar), Koprivnica (71 km from Ščitarjevo, 70 km from Daruvar), Majur (47 km from Ščitarjevo, 55 km from Daruvar) and Bjelovar (58 km from Ščitarjevo, 45 km from Daruvar). For want of more concrete indications, such as land or water communications, all of these sites were tentatively ascribed to Andautonia or Aquae Balissae exclusively on account of the distance between them: Novigrad Podravski and Bjelovar to Aquae Balissae, and Križevci, Majur and Koprivnica to Andautonia.[17] With such division, arbitrary as it is, the border between the *agri* of Andautonia and Aquae Balissae supposedly ran along a north-south line that stretched from somewhat west of Novigrad Podravski straight down to Moslavačka gora (Mt Moslavačka); from that point towards the south ran a hypothesized border between the *agri* of Andautonia and Siscia (Figure 3).[18] On the other hand, bearing in mind that *civitas* Iovia must have had some territory assigned, all of the above sites could have belonged to it, as they are all situated nearer to Ludbreg than to either Ščitarjevo or Daruvar. Indeed, the insertion of the *ager* of Iovia in the territorial reconstruction would conveniently overcome a gap created by the too large distances of the above-mentioned five sites from both Ščitarjevo and Daruvar, and too large territories of Andautonia and Aquae Balissae in comparison with the much larger and definitely more important city of Siscia. Unfortunately, we cannot be sure when Iovia acquired the status of a *civitas*. In Late Antiquity the definition of the *civitas* corresponded to the earlier concepts of autonomous status (*municipia* and *coloniae*), but this type of municipality, different from native tribal *civitates*, was introduced at the beginning of the 3rd century, with some cities figuring as *civitates* already in the early Imperial period.[19] There is no knowing when Iovia acquired this status, as it was first mentioned as such in the Jerusalem Itinerary (*Itinerarium Hierosolymitanum/Burdigalense*) from AD 333–334. In the Antonine Itinerary (*Itinerarium Antonini*), dating from the 3rd century, it figures without the apposition *civitas*, suggesting its acquisition not earlier than the later 3rd century.[20] Still, we cannot be completely sure that this did not happen earlier, for two reasons. Firstly, there is always the possibility of a mistake in literary sources, and more than one has been observed in the Antonine itinerary.[21] This means that an accidental omission of the apposition *civitas* with Iovia cannot be excluded, the more so as the *Itinerarium Antonini* was presumably composed in the early 3rd century and updated in the later 3rd or the early 4th century.[22] Finally, as László Borhy persuasively posited, the core of the late Roman concept of the *civitas* was a walled settlement, and Iovia was a small town presumably walled in the early Roman period.[23] Notwithstanding all this, there is no certainty about the existence of Iovia as a *civitas* with an associated territory before Late Antiquity. Therefore, this city was not included in the catalogue databases, and the sites potentially situated in its territory were hypothetically assigned to the *agri* of Andautonia or Aquae Balissae, respectively. In addition to the previously adduced reasons, it was decided not to operate with the *ager* of Iovia in the present context because of the lack of diagnostic monuments from the town of Ludbreg itself. Stone sarcophagi have been mentioned vaguely as stemming from one of the town's cemeteries, but they were never illustrated and could not be traced in the museums or anywhere on the ground.[24]

The southern border of the Aquae Balissae *ager* should have run along the border between the provinces of Pannonia and Dalmatia, given that no municipalities existed south of Aquae Balissae in the given stretch of southern Pannonia, or northern Dalmatia for that matter; the nearest such place in Dalmatia was *municipium Baloie* (modern-day Šipovo, BA), 147 km south-west of Daruvar.[25] The provincial border here was fixed by various authors at a larger or smaller distance from the Sava River. According to the former, the settlement Castra (modern-day Banja Luka, BA) was on the border, but still in the province of Pannonia, which was presumed on account of an inscription of a *beneficiarius consularis Pannoniae Superioris* found in

[17] Koprivnica is situated at an equal distance from Daruvar (70.39 km) and Ščitarjevo (70.59 km), but was tentatively ascribed to Andautonia because it produced a marble lion top (A[ptNE] I.1) of the same type as the one from Križevci (A[t] I.9), also presumably in the territory of Andautonia; it goes without saying that this argument is arbitrary.

[18] Lolić 2003: 132, fig. 1.

[19] Leveau 1993: 463; Borhy 2003: 4166–418.

[20] Bojanovski 1988: 325–326; Migotti 2012c: 4 (with further reading in fn. 24).

[21] Reed 1978: 251; Nagy 2004: 77.

[22] Reed 1978: 229; Nagy 2004: 77.

[23] Borhy 2003: 416–417; Gregl and Migotti 2004: 136–137.

[24] Cf. Migotti 1997: 25.

[25] Bojanovski 1988: 287–292; Grbić 2014: 128.

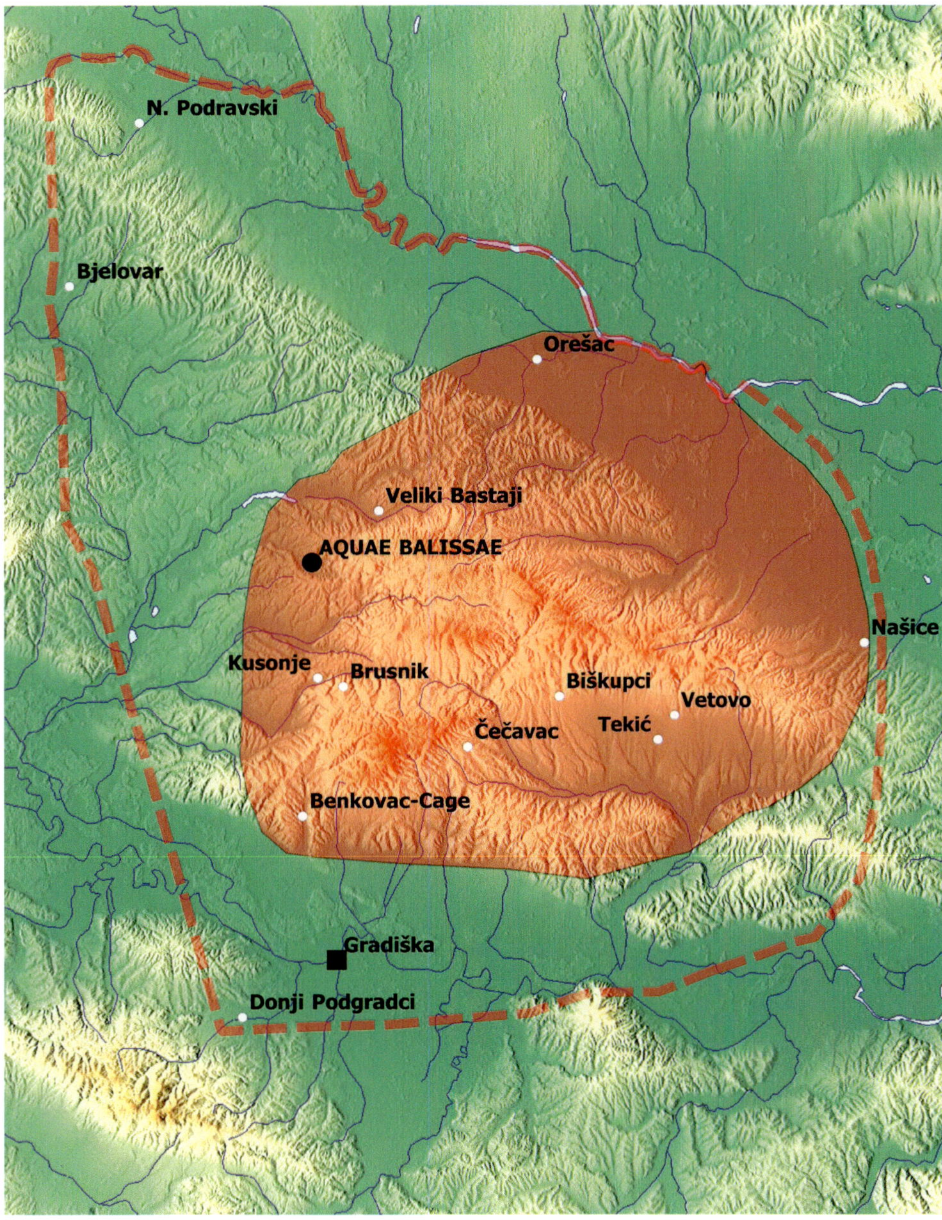


Figure 3. Map of the territory of Aquae Balissae (drawn by: D. Fudurić and T. Leleković)

Banja Luka.[26] Given the official nature of the inscription that was dedicated to Jupiter and *genius loci*, at first sight such a suggestion seems to be well founded. However, Castra was recently located in Dalmatia, on account of three inscriptions of *beneficiarii consularis* of Pannonia Superior found in places that were undoubtedly situated in the province of Dalmatia.[27] Therefore, it is more likely that the border was near the Roman settlement of Ad Fines, which is located at the sites of Laktaši or Mahovljani (BA) respectively; the former is 17 km and the latter 20 km to the north-east of Banja Luka.[28] This is still within the option that fixes the border between Pannonia and Dalmatia farther inland from the Sava River towards the south. Significantly, the only monument stemming from this area, a lion top from Mahovljani (Catalogue, Peripheral areas III.1.), suggests a Pannonian origin.[29] As for Castra, irrelevant of its affiliation with Pannonia or Dalmatia, there is no reason to believe that it was a *municipium*, as was tentatively suggested by Ivo Bojanovski on account of the above-mentioned *beneficiarius*' inscription. Admittedly, *beneficiarii consularis* were customarily present, among others, in places characterized as *castra* and *castella*.[30] Therefore, the name of Castra probably

[26] Bojanovski 1988: 302. On the provincial border see also Bojanovski 1988: 325–330; Šašel 1989: 60; Migotti 2012c: 5, fn. 27; Šačić Beća 2017: 130–133. On the location of Castra see Talbert 2000: 20 E5.

[27] Cf. *CBI* 439, referencing to nos 463, 470 and 475.

[28] On the location of Ad Fines, see Bojanovski 1974: 97–98; Bojanovski 1988: 326; Talbot 2000: 20 E5.

[29] This type of monument probably penetrated to the inner province of Dalmatia (Paškvalin 2012: 146–147 [nos 82–85], 152–154) from Pannonia, because no such pieces are known from the Dalmatian coastal areas.

[30] France and Nelis-Clément 2014: 119. On Castra, see also Šačić Beća 2017: 138–140.

derived from its original character as a military settlement. As stated above, the southern border of the Aquae Balissae *ager* should have run along the border between the provinces of Pannonia and Dalmatia. This, however, seems to be too large a space for an *ager* of a *municipium* of no great administrative or political significance, so the territory of Aquae Balissae more likely extended only as far as the Sava River. In that case, the area between Castra and the Sava could have been in the category of unadministered land, or it could have been a military reserve. The third possibility, adduced recently by Amra Šačić, is that Roman Servitium, mentioned in several itineraries without any specific administrative determination and located at the town of Gradiška (BA) on the Sava River, 50 km south-east of Daruvar (Figure 3), was actually an autonomous town with its own *ager*.[31] Although this is a well-founded hypothesis, it was not applied to the conception of the catalogue database in this research, because it is not finally proved.

There are only two published funerary monuments from Banja Luka, or its immediate surroundings, now in the Museum of Republika Srpska in Banja Luka.[32] One of them is a completely preserved lidded sarcophagus with a plain chest and a gabled lid holding large corner *acroteria*, without revealing a secure structural origin from either Dalmatia or Pannonia.[33] The other is a fragment of the front side of a sarcophagus chest from an unknown site, but probably from Banja Luka or its surroundings. It features a circular inscription panel supported by (possibly) floating Erotes.[34] Only one example of Erotes, though standing ones, supporting a rounded inscription panel is known to this present author from the coastal stretch of Dalmatia, and none with floating Erotes. Erotes (especially floating ones) supporting a circular inscription panel is a motif characteristic of Norico-Pannonian funerary iconography. Therefore, those rare examples with floating Erotes by a circular portrait panel on funerary buildings in the interior of the province of Dalmatia should probably be ascribed to Norico-Pannonian artistic influences.[35] The motif of the circular inscription or circular portrait panel is in its own right reminiscent of Norico-Pannonian funerary medallions, whose origin should be sought in North-Italian funerary monuments with portraits in rounded frames, and is therefore associative of North-Italian and Norico-Pannonian artistic prominence in regions bordering Dalmatia.[36] Significantly, some of the funerary evidence from the nearest Dalmatian *municipium* to the south of Banja Luka (Baloe, modern-day Šipovo, BA) reveals an interdependence of Dalmatian and Pannonian iconographic traits, reminiscent of a similar situation concerning a Dalmatian *municipium Metulum* south of Siscia (see below).[37]

The northern border of the Aquae Balissae territory is here taken arbitrarily as running along the Drava River, that is, following the modern-day state border between Croatia and Hungary. This means that the indications of the stretching of Aquae Balissae *ager* across the Drava, based on the presumption that the tribe of the Iasi lived there as well,[38] was ignored here. This was done on account of the lack of data on funerary monuments from the area covering some 30 km inland Hungary from the Drava.

Fixing the eastern border of the Aquae Balissae *ager* is constrained in a similar way as that on the south, because it should have followed a still contested southern stretch of the border between Pannonia Superior and Pannonia Inferior. Some authors see it as stretching along the north-south line from Donji Miholjac on the Drava River to Slavonski Brod on the Sava, thus cutting the eastern slopes of the mountains surrounding the Požega Valley (the Slavonian mountains) on the east side and probably leaving only the easternmost mountain fringes in Pannonia Inferior; this is the solution pursued here. Others, however, fix this line some 15 or 20 km to the west, leaving the easternmost part of the Požega Valley outside Pannonia Superior, and, by implication, outside the *ager* of Aquae Balissae.[39] An inscription to Jupiter was allegedly found in Roman ruins in the mountain west of the town of Našice (69 km SE of Daruvar). It could bring more light into this issue were its authenticity not questionable and were it not missing.[40] Its record in the 19th century manuscript reveals a mention of Pannonia Inferior (see the illustration at Catalogue AB[t] VIII.1), suggesting that the border between Pannonia Superior and Pannonia Inferior ran somewhat more to the west, leaving some of the Požega Valley in the latter. Indeed, the distance between Našice and Daruvar seems to be too large for the inclusion of the former into the *ager* of Aquae Balissae, but the alleged findspot of the inscription in question is some 10 km to the south-west of Našice, on the slopes of Mt Krndija. On balance, this issue remains open, but only in details which do not significantly blur the eastern boundary of the *ager* of

[31] Šačić Beća 2017: 123–141.

[32] According to a personal communication of Ljubica Srdić of the Republic Institute for the Protection of Cultural, Historical and Natural Heritage in Banja Luka, no other such monuments exist in the Museum.

[33] Paškvalin 2012: 445, no. 11.

[34] Paškvalin 2012: 443–444 (no. 2), 447, 451. They were described as floating Erotes by Paškvalin, but the fragment is too small to prove this beyond doubt.

[35] Cambi 2010: 126–127, no. 150; Paškvalin 2012: 471–472, nos 1and 3. See also fn. 90 in Ch. II.3.1.

[36] Kranz 1986: 194; Kremer 2001: 346–347; Pochmarski 2011: 24–25; Paškvalin 2012: 447, 451; Cambi 2017: 153. For North-Italian monuments with portraits in rounded frames see Pflug 1989: 161 ff., nos 25, 26, 58 ff; *lupa* 2390, 3699, 17236, ff.

[37] Cambi 1982: 97; Paškvalin 2012: 447–451.

[38] Kovács 2014: 6.

[39] Cf. Migotti 2012c: 2–4, map 2.

[40] Ilić 1852: 363.

Figure 4. Map of the territory of Siscia
(drawn by: D. Fudurić and T. Leleković)

Aquae Balissae, bordering the *agri* of Mursa (present-day Osijek) and Cibalae (present-day Vinkovci) in Pannonia Inferior. The same cultural feature as appearing in other boundary areas can be observed here, exemplified by the funerary medallion from Biškupci (AB[t] VI.1), 35 km to the south-east of Daruvar. This find namely reveals art-historical traits of the production of this kind of monument in Pannonia Inferior, while the site was undoubtedly located in the *ager* of Aquae Balissae, and, by implication, in Pannnonia Superior (see at Ch. II.3.1).

*Siscia*

If the assessment of the Siscian *ager* at some 50 km in diameter is accepted,[41] this would exclude not only its *possible* territory, but also parts of its *presumed* territory as given in the present Catalogue (Figure 4). However, the 50 km diameter seems to be a conventional orientation in space and not a specific distance, as results from the fact that at the same time the surroundings of Dubica, at a distance of 47 km to the southeast of Sisak, were included in the *ager* of Siscia on account of a milestone found there.[42] On balance, the starting point for the discussion of the Siscian *ager* is *the presumed territory of Siscia* (S[t]), reaching to the south at least as far as Topusko, at a distance of 38 km, while monuments that would produce similar indications in the north-north-west direction, towards the cities of Andautonia and Aquae Balissae, are missing. The town of Topusko was earlier identified with Roman Ad Fines, and thus placed on the border between Pannonia and Dalmatia, leaving the question of its provincial affiliation open. Recently, Roman Quadrata was located in Topusko, while Ad Fines was tentatively placed at Mali Gradac, 20 km to the southeast, presumably fixing the provincial border there.[43] On a funerary stele from Topusko (S[t] II.1) a veteran gave his origin as *domo Siscia*, which was interpreted as a possible indication that Topusko was not in the *ager* of Siscia, and therefore not in the province of Pannonia.[44] This, however, cannot be taken as a conclusive argument, because similar examples are known without proving a foreign origin for the respective tomb owners.[45] Furthermore, the monuments from Topusko clearly reveal Pannonian structural typology and iconography (see Chapters II.3.1 and II.3.3).

In the west, the *ager* of Siscia bordered the territory of the westernmost south Pannonian town – Neviodunum

[41] Lolić 2003: 132–133, fig. 1.
[42] Lolić 2003: 133.
[43] Gračanin 2010: 15, 49; Lolić and Wiewegh 2012: 195.
[44] Šegvić 1988: 61.
[45] Cf. inscriptions from Budapest (Aquincum) giving *domo Aquinco* as the origin: *lupa* 2911, 10562, 10573.

(Drnovo, SI). This border should have been on the Kupa River, although some historical and geopolitical circumstances leave open the possibility that the modern-day Bela krajina (White Carniola) on the left bank of the Kupa River in Slovenia belonged to the Siscian territory. While Viktor Hoffiller and Balduin Saria only claimed that Bela krajina did not belong to the *ager* of Neviodunum, Milan Lovenjak specifically mentioned the possibility that it was part of the Siscian *ager*.[46] The fact that Bela krajina most probably belonged to the early Christian bishopric of Siscia and certainly to its medieval and modern successor, the bishopric of Zagreb, only indicates without proving it, that previously it had belonged to the *ager* of *colonia Siscia*.[47] Because of this uncertainty, and the present territorial affiliation of the Bela krajina region with Slovenia, this area has not been included in the present research. Otherwise, its funerary *stelai* represent a closed and specifically unique art-historical whole, with a rustic touch about them and with no close comparisons in the Siscian evidence.[48] On balance, the western border of the Siscian territory may have been on the River Kupa but may have also stretched on its left bank to comprise some of the Bela krajina region.

Defining the *ager* of Siscia in the south is equally debatable and is related to the issue of the border between the provinces of Pannonia and Dalmatia in its south-western stretches. The line of this border is usually described vaguely as (at a small distance) *south of the Sava*. From the accompanying maps it transpires that some authors put it some 30 km south of the Sava, while others see it as lying further south, nearer to the Adriatic hinterland.[49] The latter assumption seems to be realistic, at least as concerns the western (Croatian) part of the border in the regions of the Pokuplje and Kordun to the south of the town of Karlovac (64 km to the SW of Sisak). These comprise the valleys of the River Kupa and its southern tributaries Dobra, Mrežnica, Korana, and Glina, which, serving as traffic routes between the provinces of Dalmatia and Pannonia, decisively shaped the political geography of the region, with, as it seems, the prevalent orientation towards Siscia. Earlier research into the stone quarrying and stone cutting in the Kordun has revealed that some of these activities were directed towards Siscia, and that both stone blocks and half-finished products were transported via the River Kupa and its tributaries, especially the Mrežnica and Korana.[50] This has been further substantiated through characterizations of some of the monuments from Siscia, conducted within the present project (see Ch. II.3.3). Another (indirect) indicator of this area's affiliation with the province of Pannonia, and, by implication, the possible/hypothetical *ager* (or the area of influence) of Siscia, comes from the pre-Roman material culture of its population, an Illyrian-Pannonian ethnic community of the Colapiani, which reveals the stretching of the future province of Pannonia further south, towards the upper course of the Kupa.[51] Hoffiller and Saria considered the area of Pokuplje and Kordun as not belonging to the Siscian territory, while not mentioning any other municipality in this role.[52] The nearest autonomous town in the south was *municipium Metulum* in the province of Dalmatia, hypothetically located in the vicinity of the village of Josipdol.[53] Thus, the southern border of the Siscian *ager* and that of *municipium Metulum* evidently coincided with the provincial border at the respective stretch, but where exactly this line ran, remains debatable. The monuments do not give a decisive clue, as transpires from the evidence from the sites both north and south of the presumed borderline. Both funerary monuments from the site of Josipdol (Catalogue, Peripheral areas, Metulum, I.1, 2) reveal more traits of art-historical influences from Pannonia than Dalmatia. A fragment of tombstone I.1 holds the motif of *dextrarum iunctio*, which frequently appears on various funerary monuments in Rome and northern Italy, as well as Noricum, and quite rarely in Pannonia, but is absent from Dalmatia.[54] Neither the shape of the dress of the woman on this stele nor its embroidered belt point exclusively to one of those provinces. Embroidery on female clothes is very rarely found in the province of Pannonia, and is connected nearly exclusively to the sashed robes introduced in the 3rd century under the influence from the eastern provinces.[55] Still, the belt decorated as on the tombstone in question can be associated with some of those worn by Norican servant girls, decorated with metal appliqués.[56] On the other hand, embroidery decorations on clothes are more widespread in Dalmatia, and without being limited to sashed robes, but also without presenting so decorated belts.[57] On

[46] *AIJ*: 218–219; Lovenjak 2003: 281–283. Otherwise, the uncertainty between the two territories can already be observed in *CIL* III, Suppl. 1, in its wavering between attributing the stones from Černomelj in Bela krajina to Neviodunum and Siscia alternately (cf. nos 10.823-10.825/Neviodunum vs. 10.826 and 10.827/Siscia).

[47] Migotti 2011b.

[48] Cf. *lupa* 3791–3794, 3796–3798, 9143, 9209. M. Lovenjak (2003: 283) observed a basic primitive execution and some iconographic traits in funerary stones of both Bela krajina region and the area along the Sava and Kupa rivers in the wider surroundings of Sisak. This, however, is a general observation, first put forward by J. Šašel (1985: 332–333), but it does not impose on the impression of the art-historical uniqueness of the Bela krajina funerary evidence within the wider regional surroundings.

[49] See fn. 26.

[50] Migotti 2005b: 371; Perkić 2012: 118, 216–217; the latter author suggested a probable administrative affiliation of the Kordun with Siscia.

[51] Čučković 2008: 80.

[52] *AIJ*: 219.

[53] Olujić 2007: 54.

[54] Cf. Gabelmann 1973: 51, 131, Pl. 33: 2; Koch and Sichtermann 1982: 98–99; Pflug 1989: 104–105; Jaeger 2003: 476; Pochmarski 2011: 21; Birk 2013: 29, *passim*; Facsády 2014: 197.

[55] Cf. *lupa* 681, 3823. A very decorative embroidered robe is represented on a statue of a presumed empress from Carnuntum, dating from the turn of the 3rd and 4th centuries (Landskron 2007).

[56] Garbsch 1985: 555, fig. 1: 1, 2; Rothe 2012: 219–220.

[57] Maršić 2014: 203–207, figs 1–6.

balance, the depiction on the stone from Josipdol can be considered as an individual local rendering revealing elements of Norico-Pannonian and Dalmatian fashions, the latter further suggested by the grooved line cutting diagonally across the lower part of the woman's dress.[58] The other piece, a fragment of a funerary building I.2 is more specific in terms of Pannonian traits, which are reflected in the type of monument and the scene with an Eros holding a down-turned torch pulled through a wreath. Funerary *aediculae* are typical of northern Italy and the northern provinces, but again, they are lacking in the funerary art of Dalmatia.[59] As for the motif, while mourning Erotes holding wreaths are ubiquitous in the funerary art of Dalmatia and north Italy, the iconography as witnessed in the piece from Josipdol is not found there, but it appears sporadically in the province of Pannonia.[60] Another possibility, but less likely, is that the artistic features foreign to Dalmatia and visible in piece I.2 from Josipdol came directly from northern Italy, as Aquileia abounds in the motif of mourning Erotes on various kinds of monuments.[61] It was pointed out recently by Friederike Harl that northern Italy, notably Aquileia, has been unduly considered as a non-negotiable, predominant source of the majority of Norican (and indirectly also Pannonian) art, as was often claimed by various commentators. While some influences, especially in the 1st century AD are indisputable, other authors have introduced a more balanced view, pointing to the role of Greek art in shaping the artistic and cultural profile of the two provinces, especially from the 2nd century.[62] On the other hand, the role of Aquileia in spreading artistic and cultural influences in Noricum and Pannonia is securely anchored in the fact that the earliest occupation of those regions started from Aquileia along the already existing trade routes, as proved by written sources and archaeological evidence.[63] Therefore, a cultural ascendancy of Aquileia over its north-eastern neighbouring regions cannot be easily dismissed and probably lasted longer than the 1st century.

The circumstances on the northern side of the presumed border in the area of Kordun, which was tentatively ascribed to the Siscian *ager*, are similar. The monuments there are mostly undecorated, but some hold circular inscription panels (I.2, III.9, III.12, III.14, III.27, III.35 and III.49), which are associative of Pannonian rather than Dalmatian funerary iconography. On the other hand, one of the decorated pieces from the Kordun area (S[ptSW] III.48) points to a model from Salona (Dalmatia).[64] On balance, the present research seems to be in tune with the hypothesis posited long ago by Nenad Cambi, who suggested that there was a stone cutting workshop somewhere in the area of Kordun, which was outside of the province of Dalmatia and which worked on Norico-Pannonian models.[65] Such reasoning excludes Metulum (Josipdol) because it is situated on the border between the modern regions of the Lika and Kordun, both within the Roman province of Dalmatia. Therefore, the hypothesized workshop should either be looked for in the area of the dispersed Kordun quarries or we should leave open the possibility that it was indeed in Metulum but was oriented towards Siscia in terms of artistic production and economic relations. On balance, what we have in the funerary evidence from the site of Josipdol is the administrative appurtenance to the province of Dalmatia and artistic affiliation with Pannonia.

Fixing the south-eastern border of the Siscian *ager* is even more problematic than that on the south-west, because of the lack of points of reference such as *municipia* Neviodunum and Metulum in the west and south respectively. According to the most likely hypothesis, the border between the province of Dalmatia and Pannonia ran south of Mts Zrinska Gora and Kozara.[66] There were no municipal towns in this area in Pannonia, unless we take into account the above-mentioned Servitium, and *municipium Faustinianum*, which, according to the epitaph from a Siscian sarcophagus, should have been situated in the vicinity of Siscia; the sarcophagus belonged to Pontius Lupus, an *augustalis* in Siscia and a *scriba* of *municipium Faustinianum* (S II.2).[67] Although the other two inscriptions mentioning this city fail to produce any geographic indication, *municipium Faustinianum* must have been situated to the east of Siscia. However, its presumed territory was not included in the Catalogue, as its precise location has not been archaeologically confirmed.[68] Furthermore, all three inscriptions mentioning *municipium Faustinianum* stem from the turn of the 2nd and 3rd centuries, with sarcophagus S II.2 as possibly a little later, while the city's name suggests its probable municipalisation in later Antonine times.[69] On balance, before its administrative elevations, the settlement was either situated in the Siscian *ager* or depended on Siscia.

[58] A fairly close analogy is found on a stele from Vašarovine, in the immediate hinterland of the Adriatic coast (Čremošnik 1963: 112–116, fig. 6).
[59] Cf. Kremer 2001; Ertel 2010; Scholz 2012; F. Harl 2016: 20–22.
[60] Gabelmann 1977: 206–208, fig. 9, *passim*; Kolšek 1991; Cambi 2010: 47–50/282–284; Migotti 2016b: 228.
[61] Gabelmann 1973: 64, *passim*; Gabelmann 1977: 237–238; Kolšek 1991: 139–140.
[62] Gabelmann 1977: 234, 238; Koch 1993: 60–61; Djurić 2005: 77–78; Djurić 2008: 161–163; M. Scholz 2012: 94–95, *passim*; F. Harl 2016; Cambi 2003: 19–22; Cambi 2005b: 17–28, 30; Pochmarski 2016c; Djurić, in press. See also fn. 219 in Ch. II.3.2.
[63] Verzár-Bass 2006: 245; Domić Kunić 2012: 38, *passim*; Radman-Livaja 2012: 163; Guštin 2015: 221–223.
[64] Paškvalin 2012: 447, 451; Cambi 2017: 153.
[65] Cambi 1975: 79.
[66] See fn. 26.
[67] Migotti 2016b.
[68] Póczi 1980: 262; Kovács 2014: 102; Šačić Beća 2017: 118–123. The possibility that the *municipium Faustinianum* was established by the Flavian emperors (Šačić Beća 2017: 122) cannot be straightforwardly rejected, but it has no firm support either.
[69] Šačić Beća 2017: 118–123, 149–154.

With Servitium excluded on account of its uncertain administrative status, the nearest municipality to the south of the presumed provincial border was deep in the territory of Dalmatia (*municipium Baloie*, modern-day Šipovo).[70] With the provincial border as presumed above, the north-westernmost Dalmatian territory comprised the valley of the middle course of the Sana River, before its winding westwards to discharge into the Una. Significantly, it was claimed that this area was culturally under the influence of Siscia.[71] The monuments from the vicinity of Prijedor in the upper courses of the rivers Sana and Japra indeed show predominant, though not absolute, Pannonian artistic traits (see at Ch. II.3.1). The complexity of the issue of the southern border of the Siscian *ager* in its relation to the provincial border in the same area is reflected in the otherwise well-argued reasoning of Ivo Bojanovski, who was vague about the mining area in the valley of the Japra River in the surroundings of Prijedor. He ascribed it alternately to the Siscian territory and, together with another mining area in the valley of the Sana River to the south, to the territory organized as an Imperial *territorium metalli*, remaining undecided about the possibility of a municipalisation of those areas at some point in the Imperial period.[72] To sum up, the Siscain *ager*, or rather, its extended area of influence in the south-east, was identical with the Pannonian-Dalmatian border along the respective stretch. Its border with the *ager* of Aquae Balisasae along the same stretch must have run between the modern-day towns of Gradiška (BA) and Prijedor (BA), with the site of Donji Podgradci (AB[t]) I.3) probably belonging to the *ager* of Aquae Balissae on account of the distances: 58 km from Daruvar and 73 km from Sisak. This, of course, is valid only if the settlement Servitium was not a municipality, in which case both Mahovljani (Catalogue, Peripheral areas III.1) and Donji Podgradci would have fallen under its administrative authority. If this were not the case, all of the Siscian extended *ager* (the *possible territory* in the Catalogue) corresponds to the concept of the unadministered land as discussed above, while its *presumed territory* is tentatively identified with its *ager* in the narrow sense.

Finally, it remains to comment on the possibility of the existence of another Roman autonomous town in north-western Croatia – *municipium Halicanum*. Halicanum, mentioned in an inscription from Budapest and in two Roman itineraries, has been customarily located at the settlement of Sveti Martin na Muri in the north-westernmost part of Croatia where it borders Slovenia. However, a municipal status of this place has not been conclusively proven.[73] On the other hand, the only monument considered to have been found in

Figure 5. Stele from Čakovec (photo: T. Leleković)

its hypothetical territory – a Roman stele now in the Museum of Međimurje in Čakovec (Figure 5), most probably does not stem from those surroundings at all.[74] Its original findspot was probably in the territory of Savaria (Szombathely) in western Hungary, for two reasons. Firstly, the stele was allegedly found in the ruins of the Zrinski Castle in Čakovec, which now houses the above-mentioned museum; the suspicion is raised by the fact that it is the only Roman funerary monument from the wider surroundings of Čakovec and Sveti Martin na Muri.[75] Secondly, its alleged findspot, the Zrinski Castle in Čakovec, was in the 16th and 17th centuries owned by the Zrinski counts, who also had a possession at modern-day Vép in the surroundings of Szombathely, and who were responsible for bringing

[70] See fn. 25.

[71] Bojanovski 1988: 288.

[72] Bojanovski 1988: 277, 281, 337.

[73] Ragolič 2014: 331, fn. 65 (with literature).

[74] Migotti 2013a: 307, no. 18, 315, fig. S 18; *lupa* 3786.

[75] *AIJ*: 457.

two more *stelai* from Savaria to north-western Croatia.[76] Also, information on a 'Good Shepherd' sarcophagus, allegedly found in Štrigova 6 km south-west of Sveti Martin na Muri and now in the Hungarian National Museum,[77] is completely unfounded as no such monument exists there.

*Concluding remarks*

As observed several times in the above discussion, the so-named *possible territories* of the cities can be taken as such in the wider sense. Unfortunately, no precise determinations exist of the borders between the three cities discussed here; in two cases such borders were actually boundary lines between the provinces of Pannonia and Noricum on the north-west, and Pannonia and Dalmatia on the south. It has often been claimed that Roman stone monuments within individual provinces contain particular distinguishing traits in comparison to other provinces. This comes from a much contested and in recent scholarship mostly rejected, but in its essence still justified, interconnection of the concepts of culture and ethnic affiliations, which served as a broad basis for the Roman provincial organization.[78] In view of this controversy, it was hoped that the archaeological evidence could produce some solutions towards establishing the administrative borders between the three cities under research. Such expectations, however, turned to be unrealistic, at least as far as the funerary monuments are concerned. When the monuments from both sides of the presumed borders are viewed in tandem, what we get is boundary areas in which artistic and workshop influences from both provinces, or from two bordering cities, are perceptible, leaving the precise fixing of the city and provincial boundaries undecided. If the north-west border of the Andautonia territory was set realistically, that is, with the site of Lobor included, then this part of Pannonia was thoroughly under the artistic influence of Norico-Pannonian workshops established in Celeia and Poetovio on the basis of marble quarrying on Pohorje.[79] This is in tune with a long-established existence of a trans-provincial Norico-Pannonian cultural circle.[80] The same can be observed in the bordering areas south of the river Sava, which, according to the previous model, could tentatively be labelled as a Pannonico-Dalmatian cultural circle.

II.2. THE FINDSPOTS AND NUMBER OF FINDS

*Introduction*

Before tackling this subject matter individually for each city, it should be remembered that the quantification of Roman stone monuments is seriously hampered by the haphazard nature of the preserved evidence, mainly for two reasons. Firstly, much of it remains unrecovered, and secondly, enormous quantities of stone monuments, marble even more than limestone, ended up as recycled raw materials for building purposes.[81] The total of the preserved stone monuments in various parts of the Roman Empire has been tentatively estimated at 2.5–5.1%.[82] If sarcophagi are taken as an example, the preserved number of them in the whole of the Roman Empire was estimated at between 12,000 and 15,000 pieces; this must have been a very small percentage of the original output, the number of which is hopelessly conjectural.[83] In all probability, the number of lost pieces from the territory of the present research is even larger, meaning that less than 2% has been preserved; this transpires from many archival and other data on missing pieces, some of which have been included in the Catalogues if the minimum data on them existed. Archival research and personal enquiries for Siscia, Aquae Balissae and Andautonia suggest that all three cities, with Siscia leading, have been hard hit by a subsequent loss of monuments. In combining data from the archives in the Archaeological Museum in Zagreb with secondary literature, and supplementing them with those gained through personal communications from colleagues in museums and other people in the field, it transpired that not only monuments mentioned in the very old literature, but even quite a lot of those which survived into the second half of the 20th century, have gone missing in the meantime. This is especially true of the undecorated and/or uninscribed pieces. The example of Siscia is the most illustrative in this regard, because the archival data and literature abound. Even prior to the mid 19th century, when the urban development of Sisak led to discoveries of Roman architecture and its immediate destruction for pragmatic reasons, Roman stones and other finds had been mercilessly devastated by whoever found them while working on their land.[84] Just a few illustrating examples of unfortunate losses of archival data and/or the respective monuments

[76] Migotti 2012d.

[77] Vidović 2003: 85.

[78] The concept of ethnically based culture(s) in Roman archaeology was widely discussed in literature in the second half of the 20th century, so here is just a selection of works with further literature included (von Schnurbein 1982; Dally 2000: 87–88; Rothe 2009; Migotti 2012c: 7–9), as well as those comprising provincially based research of funerary monuments: O. Harl 1991a; Pochmarski 1991; Kremer 2001; Starac 2002; Konrad 2004; Scholz 2012; Handy and Pochmarski 2014; F. Harl 2016. On the critique of such conception, see Djurić, in press.

[79] Cf. Kremer 2001: 27–52; Djurić and Müller 2009: 116; Scholz 2012: 137.

[80] Garbsch 1965; Gorenc 1971; Garbsch 1985; Kranz 1986; Kastelic 1998: 27–29, 219–220; Lehner 2012: 46; Migotti 2013a: 305; Ubl 2013.

[81] Koch1993: 60; O. Harl 2003: 338; Cambi 2010: 15.

[82] O. Harl 1991a: 12–13; Pochmarski 2011: 10; Pochmarski and Weber-Hiden 2016: 12.

[83] While there seems to be an understanding on the number of preserved pieces, the estimations of what percentage of the original output these numbers represent, wavers from 2% to 20% (Koch 1993: 1; Russell 2011: 127–128; Birk 2013: 9; Russell 2013: 257; Koch 2015: 148–149).

[84] One of the earliest reports on such practices, described as *vandalism*, comes from the 18th-century learned lady and travel writer Therese von Artner (1830: 9–10). See also Vukelić 2011: 60, *passim*.

will be adduced here. In 1876 in the area of the south-east cemetery a sealed sarcophagus was found holding three skeletons, and was sent to Zagreb.[85] Since none of the sarcophagi and ash-chests now in the Archaeological Museum in Zagreb (S II.8, 10, 11, 22; S III.1, 2.) corresponds to the date of this discovery, such an extremely important find was long lost. Another such example is the epitaph from a sarcophagus found in the area of the south-west cemetery and remodelled in 1902 into a bench. Since the inscription has not been recorded in the *CIL*, all that remains is its Croatian translation, reporting that one Iuliana, a member of a *collegium virginum*, put up a sarcophagus for her brother, a carpenter and a veteran of the 7th legion (*Legio septima Claudia*).[86] Finally, a sarcophagus which was seen in the courtyard of a private house in Sisak in 2004 (S II.5) was no longer there in 2015, and the same happened with a tombstone from Petrinja (S[t] VIII.1), last seen in 1995. Similar data can also be found for Aquae Balissae, although not in such abundance. For example, a sarcophagus base similar to that of the sarcophagus from Veliki Bastaji (AB[t] II.1) was seen long ago in the village of Badljevina, 10 km south-west of Daruvar.[87] The importance of this find is evident, because the sarcophagus from Veliki Bastaji is the most sumptuous example in north Croatia, and the only one taking a separately carved elaborate base. Finally, the number of types of *stelai* and sarcophagi, which is too large if compared to the total number of finds, testifies to a high percentage of the loss of evidence in this way or the other. On balance, any statistical quantification conducted on such a database cannot be more than an indication towards quantitative relationships at various levels of comparison for individual cities and for all three of them viewed as a whole.

*Siscia*

The funerary topography of Siscia has been drafted in general lines in the literature, revealing that there were cemeteries on all four sides along the roads leading off the city. This was done more through archival data than through archaeological research, resulting in quite a few instances of the finds that cannot be precisely mapped. This is especially true of the western and eastern cemeteries, which possibly stretched all along the city perimeter, and not only along the (presumed) lines of the roads. Therefore, the determination *east cemetery* tentatively comprises the finds from the eastern stretch outside the city walls, while *south-east cemetery* includes finds from the south-eastern and southern stretches, as they seem to have been separate wholes. It should be observed that in the literature the terms *eastern* cemetery and *south-eastern cemetery* were used alternately and alternatively, that is, in a confusing manner. Similar circumstances refer to the western cemetery, which is denoted as south-western in the Catalogue because the majority of the finds stem from its southern stretch, while there have been indications, both from the archives and recent research, of funerary finds along the entire western fringe of the city. The area of the northern cemetery is also somewhat elusive in terms of spreading, with only indications of an eastern 'branch' and the majority of graves taking the area to the north-west of the city. This is why in the most recent literature this cemetery has been labelled as north-west[88] (Figures 6 and 7). Because of such uncertainties, stele S I.1 remained tentatively placed outside the cemetery (see below).

On the basis of secure and tentative affiliations of the finds with the urban cemeteries, the city itself and its presumed *ager* have produced 56 pieces, of which eight *stelai*, 22 sarcophagi, two ash-chests, four *tituli*, three funerary buildings, and 17 unclassifiable monuments. In terms of microtopography, the finds were recovered from the following places: within the city walls (S I.5; IV.3); north cemetery (S I.1[?], 2, 3; S II.8[?], 21; S VIII.1, 16[?]); east cemetery (S I.4; S II.19[?]; S VII.2[?]; S VIII.8[?]); south-east cemetery (S I.7, 8; S II.4, 10, 13, 14, 15, 16, 17, 18, 22; S III.1; S VIII.2, 3, 9[?], 15); south-west cemetery ( S II.1, 2, 3, 5, 6, 7, 9, 12, 20; S III.2; S IV.1, 4; S VIII.11, 13); unknown sites (S I.6; S II.11; S IV.2; S VII.3; S VIII.4, 5, 6, 7,10, 12, 14, 17). While the monuments within the city were found in unknown contexts, only six of them were recovered in archaeological (rescue) excavations in the south-east cemetery (S II.4, 13, 14, 15, 16, 17).[89] If the monuments from the possible *ager* are added, they give a total of 127 pieces.

All but one (S[t] VII.1) of the 14 monuments (seven *stelai*, two each ash-chests and altars, and one each sarcophagus, *aedicula* and unclassifiable fragment) from the presumed *ager* of Siscia (S[t]) were chance finds, which probably represents a very small percentage of its original evidence, as the territory of Siscia is notorious for the lack of systematic excavations.[90] Only two sites have yielded more than one find: Topusko: (S[t] I.1, 2, 3, 4, 5; S[t] III.1, 2; S[t] VII.1) and Odra Sisačka (S[t] I.6; S[t] II.1; S[t] V.1, 2), while one each piece was found in Petrinja (S[t] VIII.1[?]) and Sela (S[t] I.7). The Roman spa settlement at the site of modern-day Topusko was located on the important road leading from Senia (modern-day Senj on the north-eastern Adriatic coast) towards Siscia. Its main cemetery stretched along the road towards Siscia, with all of the

[85] Vukelić 2011: 136.

[86] Vukelić 2014: 173; *ILJug* 1128.

[87] Schejbal 2004: 114, 120, fn. 136.

[88] Here is a selection of literature on the cemeteries of Siscia or the literature comprising chapters on this theme, with archaeological and archival aspects included: von Artner 1830; Burkowsky 1996; Buzov 2002; Wiewegh 2003; Vukelić 2011; Leleković 2011; Leleković 2012; Lolić 2014; Baćani 2018.

[89] Tomaš Barišić 2018: 65.

[90] Leleković 2012: 331; Vukelić and Glazer 2015: 569.

Figure 6. Plan of Siscia with cemeteries
(after Leleković 2012)

monuments represented in the Catalogue (S[t] I.1[?], 2, 3, 4, 5; S[t] III.1, 2; S[t] VII.1) most probably stemming from there.[91] Recently, Roman Quadrata was located in Topusko, while Ad Fines presumably stood on the site of the village of Mali Gradac, 20 km to the south-east.[92] Apart from the monuments in the Catalogue Siscia, the village of Odra Sisačka has produced one or even two more funerary monuments, now missing. One was the portrait niche of a stele figuring a 'nicely carved' couple, and another was a fragmentary funerary inscription *CIL* III 3797. The former was not included in the Catalogue because of insufficient data and the latter because it has been presumably mistaken for an inscription from Aelium Cetium (*CIL* III 5663), which remains a contested issue (see Chapter IV).[93] Given, however, that Odra Sisačka is situated at a distance of only 3 km north-west of Sisak, and even less from the northern perimeter of Siscia and its north cemetery, the Roman finds there could have belonged to a villa in the vicinity of the city, but they can equally be stemming from the western extension to the north cemetery; the same is true of an *aedicula* pediment (S VII.1) recovered from the site that was tentatively explained as a suburban villa, just off the north-west fringe of the north cemetery.[94] A similar uncertainty applies to stele S I.1. It was interpreted as a stray find resulting from a failed attempt of unloading from a boat which should have brought it via the Sava River to Siscia. Presumably, it got broken and was never used, remaining in the ground some 20 m from the riverbed, at the site of Štakori. However, this findspot is about 1 km from the northern periphery of Siscia, and therefore close enough to its north cemetery, and possibly originally stemming from the cemetery area.[95] In that case, however, its findspot should have been secondary, for several reasons. Firstly, it was researched in more detail immediately, but no associated grave or any other Roman finds could be spotted; also, no other traces of a funerary ground are known from the vicinity. Finally, stele S I.1 is the earliest tombstone of Siscia and therefore probably set up somewhere closer to the northern city walls and not in its periphery and in the immediate vicinity of the river. If, however, the site of Štakori is the original findspot of the stele then it should be presumed that the river changed

[91] Čučković 1986: 14–15; Šegvić 2006; Lolić and Wiewegh 2012: 195, 217.
[92] Gračanin 2010: 15, 49; Lolić and Wiewegh 2012: 195; Talbert 2000: 20 D4.
[93] Kuntić-Makvić 1993: 265–266.
[94] Vukelić and Glazer 2015.
[95] Slukan-Altić 2012, map on p. 192.

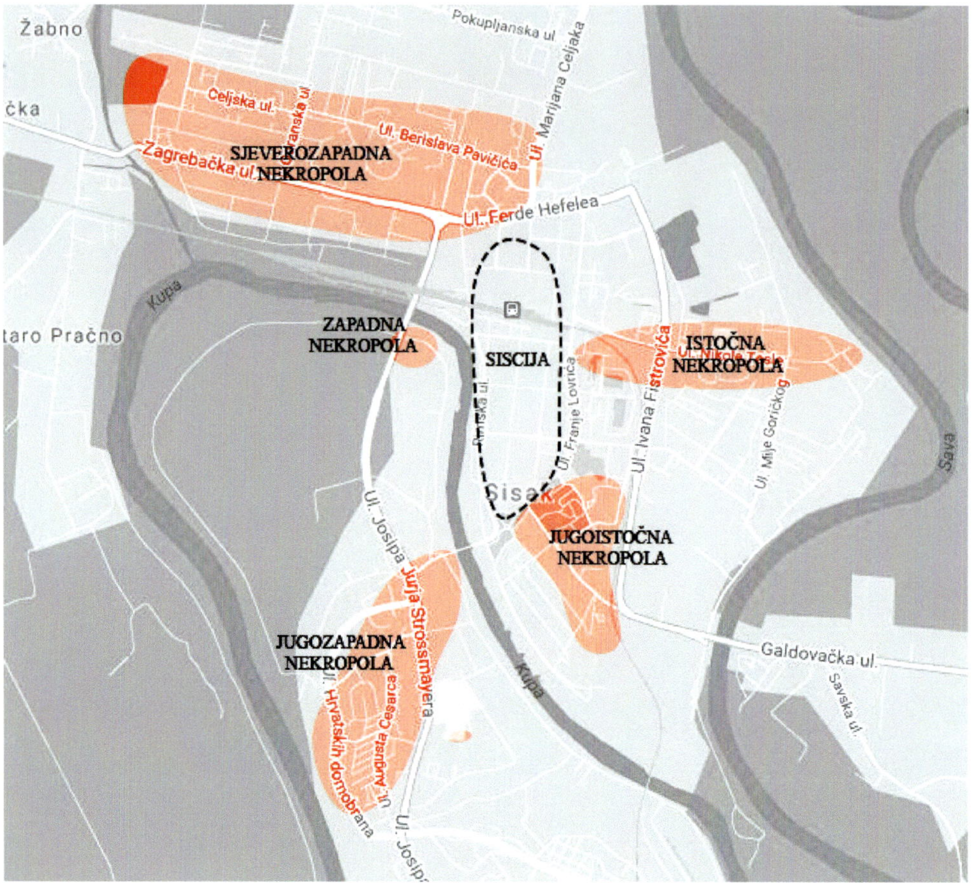


Figure 7. Plan of Siscia with cemeteries (courtesy of I. Baćani 2018)

its course over the centuries.[96] The area of the town of Petrinja (S[t] VIII.1) has never been researched archaeologically, but its Roman predecessor can be presumed on account of several substantial indications: a chance find of a *beneficiarius* inscription (*CBI* 287), presumed Roman quarrying in the village of Hrastovica, 11 km south-west of Sisak (see Chapter III), and abundant remains of Roman building materials in the city surroundings.[97] The village of Sela (S[t] I.7) also has never been researched archaeologically, but a lot of chance finds, mostly building materials, as well as remains of a road, were found there.[98] These modest data on the Roman finds from Petrinja and Sela are insufficient for a suggestion on either a villa or other type of settlement. If the fact that practically all of the monuments from the presumed Siscian *ager* stem from the Pokuplje to the south of Siscia is not accidental, it suggests that the majority of the city's *ager* spread on the south.

As explained in Chapter II.1, the region of Kordun can hardly be imagined as belonging to the immediate *ager* of Siscia because of the large distance; this is why it is here defined as a *possible territory in the south-west* (S[ptSW]). It takes the area of some 5 x 20 km and was situated in the border region between the provinces of Dalmatia and Pannonia, probably gravitating towards the latter, although the archaeological evidence does not prove it finally. This is because the monuments found there constitute a rounded whole with specific local characteristics and with no clear marks of Dalmatian or Pannonian structural typology or iconography. Nevertheless, circular inscription panels appearing on one stele and six ash-chests (S[ptSW] I.2; III.9, 12, 14, 27, 35 and 49) suggest a possible Pannonian influence.[99] On the other hand, a pelta-type of *tabula ansata* as found in an ash-chest from Erdelj (S[ptSW] III.48) points to a Dalmatian inspiration. Although various forms of such framings are known from Italy and northern provinces, Pannonia included, the only close analogy to the piece from Erdelj, known to this present author, comes from Salona in Dalmatia.[100] Apart from several old chance finds (S[ptSW] I.1, 2, 3, 4;[101] V.1; VIII.1, 2, 3), the remainder

[96] The Sava River is notorious for its meandering and changing its course in the past (Nemeth-Ehrlich and Kušan Špalj 2007: 31; Slukan Altić 2012: 11).

[97] Kristina Brkić and Aleksandar Durman, personal communications.

[98] These finds have never been published (Spomenka Jurić, personal communication).

[99] Paškvalin 2012: 447, 451; Cambi 2017: 153. See also fn. 35 in Chapter II.1.

[100] Spieß 1988: 272, fig. 7; Ciliberto and Mainardis 2015: 63, fig. 1; *lupa* 2924, 3263, 3470, 3044, 9725 ff. On the piece from Salona see Cambi 2010: 109. no. 60, Pl. XXXVI: 1.

[101] According to the original report of the find, there were two graves, each marked by a similar stele, but the other one was destroyed during the excavation (Migotti 2010a: 94–95).

of the evidence (S[ptSW] II.1–8; III.1–50), although partly known long ago, was researched anew through field surveying, trenching, and even some small-scale test excavations, and systematically documented in the 1997–2001 researches by Domagoj Perkić.[102] A total of 64 finds (four *stelai*, eight sarcophagi, 50 ash-chests, one altar, and three indeterminable pieces) are given in the Catalogue, which is less than listed in Perkić (2012), while at the same time more finds than comprised in that work can be found in the field.[103] Nevertheless, more important than the exact number of finds in this context is the fact that the whole area was occupied by surface quarries and many scattered pieces, some unfinished and/or discarded, testifying to a lively stone-cutting activity; as many as 29 smaller quarry sites were registered, grouped into 11 clusters.[104] The distribution of the finds by sites is as follows: Barilovečki Cerovac (S[ptSW] II.1, 2), Blagaj (S[ptSW] II.6, 8), Generalski Stol (S[ptSW] III.16, 17), Erdelj (S[ptSW] III.48), Gorinci (S[ptSW] II.7), Hrnetić (S[ptSW] II.3), Jankovo Selište (S[ptSW] III.5), Mala Crkvina (S[ptSW] III.13–15), Mateško Selo (S[ptSW] III.1–4, 18–47), Mrežnički Varoš (S[ptSW] V.1), Perjasica (S[ptSW] II.4), Ponorac (S[ptSW] II.5; III.6–10), Srednji Poloj (S[ptSW] III.49), Sveti Petar Mrežnički (S[ptSW] I.3; VIII.1, 3), Svojić (S[ptSW] I.4), Točak (S[ptSW] I. 2; III.11, 12), Vali (S[ptSW] II.50 and probably III.16, 48; VIII.2), and Vukova Gorica (S[ptSW] I.1). A combination of older finds and recent research, including not only funerary monuments but also small finds and building materials, suggests that Roman settlement agglomerations of the area should be looked for in the villages of Svojić, Mateško Selo, and Sveti Petar Mrežnički, with the latter tentatively identified as Roman Romula Quadrata.[105] All of the sites in the possible *ager* of Siscia in the south-east (S[ptSE]): Japra, Blagaj, Cikote and Rakanske Barice, are situated between the modern-day cities of Novi Grad and Prijedor in the valleys of the Sana and Japra Rivers (BA). As an important mining area, this territory witnessed some excavations in the past, but none of the funerary monuments was recovered in such excavations.[106] The distribution of the seven finds (six *stelai* and one *titulus*) by sites is as follows: Japra (S[ptSE] I.1), Blagaj (S[ptSE] I.2, 6), Cikote (S[ptSE] I.3, 4, 5), and Rakanske Barice (S[ptSE] IV.1). Roman Clandate has been tentatively located in the vicinity of Novi Grad and/or the village of Rakanske Barice.[107] Quantified together, the hypothetical Siscian territories on the south-west and south-east produced a total 73 finds (ten *stelai*, eight sarcophagi, 50 ash-chests, three unclassifiable fragments and one each *titulus* and altar).

*Andautonia*

Three cemeteries have been established in Andautonia, two on the south (following two phases in the urban development) and one on the east side (Figure 8). Despite the fact that systematic archaeological excavations in Ščitarjevo have had a long history starting in 1969, all that the city's cemeteries have produced is one stele and two sarcophagi (A I.1; A II.1); one of the sarcophagi, found in 1928, went missing in the meantime, as did the stele. All three finds were recovered from the south cemetery, but none in archaeological excavations.[108] Such dearth of evidence can at least partly be explained by the fact that quite a few funerary monuments, mostly fragmented, had been found in the cemeteries, but went missing in the meantime.[109] With the 16 finds from its presumed *ager* (11 *stelai*, one each sarcophagus and a funerary altar, and three unclassifiable pieces) and a further 12 from its hypothetical territory (three *stelai*, seven *aediculae* and two unclassified fragments), the evidence from Andautonia amounts to a modest number of 30 monuments.

Contrary to the city itself, its presumed *ager* has furnished more abundant funerary evidence, not least due to its dense settlement. This was proved by the state of research so far, and has been further substantiated by the evidence of funerary monuments compiled here, which amount to 16 finds distributed across 11 sites.[110] Villa estates have been presumed at Vugrovec (A[t] VIII.2) and Donji Čehi (A[t] I.6).[111] In the village of Samoborski Novaki (A[t] II.1), either a *villa rustica* or a *vicus* has been hypothesized.[112] The same is true of some other sites that do not furnish enough diagnostic data on the nature of Roman settlement, such as Odra (A[t] I.3),[113] Lekneno (A[t] I.8),[114] Kerestinec

[102] This was done within a wider research on the population and settlement of the region (Perkić 2012). Therefore, the work of Perkić is given as a source for all those pieces, with the earlier literature adduced for the rare monuments that had been published before.

[103] The number of finds as given in Perkić (2012: 5) is two *stelai*,46 sarcophagi and 64 ash-chests.

[104] Perkić 2012: 5.

[105] *AIJ*: 218–219; Gračanin 2010: 13; Perkić 2012: 220–227, 236–237, *passim*.

[106] On the archaeological finds and excavations in the mining area of the Japra (and Sana) Valleys, see Bojanovski 1988: 273–283; Škegro 2006: 151–152, 155.

[107] Bojanovski 1974: 231; Bojanovski 1988: 373. See also Talbert 2000: 20 D5.

[108] Pintarić 1998: 2; Nemeth-Ehrlich and Kušan Špalj 2007: 18–19, *passim*.

[109] Degmedžić 1958; Pintarić 1998: 2.

[110] On the settlement of the Andautonia *ager*, see Gregl 1997: 14–18; Nemeth-Ehrlich and Kušan Špalj 2003: 108–110; Leleković 2012: 331; Leleković and Rendić-Miočević 2012: 304.

[111] The villa was actually found at the site of Gornji Čehi, while the site of Donji Čehi produced the associated *tumuli* cemetery. On all three sites, see Gregl 1991: 14–21; Leleković and Rendić-Miočević 2012: 282.

[112] Galić and Radman-Livaja 2002: 242.

[113] A. Rendić-Miočević 2013: 343–348.

[114] The site has not been published yet. According to a personal communication by Tanja Pintarić, some walls have been identified in the vicinity of the stele, indicating a funerary enclosure.

Figure 8. Plan of Andautonia with cemeteries (after Lolić and Wiewegh 2012)

(A[t] I.11)[115] or Čučerje (A[t] VIII.1).[116] The evidence from the city of Zagreb (A[t] I.1, 2 [(?], 5 [?)], 7, 10) is not particularly helpful in detecting the nature of the Roman settlement(s) there, as only one piece (A[t] 1.7) comes from a researched cemetery. Given that graves have been found in many other places within the city and its outskirts, various types of settlements, including *villae* and perhaps *vici* can be hypothesized there.[117] The cemetery that produced stele A(t) 1.7 is quite diversified in types of graves and was long in duration (1st–4th centuries), without revealing clearly the nature of the associated settlement.[118] While no data on Roman finds for the sites of Majur (A[t] VI.1) and Hraščina (A[t] I.4) exist, the only evidence for the Roman presence at the place of the town of Križevci (A[t] I.9) is Roman coins.[119] Both at the site of Repišće (A[t] VIII.3) and Donji Čehi (A[t] I.6), a *tumuli* cemetery has been established, but with no traces of the associated settlement.[120]

Of the sites in the territory possibly gravitating towards Andautonia, the most important is Lobor (A[ptNW] I.1, 2; VII.1–7). It is therefore necessary to observe that the majority of commentators, including the most outstanding researcher of the Lobor site, Krešimir Filipec, believe that the southern slopes of Mt Ivanščica belonged to the territory of Poetovio.[121] Nevertheless, the possibility that the border between the *agri* of Poetovio and Andautonia ran along the ridge of Ivanščica, as supposed by other researchers, has been considered here.[122] While not completely rejecting the border as fixed by the former authors, I believe the latter's hypothesis should be allowed for also on account of the finds from Lobor. This is because some of them, and especially funerary-*aediculae* fragments, reveal a structural and iconographic affiliation with the Celeian production, as particularly evidenced in the Šempeter cemetery (see Chapter. II.3.1). Admittedly, this can also be a false impression produced by the fact that owing particularly to the Šempeter cemetery, funerary buildings are much better documented

[115] The stele was found within a cemetery by the road Emona–Siscia, possibly suggesting a type of *vicus* for the associated settlement (Gregl 1991: 25).
[116] Bekić and Lovenjak 2002: 201–202; Leleković and Rendić-Miočević 2012: 283.
[117] Gregl 1991: 21; Leleković and Rendić-Miočević 2012: 282.
[118] Gorenc 1960: 22–27.
[119] Demo 1982: 76–77.
[120] Gregl 1997: 21.

[121] Nemeth-Ehrlich and Kušan Špalj 2003: 107–110; Ragolič 2014: 329–333, figs 2, 3 and 4; Filipec 2017: 121–122.
[122] In the older literature, such an hypothesis was favoured by Hans Pirchegger, Mihovil Abramić, and Balduin Saria, and more recently by Jana Horvat (Horvat *et al.* 2003:160).

Figure 9. Plan of Aquae Balissae (after Schejbal 2004)

for the Celeian workshop circle than Poetovian.[123] Nevertheless, it is possible that Andautonia was under the artistic influence of Celeia rather than Poetovio for practical reasons. This is because Celeia and Andautonia were connected by the waterway traffic routes via the Savinja and Sava Rivers, preferred for the transportation of monuments in Roman times, while there was no waterway connection between Poetovio and the site of Lobor, or Petovio and Andautonia for that matter. A prevailing (though not exclusive) Norican influence, in this case presumably transmitted through Celeia, can be observed in autochthonous clothing represented on some funerary monuments form the *ager* of Andautonia (A[t] I.6; I.8; A[ptNW] I.1). Certainly, the lack of a conspicuous artistic influence from Poetovio does not finally prove that Lobor was *not* in the *ager* of Poetovio, and a further weakness of this theory is the lack of Norican-type funerary buildings in the city of Andautonia, which would connect it with its presumed territorial unit at the site of Lobor. Notwithstanding all that, a territorial affiliation of Lobor with Andautonia cannot be discarded with certainty. Otherwise, the monuments from Lobor included here represent only a limited choice of those published thus far, while there are also a lot more unpublished pieces made of Pohorje marble and local limestone.[124] No traces of a Roman settlement have yet been found in the valley at the foot of Mt Ivanščica, where one stele was recovered (A[ptNW] I.1), while the remainder of the evidence was used as *spolia* in the hilltop medieval church and its premises. Given that the monuments are all fragmentary and many of them cut into thin slabs for reuse in the medieval church pavements and other structural parts, it is not possible to assess the original number of gravestones or *aediculae*, especially as only a small part of the evidence has been published. Therefore, the cemetery that has yielded all this evidence could have been associated with a country estate, but also with another type of settlement, such as a *vicus*, possibly existing in connection with quarrying on Mt Ivanščica (see Chapter III).

*Aquae Balissae*

The urban topography of Aquae Balissae, including the spreading of its cemeteries, is insufficiently known. So far, those have been located to the south of the hypothetically fixed city perimeter, and also on the north side (Figure 9). Both these funerary grounds allegedly produced several lavishly furnished underground funerary chambers, of which no traces or finds survive.[125] However, several funerary monuments stem from a wide area stretching from the north-west to the south-west of the city, indicating at least one cemetery on the western fringes of the city; some of them were probably found in secondary positions in the inner urban periphery on the west side. This is less likely for its eastern side, as only one funerary stone was found there, but in a secondary position as built in the city walls (AB VIII.1). The distribution of the

[123] The evidence of *aediculae* in Poetovio and its *ager* is fragmentary, with no possibility for complete reconstructions; nevertheless, use of the same types as in Noricum has been presumed (Maver 2008).
[124] Cf. Filipec 2017: 98–99.
[125] Schejbal 2004: 104 (map 4: 11–13), 112, 118.

remainder of the total of eight finds (four *stelai*, two sarcophagi, and one each building and unclassified piece (AB I.1, 2, 3; II.1, 2; V.1; VII.1; VIII. 1) is as follows: west (?) cemetery (AB I.1–4; II.2; V.1); south cemetery (AB II.1). Funerary altar AB V.1 was found 2.5 km to the west of the city, which is near enough for a city cemetery, but at the same time far enough for a private estate in the urban periphery. None of the finds was recovered in archaeological excavations. If to the eight finds from the city a further 19 from its presumed territory are added, they amount to a very modest total of 27 monuments.

In discussing the sites of the presumed *ager* of Aquae Balissae (AB[t]), it is first necessary to pay attention to a contested issue of fixing the border between this city and Andautonia on the south-west and possibly even Iovia on the north-west. Similarly, the southernmost site of Donji Podgradci (AB[t]) I.3) can only tentatively be ascribed to the city's *ager* on account of a large distance (58 km). On the other hand, this site can be hypothetically associated with the site of Mahovljani (Catalogue, Peripheral areas III, Mahovljani) in terms of their common belonging to an unknown municipality situated between Aquae Balissae in Pannonia superior and Baloie in Dalmatia, if such existed; one such possibility is Servitium (see Chapter II.1). In the total of 19 finds, there are nine *stelai*, two or four sarcophagi (two, if pieces II.3 and II.4 are discarded as not from the Roman period), one funerary building (with an unknown number of inscribed slabs), one medallion, one altar, one ash-chest and three indeterminable pieces, one of which (AB[t] VIII.1) is possibly not authentic. The distribution by sites is as follows: Biškupci (AB[t] VI.1), Bjelovar (AB[t] I.4), Brusnik (AB[t] I.5), Cage (AB[t] II.3, 4), Čečavac (AB[t] I.2, II.2), Donji Podgradci (AB[t] I.3), Kusonje (AB[t] I.6), Našice (AB[t] VIII.1), Novigrad Podravski (AB[t] III.1), Orešac (AB[t] I.1), Tekić (AB[t] I.8; VIII.2, 3), Veliki Bastaji (AB[t] I.7; II.1; VII.1), Vetovo (AB[t] I.9). None of the finds was recovered in archaeological excavations. The Roman settlement Incerus/Incerum (recorded as *Incero* in the Antonine Itinerary) has been located at various places in the Požega Valley, among others on the site of Tekić (AB[t] I.8, VIII.2, VIII.3), which has also yielded Roman baths.[126] Another candidate for the identification with Incerus/Incerum is the village of Vetovo (I.9), where undetermined Roman architecture was found.[127] Curiously, inscription AB(t) VIII.1, which was allegedly found in the vicinity of the town of Našice, but whose authenticity is unfortunately debatable, brings the name *Incero*. Otherwise, Roman Stravianae, mentioned in the Antonine Itinerary as *Stravianis*, has been tentatively located in the vicinity of Našice.[128] In the village of Orešac AB[t] I.1) Roman Bolentius/Bolentium (recorded as *Bolentio* in the Peutinger Table) has been hypothetically located.[129] The site of Cage has been confirmed as the place of a Roman villa, with no medieval finds in the immediate vicinity.[130] Nevertheless, a Roman origin for the two presumably Roman fragments (AB[t] II.3, 4) cannot be ascertained. An upland settlement on the site of Čečavac (AB[t] I.2; II.2) has been securely proven as Roman through abundant remains of building materials.[131] The city of Bjelovar (AB[t] I.4) has long been considered as completely bare of Roman finds, but this has been proven wrong recently, when the location of a cemetery has been identified close to the city centre.[132] The village of Biškupci has never been researched archaeologically, but there is a local tradition of the village's Roman (specifically, early Christian) origin, and its only funerary piece (AB[t] VI.1) seems to have been found within a funerary precinct.[133] In the village of Brusnik (AB[t] I.5), a luxurious villa or a smaller-size settlement was located.[134] The findspot at Kusonje (AB[t] I.6) had allegedly yielded another funerary stone, which was subsequently built in the foundations of a village pigsty.[135] The site of Crijepci in the village of Veliki Bastaji (AB[t] I.7; II.1; VII.1) most probably held a luxurious villa with at least two subterranean and lavishly furnished vaulted tombs.[136] In the immediate vicinity of the town of Novigrad Podravski (AB[t] III.1) a Roman baths has been partly excavated in 1925, but the location of the cemetery of the respective settlement (a presumed villa) remains completely unknown, because ash-chest III.1 is walled in a secondary position.[137]

### *Concluding remarks*

If now we compare the three cities by sites and finds, the differences both in numbers and nature of the evidence can be observed (Figure 10). Of the 70 funerary stones from Siscia (the hypothetical *ager* excluded), as many as 56 pieces stem from the cemeteries and only 14 from its presumed *ager*; the latter are distributed among four sites. To this, the area of Kordun should be added as a hypothetical Siscian territory, or at least the area gravitating towards Siscia, and furnishing as many as 66 finds distributed among several smaller and larger sites, with only seven monuments stemming from the hypothetical territory in the south-east. Owing to the particular stone-cutting nature of the Kordun region,

[126] Talbert 2000: 20 E4; Sokač-Štimac and Korać 2007; Gračanin 2010: 33–35, 53.

[127] Schejbal 2004: 121; Sokač-Štimac 2005: 2. See also M. Šašel Kos in the respective Catalogue entry.

[128] Gračanin 2010: 35, 53.

[129] Salajić 2001: 11; Gračanin 2010: 23, 50.

[130] Leleković and Rendić-Miočević 2012: 280, 287.

[131] Migotti 1997: 27–28.

[132] Jakovljević 2012: 9.

[133] Data from the Museum Archives (Arheološke bilješke 171/139, 24 December 1954).

[134] Leleković and Rendić-Miočević 2012: 303.

[135] Siniša Njegovan Stárek, personal communication.

[136] Migotti 1997: 43–44, 47–49; Schejbal 2004: 119–120.

[137] Hoffiller 1937b: 296; Leleković and Rendić-Miočević 2012: 279–280.

| | I. Stelai | II. Sarcophagi | III. ASH-Chests | IV. Tituli | V. Altars | VI. Medallions | VII. Buildings | VIII. Unclassifiable | |
|---|---|---|---|---|---|---|---|---|---|
| Siscia S | 8 | 22 | 2 | 4 | | | 3 | 17 | 56 |
| Siscia S(t) | 7 | 1 | 2 | | 2 | | 1 | 1 | 14 |
| Siscia S(pt) | 10 | 8 | 50 | 1 | 1 | | | 3 | 73 |
| Total | 25 | 31 | 54 | 5 | 3 | | 4 | 21 | 143 |
| Andautonia A | 1 | 1 | | | | | | | 2 |
| Andautonia A(t) | 11 | 1 | | | | 1 | | 3 | 16 |
| Andautonia A(pt) | 3 | | | | | | 7 | 2 | 12 |
| Total | 15 | 2 | | | | 1 | 7 | 5 | 30 |
| Aquae Balissae AB | 3 | 2 | | | 1 | | 1 | 1 | 8 |
| Aquae Balissae AB(t) | 9 | 4 | 1 | | | 1 | 1 | 3 | 19 |
| **Total** | **12** | **6** | **1** | | **1** | **1** | **2** | **4** | **27** |
| **Sum Total** | **52** | **39** | **55** | **5** | **4** | **2** | **13** | **30** | **200** |

Figure 10. Numerical survey of the monuments by kinds and cities (made by: B. Migotti and Lj. Perinić)

the total amount of the finds from the hypothetical Siscian *ager* is, so to say, artificially augmented, and this should be kept in mind in the comparative analysis of the entire area of study. Eight of a total of 27 funerary stones from Aquae Balisssae stem from the urban cemeteries, while 18 were recovered from the presumed *ager*, with one of the sites (Donji Podgradci) only tentatively belonging to, or just gravitating towards, Aquae Balissae. Of a total of 30 finds from Andautonia, only two (three, if a missing sarcophagus, not included in the Catalogue, is counted) stem from the urban cemeteries, while its presumed *ager* furnished 16 finds distributed across 11 sites. The hypothetical *ager* of Andautonia has furnished an additional 12 funerary monuments, with as many as nine of them stemming from the site of Lobor, which actually holds many more finds.

A distribution by sites reveals various relationships at two levels: between each city and its territory individually, and between the three cities as compared. The large number of finds from the Siscian cemeteries is realistic, while the extremely small number of finds from its presumed territory undoubtedly results from an inadequate state of research. The circumstances in Andautonia are comparable, but in the opposite direction, yielding an extremely small number of monuments from the urban cemeteries, which can only be explained as accidental, and producing an abundance of rural finds. The latter may be pointing to the fact that the majority of the population of Andautonia, who were able to afford stone monuments, and especially marble ones, lived in their country estates in the city's *ager*. The ratio of urban and rural finds from Aquae Balissae seems to be slightly more balanced than in the previous two cities, but the original number of urban finds should have been much larger in proportion to those from the *ager*. Otherwise, the larger number of sites in the territory of Andautonia as compared to that of Aquae Balissae is realistic if the geography of the areas is concerned, with a fertile valley taking a considerable part of the *ager* of Andautonia and a mountainous landscape prevailing around Aquae Balissae.

Finally, if the number of finds is juxtaposed with the hypothetical territories, Siscia expectedly emerges as the site with the largest number of monuments (143), followed by Andautonia (30) and Aquae Balissae (27). It seems that such a huge discrepancy is realistic, not only on account of the historical importance of Siscia, but because the loss of Siscian monuments has been well substantiated. In other words, the possibility that the discrepancy is due to a heavier loss of monuments in the remaining two cities is practically negligible. If now the approximate surfaces of the city territories are juxtaposed with the respective number of finds, we see that the Siscian *ager* is not four of five times larger than the other two, as the number of finds would suggest. As a matter of fact, the territories of the three cities seem to be surprisingly levelled (Figure 1). However, more balanced spatial relationships would probably be gained by the knowledge of the territory of Iovia, situated between Andautonia and Aquae Balissae, as well as the location and territory of *municipium Faustinianum*, which must have been somewhere to the east of Siscia. The former would have probably reduced the territories of both Andautonia and Aquae Balissae, while the latter would have presumably further reduced the *ager* of Aquae Balissae, and would have also affected the area of the Siscian *ager* in the south-east. Admittedly, such 'restructuring' would be even more nuanced if the hypothetical autonomy of Roman Servitium is considered. With all these calculations in mind, the predominance of the evidence from Siscia still seems to result from the importance of this city rather than from its territorial scope.

## II.3. STRUCTURAL TYPOLOGY, ICONOGRAPHY AND CHRONOLOGY OF THE MONUMENTS IN THEIR MATERIAL, SOCIAL AND RELIGIOUS CONTEXTS

### *II.3.1. Structural typology, iconography and chronology*

*Introduction*

Only six epitaphs have furnished original words naming the monuments: *arca* (S II.9, 22), *memoria* (S II.10; AB[t] I.5) and *titulus* (S[t] I.1; S[ptSW] I.4). The word *arca* was expectedly used to denote sarcophagi, while the other two expressions are not precise enough, as transpires from the fact that *memoria*, as a multiple-meaning word, was used for both a stele and a sarcophagus.[138] Finally, the word *titulus* was indeed used for the monument combining structural traits of both *stelai* and *tituli*. Still, there is no doubt that this word did not specifically denote the kind of monument of the same name as employed in the Catalogue, as *titulus* is another multiple-meaning word.

When a structural typology of the *stelai* is concerned, what first strikes us is a comparatively large number of types to a small number of pieces. A differentiation between architectural and non-architectural *stelai* has been neglected here, as such classification varies among authors, and is ultimately not functional enough; in spite of their relative diachronic sequences, both types were used from the 1st to the 4th centuries.[139] Subtypes have been established of necessity, but such affiliations are often arbitrary and inconclusive due to the fragmentary state of many monuments, so that quite a few remain hypothetically attributed to one (sub)type or another. Except in the case of type 4, types were not devised on account of tentatively restored structural elements, even if such restorations would have been probable. On the other hand, subtypes were devised as broadly as possible, that is, including variants without a further subtyping; particular cases will be commented below, within the discussions of individual cities' evidence. Subtypes and variants are indicative of the presumed model of production, in which local workshops using stone other than marble probably imitated and/or slightly adapted original marble models (see Chapter II.3.3). In addition, there are monuments that should fall into amalgamated forms of *stelai* and other kinds of funerary stones, but are for convenience classified among *stelai*, such as S(t) I.6, AB I.3, AB(t) I.6 or S(ptSE) I.2.[140] On balance, the typology as given here contains nine types of *stelai* with subtypes (types 1, 2, 3a and 3b, 4, 5a and 5b, 6, 7, 8, and 9), the majority of which are represented by single pieces (types 1, 2, 4, 6, 7, 8, and 9) (Figure 11). The types are arranged in a roughly chronological order, with the following basic structural components: type 1 (plain inscription panel, free pediment); type 2 (moulded pediment in a pentagonal upper part, columnar inscription panel, unframed base/pedestal); type 3a (moulded lion pediment, frieze, architrave, columnar inscription panel, unframed base/pedestal); type 3b (moulded pediment within a rectangular upper part, frieze, columnar inscription panel, base/pedestal); type 4 (moulded lion pediment, frieze, portrait niche, decorative band, inscription panel, continuous columns flanking the portrait niche and inscription panel, base/pedestal?); type 5a (moulded lion pediment, frieze, columnar portrait niche, decorative band, columnar inscription panel, decorative band, framed base/pedestal); type 6 (detached lion top, simply framed pediment in a rectangular upper part, columnar portrait niche, simply framed inscription panel); type 7 (top of an unknown shape, moulded inscription panel and base framed by a continuous decorative band, tenon base); type 8 (linearly incised structural elements: pediment in a rectangular upper part, blind frieze, rectangular body with a circular inscription panel); type 9 (concave pentagonal upper part with an arched niche, blind frieze, columnar inscription panel, unframed base/pedestal).

Of other kinds of monuments only sarcophagi and ash-chests come in sufficient numbers to produce a coherent typology. These two kinds have been classified into the same typological category because they share shapes and tectonics, which at times makes it impossible to differentiate between them despite some discriminating features, such as larger dimensions of sarcophagi and more substantial bottoms of ash-chests. True, sarcophagi should be larger, but the problem occurs with the juxtaposing of an ash-chest and a child sarcophagus. On the other hand, the bottoms in ash-chests are typically much thicker than in sarcophagi.[141] However, several chests in the Siscian evidence display dimensions proper to sarcophagi, but with disproportionally thick bottoms and without a securely identifiable stone 'cushion' on one of the short sides to suggest an inhumation. In the Catalogue they are marked with an asterisk denoting hesitancy about their classification. This phenomenon has been noticed among researchers, who mostly failed to discuss it, but Fulvia Ciliberto did posit that a massive bottom was a secure marker of an ash-chest despite the measurements of the box.[142] Nine types (Figure 12) have been established for the sarcophagi and ash chests: type 1 (moulded inscription field and plain lateral panels of various width); type 2 (moulded inscription panel,

[138] The word *arca* to denote sarcophagus was often used in Dalmatia (Cambi 2010: 21–22), while one funerary altar from Rome was named *memoria* in the epitaph (Boschung 1987: 13).
[139] Migotti 2013a.
[140] Amalgamated structural typologies abound in Roman funerary monuments (Pflug 1989: 47–50; Cambi 2010: 7/253; M. Scholz 2012: 3).
[141] Migotti 2013b: 181.
[142] Rebecchi 1978: 229; Koch 1993: 154–155; Ciliberto 2007: 328; Cambi 2010: 25.

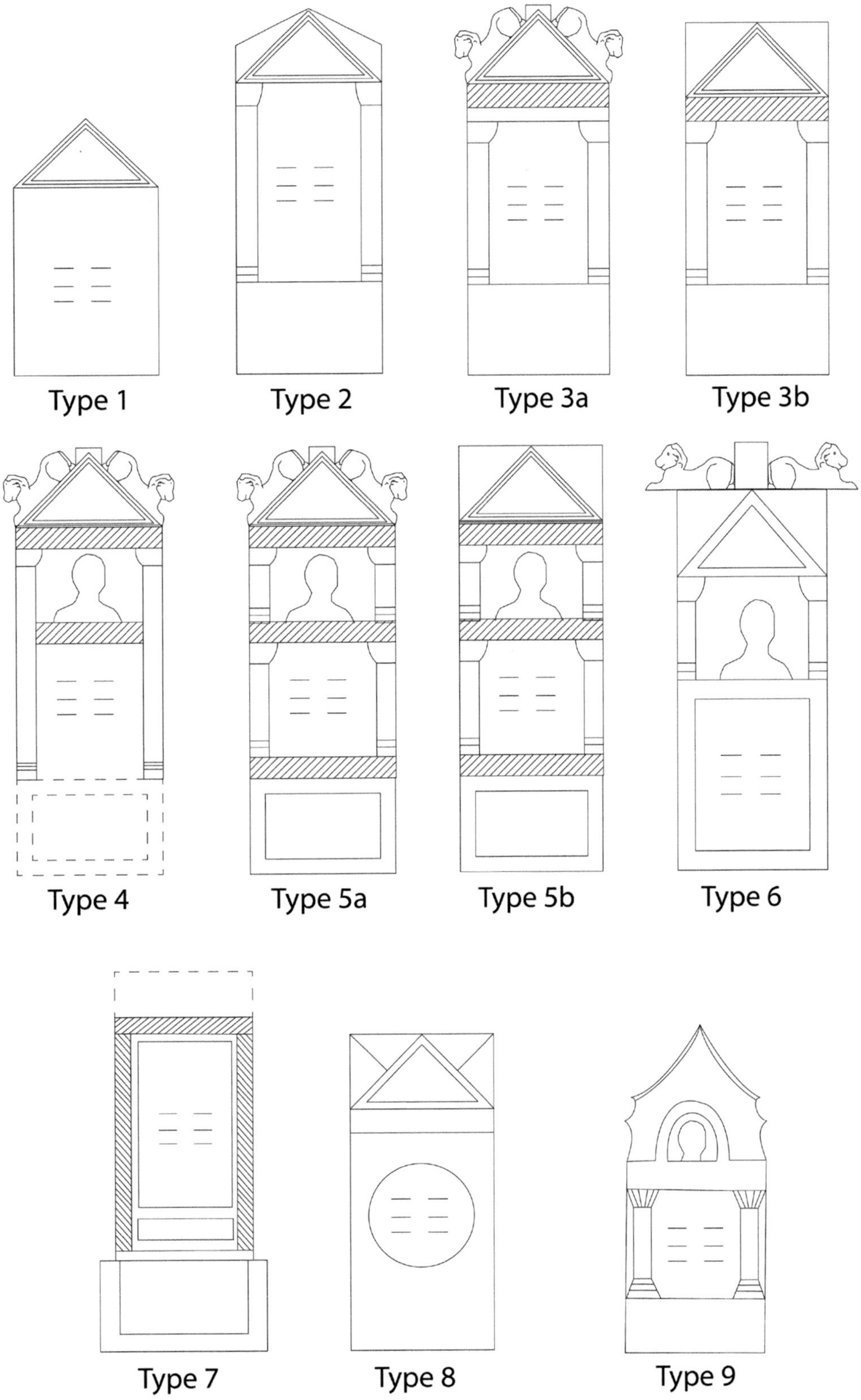


Figure 11. Typology of *stelai* (made by: D. Fudurić and T. Leleleković)

architectural arched lateral niches, moulded short sides); type 3a (moulded inscription panel flanked by moulded arched lateral niches); type 3b (the same as type 3a, but provided with a detached base); type 4 (moulded inscription panel within a simple outer frame); type 5 (framed inscription panel flanked by framed rectangular lateral niches); type 6 (framed front side with a central inscription panel in the form of a *tabula ansata*); type 7 (plain); type 8 (plain front with a central circular inscription panel). Lid types (Figure 13) are the following: type 1a (gabled with corner *acroteria* – higher); type 1b (gabled with corner *acroteria* – lower); type 2 (gabled and tiled); type 3 (gabled with middle and corner *acroteria*).

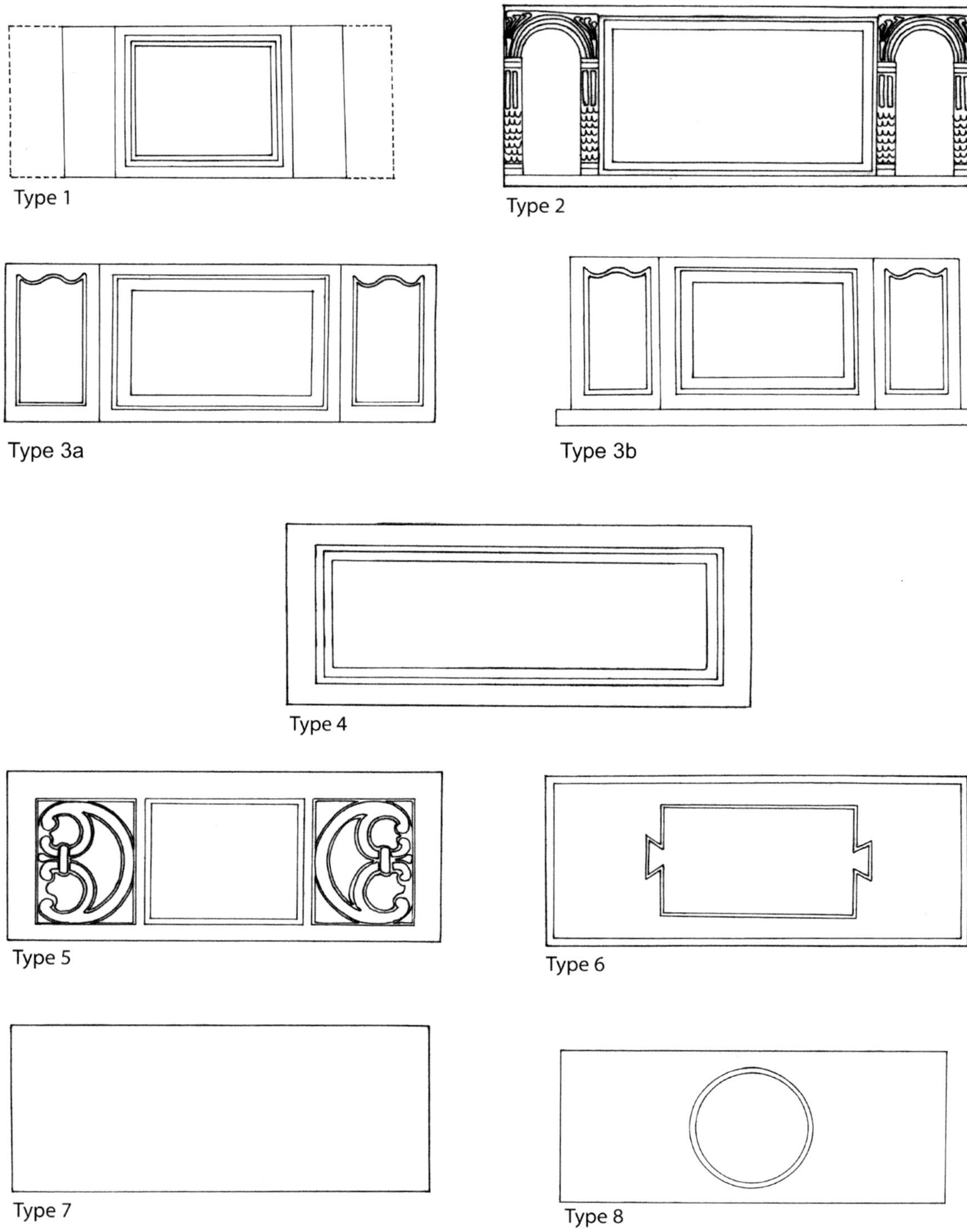


Figure 12. Typology of sarcophagi
(made by: M. Galić and Lj. Perinić)

*Tituli* make an important kind of funerary monument, because they indicate the existence of buildings, be that underground chambers, tumuli, grave enclosures, or the like. Generally, they should be defined as relatively thin slabs taking the inscription encircled by a frame (although not considered as indispensable by all commentators) and should be recognized by less carefully worked sides prepared for walling, as well as the lack of devices for fitting onto abutting structural components, as typical of a specific category of Norico-Pannonian funerary buildings (*aediculae*). Unlike *aediculae* (floored structures composed of individual relief stone blocks, fitted together by various devices), funerary buildings and enclosures were constructed from any kind of mortared building materials.[143] However, the severest obstacle to recognizing a funerary *titulus* is the fact that, when fragmented, it can

[143] On such monuments in Noricum and Pannonia, see Kremer 2001; Ertel 2010; M. Scholz 2012.

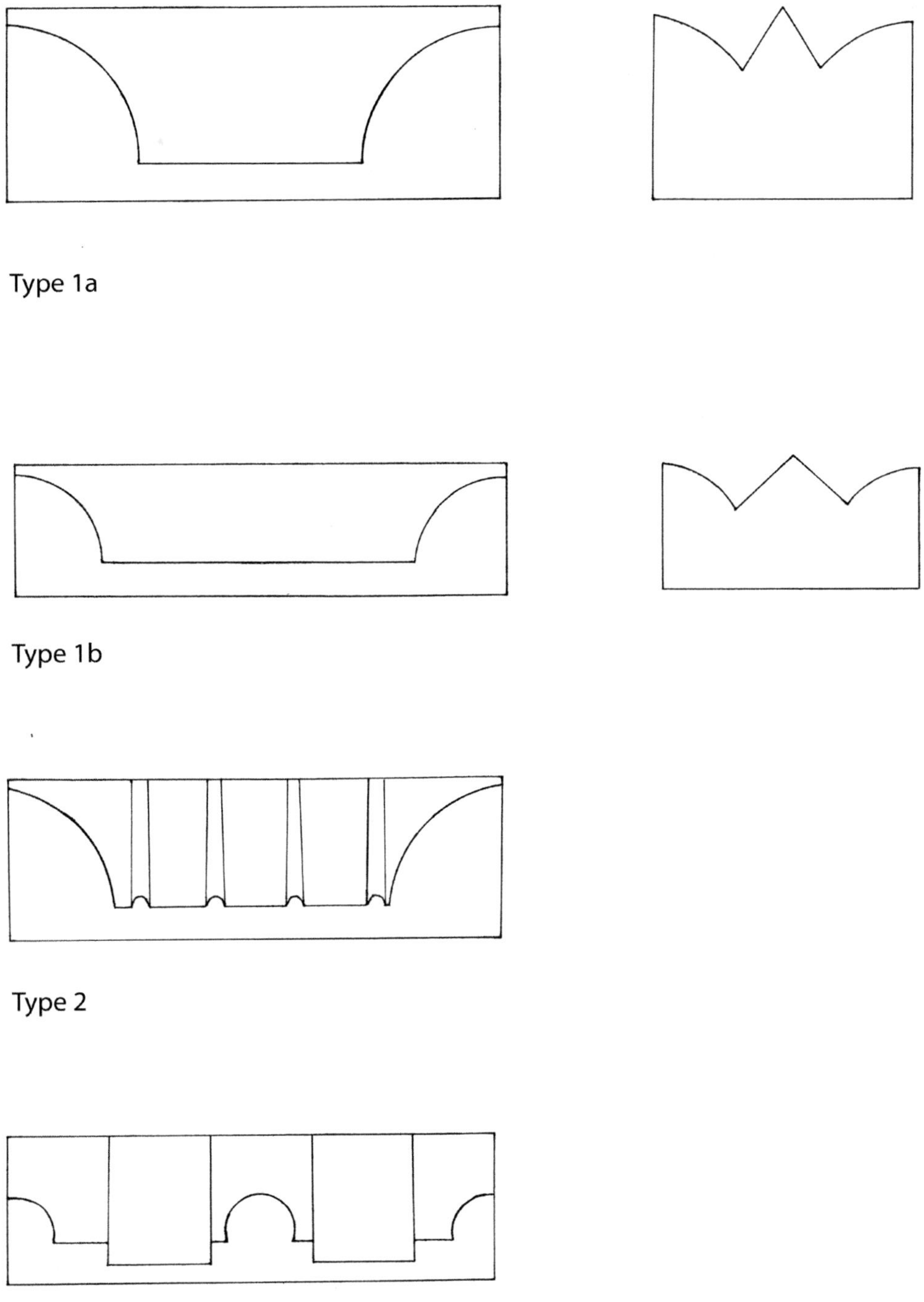


Figure 13. Typology of sarcophagi lids
(made by: M. Galić and Lj. Perinić)

be mistaken for an inscription belonging to any kind of monument.[144] While *tituli* are generally numerous in Noricum in Pannonia, they are under-represented in the present evidence, which can be explained in two ways. Firstly, they are most easily lost by destroying and reusing owing to relatively small dimensions, and secondly, it is very likely that a considerable part of the pieces which remain unclassified (Category VIII) are actually *tituli* fragments.

A similar uncertainty is inherent in a further two monument kinds that indicate funerary buildings: a portrait relief from *aediculae* and a portrait relief from other types of funerary structures. The nomenclature of these monuments varies greatly, especially in German and Italian, and among others they are often

[144] On *tituli*, see Schober 1923: 4, *passim*; Kremer 2001: 153–160, 363; Mosser 2003: 230–236, *passim*; Ertel 2010: 149; Lazar 2013: 263–264. See also *lupa*, s. v. Titulus.

alternatively called *portrait niches*; therefore, it is important to explain the terms used in each individual discussion.[145] There are several structural differences between two kinds of portrait reliefs (niches), relative to the way they were incorporated into their architectural settings, the most important of them being in fitting devices. Portrait reliefs from *aediculae* should be provided with fitting devices on the short sides and/or the upper and bottom surfaces, such as clamps and dowels, while reliefs from other buildings lacked such mechanisms, as they were inserted by using mortar; the former were also generally of a higher quality than the latter.[146] Neither of these two kinds of monuments is securely evidenced in the present material. Portrait reliefs of an *aedicula* can be presumed for Aquae Balissae (AB VII.1 – probably; AB I.3 – a slight possibility), while only one such building (S VII.1) is ascertained for Siscia, with several more possibly belonging to this kind of monument (S VII.2, 3; S[t] VII.1). As for the building other than *aediculae*, they can only be tentatively presumed on account of *tituli* in Siscia, while a funerary enclosure may have existed in Andautonia, as suggested by a shaft block (A[t] VIII.3).

A further two monument kinds that were widespread in Noricum and slightly less in Pannonia, are scarcely present in the area of this current research: funerary medallion[147] and funerary altar. A poor representation of funerary altars in the present evidence is in tune with the general scarcity of this kind of monument in Pannonia, contrary to Noricum where it abounds.[148] The biggest obstacle to a reconstruction of funerary altars in the Norico-Pannonian area is the fact that they are mostly found as disintegrated parts.[149] The majority of pieces are represented by topless and baseless shafts, which has resulted in a slightly contradictory approach towards the terminology of the funerary altar. The question is whether to consider as an altar each such parallelepiped block featuring a frontal inscription and relief-decorated sides but lacking a crown and/or a base. While Gabrielle Kremer leaves open the possibility that such shafts took different tops, which call for a different typological classification, recommending therefore their naming as shaft-blocks and not altars, Erwin Pochmarski consistently names such pieces as altars.[150] Those few pieces that appear in the present evidence are considered under the title of altars.

The three main categories of iconographic subject matters that will be considered in this discussion are: portraits with clothes, jewellery and attributes included; mythological and genre scenes, the former including scenes in the proper meaning and single motifs, such as mythological persons and animals depicted individually or in various relationships between themselves or with floral motifs; floral and geometrical decorative motifs. In regard to the last-mentioned category, it is necessary to point out that the term *decorative* is here used conventionally, while it has been maintained that each and every motif used in a funerary context had its specific original symbolism, even if it was not straightforwardly disclosed to all the participants in funerary rites and rituals at all times.[151] The same is even truer of individual mythological motifs, such as a *gorgoneion* or pediment lions, which are otherwise often perceived as purely conventional and/or decorative.

*Siscia*

The city itself has furnished 8 *stelai*, 22 sarcophagi (20 chests and 2 lids), 2 ash-chests, 4 *tituli*, three fragments of funerary buildings, and 17 fragmented and unclassifiable monuments (a total of 56 pieces). From its presumed *ager* come seven *stelai*, one sarcophagus, two ash-chests, two altar pieces, one each *aedicula* fragment and an unclassifiable piece (a total of 14 pieces), while its possible (hypothetical) *ager* produced ten *stelai*, eight sarcophagi (seven lids and one chest or lid), 50 ash-chests, one *titulus*, one each altar and *aedicula* piece, and three unclassifiable fragments (a total of 73 pieces). Taken as whole, Siscia and its territory have produced 25 *stelai*, 31 sarcophagi (21 sarcophagi, nine lids and one undetermined piece), 54 ash-chests, five *tituli*, three altars, four *aediculae*, and 21 unclassifiable fragments, that is, altogether 143 pieces.

Of the 25 *stelai*, only 12 examples can be typologically classified: type 1 (S I.1.); type 2 (S I.2); type 3a or 3b (S[t] I.4); type 3b or 5b (S[t] I.5); type 5a (S I.8; S[t] I.2, 3); type 5a or 5b (S[ptSW] I.1; S[ptSE] I.3, 5); type 8 (S[ptSW] I.2); type 9 (S[ptSW] I.4). Only three *stelai* are marble (S I. 4, 6, 8), all originating from the city. In so fragmentarily preserved evidence it is not possible to discern hypothetical marble models for local limestone pieces. In the case of the most frequently represented

[145] Cf. Migotti 2012a: 389. The above differentiation is not straightforwardly expressed in English, as the buildings providing contexts for the two types of reliefs cannot be so clearly distinguished as in German (*aediculae* vs. *gemauerte Grabbauten*).

[146] Cf. Kremer 2001: 24–25, pl. 8: I,24; I,26 (portrait reliefs from *aediculae*); 160–161: pl. 18: 1,82; I,83 (portrait reliefs from other buildings); see also pp 359–360.

[147] Cf. Schober 1923; O. Harl 1991: 32–34; Palágyi 1991; Kremer 2001: 344–347; Pochmarski 2011: 22–26; Lazar 2013: 262. See also *Lupa*, s. v. Grabmedaillon (Noricum, Pannonia).

[148] Ertel 2010: 161–163. According to the evidence on the *Lupa* (last accessed on 25 June 2018), there are 68 funerary altars in Pannonia and 170 in Noricum. These numbers should be taken roughly, as there is no knowing how many pieces remained outside this database. Nevertheless, they can be considered as a reliable framework for a juxtaposition of two provinces about equally covered by the *Lupa* database.

[149] Kremer 2001: 349; Handy and Pochmarski 2014: 57, 83; Ertel 2010: 151–154; Pochmarski and Weber-Hiden 2016: 115–144, nos 66–91.

[150] Kremer 2001: 351; Handy and Pochmarski 2014: 57, 83–84; Pochmarski and Weber-Hiden 2016: 18–19.

[151] Cf. Koch and Sichtermann 1982: 601; Dexheimer 1998: 16; Ortalli 2005: 267–279; Fejfer 2008: 106; Chioffi 2015: 641–642; Mainardis 2016.

type (5a), stele S I.8 could not have served as a model for two limestone *stelai* from Topusko in the Siscian *ager* (S[t] I.2, 3), and for further two in its possible south-east territory (S[ptSE] I.3, 5). This is not only on account of the lack of a frieze and dividing decorative bands between the panels in stele I.8, which makes it a simplified variation on the original type 5a, but also on account of its later date, as all five limestone examples are earlier. Therefore, the limestone *stelai* must have been fashioned as reworked examples of the original type 5a, probably imported to Siscia from one of the Norican workshops, as this type has not been identified in the territory of Poetovio.[152] How exactly the local workshop became acquainted with this model escapes us, but two *stelai* from Topusko need to be commented upon, as they are from the same place and evidently from the same workshop, displaying curious variations on type 5a. Stele I.3 is closer to type 5a in the structure of its pediment, while the affiliation with this type depends on the presumed restoration of its missing middle and lower sections. Stele I.2 provides a better clue for its model, because its structural analogy can be found in Piber, a site in the vicinity of Salla and Kainach, the quarries supplying Flavia Solva (Leibnitz, AU).[153] From this place comes the portrait niche of a marble stele featuring a stylized shell background and flanked by columns sitting on bases in the form of high shafts.[154] However, the pediment of this stele is missing, so the shape of its upper part remains unknown and cannot be juxtaposed with a particular variation of the pediment of the stele from Topusko, which is considerably higher and shorter (stopping short of the architrave's ends) than usual for type 5a. While a general model for the pediment whose triangle is slightly narrower than the architrave supporting it can be presumed in *stelai* with accentuated corner *acroteria*, the examples of virtually narrower pediment triangles are rare, and do not furnish a convincing clue for the origin of stele S[t] I.2. Even though, a marble monument from the territory of Poetovio, carved in Pohorje marble, needs to be taken into account.[155] On the other hand, the above-mentioned tombstone from Piber furnishes a good analogy for both the portrait niche flanked by columns sitting on very high parallelepiped pedestals and for the background of the portrait niche in the form of a simplified shell.[156] Therefore, it suggests that the model for stele S(t) I.2 was created in a Norican workshop, most probably in Flavia Solva, to be imitated locally. As suggested above, the original marble model, which should have existed in Siscia, has not reached us.

With its thickness of 33 cm, stele S(t) I.6 stands out among the remainder and was tentatively defined as a pillar (*Pfeiler*).[157] However, the ratio between its width and thickness (88:33 cm) rather speaks in favour of a large stele with possibly a lion top and an inscription field below the two relief panels. The remaining 13 *stelai* are either insufficiently preserved to be classified typologically, or are missing without leaving an inscriptional record (S I. 3–7; S[t] I.1, 6, 7; S[ptSW] I.3; S[ptSE] I.1, 2, 4, 6). Nevertheless, one of them (S I.3) most certainly represents an additional type not given in the list; it was not classified as a separate type because its full shape remains unknown. Judging from the comparative evidence, the top of this stele could have been straight or arched, and there could have been another panel between the inscription field and the pedestal.[158] Stele (S[ptSE] I.2) is of amalgamated form: its thickness, corner *acroteria* and vertically elongated form suggest a stele, while its florally decorated border on all four sides is characteristic of *tituli*. Only one stele from the *ager* of Siscia was found together with the tenon base, accompanied by presumed remains of a pottery urn.[159] The earliest stele (S I.1) is dated to the first half of the 1st century, which makes it the earliest example in the whole of northern Croatia. Near to it in terms of chronology is the only example of type 2 (S I.2); its tentative later date has been refused on the grounds explained in the Catalogue. The latest this stele can be dated is the last quarter of the 1st century, on analogy of similar *stelai* from Carnuntum.[160] *Stelai* of type 3 are generally dated from the late 1st to the mid 2nd century, while the only Siscian example (S[t] I.4) falls in the beginning of the 2nd century.[161] The most numerous *stelai* are those of type 5 (S I.8; S[t] I.2, 3; S[ptSW] I.1; S[ptSE] I.3, 5) spanning the time period from the early 2nd century to the late 3rd. The latest *stelai* are from the later 3rd, or the turn of the 3rd and 4th centuries (S I. 8 – type 5a; S[ptSW] I.2 – type 8; S[ptSW] I.4 – type 9).

All of the 20 sarcophagi chests can be classified typologically: type 1 (S II.1, 6*, 12); type 2 (S II.2); type 3a (S II.4*, 7, 8, 9, 10; S[t] II.1); type 4 (S II.3, 5*); type 5 (S II.11); type 6 (S II.22); type 7 (S II. 13, 14, 15, 16, 17, 18*, 19*). The same is true of all but one (S II.12) of the 12 lids (nine individual lids and three coming with chests): type 1a (S[ptSW] II.1, 2, 4, 5, 6, 7, 13); type 1b (S II.8); Type 2 (S. II. 21, 22); type 3 (S II.11). Only one chest is marble (S II.10), and one was possibly marble (S II.9); both are from the city. Sarcophagus S II.2 is most curious as the

[152] Cf. Djurić 2010; Djurić 2016, especially fig. 3 on p. 181.

[153] Salla and Kainach are quarries in Oswaldgraben (some 60 km to the northwest of Leibnitz), which are alternately (and alternatively) used in literature to denote a local marble source for Flavia Solva, without discriminating between the two. See Hudeczek 2008: 11; Djurić and Müller 2009: 115–116.

[154] *Lupa* 1609.

[155] Djurić 2008: 161, 165, fig. 2.1 (*lupa* 3102). Further examples: *lupa* 374 (Noricum), 2846 and 3053 (Pannonia).

[156] Another example of a high base stems from Klagenfurt (AU), but it is segmented and applied to a different type of stele (3b) – see *lupa* 1144.

[157] M. Scholz 2012: 209–211.

[158] Škrgulja and Migotti 2015: 31–33.

[159] Brunšmid 1907: 184. The base was subsequently acquired for the museum but has gone missing in the meantime.

[160] Weber-Hiden 2014: 72, fig. 1 (type D 1–3).

[161] On the chronology of the type, see Djurić 2010: 208–210.

only one of architectural type, which was evidently produced locally on the model that remains unknown, as it could not be found in the available evidence from either Noricum or Pannonia.[162] Thus, these monuments seems to have been carved under the influence of North-Italian (most likely Aquileian) production, with no clear trace of its possible marble model transmitted through Norican workshops. As suggested also by its iconographic scheme (see below), this piece can be perceived as a local rendering fashioned under various influences to produce a hybrid and possibly unique example.[163] The same is true of sarcophagus S II.22, which is a unique piece in combining a *tabula ansata* as a common feature of funerary chests everywhere, including north Italy, but not in combination with bucolic motifs in the place of Erotes. If not from north Italy or from the local tradition of funerary art, the inspiration for the sarcophagus in question could have come from Salona, where the sarcophagus with the *tabula ansata* was a standard type from the 3rd century, while the free space on the front was from the start of the production often filled with floral motifs.[164]

Contrary to sarcophagi represented by one to three pieces to a type (types 1, 2, 4, 5 and 6), types 3a and 7 predominate heavily among the chests, while the same is true of type 1a for the lids. Sarcophagus II.19 should be a variant of type 7, as its slightly rounded inside corners remind of tub-shaped sarcophagi (*lenoi*), typical of Rome and Italy, and appearing only sporadically in northern provinces;[165] no such example from Noricum or Pannonia is known to this present author.

Owing to the 50 ash-chests from the Kordun (S[ptSW] III. 1–50), the overall number of this type of funerary monument is disproportionally large. However, the city of Siscia and its presumed *ager* have furnished only four examples: two lidded chests (S III.2; S[t] III.1) and one each lid (S[t] III.2) and chest (S III.1). None of the 50 ash-chests from the Kordun is complete; there are 28 chests, 20 lids and two undetermined pieces. The chests from Siscia and its presumed territory can be classified typologically as follows: type 4 (S III.1); type 3a (S III.2) and type 7 (S[t] III.1), while of the three lids one falls under type 1a (S III.2) and two under type 1b (S[t] III.1, 2). All of the lids from the Kordun fall under type 1a, while the majority of the chests fall under type 7; there are only six exceptions, belonging to type 8 (S[ptSW] III.9, 12, 14, 27, 35 and 49). In sum, a small number of examples from Siscia and its presumed *ager* do not allow any meaningful observations on structural typology, while the south-west hypothetical Siscian territory is characterized by a clearly distinguished typology falling under two types. Type 7 represents the most basic shape of sarcophagi and ash-chests from Siscia and its immediate and wider area. Type 8, featuring a circular inscription panel and appearing only in the Kordun evidence, represents a specific trait of one of the two lines of local production (the other represented by type 7).

The ratio of ash-chests and sarcophagi should indicate a shifting from incineration towards inhumation in the funerary ritual. Given that the ratio between sarcophagi and ash-chests in Siscia and its presumed *ager* is in favour of the former (23:4), while the situation is opposite in the city's wider surroundings (8:50), an hypothesis can be put forward that the rite of inhumation entered the urban society earlier than the rural one, which could have been more traditional and adhering to the older ritual. The prevailing number of ash-chests was indirectly but convincingly explained by the main researcher of the area through a sudden stop of stone production in the 270s.[166] Nevertheless, sarcophagus S II.22 and ash chest S III.2, most probably made of the stone quarried in the Kordun, testify that production there continued, at least in a very reduced form, possibly just for the needs of Siscia. As for the above-mentioned ratios between sarcophagi and ash-chests, they would change for the city and its presumed *ager* if the thick-bottomed sarcophagi marked with an asterisk are considered to be ash-chests.[167] None of the five pieces marked with an asterisk (S II. 4*, 5*, 6*, 18* and 19*) have a thickening on one of the short sides, which would qualify them as sarcophagi; only in chest S II.4 does the bottom rise gradually from the left to the right side. In chest S II.5, now missing, the inner of the bottom was not accessible, while the remainder (S II.6, 18 and 19) do not show any cushion-like thickening. If, therefore, these five pieces are considered as sarcophagi, the ratio between sarcophagi and ash-chests in the city and its presumed territory becomes more balanced (18:9), but still in favour of inhumation. However, these calculations need to be put in a perspective that changes them considerably, again only for the city and its presumed *ager*. This perspective concerns the fact that only above-ground stone monuments have been considered here, and not those that were not intended for viewing.[168] An illustrative example is an ash-chest from the village of Gola, 87 km NE of Ščitarjevo and 70 km NW of Daruvar, and therefore tentatively in the *ager* of Andautonia, Aqauae Balissae, or Iovia (Figure 14).[169] It was not included in the evidence because it had

[162] The only three (marble) architectural sarcophagi known to the present author from the Norico-Pannonia region are structured differently; two of them stem from Sirmium, Pannonia (*lupa* 4329 and 4358), and one from Teurnia, Noricum (*lupa* 3660).

[163] Migotti 2016b.

[164] Migotti 213b: 198–199.

[165] Cf. Koch and Sichtermann 1982: 80–82; Ciliberto and Mainardis 2015; Sinn 2003: 308: fig. 1.

[166] Perkić 2012: 189–190.

[167] See fns 4 and 5.

[168] Cf. Fejfer 2008: 132. The verb *viewing* here implies anyone, and not everyone, meaning that receptacles held in underground chambers were presumably accessible only to the family and friends, while those in the open were visible to everyone.

[169] Kolar-Sušanj 1971.

Figure 14. Ash-chest from Gola (photo: Lj. Perinić).

been buried inside a tumulus and was not intended for viewing, which is further corroborated by its extremely course craftsmanship. Otherwise, this piece is only slightly smaller (chest: length 56–63 cm, height 38 cm) than the average ash-chests, while sharing with them a structural typology. Finally, the insight into the chronological aspect of the shift from incineration to inhumation is much hampered by inconclusive chronologies of some of the monuments, as well as the fact that both kinds were used during the 2nd and 3rd centuries, while a larger group of plain sarcophagi (S II.13–20) cannot be dated more precisely than within the broad time-span of the 2nd to 4th centuries. It is generally considered that the height of sarcophagi can be indicative of chronology, growing in time from quite low and elongated receptacles towards those considerably higher. Accordingly, the ratio between the length and height of approximately 4:1 is typical of early specimens of Roman sarcophagi in various areas, while the approximate scale 3:1 or less takes over from the beginning of the 3rd century.[170] The majority of the Siscian receptacles reveal the latter scale either closely or approximately, the latest of them, securely dated to the early 4th century (S II.22), showing the ratio of nearly 2:1 and thus theoretically speaking in favour of the dating criterion in question. Curiously, the majority of the chests of type 7 (S II. 13–18) reveal the scale of 4:1 or slightly less, which should be pointing to an earlier date of their manufacture, that is, the turn of the 2nd and 3rd centuries. Nevertheless, in the lack of additional arguments, those pieces were dated more broadly.

All of the four *tituli* from Siscia (S IV.1–4), as well as the one originating from its hypothetical territory (S[ptSE] IV.1), are actually Late Roman unframed (or only slightly framed by incised lines) slabs, which means that the present evidence has failed to produce any traces of earlier funerary buildings and enclosures as known from Noricum and Pannonia. All but one (S IV.1) of the five *tituli* were made of Pohorje marble. Given the meagre evidence and the lack of context, it cannot be hypothesized with any certainty that those *tituli* were fashioned from reused earlier stones, as was the case in Sirmium, where early Christian inscriptions were mostly made of reused Mediterranean marbles, originally imported for Late Roman Imperial buildings.[171] Such reusing can nevertheless be presumed for Siscia on account of a marble fragment inscribed on both sides (S VIII.1). Generally, the Siscian *tituli* possibly suggest a continuation of the import of marble from the Alpine region in Late Antiquity, but this cannot be established with certainty. On balance, the evidence for funerary buildings substantiated by *tituli* is meagre, but still probably larger than can be proved. The evidence for *aediculae* is equally scanty, with some of it questionable either in terms of provenance or funerary use. Two of the three pieces from Siscia probably testify to luxurious marble *aediculae* (S VII.1, 2), while the third (S VII.3) is even more questionable because its findspot is unknown. The only fragment from the presumed territory (S[t] VII.1) can be perceived as a reduced simplified form of Norico-Pannonian *aediculae* of mausoleum type; if their inner architectonic elaboration is neglected, their upper parts receive pentagonal contours.[172]

The only two funerary altars from Siscia are from the same site (Odra Sisačka) in its presumed territory (S[t] V.1, 2). The first one is a now missing shaft, with only one of the sides decorated, if the preserved drawing is faithful. However, it has been posited that both sides of funerary altars were originally carved in relief, and that the lack of such depictions is due to the state of preservation.[173] The other piece is a pyramidal top, characteristic of one type of altar structure.[174] These two pieces comply chronologically with funerary altars in Noricum and Pannonia, which are dated from the end of the first to the second half of the 3rd centuries.[175]

Two of the 17 unclassifiable fragments from the city of Sisak need to be commented on. The first is a statuette

[170] Migotti 2013b: 203–204.

[171] Djurić, Müller and Filipović 2010: 10. The reuse of marble in Late Antiquity was customary, cf. Rebecchi 1978: 202, 254–255; Fejfer 2008: 65, *passim*; Barker 2012; Russell 2013: 35; Previato and Mareso 2015: 301.

[172] Kremer 2001: 28–29, 36–37, 50–51, figs 6, 10, 20. The same shape can be identified in some *stelai* (*lupa* 361, 1539, 2635).

[173] Dexheimer 1998: 4; Handy and Pochmarski 2014: 84.

[174] Kremer 2001: 267–270, 351–352; Ertel 2010: 81–82; M. Scholz 2012: 233. Contrary to pyramidal altar tops, pyramidal roofs of *aediculae*, typical of north-western provinces and north Italy, do not appear in Noricum and Pannonia (M. Scholz 2012: 128). On the other hand, such altar tops, mostly decorated with two dolphins by a trident, are frequently found in Aquileia (*lupa* 14030, 14125, 14126, ff).

[175] Kremer 2001: 343–344; M. Scholz 2012: 233–238; Handy and Pochmarski 2014; Pochmarski and Weber-Hiden 2016.

of a lion (S VIII.16), which can be hypothetically interpreted as one of a pair watching the entrance to a grave, as witnessed in several Norico-Pannonian cities.[176] However, the dimensions of the Siscian example, which is far smaller than is usual for such sculptures, leaves open the question of its funerary architectural setting. Therefore, the possibility that it was a fragmented lion top of a stele cannot be excluded, the more so as its base is broken along the whole length, which means that its long side could have been the short side of the base of a lion top.[177] The second is an unusual stone fragment, looking like a small altar but undeterminable precisely (S VIII.17). Neither this piece's structure nor its inscription have parallels in Norico-Pannonian funerary monuments, while both the sphinx and the exclamation *zaesis* (= *zeses*) / *vivas* are in accordance with funerary milieus, the latter possibly suggesting a Christian context, as it often comes with an adverbial phrase *in deo*.[178] Otherwise, the column on an inscribed basis from Siscia is associative of Greek archaic grave *stelai* topped by statues of sphinxes.[179] Perhaps this can be brought in connection with the fact that, in spite of the Latin name of the presumed deceased (Ianuaria) and the Latin script of the epitaph, the Greek verb *zaesis* was used. Curiously, despite its Latin descent and its presence across the Roman Empire, the name Ianuarius in some contexts possibly reveals Oriental connections.[180] This stone probably stood within a family grave enclosure or a funerary chamber.

All of the above-mentioned iconographic components (portraits with clothes, jewellery and attributes included; mythological and other scenes; decorative motifs) are evidenced in Siscian monuments, although not abundantly. Portraits are completely or partly preserved in 15 examples (S I. 7, 8; S[t] I.2, 3, 6; S[ptSW] I.1, 3, 4; S[ptSE] I. 3, 4, 5, 6; S II.10; S III.2; S[t] VIII.1), and possibly on S II.9. A married couple is represented on four monuments (S[t] I.2, S[ptSE] I.5; S II.2), as is a family with children (S I.7, 8; S[ptSW] I.1; S[ptSE] I.6). One person is depicted on two *stelai* (S I.3; S[ptSE] I.3), three persons on three (S[t] I.6; S[ptSW] I.3; S[ptSE] I.4), while on the front of one sarcophagus, otherwise reserved for the portraits of a married couple, a servant boy and a servant girl are depicted (S II.10), yet not in a customary iconography of servants, but with portrait features of the era. Only two of the married couples were depicted in the gesture of *dextrarum iunctio* (S[t] I.2; S[t] VIII.1), while parents hugging their children can be presumed in all four cases, although obscured on stele S(ptSW) I.1.

On the majority of the tombstones female hairdos can be only tentatively associated with Imperial fashions, or not at all, for various reasons: heavy damage, a possibly unfinished state of execution, or extremely rough workmanship. The same is true of wrapped hairdos or hairdos worn with bonnets/hats, which can sometimes be only roughly identified as types established in the well-researched Norico-Pannonian evidence. Even in the examples which are not completely disfigured, recognizing the hairdo and dating the portraits is problematic in funerary sculpture, because the back of the heads remains invisible. On stele S I.8 all three female hairdos are the same, featuring a typical forehead arrangement in voluminous locks on both sides of the central parting, uncovered ears and short geometricized strands below them, which represent the hair braided and folded at the back and raised towards the crown to form a chignon. In its front view, this hairdo corresponds to that of the Severan empresses as well as those of the wives of the so-called soldier emperors, covering the entire time span of the first half of the 3rd century,[181] while its precise appearance and dating remains hidden under the headwear. Even such broad chronological span is in discrepancy with the date of the stele in the second half of the 3rd century (based on the type of brooch worn by the men), thus exemplifying a fairly widespread feature of the lag in female fashions in Roman funerary portraiture in the provinces. This happened for various reasons, such as the distance from the trend-setting centres, the preference for the hairdos of their youth in older women, and the use of tombstones only long after their carving.[182] While concealing the full appearance of the hairdos, the headwear in question is the closest in appearance to type H 7 as established for the Middle Danube region by Ursula Rothe (Figure 15). Significantly, this type is dated from the 1st century to the Severan period, that is, correspondent to the date of the hairdo depicted on stele S I.8.[183] Unfortunately, the headwear on *stelai* S(t) I.2 and 3 is mostly illegible due to extreme wear or, rather, the unfinished state. In stele S[t] I.3 it resembles Norican type H 1.1.1. (Figure 16), with the hairdo which customarily matches this type remaining obscured on the forehead, but also differing in the short strands of hair behind and off the ears, which do not appear

[176] Hudeczek 2008: 41.

[177] Cf. Dautova-Ruševljan 1983: 31, no. 224.

[178] On the sphinx, see Nicole 1873, and on the exclamation *zeses/vivas*, see Migotti 2001: 5, and also *EDR*139234, 139511, 149097, 267511, 006603, 247206 ff.

[179] Charbonneaux, Martin and Villard 1969: 153–154, fig. 190.

[180] Alföldy 1969: 220.

[181] M. Bergmann 1977: 91–99, pls 7 and 8; von Kersauson 1996: 414–415, 422–423, nos 191, 195; Ziegler 2000; pl. 26: 80–82; pl. 27: 87, ff; Croom 2002: 104, fig. 48.17.

[182] Braemer 1959: 17; Pflug 1989: 13; Pochmarski 1996: 133; Braemer 2005: 32; Fejfer 2008: 280–284, 356; Cambi 2010: 66, fn. 378/296; Rothe 2012: 162; Birk 2013: 17–19, 77, fn. 328; Skupińska-Løvset 2014: 41.

[183] Rothe 2012: 205–209, fig. 54. Presumably, south-Pannonian examples were not included in this typology, not because they would not have been recognized as such, but because the research comprised only the Middle Danube provinces. On a stele depicting the same headdress in combination with the hairdo of (probably) theempress Faustina the Elder, with the hairstyle on the forehead the same as in the Siscian example, see *lupa* 2879 (Budapest/Aquincum, Pannonia). On Faustina's hairdo, see Croom 2002: 102, fig. 47.10; von Kersauson 1996: 210–211, no. 90.

Figure 15. Types of women's head-wear in the middle Danube region (after Rothe 2012: fig. 54)

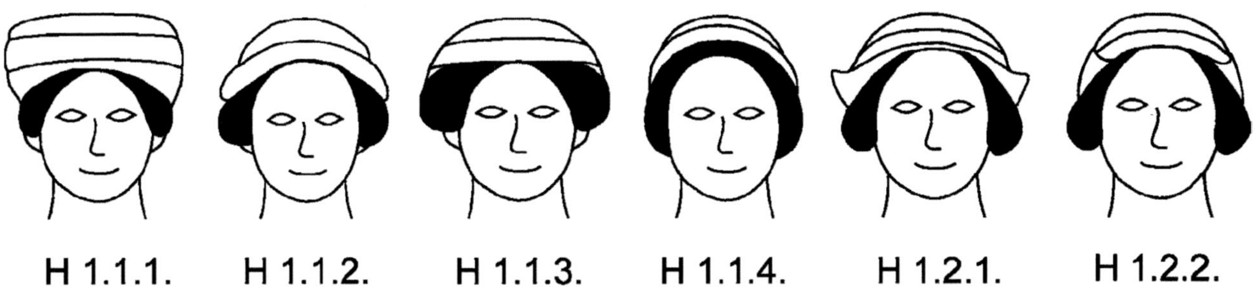


Figure 16. Types of Norican bonnets (after Rothe 2012: fig. 48)

in the Norican typological model.[184] If indeed these strands represent a hairdo, they push the date of the stele towards the 3rd century as on stele S I.8 above, which is in strong discrepancy with the early date of the headwear type H. 1.1.1 (mid 1st century to the early 2nd century). However, the dating in the present example remains further insecure because the hairdo on the forehead is not visible. It is virtually impossible to establish the female headwear and hairdo in stele S[t] I.2, as its hairdo evidently contains a central parting, but with the rest of the hair on the temples seemingly peeping underneath as if covered by a headwear whose structure is not clear due to the harsh work. Its shape does resemble Norican type H 1.1.4. (Figure 16), dating from the late Antonine to the late Severan period, but it should be observed that this type has not been evidenced in combination with the hair parted on the forehead.[185] On stele S[t] I.6 the female hairdo also remains unclear, although it should probably be interpreted as following the types of the middle and second half of the 3rd century, with a high chignon and thick strands of hair sticking off the sides of the face.[186] A slightly less likely possibility is that this is a veiled headdress of type V 2, typical of south-east Noricum and featuring a veil shorter than in such headdresses in northern Pannonia (Figure 17).[187] It is not possible to determine with certainty whether the woman on stele S(ptSE) I.5 is bareheaded or whether she is wearing a head-cover. Judging from small strands of hair sticking from the sides of the lower cheeks, in both cases she is featuring a hairdo of the first half of the 3rd century.[188] If there is any headwear, it should be of Norico-Pannonian type H 7, dating from the 1st century to the Severan period, suggesting a time lag as with stele S I.8.[189] In four or five of the interpretable examples the women in the portrait niches are bareheaded. Despite heavy wear, the woman on stele S(ptSW) I.3 features a fairly clear shape of the hairdo of the middle and the second half of the 3rd century, notably of the empress Salonina,

[184] Rothe 2012: 200, fig. 48.
[185] Rothe 2012: 200–203, fig. 48.
[186] M. Bergmann 1977: 96–97, pl. 29: 3, 4, pl. 30: 4; Croom 2002: 104, fig. 48.19. A close parallel is found on a stele from Celje (Celeia, Noricum; *lupa* 3610).
[187] Rothe 2012: 195–197, fig. 44.
[188] M. Bergmann 1977: 29–30, 39, pl. 7: 1–4; von Kersauson 1996: 414–415, 422–423, nos 191 and 195; Croom 2002: 104, fig. 48.17.
[189] Rothe 2012: 205–209, fig. 54.

Figure 17. Types of separate veils in the middle Danube region (after Rothe 2012: fig. 44)

with a chignon pushed towards the forehead.[190] The hairdo of the woman on sarcophagus S II.10 is certainly from the mid or second half of the 3rd century.[191] On the other hand, the Imperial model for the hairstyle of the lady on ash-chest S III.2 is extremely difficult to establish for two reasons. Firstly, it is damage to the stone which produces a false impression of the hair, taking three superimposed horizontal layers instead of a central parting with a 'melon-style' arrangement at the front. Secondly, women sometimes wore hairstyles without revealing recognizable prototypes, or with an emphasized personal touch to original models.[192] Therefore, it is possible that the hairdo in question was a crudely executed imitation of the hairstyle as established for the previous example (S II.10). However, it can also be an imitation of a hairstyle from the turn of the 3rd and 4th centuries.[193] Although stele S(ptSE) I.6 is missing and is accessible only through a low-quality illustration from the literature, the Late Roman hairdo of the man is clearly discernible, while that of the woman is heavily damaged and therefore not so readily recognizable; still, she seems to be wearing a 'tiered' hairstyle, as sported by the empresses Helena and Fausta at the beginning of the 4th century.[194] Finally, if we presume that on sarcophagus S II.9 four female mythological figures depicted with identical hairdos impersonate the deceased Maximiana Aemilia, as is often the case in funerary iconography,[195] her hairdo can tentatively be interpreted as reflecting a hairstyle of the period. In view of that, Maximiana Aemilia's hairstyle can be perceived as an imitation of the empress Iulia Domna's wigged style.[196]

Of the extremely small number of examples showing discernible female clothing, Roman *palla* seems to be worn by a girl on S I.7 and women on three *stelai* S(ptSW) I.1, 3 and S(ptSE) I.6, but such interpretation remains insecure due to a bad state of preservation. In all other cases the women are wearing native clothes, in most cases combined with hairdos of the era, while in the remainder native headdresses come with matching clothes. The most curious example in terms of blended styles is produced on stele S I. 8; the Roman fashion is displayed in the hairdos of the female deceased and their jewellery, but the mother is still wearing a native cloak hanging on her shoulders and falling in straight vertical folds down the sides of her upper body. Arguably, it corresponds to a pre-Roman overcoat (Figure 18), but combined with a wide-sleeved tunic and not a tight-sleeved bodice.[197] The research of Ursula Rothe has established that in the Middle Danube region pre-Roman dress was worn well into the 3rd century.[198] Nevertheless, native clothes should not be expected at the end of the 3rd century, and in a family of Roman knights, but it seems that at least some items of native dress were cherished much longer in southern Pannonia.[199] Such blend of native and Roman fashions is possibly employed on stele S(t) I.6, on which the woman is dressed in native clothes, while her hairdo may be interpreted as revealing a hairstyle of the era and not a native, veiled type. Given that the women on two *stelai* from Topusko (S[t] I.2, 3) are depicted wearing native headdresses, no doubt they sported native clothes as well, but this is ascertained only in stele no. 3, on which two large brooches of a wing type on the shoulders are discernible, most probably holding a sleeveless overtunic.[200] This can be considered as the easternmost evidence for a Norican fashion in the entire area of research, as other such finds are limited to the territory of Andautonia. Possibly, the owners of the stele from Topusko were immigrants from Noricum. The remainder of the portrayed women wearing local clothes are represented on three *stelai* from the possible Siscian territory in the south-east (S[ptSE] I. 3, 4, 5). One of them (no. 3) is the only other example depicting brooches as a mark of local female fashion; unfortunately, they are fragmentarily preserved and not represented clearly enough. This is possibly due to a shortened perspective, so that the dress depicted associates a bulbous overtunic of Rothe's type O 1, whose belt is hidden behind the folded lower arms; this type was spread over eastern Noricum and the neighbouring parts of Pannonia (Figure 19).[201] The type of brooches depicted on this stele cannot be established

[190] M. Bergmann 1977: 98–100, pl. 27: 2, pl. 28: 3, 4; Öğüş 2014: 139–140, fig. 5.

[191] Kleiner 1992: 378–380, fig. 349; von Kersauson 1996: 492–493, no. 232; Ziegler 2000: 98–125, pl. 34: 102, 103; pl. 35.

[192] Braemer 1959: 17; Pflug 1989: 72–73; B. Scholz 1996: 169; Fejfer 2008: 356.

[193] M. Bergmann 1977: 194–195, pl. 58: 5, 6; pl. 59: 1, 2.

[194] Wessel 1949: 72. fig. IV; von Kersauson 1996: 524, no. 25; Croom 2002: 104, fig. 48.22; Donati and Gentili 2005: 212, no. 8.

[195] Koch and Sichtermann 1982: 615–617; Koch 1993: 50; Kremer 2006: 47–52; Gabler 2007: 22; Fejfer 2008: 124–127; Lo Monaco 2011; Birk 2013: 94–95, *passim*.

[196] Kleiner 1992: 326–328, fig. 291; von Kersauson 1996: 370, no. 170; Ziegler 2000: 73–82; pls 20 and 21; Croom 2002: 102, fig. 47.15.

[197] Rothe 2012: 178, fig. 26: 4.

[198] Rothe 2012: 174. According to other commentators, this happened during the Severan period (Gabler 2007: 26; Facsády 2014: 199).

[199] This can be related indirectly to the phenomenon of a 'renaissance' of various indigenous traditions in later Antiquity. See Šašel Kos 1998: 20; Reuter 2003; Adam 2007: 289; Russell and Laycock 2011: 87–88.

[200] Rothe 2012: 178, fig. 26: 3.

[201] Rothe 2012: 183 –184, fig. 31.

Figure 18. Schematic drawing of the main components of native women's dress in central and north-western Europe (after Rothe 2012: fig. 26)

with certainty because only very elongated heads are preserved, but they can hardly be identified as either of a wing or of a *Doppelknopf* type, both of which are customarily matched with the native dress. The brooches on the stele in question should sooner be recognized as of a knee or a *kräftig profilierte* type, which were also worn with a pre-Roman garment, extending a possible date of such depictions to the turn of the 3rd and 4th centuries.[202] Hanging rings as seen in our example are sometimes found with brooches matched with a native dress, but this habit is rarely documented on funerary monuments, and seems to have been limited to Pannonia.[203] The women depicted on two *stelai* from the same site (Cikote) in the possible south-eastern Siscian territory are wearing local clothes, but with no clear parallel in the known Norico-Pannonian repertoire. The one on stele S(ptSE) I.5 seems to be dressed in a tunic with extremely wide sleeves and a cloak draped on the breast in thick, oval and V-shaped folds in the manner more resembling the Roman *palla* than a native dress; still, it can hardly be interpreted as a Roman piece of clothing. The clothes of the woman on stele S(ptSE) I.4 is even more unusual, as she is wearing a tunic with long and tight sleeves and above it apparently a stiff and heavy lapelled cloak.[204] Two examples of the female dress attract special attention for their unusual shapes, the first being the toga worn by the woman (probably a servant girl depicted in the guise of her mistress) on sarcophagus S II.10. In spite of some inconsistent details, such as the lack of a rounded lower edge and *lacinia,* and the fact that the contabulated *umbo* is covered by the upper part of the *sinus* on the left shoulder, this robe should be designated as toga, above all on the basis of its clearly depicted *sinus* and a contabulated *umbo.*[205] Therefore, it should not be confused with the so-called contabulated *palla*, because none of the variants of contabulated garments, which appeared in the 3rd century, under the influence of oriental priestly robes, to become characteristic of the female dress, possesses a sinus.[206] Otherwise, differentiating between those garments is much hampered by the fact that mostly busts are depicted on funerary monuments. Significantly, the garment worn by two ladies depicted as full-standing figures on a sarcophagus from Brescia was named an *eigentümliche Togatracht*, in spite of featuring a sash instead of a clearly executed *contabulated umbo*, as is seen on the sarcophagus from Siscia.[207] Although at first glance the stately *togata* associates the owner of the sarcophagus in question, this figure actually represents a servant girl in the guise of her mistress, as transpires from many depictions of paired servants on Norico-Pannonian funerary monuments. This is further substantiated by a servant girl on a funerary altar from Stallhofen (Noricum), depicted as wearing a toga-like garment and a hairdo of the era.[208] The other of the two unique cases of female dress is represented on ash-chest S III.2, where the lady seems to be wearing a sashed robe, mentioned in the previous example as *not* representing the toga. Three items of clothes are discernible in this depiction. First, there is a tunic with

[202] On the types of brooches worn with a native dress in the Middle Danube provinces, see Rothe 2012: 213–215, *passim*. On dating of knee brooches to as late as the 4th century, see Riha 1979: 85.
[203] Rothe 2012: 189, fig. 39. See also *lupa* 76, 734, 760 and 2255.

[204] The only two similar examples the present author could find stem one each from Pannonia Superior (*lupa* 2273) and Noricum (*lupa* 349).
[205] Possible misunderstandings in the arrangement of the toga, as well as local or individual details in its draping, could have resulted in some inconsistencies in the carving of togate statues in regional or local workshops. Cf. Fejfer 2008: 197.
[206] On contabulated garments, see Gabelmann 1973: 103, pl. 41: 1; Kranz 1986: 231, fn. 138; Pflug 1989: 98; Glaser 1997: 128–129; Croom 2002: 82–83, fig. 34.2; Migotti 2007b: 66–69; Fejfer 2008: 251–252.
[207] Gabelmann 1973: 103, pl. 41: 1.
[208] Pochmarski 2005: 361–363, fig. 6.

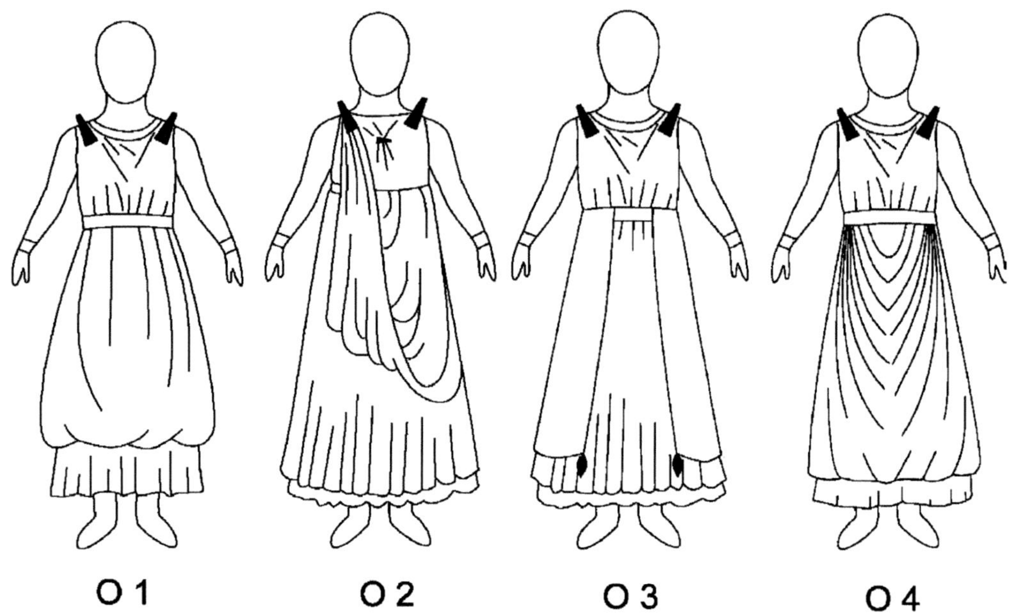


Figure 19. Types of over-tunics in the middle Danube region (after Rothe 2012: fig. 31)

an oval neckline and long sleeves whose shape is not readily recognizable because of the damage at the area of the right lower arm and wrist; either the sleeves were not very wide or they were wide but possibly tightened at the wrists. Then there is a *palla* whose diagonal folds stretch from the left hip to the right knee, resembling the *sinus* of the toga and falling also over the woman's left lower arm. The third item consists of a scarf-like piece of cloth whose upper part falls over the left shoulder perpendicularly to the left knee, ending in fringes, while the remainder stretches diagonally from the left armpit to above the right hip. Although quite resembling a contabulated *umbo* of a toga, the lack of a proper *sinus* favours its interpretation as a sash, possibly indicating a religious or some other ceremonial dress.[209] All in all, cloaks worn by the women in the Siscian territory are quite varied, while tunics with the neckline invariably oval to rounded come in two variants regarding the sleeves; these are always long but can be either tight or wide. Of the jewellery, only necklaces and armbands (or bangles) are discernible in several examples. However, the latter can never be securely distinguished from cuffs in cases when long- and tight-sleeved tunics or bodices are depicted. Armbands are more easily recognizable when there is more than one band around the wrist, while cuffs are clearer when they are rolled back, but the latter come with male, notably military, fashions.[210] On the whole, hand jewellery is clearly discernible on only two monuments. On stele I.8 a beaded bracelet is carved, matching the same type of necklace as the only one such item of jewellery in the evidence from the Siscian territory.[211] Two armbands seem to be depicted on the right hands of the women on *stelai* S(t) I.6 and S(ptSE) I.3. However, in the former case this remains debatable, because the wide, hanging sleeve seems to be gathered at the wrist, leaving open the possibility that at least one of the two bands represents a cuff and another perhaps an armband. Three kinds of attributes are found with the portrayed women. The first is a roundish fruit held in their right (S. I.8; S[t] I.6; S[ptSE] I.3, 4; S II.10; S III.2) or left hand (S[ptSW] I.3). A scroll, held in the left hand, is evidenced three times (S[t] I.6; S II.10 and S III.2); in the last example, the woman is pointing at it with her right hand. In only one instance (S[ptSE] I.4) is a small bulbous vessel depicted, probably a toilet bottle, as a fairly rare attribute in Noricum and Pannonia, contrary to some northern provinces.[212]

Hairdos in the male portraits show Imperial models, or presumably so, of the time span from the Antonines or the early Severans (S[t] I.2), to the soldier emperors (S I.8; S[t] I.6; S[ptSW] I.3; S II.10; S III.2), to the early 4th century (S[ptSW] I.4; S[ptSE] I.6). Due to the bad state of preservation there are some ambiguities of dating between the emperors of the 2nd and 3rd centuries (S[t] I.3; S[ptSW] I.1), as well as the lack of a more refined dating within the 3rd century (S[ptSE] I.5).[213] Also, hindrances to dating provincial funerary stones on the basis of portraits and imperial fashions are true for male portraits as well as female, but for slightly different reasons; they refer to the way of carving

[209] Migotti 2005b: 377–378.

[210] Cf. Sumner 2009: 50, fig. 39; Rothe 2012: 145, 181.

[211] This type of jewellery is worn on 3rd- and 4th-century funerary stones (Maršić 2014: 216; *lupa* 492, 1204, 1223, 1355, 4613, ff).

[212] Rothe 2009: 125, T 33, ff.

[213] On the basic styles of male hairdos from the Antonines to Constantine, see Wegner 1939; M. Bergmann 1977; Kleiner 1992: 237–463; von Kersauson 1996: 117–543; Croom 2002: 65–67, figs 23 and 24.

rather than to a time lag or specific variants of Imperial hairdos.[214] Therefore, establishing Imperial models was completely unrealistic in some cases while in others it remained questionable, resulting in only a cursory discussion of male portrait features. Nevertheless, portraits whose Imperial models are more easily perceived should be noted: S I.8; S(t) I.2; S(ptSE) I.6; S II.10; S III.2; curiously, all date from the later Roman period. All men whose clothes are discernible are wearing tunics and *saga*. Where discernible, the former have long sleeves that are either wide (S I.8; S[t] I.6) or tight (S[ptSW] I.3; S[ptSE] I.4 and possibly 5). The *saga* are fastened with brooches of various types on the right shoulder; these are round on *stelai* S I.8; S(t) I.6; S(ptSW) I.1, 3 and S(ptSE) I.4, possibly of a knee type on S(ptSE) I.6, and indistinguishable on S(ptSE) I.5. The *sagum* has fairly clearly determined social and chronological implications as an item of clothing in the Norico-Pannonian region. Depicted on funerary monuments, it generally points to soldiers from the turn of the 2nd and 3rd centuries to the later 3rd century. This, however, cannot be taken as an absolute criterion for establishing the nature and date of individual funerary portraits, as correctly stressed by Ursula Rothe and some other commentators. This is because the *sagum* was worn in the 1st and 2nd centuries, and was, throughout, depicted also on civilian monuments.[215] According to numerous analogies from both Germanias, the veteran on stele S I.3 was clothed in a tunic and a cloak or even a toga.[216] The attribute appearing on male portraits is nearly always the scroll, which is regularly associated with persons wearing *saga*, with only one exception (S[ptSE] I.6). Men are holding the scroll in their left hand and are pointing at it with their right in all examples but one. In this exception (S[ptSE] I.4) one of the two men is holding the scroll with both hands, while the other is holding it with his left hand, but is not pointing at it with his right, with which he is touching the end of his *sagum*. Curiously, and possibly significantly, in one case (S I.8) one of the men holding the scroll is pointing at it, while the other is not. The only other attributes apart from the scroll can be found on *stelai* S(t) I.6 and S(ptSW) I.3. On the former, one of the men seems to be holding a sword tucked under his left arm, while at the side of the other a diptych and a bunch of scrolls is depicted. On the latter, one of the two men is holding a scroll, while the other one is most probably represented with a diptych and a *stilus* in his hands. Finally, there is a sarcophagus holding depictions of two servants (S II.10), which should normally be discussed within genre scenes. Nevertheless, an exception is made here because the servants are depicted in the place reserved for the portraits and in iconography that suggests a particular symbolism of the servant girl (see above). On the other hand, the servant boy fits in with the customary 'servant iconography', but still with nuances, given that a towel in his left hand is ubiquitous in such contexts, while a skillet in his right is not.[217] A similar doubt concerning the nature of the horseman on monument S(ptSW) VIII.3 comes from the fact that the full shape of the stone remains unknown, and the horseman motif occurs often on various parts of Roman of funerary monuments: bases, pediments, and portrait fields. Presumably, it was a kind of a generic portrait of the monument's owner related to its status, such as on stele A(t) I.3.[218]

In terms of hairdos and clothes, children's portraits are discernible in only two examples, in one of them clearly (S I.8), and in another (S[ptSE] I.6) less so. In both cases the children are 'small copies' of adults, wearing the same kind of clothes and hairstyles. A curious detail is perceivable in the latter example, in which one of the daughters is wearing the same classic form of *palla* draped sinuously across the chest, as does her mother, while another one is wearing a typical Late Roman *palla* crossed diagonally on the chest.[219] No conclusions can be drawn on such meagre evidence, but these two examples are in accordance with the fact that children in funerary iconography were everywhere treated as 'small adults' and not as an individual social category.[220] Only on one stele (S I.8) are objects in the two girls' hands discernible: a small wreath with the younger and writing implements with the older.

On two *stelai* and one sarcophagus mythological scenes in the proper meaning of the word are depicted. When first published, the scene on the base of stele S I.6 was interpreted as a centaur being pierced in the chest with a spear. Such interpretation remains insecure because a scene like that has not been evidenced in Roman art, but it can still be a reminiscence of a Greek iconography that comprises episodes with centaurs speared by hoplites.[221] The base of stele S(ptSE) I.3 holds a most curious scene, which was interpreted as a Sabine woman separating the Romans and Sabines. Another tentative explanation is the goddess Athena separating Achilles and Hector in a scene from the Iliad. Still, the former interpretation seems more plausible, because the woman separating the two warriors is not

[214] Fejfer 2008: 279–284, *passim*; Pochmarski 2011: 17, *passim*; Öğüş 2014: 138; Skupińska-Løvset 2014: 43; Pochmarski and Weber-Hiden 2016: 22.

[215] Here is a selection of abundant literature on (or mentions of) the *sagum* on tombstones in Noricum and Pannonia: Pochmarski 1996: 129–135; Pochmarski 2011: 19–21; Rothe 2012: 159–160, *passim*; Migotti 2013c; Ubl 2013: 305–317; Carroll 2015: 506; Hainzmann 2015; Handy and Pochmarski 2016; Pochmarski and Weber-Hiden 2016: 23.

[216] *Lupa* 6975, 6983, 15538, 15539, 15543, 6975, 6983 ff.

[217] Cf. Piccottini 1977; Handy and Pochmarski 2014: 85–87.

[218] On many examples of horsemen on Roman funerary monuments, see *Lupa* s. v. Reiter.

[219] For the latter type, see Howells 2015: 128–129, nos 38 and 39.

[220] Pflug 1989: 73; Jaeger 2003: 477; Jaeger 2007: 32; Birk 2013: 157; Facsády 2014: 198; Carroll 2015: 504–505.

[221] Leventopoulou 1997: 686, no. 175, ff; Sengelin 1997. On the translation of iconographic schemes from Greek into Roman art, see Vacinová 2017: 53.

depicted in the guise of Athena. No parallels for either of these two episodes are known to the present author in Roman funerary iconography, while both are found in modern sculpture and paintings; presumably, both were sourced from Classical art.[222] Although the episode of the punishment of the unfortunate Sabine Tarpeia is known from Greek and Roman coins,[223] an unknown (or lost) Roman prototype that presumably inspired the Classical relief by the French painter Jacques-Louis David, seems to be a more likely model for the relief on stele S(ptSE) I.3. This is because the scene depicted on it is much more similar to the one on David's painting than to any represented on coins. The most lavish use of Classical myth was employed on sarcophagus S II.9, on which various episodes connoting death were depicted: Eros with Venus or Helena, Apollo and Daphne, and Hercules and Hesione.[224] Individual mythological motifs can be classified in two groups, the first comprising those employed repeatedly as a part of a workshop 'package offer' with no apparent individualised approach, such as a *gorgoneion* (or a rosette probably with the same symbolism) flanked with birds (S I.8; S[t] I.3), dolphins (S I.4; S[t] I.2; S[ptSW] V.1), sea monsters (S I.2), lions (S I.8; S[t] I.2, 3), an eagle (S[ptSE] I.6),[225] and Erotes, as the most numerous in the present evidence (S[t] I.2, 5; S II.2, 4, 7, 8; S III.2). In funerary art Erotes appear in various guises, with two dominant types: mourning Erotes and Erotes supporting a garland. The most widespread in the present evidence is the former type, depicting Erotes with a reversed torch and the head bowed and supported on one hand, sometimes holding a wreath in the other hand.[226] Floating or striding Erotes supporting a garland (S III.2) or the sides of an arched niche (S[t] I.5) are each represented only once.[227] A very important aspect of the motif of Erotes in Norico-Pannonian funerary art is its chronological sensitivity. This aspect was amply discussed by Erwin Pochmarski on a number of occasions, starting from the study of Roman models and leading to a categorical conclusion that in Norico-Pannonian funerary art the 'baby-type' Erotes (small boys) lasted from the mid Antonine to the early Severan period, when they began giving way to somewhat bigger and more grown-up boys.[228] This was observed also by earlier commentators as a general phenomenon presumably related to the enlargement of the sarcophagi's dimensions and the altered ratio between the chests' height and length.[229] While these observations are generally true and well substantiated archaeologically, the thus established chronological rule cannot be taken as absolute in every individual case, for various reasons. One of these is a difficulty in distinguishing between various ages of the depicted figures.[230] Another reason, inherent to all chronological sequences, is the lack of a sharp time border between the two types, which were evidently in use contemporarily for some time. A further three observations should be taken into account when estimating the chronological sensitivity of the Erotes' age. First, it is a fact that in north Italy sarcophagi production 'baby' Erotes were taken from Attic sarcophagi and those larger ones from Asia Minor workshops.[231] Also, in the local Dalmatian production of sarcophagi Erotes remained styled as little boys from the 1st to the 4th centuries.[232] Further, the Erotes on the early 2nd-century stele from Andautonia (A[t] I.3) are shown as grown boys, while on only one Siscian example are the Erotes represented as personifications of seasons (the so-called season *genii*) looking like older boys (S II.8). The remainder are depicted in the guise of small boys, but whose age is not easily discernible due to a course execution in all examples but one (S II.2). According to the established chronology, they should be dated to the late Antonine period, or the early Severan period at the latest. Nevertheless, they were dated here exactly to the Severan period, supported on the examples provided with other chronologically sensitive elements: S II.2 and 7, and III.2. Sarcophagus S II.2 was dated to the later 2nd or, rather, the early 3rd century on account of several indicators, such as onomastic traits (the lack of a *praenomen*), size and shape of the receptacle, stylistic-iconographical features of its reliefs, as well as the choice of motifs of the Dioscuri on its short sides.[233] While the Erotes are an impersonal subject-matter preceding portraits, the motif of the Dioscuri in this case bears on the religious activity and social status of the deceased

[222] On the scene from the Iliad by the 19th-century Italian artist Giovanni Maria Benzoni, see http://metmuseum.org/art/collection/search/209060, and on the scene from the Roman myth by the 18th and 19th century French painter J-L David, see https://www.google.com/search?rlz=1C2LENP_enHR513HR537&ei=HmfcWpqVA5KmwQL03JOwCw&q=David%2C+sur+l%27enl%C3%A8vement+des+Sabines+&oq=David%2C+sur+l%27enl%C3%A8vement+des+Sabines+&gs_l=psy-ab.3..0i22i30k1l2.2986.6972.0.8115.14.14.0.0.0.0.104.1257.12j2.14.0....0...1c.1j2.64.psy-ab..0.2.193...33i21k1.0.QPaCY9n34xg (last accessed, 21 June 2018).

[223] Vacinová 2017: 45, figs 2 and 3.

[224] For a more detailed discussion of those scenes and vagaries of their interpretation, see Migotti 2017b: 504–506.

[225] Although of worldly fauna, birds, dolphin and lions in funerary contexts are considered to be mythical creatures with various roles in transferring the deceased to the Underworld and/or watching over them.

[226] The mourning Eros is a motif that spread from northern Italy to the provinces of Noricum and Pannonia (Gabelmann 1973: 64). More on various types of funerary Erotes, see Cambi 2010: 47–51/282–284.

[227] On floating Erotes supporting either rounded niches or pediment sides, see Gabelmann 1977: 69–70, pl. 12: 1; *lupa* 18508 (Aquileia), 674 (Celje), 3107 (Šmartno na Pohorju), 3106 (Poetovio), 4338 (Sirmium).

[228] Kranz 1977: 352–354; Rebecchi 1978: 244; Koch and Sichtermann 1982: 206; Pochmarski 1986: 249, *passim*; Pochmarski 2015; Pochmarski 1998: 196–200; Pochmarski 2009: Pochmarski and Pochmarski-Nagele 2013: 288–290.

[229] Koch and Sichtermann 1982: 206, 220; Kranz 1986: 204–208, 233; Pochmarski 2015: 212.

[230] A good case in point are two pairs of Erotes designated as younger and older boys, respectively, while they actually look very much the same. Cf. Pochmarski 2015: 213–214, figs 1 and 2.

[231] Gabelmann 1973: 121.

[232] Cambi 2010: 48–49/283. The same was observed for some of the Norican production (Kranz 1986: 234, fn. 154).

[233] For a more detailed discussion, see Migotti 2013b: 185.

(see below). Thus, it brings its iconography closer to personal themes, otherwise fully represented by portraits of the deceased and finally marking funerary iconography from the Severan period at the latest.[234] Theoretically, the lack of portraits on this piece should date it to the 2nd century, but it seems that the Siscian workshop(s) lagged behind the iconographic trends in Rome, north Italy and Dalmatia, as the only two chests featuring portraits are dated to the second half of the 3th century (S II.10; S III.2). On chest S III.2 the motif of Erotes was relegated from the front to the short sides, as an indication of a replacement of an out-dated motif by the one becoming fashionable (a portrait). At the same time, this means that Erotes depicted on the front can be dated to as late as the mid 3rd century, especially season Erotes, whose appearance on Roman sarcophagi started only at the end of the 2nd century.[235] On the whole, the lack of a portrait on sarcophagus II.2 need not be an obstacle to its later dating, the same being true of the motif of the Dioscuri on its short sides, despite its relatively early prototype in Roman art. The city-of-Rome sculpture in the round of the Dioscuri in the *contrapposto* stance, such as on the Siscian example, dates from the mid 2nd century, while the sculpture showing them as striding, again from Rome, is later. Nevertheless, the former motif does not seem to have entered funerary art before the late 2nd century, so it does not contradict the first half of the 3rd century as the date for the Siscian sarcophagus.[236] Except for its chronological aspect, the motif of the Dioscuri on sarcophagus II.2 is worth mentioning for its peculiar iconography marked by *vexilla* instead of spears in the youths' hands. No parallels in Roman art have been found for this motif; it was tentatively related to the sarcophagus owner's role as an *Augustalis*.[237] The last, however feeble, argument in favour of a 3rd-century date for sarcophagus S II.2 is the fact that the other two inscriptions mentioning *municipium Faustinianum*, both from Rome, date from the 3rd century.[238]

Among the remainder of non-standard mythological motifs, one of the two pairs of Erotes on ash-chest S III.2 should be mentioned. While the Erotes supporting a garland, represented on one of the short sides, is a widespread motif, the picture of those on the opposite side is uniquely reminiscent of water nymphs. They are depicted as wingless boys holding a huge shell between them, and are roughly comparable to a pair of Erotes depicted in the pediment of stele S(t) I.2. These are supporting a circular object with a wide hole in the middle, which should be interpreted as a wide-mouthed receptacle of indiscernible shape. Although this picture fits well with the thermal nature of the monument's findspot at Topusko, the question remains why nymphs were not depicted instead of Erotes, as the former were customary in spa milieus while the latter were not, despite their connection with the marine world, possibly effectuated through Venus.[239] Whatever the answer, those two non-standard Erotes scenes are close to the second group of mythological motifs, those that appear far less frequently than mourning Erotes, possibly suggesting individualized reasons behind the choice of funerary themes. Apart from the already commented Dioscuri, this group comprises Hercules (S[t] V.1), an indistinct and probably syncretised female mythological character (S[ptSW] VIII.2), and an elaborately and singularly decorated pyramidal top of altar S(t) V.2, featuring a dolphin perched on a trident and flanked by two snakes and two lions with prey in their paws.[240]

Three kinds of genre scenes are depicted on the Siscian monuments: a hunt (St I.6), a bucolic background (S II.22); and servants (S I.3; St I.6). On stele S I.3 the servant boy is given within a well-known military iconographic scheme, typical of soldiers' tombstones in both Germanias and produced in the workshops in Mogontiacum (Mainz) and Colonia Agrippina (Köln).[241] On the other hand, servants on the short sides of stele S(t) I.6 are executed in a non-standard way, comparable to that on the front of sarcophagus S II.10. In both cases the servant boy is depicted in a regular servant iconography, that is, dressed in a tunic and holding a vessel in one of his uplifted hands. Contrarily, their female counterparts are shown as dressed in long robes and holding a scroll as an utterly irregular attribute for a servant; both girls seem to be depicted in the guise of their mistresses, probably connoting the latter's social position. As for individual decorative repertoire, the majority of monuments hold customary geometrical and floral motifs, as well as Christograms, sometimes fused on one and the same piece, which appear on funerary monuments throughout the Roman Empire (S I.1, 4; S II.22; S[ptSE] II.11; S[ptSE] I.1, 2, 5, 6; S VII.2, 3; S[t] VII.1; S[t] VIII.1). Apart from that, there are several exceptions that can be estimated as slightly more individualized, at least in the present context, such as a pedimental wreath with a central gem (S VII.1), or Christian pictograms applied to inscription S VIII. 15, if those were correctly copied from the now lost original. Finally, plain circles framing inscription panels on six ash-chests and one stele from the south-west Siscian

[234] This knowledge is well established for Rome, north Italy and Dalmatia (Gabelmann 1973: 52–56, 121–130; Rebecchi 1977: 117–122; Koch and Sichtermann 1982: 255–256; Koch 1993; 51–52; Cambi 2010: 57/290, 61/292; Birk 2013: 14–15), while its relevance for the Norico-Pannonian region has not been sufficiently studied.

[235] Gabelmann 1973: 47, *passim*, Rebecchi 1977: 115–122; Rebecchi 1978: 242, 249; Cambi 2010: 53.

[236] On the Campidoglio and Quirinal sculptural models for the Dioscuri in funerary art, see Kranz 1986: 204–208.

[237] Cf. Migotti 2005a.

[238] Šačić Beća 2017: 149–152 (nos 3 and 4).

[239] Cf. Kušan Špalj 2014: 58–61. On Erotes in the context of the marine world and Venus, see Piganiol 1892; Koch and Sichtermann 1982: 196, 238–240.

[240] For similarly decorated tops see Kremer 2001: 267–268, nos II.362–366.

[241] Škrgulja and Migotti 2015.

territory can be taken as unique, as they evidently represent a single workshop characteristic that is absent from the rest of the region (S[ptSW] I.2; III. 9, 12, 14, 27, 35, 49).

*Andautonia*

All that the city's cemeteries have furnished is one each stele and sarcophagus chest. From its presumed *ager* 15 *stelai* (of which two are individual lion tops), two sarcophagi, one funerary medallion, and three unclassifiable fragments have been recovered, while its possible *ager* produced three *stelai*, seven *aediculae* fragments and two unclassifiable pieces. They give a total of 15 *stelai*, two sarcophagi, one medallion, seven *aediculae* and five unclassifiable pieces, i.e. 30 monuments altogether.

Only six of the 11 *stelai* are preserved sufficiently for a secure or tentative typological classification: type 3a (A[t] I.3); type 3a or 3b (A[t] I.7); type 4 (A[t] I.5); type 5a (A[t] I.6, 8). Stele A(t) I.3 is the only certain example of its type in the entire territory, and the only one with an elaborate tripartite base (see Chapter II.3.3), and also the first example decorated with the so-called Norico-Pannonian volute, which is earlier than previously presumed for this motif.[242] Stele A(t) I.4 possibly falls under types 5a or 5b, but can also be an example of an individual type that was not defined as such because the shapes of its upper and lower parts cannot be restored with complete certainty. It is one of the three examples in the entire evidence of a portrait stele featuring a narrow architrave instead of a frieze, and possibly topped by a pentagonal upper part.[243] The other two examples of portrait *stelai* provided with a narrow architrave instead of a frieze come from Siscian territory (S[t] I.2, 3), which makes them exceptions from the given typological structure of types 5a and 5b, whose essential element is a decorated frieze. Nevertheless, their affiliation with type 5a is ascertained through lion pediments. *Stelai* A(t) I.6 and 8 both fall under type 5a, in spite of the fact that the former has a narrower frieze and a less steep pediment. Given that no. 8 is marble and no. 6 limestone, it is likely that the latter was sculpted in the local workshop on the model of the former. Stele A(ptNW) I.1 most certainly represents an additional type, as it is the only example in the evidence of all three cities featuring standing figures and exhibiting a structural coherence sufficient for a typological classification. Another example with standing figures (A[t] I.11) is a very primitive improvisation, which cannot be considered as a representative of a type. Stele A(ptNW) I.1 was not classified as a type because its full shape remains unknown. It probably had a lion top, but there is no knowing whether there was another panel figuring as a base below the inscription field.[244] One of its structural-iconographical details indeed testifies its production to order, and not a choice from stock: the male figures on the sides slightly cut into the portrait niche's frame. Another important aspect of this monument is its possible affiliation with the Celeian production, suggested by a combination of the Norico-Pannonian volute and a rectangular form of *stelai*.[245] The earliest stele (A[t] I.1) is dated to the second or the first half of the 1st century to the 2nd century on account of both inscription and iconography, since its full structural type is unknown. Nevertheless, portraits comprising only heads and necks (possibly with a small part of the collarbone area, but with no clothes depicted) abound in north Italy, Noricum and Pannonia, while an oval lower line of the portrait niche has an analogy in Istria.[246] Type 4 (A[t] I.5) is structurally unique in the entire Norico-Pannonian production, whose models, mostly dating from the 1st century or from the turn of the 1st and 2nd centuries at the latest, are found in north Italy, Istria and Dalmatia.[247] What makes this piece unique in the Pannonian context is its shaft flanked by continuous columns, as an Adriatic feature, amalgamated in this example with a typical Pannonian lion pediment. Both the lion pediment and the motif of laurel leaves tied with a ribbon point to north Italian rather than Istrian or Dalmatian inspiration for this stele.[248] The majority of the remaining pieces fall under types 3 and 5 (A[t] I.3, 4, 6–8), and are dated from the late 1st to the end of the 2nd century, or possibly the 3rd century.

Both sarcophagi chests from Andautonia fall under type 7 (A II.1*; A[t] II.1). The one marked by an asterisk is more problematic in terms of its function as a sarcophagus or an ash-chest, than such examples from Siscia. On the one side, this is because it has a very thick bottom, but on the other it (unexpectedly) holds cushion-like stone bolsters on both ends; this feature is

[242] On the Norico-Pannonian volute, see Pochmarski and Nagele 2013: 289–292.

[243] This type is rare in the Norico-Pannonian material; cf. Djurić 2008: 161, 165, fig. 2.3 (*lupa* 3778).

[244] This cannot be inferred either from a fairly modest evidence of Norico-Pannonian military and/or family *stelai* with standing figures, with early military *stelai* of individual soldiers excluded. Cf. *lupa* 3596, 3598 (Noricum); *lupa* 348, 3095, 3596, 4364, 4981, 10456, 10685 (Pannonia).

[245] On the *stelai*, see *lupa* 673, 3118, 3119, 3594, 3613, 3615, etc., and on the volute, see *lupa* 11198, 11200, 13256, 4199, 4122, 4139, etc. See also Djurić 2007: 896 and Migotti 2013a: 332, fn. 36.

[246] On head-neck portraits see Pflug 1989: cat. 8, pl. 1: 1, ff; *lupa* 93, 94, 2274 ff (Pannonia); *lupa* 355, 356, 2390 ff (Noricum). For the form of the portrait niche, see Starac 2002: 68–69 (type IId1), map 11 (Novigrad).

[247] Pflug 1989: cat. 7, T.2: 1, ff; Mercando and Paci 1998: 83–84, pl. LI: 31 ff; Starac 2002: 68–69, IId1, IId3, map 11, map 13; Cambi 2005a: 38 (fig. 45), 52 (fig. 68).

[248] Lion pediments, although slightly different from Norico-Pannonian pieces, are a usual motif of *stelai* from north Italy (Pflug 1989: cat. 39, pl. 8: 2; Mercando and Paci 1998: 186–189, nos 113–115, pl. XCIX: 113, 114, ff.), which also sometimes feature columns decorated with laurel leaves (Pflug 1989: 57). To this author's knowledge, none of the two features appears on *stelai* from Dalmatia and Histria.

otherwise indicative of sarcophagi when appearing on one of their ends. There are no elements for refining the chronology of these two chests within the time-span from the 2nd to the 4th centuries. The only funerary medallion (A[t] VI.1) is of a customary Norican type, and obviously imported from Noricum, as results also from its material (marble). Otherwise, there are some general differences between Norican and Pannonian funerary medallions. The former are almost exclusively made of marble and are normally larger than the latter, which are customarily made of stone other than marble; the time-span of both is from the early 2nd to the early 4th century.[249]

All of the *aediculae* fragments (A[ptNW] VII.1–7) are from the site of Lobor, hypothetically ascribed to the *ager* of Andautonia. Only a selected number of pieces have been presented here, all fragmentary, and therefore insufficient for a trustworthy restoration of the precise types of the respective funerary buildings. Nevertheless, various subtypes of Norican *aediculae* of the mausoleum type (A.1–3. after G. Kremer), known from the Šempeter cemetery near Celje (Roman Celeia) and some other Norican sites, seem to be identified among the evidence from Lobor.[250] The fact that pieces 1 and 7 are marble, and the remainder limestone, indicates that the customers purchased both from Norican[251] marble-quarry workshops and from local stonecutters. The latter at least partly worked on the imported models, as is best seen from pieces 3, 4 and 7; the first two are fragments of an *aedicula* type A.2. (with a relief niche and lacking columns), while fragment no. 7 could have belonged to any of the three above-mentioned *aedicula* subtypes. Significantly, fragment no. 4 has direct constructive (though not iconographic) analogies in marble pieces from Pöchlarn and Enns, both sites situated on the Danube in north-eastern Austria, while limestone examples mostly come from north-Pannonian sites on the Danube.[252] Otherwise, the fragments of *aediculae* from Lobor are chronologically sensitive to a certain measure, because such buildings of mausoleum type started spreading in Noricum from the second half of the 2nd century.[253]

Of the five unclassifiable pieces (A[t] VIII.1–3; A[ptNW] VIII.1–2), two should be given prominence: a marble statue of Icarus (A[t] VIII.2) and a shaft with a relief depiction (A[t] VIII.3). Icarus statues have been variously restored as regards their function in funerary contexts – as monument finials, funerary buildings and enclosures, or even tumuli, although none of the pieces was ever found in its precise primary context.[254] The Icarus statuette from Andautonia is very similar to the one from Leibnitz (Flavia Solva), proving that the former was certainly imported from a workshop there, as is further proved by the marble used (see Chapter II.3.3). Shaft-block A(t) VIII.3 was recovered from a tumulus cemetery site and was possibly part of a tumulus structure (enclosure or entrance) or some other building that was occasionally found in such contexts.[255] It was mentioned in the literature as a stele with the depiction of a crested helmet.[256] The relief is quite worn out, but its motif is still recognizable as a footed vase flanked by snakes or springing tendrils. Vases springing plants, mostly vines, or holding fruits, are customary symbolic depictions in Norico-Pannonian funerary art.[257] The only other monument from a tumuli cemetery in the entire area of study also comes from Andautonia (stele A[t] I.6). Unfortunately, it was found in building works, so its precise position in the given context is not known; the same applies to another stele of which only a tenon base was found in the vicinity.[258] On the basis of such finds from northern Pannonia, these two *stelai* were presumably placed in front of a tumulus.[259] The only other stele that was found together with the tenon base is A(t) I.8; tenon bases are usually lost with monuments recovered by chance. As for tumuli, they are customarily called Norico-Pannonian and are considered as originally a Norican feature that spread towards Pannonia; the other of the two clusters of such burial mounds from northern Croatia appear in the north-eastern part of the area of study (Podravina), possibly belonging to the territories of Andautonia and Aquae Balissae, but they have not produced funerary monuments.[260]

Portraits, preserved enough to be analysed in any measure, can be found in only seven examples (A[t] I.4, 5, 6, 8; A[t] VI.1; A[ptNW] I.1; A[ptNW] VIII.2), with

[249] Pochmarski 1991: 126; Kremer 2001: 347; Ertel 2010: 163–165; Pochmarski 2011: 23.

[250] Cf. Kremer 2001: 89–97, 327–330; M. Scholz 2012: 128–132; Lazar 2013: 259; *lupa* 13344.

[251] Poetovio is also included in the Norico-Alpine marble-producing circle, although it was a Pannonian city (Migotti 2012c: 4). This is because Poetovio was situated on the border with Noricum and close to the foot of the marble-bearing mountain of Pohorje (Djurić 2001a; Djurić 2001b; Djurić 2005; Djurić 2008; Djurić and Müller 2009: 116).

[252] Kremer 2001: 74–78, nos 16–20. On the finds from Pannonia, see Ertel 2010: 116–122; M. Scholz 2012: 136 –137.

[253] Kremer 2001: 34, *passim*; M. Scholz 2012: 139.

[254] Statues of Icarus topped various types of *stelai* (Schober 1923: 166–167; Hudeczek 2008: 25–28; Ertel 2010: 83–84, 139–141). They were certainly placed on lion tops amidst the lions (Šajn 2013: 244, fig. 4; *lupa* 3392), but possibly also as top *acroteria* on pediment *stelai*, as suggested by many pieces from north Italy holding similar statues (Pflug 1989: 167, cat. 3, ff; Berti 2006: 13–14, fig. 7).

[255] Kremer: 2001: 362–366; M. Scholz 2012: 61–62.

[256] Gregl 1990: 101–103, 106, fn. 9.

[257] On the symbolic meaning of such motifs, see Ortally 2005: 271. These vases usually show recognizable forms of *kantharoi* or craters, but in our example the upper section with the presumed handles is battered, so the type of the vase remains insecure. For the depictions of vases similar to that from Andautonia, see Kremer 2001: 201, no. II, 98a; Ertel 2010: 58, cat. 26. For a vase flanked by snakes, see *lupa* 4345 (Sremska Mitrovica).

[258] Gregl 2000: 165–166.

[259] Cf. Kremer 2001: 364; M. Scholz 2012: 59, 62.

[260] Kremer 2001: 362–366; Leleković 2012: 324–326; M. Scholz 2012: 58, 61–65.

two more *stelai* featuring indistinguishable people. A married couple is represented on two *stelai* (A[t] I.4, 6), as is a single man (A I.5; A[ptNW] VIII.2), while one each case is of a family of three (A I.8), a mother with two sons (A[ptNW] I.1), and a father and son (A[t] VI.1). On stele A(t) I.11 full figures of three people of unknown family relationship are depicted, and in a manner so schematic as to completely obscure their portrait features, while only necks of the married couple survive on stele A(t) I.1 (see above). All in all, married couples prevail in portrait iconography. In only one case (A[t] I.6) is a parent hugging a child depicted, while the mother on stele A(ptNW) I.1 is possibly resting her hand on her grown-up son's shoulder.

All but one of the five women appearing among the portrayed people are sporting local head-wear, which can be identified, or at least broadly compared, with the well-established Norico-Pannonian types. One of them is the woman depicted on stele A(t) I.4, whose fairly low, sub-trapezoidal head-wear is of type H 8 (*modius* hat), although lacking its height (Figure 15).[261] This type of hat was typical of northern Pannonia and the whole of Noricum, with the concentration around Virunum, and with its date corresponding with the date of stele A(t) I.4 (the earlier 2nd century).[262] Significantly, the woman wearing what seems to be a flamboyant version with huge side locks of the hairdo is depicted on a funerary medallion from the territory of Virunum.[263] On balance, what we have here is an individual type formed on the Norican model. The remainder of the female head-wear can also be identified among the known repertoire. The women on stele A(ptNW) I.1 is wearing a tight, plain bonnet of type H 7, widespread in the whole of the Middle Danube region from the 1st century to the Severan period (Figure 15).[264] The forehead section of the woman on stele A(t) I.8 is damaged, so that it is not certain whether she is wearing a fringe, or whether her hair is covered by a hat or bonnet in a typically Pannonian manner. Since the latter case seems to be more likely, she is probably wearing a Norican bonnet of type H 1.1.2., with the hair covered, rather than H. 1.1.4., which leaves the fringe visible (Figure 16). Both types lasted into the Severan period and were concentrated in Noricum, with isolated instances in northern Pannonia.[265] On the other hand, both women on stele A(t) I.6 are wearing bonnets of type H 1.1.4. but presenting a peculiarity which appears to be a blend with a 'layered' type H.1.1.1.; this is because the bonnets on the stele from Andautonia contains several (three or even four) wrapped layers instead of only two, as in type H.1.1.4. In addition, the woman in the middle reveals a curious detail to her hairstyle, that is, two characteristic locks at the central parting, typical for the styles of empresses Faustina Maior and Faustina Minor, while at the same time her hair framing the cheeks is arranged in a native (Norican) manner.[266] Expectedly, all of the five women are wearing local clothes, matched with jewellery that partly reveals native styles. On two *stelai* (A[t] I.4; A[ptNW] I.1) women are wearing sleeveless over-tunics held at the shoulders with large brooches (Figure 18). The brooches are typically very large and of a clearly recognizable *kräftig profilierte* type on the former stele, while on the latter they are smaller and of uncertain type. Given that only a half-bust portrait is depicted on stele A(t) I.4, no more details of clothing are visible. On the other hand, a full standing female figure on stele A(ptNW) I.1 is wearing on her chest a circular brooch which could have been purely decorative or still functional, that is, used for fastening the over-tunic to the garment underneath. Also, this stele is the only example on which a female tunic with considerably wide sleeves is depicted.[267] Overall, the stele in question is very important as the only example of full standing figures in the entire evidence, revealing a full ensemble of dress. Owing to that, the depiction of the woman on it can be tentatively perceived as an additional proof of the fact that the underskirt reaching to the ankles could have been a separate item of clothes, and not part of the under-tunic.[268] The women on stele A(t) I.6 are wearing native clothes of completely different type than in the previous two examples. Their dress comprises plaited tunics with straight necklines and above them stiff, smooth cloaks of apparently heavy fabric, with a V-cut on the chest and with long, narrow, cuffed sleeves, fastened (or just decorated) with a round, dotted brooch at the apex of the V-cut. This type of a coat-like female cloak seems to be unknown elsewhere in the Norico-Pannonian region.[269] Curiously, the woman on stele A(t) I.8 is possibly wearing the same item of clothing, but this cannot be established with certainty because the stone surface is very worn. There is a knee brooch on the left woman's right shoulder, this time evidently in a purely decorative role. Several items of native jewellery have been represented, such as twisted and plain (tubular) torques, the latter provided with lunular pendants (A[t] I.4, 6). However, a fretted necklace, matched with the same type of bracelet on stele A(ptNW) I.1 should rather be a Roman than a native type of jewellery, recalling a similar beaded set

[261] For a close comparison, classified as type H 8 by U. Rothe (2012: 211, fn. 562), see *lupa* 1607.
[262] Rothe 2012: 208, 211.
[263] Rothe 2012: 210, fig. 61 (*lupa* 851).
[264] Rothe 2012: 209–210, fig. 60.
[265] Rothe 2012: 200–203.
[266] Wegner 1939: 115, pl. 12; 220, pl. 34.
[267] Cf. Rothe 2012: 216.
[268] This possibility was fathomed by U. Rothe (2012: 180–182) on the basis of a painted example from Brunn am Gebirge (Pannonia, *lupa* 4411).
[269] The present author is not acquainted with this type of native dress in the Norico-Pannonian region. On the other hand, its V-shaped neckline bears a slight reminiscence of a male 'Gallic cape' (cf. Rothe 2009: 42–43).

on stele S I.8. The only other bracelet, an armband indeed, is worn by the woman on the on the left-hand side of stele A(t) I.6. In only one example is the woman holding something in her (left) hand – probably a fruit (A[ptNW] I.1).

Some of the male hairdos are accurate imitations of Imperial styles, such as that of Hadrian on stele A(t) I.5 or either Antoniunus Pius or Marcus Aurelius on stele A(t) I.6.[270] The remainder are less clear due to the poor state of preservation of the monuments; the men on *stelai* A(ptNW) I.1 and A(t) VI.1 are wearing hairdos of the so-called soldier emperors, starting with Caracalla and lasting throughout the 3rd century.[271] The portrait features of the man on stele A(t) I.8 cannot be grasped with any certainty because of the surface damage, which blurs the style of hair and especially beard. Still, given that the beard was certainly not voluminous, and that other details (the female portrait and the epitaph) contest a date in the 3rd century, the mail portrait should point to the early period of the emperor Hadrian at the latest. Similarly, the man on stele A(t) I.4 seems to be revealing a blend of portrait features of the emperors Trajan (the hair combed in parallel strands towards the front) and Hadrian (a beard and a block-like shape of the head).[272] Despite extreme simplification of the portrait on tombstone A(ptNW) VIII.2, the man's hairdo probably reflects the hairdo of Constantine, with a concave fringe composed of straight parallel strands.[273] Five of the seven men whose clothes are discernible are wearing *saga*, fastened with a round brooch on the right shoulder (A[t] I.6, 8; A[ptNW] I.1; A[t] VI.1), in three cases matched with long- and wide-sleeved tunics. Owing to full-figure portraits on stele A(ptNW) I.1, the men's tunics, belted by using rectangular buckles, are visible. There are two exceptions regarding the male clothes, otherwise strongly marked by the *sagum* in the evidence of all three cities. The first one is the toga, apparently of 1st-century type, worn by the man on stele A(t) I.5,[274] while in the other example (A[t] I.4) a grown man is wearing only a tunic. In the case of Andautonia, attributes are more frequent and versatile in men than in women; three of them are carrying scrolls in their left hand (A[t] I.6; A[ptNW] I.1 – two men). One of the two men on the latter stele is pointing at the scroll with his right hand, while another is holding an indistinguishable object in his right hand. This stele at the same time represents the only instance of soldiers' portraits with swords depicted. Another example of a fully armed veteran on horseback is depicted on stele A(t) I.3, but in this case it is a generic portrait placed on the base and not in the portrait niche. Finally, the slave on stele A(t) I.6 is carrying a whip in his right hand and is supporting it with his left.

No single mythological scene has been depicted on the material from Andautonia, but there are quite a few individual mythological motifs and characters, which, however, do not go beyond a classic artistic repertoire. These are a *gorgoneion* flanked with birds or acanthus leaves (A[t] I. 3, 4, 5, 6, 8), two dolphins flanking a trident (A[t] I.8), dolphins in the spandrels (A[ptNW] I.2), lions with prey in their paws (A[t] I.3, 5, 6, 8, 9, 10; A[ptNE] I.1), Icarus (A[t] VIII.2), a maenad (A[t] I.7; A[ptNW] VII.3, 4), mourning Erotes (A[t] I.3), an Eros of season (A[ptNW] VII.4) and Venus (A[ptNW] VII.4). If the reworked (?) receptacle A(ptNW) VIII.1 was originally a funerary monument, it contributes another mythological character – a mask-like Dionysian or aquatic deity. While individual servants (A[ptNW] VII. 2) can be taken as representing genre scenes with social connotation (see Chapter II.3.2), a woman milking a goat (A[ptNW] VII. 5) is an outright genre scene, and one unparalleled in Norico-Pannonian funerary art. The nature of the scene with a dog (A[ptNW] I.2) is not so straightforward, because its counterpart in the pediment is missing. Judging from the Norican evidence[275] and taking into account its place in the pediment, this scene should be presumed as being strictly mythical rather than a hunting episode with mythical connotations. As for the 'purely' decorative motifs, they mostly comprise standard floral and geometric patterns, often employed in tandem (A[t] I.3, 4, 5, 6, 7, 8; VII.2), architectural patterns (A[ptNW] VII.1, 6, 7), a vessel with plants (A[t] VIII.3), and the so-called Norico-Pannonian volute (A[ptNW] I.1).

*Aquae Balissae*

The city's cemeteries have furnished three *stelai* (of which one individual lion top), two sarcophagi chests and one each altar shaft, portrait niche from a funerary building and unclassifiable piece, while its presumed *ager* produced nine *stelai*, four sarcophagi (one lidded piece, one chest fragment, and two fragments whose Roman origin is not secure), one each ash-chest, medallion and funerary building (ascertained through several inscribed slabs), and three unclassifiable pieces. These give a total of 12 *stelai*, six sarcophagi, two (hypothetical) funerary buildings, one each ash-chest,

[270] On Hadrian, see Kleiner 1992: 238–240, figs 202–204; von Kersauson 1996: 118–133, nos 48–54. On Antononus Pius and Marcus Aurelius, see Wegner 1939: 125–153 (pls 1–9), 166–210 (pls 14–33); Kleiner 1992: 267–295.

[271] M. Bergmann 1977; Kleiner 1992: 357–378, figs 319–344; von Kersauson 1996: 449–459, nos 210–214.

[272] On the portraiture of Trajan, see Kleiner 1992: 208–212, figs 171–175; von Kersauson 1996: 68–89, nos 24–31. On Hadrian's portraiture, see fn. 133. On the uncertainties of details of the Trajanic and Hadrianic hairstyles in provincial funerary portraiture, and a possible confusion of the two, see Braemer 1959: 25, 31; Pochmarski 1996: 129; Pochmarski 2011: 17, *passim*; Rothe 2012: 160–161; Skupińska-Løvset 2014: 43–44; Pochmarski and Weber-Hiden 2016: 22.

[273] For a comparable portrait, see Djurić 2007. On Constantine's portraiture, see Kleiner 1992: 433–441, figs 392–402; Parisi Presicce 2005.

[274] Cf. Goethe 1989: 29, pl. 5: 4.

[275] Pochmarski 2016b.

altar and medallion, and four unclassifiable pieces, amounting to 27 monuments.

Only five of the 12 *stelai* are preserved sufficiently for a secure or tentative typological classification: type 3a or 3b (AB[t] I.7); type 5b (AB I.1; AB[t] I.4); type 6 (AB[t] I.6); type 7 (AB[t] I.7). Of the nine *stelai* whose material could be established, three are marble (AB I.1; AB[t] I.4, 7), while the remainder are limestone or sandstone; their time-span is from of the later 1st century (AB[t] 1.1) to the later 3rd century, or even the early 4th (AB I.3; AB[t] I.9). Scant evidence does not allow the establishment of a relationship between marble pieces and those made of local stone in terms of workshop affiliation and imitation of marble models. Stele AB(t) I.1 may fall under type 1, but we cannot trust the old sketch for accurately transmitting all structural details; the stele could have had a base and a rectangular upper part. Stele AB I.1 is a simplified variant of type 5a or 5b, as it lacks the frieze. There is a very slight possibility that stele AB I.3 was a portrait niche from a funerary building, as indicated by a chip out of the right short side, which could have possibly grown from a subsequent damage to an original dowel hole (Figure 2 of the respective Catalogue entry). Fragment AB VII.3 could be interpreted as a stele, but it was more likely a portrait niche from an *aedicula*, as suggested by a dowel hole on the upper surface; the piece seems to be too thin (14 cm) to take a lion top.[276] Most interesting in terms of a mixed structural typology is stele AB(t) I.6, as it has formal and structural characteristics of both a stele (higher than wide, a characteristically decorated base, a tenon-pedestal construction) and an altar (a considerable thickness and a characteristically framed inscription panel). The upper tenon could have supported any kind of top, either befitting a stele or an altar, but no such piece was found during the monument's recovery.[277] Funerary monuments combining shapes of *stelai* and altars are not exceptional in the structural typology of Roman funerary stones, but the above example still appears to be unique.

Only two of the six (two of which are questionable) sarcophagi can be classified typologically: AB(t) II.1 (chest type 3b; lid type 2) and AB II.1 (chest type 7). Only one is marble (AB[t] II.1), and only two could be dated with some precision, to roughly the turn of the 3rd and 4th centuries (AB[t] II.1, 2). Structurally, the only altar (AB V.1) is a typical example of Norico-Pannonian work that most usually come as separate shaft-blocks. If originally there was a base, it must have been a detached one. As there are no anathyrosis or traces of fixing devices for a crown on the upper surface, if a crown did exist, it should have been joined to the shaft by its sheer weight.[278] The only funerary medallion from Aquae Balissae (AB VI.1) is a typical representative of the Pannonian type. The evidence for the funerary monuments of *aedicula* type is meagre and inconclusive (AB VII.1 and possibly, but not very likely AB I.3), while Late Roman underground chambers are well attested in the literature and are represented by one example (AB[t] VII.1), which has furnished marble and limestone slabs inscribed with Christian hexameters. Structurally, this underground chamber was no rarity in Pannonian Late Roman contexts,[279] but in tandem with its inscriptions, it makes a thoroughly unique site, and not only in Pannonia.[280]

Portraits are depicted on eight monuments, with six of them preserved enough for an analysis (AB I.1, 3; AB VII.1; AB[t] I.5; AB[t] II.1; AB[t] VI.1). A married couple is represented on four of them (AB[t] I.5; AB[t] II.1; AB[t] III.1; AB[t] VI.1), three persons on two (AB I.1; AB VII.1), and four persons on one monument (AB[t] I.2); the remainder are fragmented and therefore not relevant for the statistics. Gestures of family affection are represented in two examples, one showing the mother (AB VII.1), and one the father (AB[t] VI.1) hugging their child.

Of seven female portraits, the only example unmistakably represented in a Roman manner, lacking any native elements either in portrait features or dress and jewellery, is depicted on funerary medallion AB(t) VI.1, on which a copy of the melon-type hairdo of the empress Plautilla is discernible.[281] A further two such examples (AB[t] VIII.2, 3), presumably originating from a funerary monument of unknown kind, are less certain as imperial imitations. These hairdos seem to copy a wig or wig-like hairstyle of the empress Iulia Domna, but in an extremely schematized provincial manner, giving an effect of a melon-like bonnet instead of a waved hairdo.[282] The remaining women sport hairdos that are either far cries from Imperial models, or are thoroughly native, both in tandem with native head-wear. The least clear example in this regard is the portrait on stele AB I.1, due to the stone wear. The woman's rounded crown with no inner elaboration of hair strands most probably represents the H 7 type of Norico-Pannonian head-wear (Figure 15), but with the covered ears as an incongruent detail in this iconography. This, however, is still not a sufficient reason to identify the head-wear in question as a bonnet of type H.1.1.4., despite their similarity and the fact that this head-wear comes with the hair covering the ears (Figure 16). As a matter of

[276] Portrait niches from *aediculae* are usually thicker, but there are exceptions. Cf. Kremer 2001: 125–126, nos 88–91 (thickness 17–19 cm).
[277] On the altar-stele, see Pflug 1989: 47–49.

[278] On such arrangements in fitting together structural elements of funerary altars, see Dexheimer 1998: 4; Ertel 2010: 157.
[279] Hudák and Nagy 2009; Visy 2016.
[280] Cf. Migotti 1997: 43–44, 47–49.
[281] Cf. Nodelman 1983: 113, fig. 15.
[282] On Iulia Domna's presumed wigged hairdo of the type with fairly short hair, see Meischner 1964: 30–42, figs 36–39. See also fn. 59.

fact, the portrait on stele AB I.1 has close comparisons in several Severan-period tombstones from Lauriacum, Virunum and Flavia Solva, matching it in clothes and head-wear.[283] As witnessed by another stele from Lauriacum, on which the woman is wearing the same hairdo but without head-wear,[284] the model for all three aforementioned hairdos should be looked for in the empress Faustina Minor's hairstyle, arranged as a blend of her mother's (Faustina Maior) and her own coiffures.[285] Since a similar hairdo was worn by the empress Iulia Domna at the beginning of her rule, and given a blurred depiction on the stele from Aquae Balissae, we cannot be sure that in this case we are dealing with a lag in fashion.[286] On the contrary, Iulia Domna's hairstyle with thick locks on the sides was most probably a model for the hairdo on gravestone AB(t) VII.1; it leaves the impression of a provincial blend of two types of empress hairstyles, one with the line of hair closing in towards the neck, and the other in which the thick hair was falling straight down the sides of the face, about shoulder length, and was thus longer than in the example from Aquae Balissae.[287] It is matched with the head cover closest to type H 7, as in the above example of AB I.1 (Figure 15). The woman on stele AB(t) I.5 is wearing a hairdo characterized by the hair thick on the sides, and reaching to approximately to ear level or covering the ears, with the central parting covered or non-existent. Such a hairdo was typically matched with the majority of types of Norican bonnets, and is dated widely from the mid 1st century to the Severan period (Figure 16).[288] In this case it is matched with the head cover that has no parallel in the established Norico-Pannonina repertoire; due to a low-cylinder shape, it mostly resembles the so-called *pileus Pannonicus*, which was a soldier cap especially popular in Late Antiquity.[289] Another possibility is that its shape is a misunderstood type of Norican bonnet.[290] In only three cases are the women's clothes discernible as a blend of Roman and native styles, or exclusively native, but not without debatable details. On stele AB I.1 a pleated tunic with an oval neckline and possibly long and tight sleeves, worn underneath a cloak that is draped symmetrically across the chest, which is not in accordance with Roman style, but is not blatantly native either. If it is a somewhat mistakenly depicted *palla*, the wide 'sleeves' are actually its folds, and if it is a native cloak, the sleeves should be of a wide-sleeved tunic.[291] On funerary medallion AB(t) VI.1 the cloak is draped across the chest similar to the previous example, except that the chest is covered all over by its swags of folds that conceal the tunic altogether, with only its rounded neckline visible above the over-sized necklace. On gravestone AB VII.1 the *palla* seems to be recognizable through the different directions of the folds on the arms, so that those on the right arm should be of the tunic, and those on the left of the *palla*. However, a wide and stiff vertical fold falling down from the right shoulder has the appearance of a native cloak. In addition, two irregularly executed necklines above the upper edge of the tunic are visible, looking as if tucked into the tunic; these cannot be explained satisfactorily, unless they are perceived as the result of rough and unskilled workmanship. What on stele AB(t) I.5 at first sight seems to be the long and wide sleeves of a tunic, is a native cloak draped as horizontal folds across the chest and diagonal ones on the arms, as suggested by two extremely large brooches on the shoulders, provided with a knobbed foot and a peaked head flanked by two knobs. In other words, the woman was wearing an over-tunic fastened with huge brooches which protruded artificially from the cloak as if being accentuated on purpose.[292] The type of brooches used to fasten the over-tunic does not correspond precisely to any of the known brooches from the 3rd century. Their heads with side knobs remind one of 4th-century crossbow brooches, but can possibly represent a 3rd-century type that immediately preceded them (T-brooches, *Scharnierfibeln*), with a central knob oval and the side ones round, although in reality the latter were not so large.[293] On the other hand, the feet of the brooches in question seem to be also ending in knobs, thus reminding of several 2nd- and 3rd-century types (knee, anchor, *kräftig profilierte*).[294] On balance, the brooches worn by the woman on stele AB(t) I.5 remain inexplicable in terms of the known repertoire of types from the archaeological record, but are not the only such example (cf. stele A[ptNW] I.1). Finally, the long- and tight-sleeved tunic and a *palla* apparently decorated with breast embroidery on sarcophagus AB(t) II.1 are of typical Late Roman forms.[295] Apart from the possibility that instead of embroidery, a lavish lunate necklace is depicted on AB(t) II.1, other items of jewellery are few: beaded necklaces on AB I.1 and AB(t)

[283] *Lupa* 495 (Lauriacum), 884 (Virunum), 1606 (Flavia Solva).

[284] Rothe 2012: 193–194, fig. 42 (*lupa* 4613).

[285] Wegner 1939: 217, pl. 36; Kleiner 1992: 308, fig. 277.

[286] On fashion lag, see fn. 45, and on the hairdo shared by Faustina Minor and Iulia Domna, see Meischner 1964: 30–42, figs 36–39.

[287] On both types of Iulia Domna's hairstyles, see fns 59 and 145.

[288] Rothe 2012: 198–204.

[289] Sumner 2009: 164–165.

[290] Such as seen on a Norican stele (*lupa* 1884), as suggested by Ursula Rothe in a personal communication.

[291] For a symmetrical draping of native cloaks and asymmetrical arrangement of the *palla*, see Rothe 2009: 45. It should be noted that a sub-symmetrical draping of the Roman *palla* across the chest could have been easily imitated in provincial art so as to produce a picture as on stele AB I.1. Cf. Kleiner 1992: 280, fig. 247; 326–327, fig. 290; von Kersauson 1996: 472–473, no. 221; Croom 2002: 91, fig. 41.2.

[292] For such draping of native cloaks, see Rothe 2009: 45. See also *lupa* 831, 1489, 1499 and 1884.

[293] Cf. Riha 1979: 166–1688, pls 50–51, nos 1434–1445.

[294] Cf. Bojović 1983: 31–61, pls V–XXII.

[295] Alternatively, it could have been a rich lunate necklace, as both these decorations are known in Late Antique female dress (von Barcley Lloyd 1990: esp. 74, pl. 6; Stout 1994: 83–96, figs 5.11, 5.18, 5.29, 5.33). See also Croom 2001: 83–92, 117–118, figs 37, 39, 42, 57; Cleland, Davies and Llewellyn-Jones 2012: 57 (on embroidery in Greek and Roman dress).

VI.1 and hoop earrings and a torque on stele AB(t) I.5. Of the attributes, one each a fruit (AB[t] II.1), a cup (AB[t] VI.1), and possibly a wreath (AB[t] I.5) are discernible.

Of the six depicted men whose portrait characteristics can be tentatively identified, but with much uncertainty due to the poor state of preservation (except in the case of AB VII.1.), the two men on stele AB I.1 suggest Severan hairdos. This is on account of fairly thick curly hair, which does not show the characteristics of cropped hairdos of the soldier emperors, but this still remains inconclusive, not least because their beards are hardly discernible.[296] Three men reflect portraiture of the soldier emperors (AB I.3; AB VII.1; AB[t] VII.1), while in one case the workmanship is so unskilled as to render the portrait's characteristics unrecognizable (AB[t] VI.1).[297] The latest portrait is from sarcophagus AB II.1, and, despite heavy wearing and damage, shows characteristics of the portraiture from the turn of the 3rd and 4th centuries (a roundish block-like head and a cap-like cropped hair).[298] Except for a man wearing a *pallium* and a long- and tight-sleeved tunic (AB[t] II.1), the remainder are dressed in tunics and *saga* fastened by a round brooch on the right shoulder. Where visible, the tunics are long-sleeved, their width mostly remaining indiscernible. Only one man (AB[t] I.5) is clearly depicted as an armed soldier (a *sagum*, a belted tunic, a sword and a shield), while at least one of the two men on stele AB I.1 was most probably a soldier, as transpires from a sword or a centurion's stick in one, while the same object seems to be depicted also in another, but this remains inconclusive. Both objects are topped by a bulbous head, which is typical of both the swords and centurion's sticks.[299] Of the other attributes only the scroll is recognizable in two cases (AB[t] I.1; (AB[t] II.1), while two objects held by the man on funerary medallion AB(t) VI.1 remain enigmatic.

Despite the fact that only one mythological scene figures among the monuments from Aquae Balissae, illustrating the episode of Iphigenia in Tauris (AB[t] I.4), there are quite a few individual mythological motifs and characters, some of which go beyond the classic artistic repertoire in Pannonian funerary art. Although the motif of lions with prey in their paws (AB I.1, 2; AB[t] I.5) counts among frequent and therefore common ones, in the latter case it is unique in featuring two unfathomable superimposed portraits or mythical being between the lions. Further common mythical motifs are a *gorgoneion* flanked with birds or acanthus leaves (AB I.1; AB[t] I.4, 5) and sea monsters (AB I.1; AB[t] I.4, 5, 6; AB[t] II.4), while several can be counted as more exceptional. One of them is a mythical figure, probably Endymion, in which case an unknown counterpart on the other end of the top of the stele should be Selene (AB[t] I.8).[300] The uncertainty of such identification comes from the fact that normally Endymion is depicted as asleep, with the left hand lifted and resting on his head.[301] Other figures from the same circle are Icarus (AB VIII.1) and the *lupa Romana* (AB[t] I.7). The most enigmatic and extremely roughly carved figures, which can be perceived as representing an oriental shepherd deity, such as Attis, Paris, Adonis, Ganymed or Orpheus, 'tainted' with Barbarian traits, are depicted on funerary altar AB V.1. Given the entire find context of sarcophagus AB(t) II.1, its rich imagery with heraldic panthers sitting by a vase filled with fruits, trees and vines with grapes, can hardly be perceived as representing a purely bucolic milieu, and should rather be seen in the light of philosophical and religious conceptions on the verge of Christian symbolism. Clearer Christian connotations can be grasped from the motif of a fish accompanied by a trefoil, carved on sarcophagus AB[t] II.2. 'Purely' decorative motifs are used modestly, comprising standard floral and geometric patterns (AB[t] I.1), as well as a frieze of running animals (AB I.1). The latter can be theoretically perceived as a genre scene associative of the hunt, but here it was applied to a decorative band and therefore less articulate in terms of symbolism.

### *Concluding remarks*

Given a huge discrepancy in the number of finds between Siscia on one side and the remaining two cities on the other (see Chapter II.2), there would be no point in comparing the components of this discussion (structural typology and iconography in chronological contexts) statistically between the cities. On the other hand, such comparisons can be more realistically conducted through an observation of conspicuous differences (or similarities) between the cities. Still, due to the scanty and fragmentary evidence, often lacking any archaeological context, it goes without saying that the remarks attempted here mostly remain hypothetical but should nevertheless be taken as directions toward a commendable future research which would include all

[296] On the Severan portraiture, see Kleiner 1992: 319–325, figs 282–287; Croom 2002: 65, no. 23.4–6.

[297] On the portraiture of the soldier emperors, see M. Bergmann 1977; Kleiner 1992: 361–378, figs 319–344.

[298] On the tetrarchic portraiture, see Kleiner 1992: 400–408, figs 364–370; M. Bergmann 2007.

[299] On the forms of centurion's sticks, including those with bulbous heads, see Pochmarski 2001: 203; Ubl 2013: 231–240, figs 400–413 and 420–421on pp 398–399. A good case in point is a depiction of a soldier equipped with a sword and a stick, both with bulbous heads (*lupa* 3632).

[300] The figure of Endymion is not a rarity in Norico-Pannonian funerary iconography (cf. *lupa* 3330, 3106, 4119), but it is still not very frequent. A possible model for its setting on a cornice-type top in tandem with a counterpart is found on a stele from Pécs, Pannonia (*lupa* 645).

[301] The only exception known to the present author is a relief from Voitsberg (*lupa* 1178), but the gesture there comes from the fact that Endymion is holding spears in his lowered left hand.

categories of monuments, and not only funerary. The sorest point in this regard is chronology; because of the so many stones dated tentatively or within the span of two or even three centuries (such as plain sarcophagi among others, dated from the 2nd to the 4th centuries), only a broad chronological orientation can be given at this stage. Siscia is the only city furnishing (scanty) evidence before the Flavian period, followed by a fairly steady distribution from the 2nd century to the Late Roman period, but still with the stress on the 2nd and 3rd centuries. Quite a lot of monuments date from the Severan period, but many of those dating from the 2nd *or* the 3rd century, together with those dating very probably from the 3rd century, suggest that no dramatic decline in their production occurred during the so-called '3rd-century crisis'.[302] Thus, the evidence for the three Pannonian cities differs from that for Virunum in Noricum, where a sharp decline in sculptural output occurred already from the mid 2nd century.[303] As the evidence for the Pannonian section is based only on funerary monuments and not all categories, as is the case for Virunum, the results are not straightforwardly comparable. Nevertheless, the situation at Virunum reminds us that the circumstances in Noricum and Pannonia did not correspond in all aspects of their historical development as a joint cultural circle. The concentration of the monuments in Andautonia is on the 2nd and 3rd centuries, with the Severan period presumably dominating, and with no single monument dating from the 1st century or the Late Roman period. The circumstances in Aquae Balissae seem to have been similar, with one difference: the city produced a luxurious Late Roman underground chamber yielding several exceptional finds. On balance, while the evidence from Andautonia failed to produce any find remotely connected with Christianity, several such monuments were from Siscia and Aquae Balissae, but their total number is still extremely moderate. As for the monument kinds, what straightforwardly catches one's eye is the difference between the distribution of sarcophagi and *stelai* in the three cities. While *stelai* are numerous in all three cities, figuring as a leading kind of monument in Andautonia and Aquae Balissae, sarcophagi are more numerous in Siscia. Surprisingly, and with no obvious explanation, only two ash-chests were found in Siscia, one from a site some distance from Daruvar, and thus just possibly in the territory of Aquae Balissae, while none are from Andautonia. The issue becomes even more perplexing if we remember that a huge number of ash-chests were produced in the hypothetical territory of Siscia in the Kordun. A further curiosity is the extremely small number of kinds of funerary monuments that were very popular in Noricum, such as *aediculae*, other funerary buildings and enclosures, as well as altars and medallions. None of those four or five kinds were particularly widespread in Pannonia, outside the area of research here, but they still come in respectable numbers, especially funerary buildings and medallions.[304] Therefore, while a small number of altars is perhaps reciprocal to their general lesser popularity in Pannonia, more funerary *aediculae* should be expected. As a matter of fact, the present research has just focussed attention on all the three monument kinds that were completely ignored in previous studies of funerary stones from south-western Pannonia.[305] On the other hand, a large number of types of *stelai*, and especially those that occur only once, suggests two conclusions. While the four types more frequently represented (3a, 3b, 5a and 5b) testify to a standard production to stock, numerous examples of *hapax* nature suggest that the original number of *stelai* was much larger in the area of study, and that they were probably the most popular kind of funerary monument from the later 1st to the end of the 3rd century. In terms of iconography, notably portraits, the three cities share several prevailing characteristics, with a negligible number of exceptions. Thus, men are mostly dressed in a tunic and *sagum*, and are rarely depicted with military equipment. Women feature a blend of native and Roman portrait characteristics and fashions, with the former predominating, and especially so in the territory of Andautonia, where a clearer affiliation with neighbouring Norican styles is detectable. The evidence for all three cities is in accordance with Ursula Rothe's claim that there is no such thing as a 'Norico-Pannonian national costume', but many local styles.[306] A further observation by the same author is substantiated through several examples from Siscia (S I.8) and Andautonia (A[t] I.6; A[ptNW] I.1). The observation in question claims that contrary to Noricum, where native head-wear was customarily blended with a local hairdo, it was typical for Pannonia in the 3rd century to merge native head-wear (as the only residue of local fashions) and a Roman hairdo by pushing back the bonnet and exposing the hair, at least partly.[307] This phenomenon is especially curious in the case of Andautonia, otherwise under a stronger influence of the Norican ethnic-cultural circle than Siscia, particularly in the light of the fact that the stele presenting such a phenomenon (A[t] I.6) dates from the mid 2nd century.

In terms of iconography other than portraits, quite a few cases of non-standard motifs can be observed, leaving open the question of whether these, too, were produced in greater numbers to create series but

[302] Cf. Migotti 2012c: 14.

[303] All categories of monuments were included, but with funerary in any case prevailing (O. Harl 1991b: 23–24).

[304] On *aediculae*, funerary buildings, funerary altars, and funerary medallions, see fns 6, 9, 10 and 11.

[305] It was Bojan Djurić who first brought attention to this subject matter in his unpublished thesis (Djurić 1991: 52).

[306] Rothe 2012: 141, 222.

[307] Rothe 2012: 193.

remained unrecovered, or whether the population of the three south-western Pannonian cities cherished a prominently individual approach to funerary art.

### *II.3.2. Social, religious and cultural contexts*

There is no doubt that in Roman times marble was a luxurious commodity in itself, especially in the cases when the cost of the material and carving was increased by the cost of transportation.[308] Since Pannonia had no marble sources except on Mt Pohorje, on the border of Noricun and Pannonia, the use of marble monuments in the latter province was an expensive luxury, generally indicative of the owners' status. In the context of reprimanding the Romans because of their yearning for the use of tombstones made of precious marbles, Pliny the Elder specifically mentioned Alpine marbles.[309] Material wealth and social position were not absolutely correlated in Roman society, as freedmen and even slaves could have been very rich, while cases of impoverished senators are known from the literary sources. In addition, archaeology testifies to high-ranking citizens who did not care for a flamboyant representation in death.[310] Nevertheless, expensive monuments were certainly in correlation with financial means. Unfortunately, the overall database, both in terms of individual cities and as a whole, is too limited to allow insight into the preferences of various social classes towards specific kinds of monuments in statistical terms. While marble monuments vary between 23% and 30% in the cities of the present study (see Chapter II.3.3), the number of those that yield information on both the kind of monument and its owners is quite small (40 pieces). Such information is preferably gained through an interaction of the epitaph and relief depiction, or the epitaph alone (S I.8; S II.9,[311] 10; S[ptSE] IV.1; A[t] I.2, 3, 4, 8; AB[t] I.7), while in some cases it can be hypothetically, and in a limited manner, restored through the depiction alone (A[t] VI.1; AB I.1; AB[t] II.1). The database is somewhat richer if the monuments made of other kinds of stone are added (S I.1, 2, 3; S II.1, 2, 3, 22; S III.1, 2; S[t] I.1, 4, 6; S[t] V.1; S[ptSW] I. 3, 4; S[ptSW] V.1; S[ptSE] I. 2, 3, 4; A I.1; A[t] I.1, 5, 6, 11; A[ptNW] I.1; AB V.1; AB[t] I.1, 5), but still inadequate for a statistical overview of the relationships between the kinds of monuments, material used, and social classes. Of the above 40 pieces, 16 were owned by the military (S I. 2, 3; S[t] I.1; S III. 1, 2; S[ptSE] I.4; S[ptSE] IV.1; S[ptSW] I.1; S[ptSW] V.1; A I.1; A[t] I.3; A[ptNW] I.1; AB V.1; AB[t] I.1, 5, 7), 20 by civilians (S I. 1; S II. 1, 2, 3, 9, 10, 22; S[t] I.4; S[t] V.1; S[ptSE] I.2, 3; A[t] I.1, 2, 4, 5, 6, 8, 11; AB[t] II.1; A[t] VI.1), and three or four by the 'military-civilian elite' (S I.8; S[t] I.6; S[ptSW] I.3(?); AB I.1). Despite the above-mentioned hindrances, an attempt has been made in this chapter to get an insight into the kinds of monument preferably employed by various classes of people, looking at the same time into the relationship between the material used and the position of monument owners in the social hierarchy.

In terms of quantification, it is necessary to bear in mind that the loss of monuments after their finding must have been enormous (see Chapter II.2). In social terms, another point should be considered, that namely only a limited number of inhabitants of the Roman Empire could afford a stone monument.[312] The cost of monuments must have varied enormously, depending on the kind, quality and availability of the stone in the first place, but also in the quantity and quality of carving. The cost range for tombstones in Italy varied between 120 and 500,000 sesterces, and in Pannonia it was estimated at between 800 and 50,000 sesterces in the 2nd and 3rd centuries.[313] The cost of two Late Roman sarcophagi from Salona of roughly the same size, type and quality, both made of local stone, cost three and 15 *solidi*, respectively, without offering a clue to such a difference in price. On the other hand, the average price of marble sarcophagi was estimated at between 40 and 50 *solidi*, while the most luxurious of these presumably cost a small fortune.[314] The only two marble sarcophagi from the area under research (S II.10; AB[t] II.1), especially the latter, were probably somewhere within this range. Two parameters are available for an insight into the financial potential of the purchasers of funerary monuments: the property census for higher social classes and military salaries. For the latter, the focus is on the 2nd and 3rd centuries, when the majority of the monuments were produced. Thus, a legionary veteran could expect 20,000 sesterces accompanying a *missio numaria*, while soldiers' payments per year varied (increasing towards the 3rd century) between 1,000 and 6,000 sesterces for auxiliaries and between 1,200 and 7,200 for legionaries, with cavalrymen getting more than infantrymen.[315] As for the higher civilian classes, notably of equestrian and senatorial ranks, their wealth and capacity to purchase a luxury monument derived from the property qualifications for their class: 400,000 sesterces for knights and 1,000,000 for senators.[316]

In terms of the hierarchy of cost, funerary buildings are presumed to have been the leading kind of funerary monuments in Noricum.[317] The only marble piece that

[308] Russell 2013: 24, *passim*; Barker 2012: 22; Lapuente *et al.* 2012: 199.
[309] Plin. *NH* XXXVI, 2.
[310] Migotti 2008b: 66, fn. 147; Mouritsen 2011: 248; Hemelrijk 2015: 14–18; Mainecke 2015: 189.
[311] Presumed to be marble on account of a sophisticated iconographic programme and epitaph.

[312] Graßl 1991: 6; Hope 2001: 3; Carroll 2006: 15.
[313] Graßl 1991: 6–7; Carroll 2006: 77–78.
[314] Cambi 2010: 177–19/261–262. Presumably, marble sarcophagi were generally not available to the lower social classes (Birk 2013: 10).
[315] M. A. Speidel 2009; M. A. Speidel 2015: 332; Škrgulja and Migotti 2015: 43.
[316] Bruun 2015.
[317] Pochmarski and Weber-Hiden 2016: 29.

certainly belonged to a mausoleum-type building is pediment S VII.1, unfortunately leaving the owner of the building anonymous. Judging from the motif of a ribbon-tied wreath with a central gem carved in the pediment, the building should have belonged to a high-ranking official. This is because, apart from emperors, only victorious generals, consuls, and magistrates offering official games were entitled to a wreath with a central gem.[318] This, however, was in theory, but it seems that this rule was taken more loosely in practice, especially in the provinces.[319] Nevertheless, there is no doubt that the family who owned the Siscian funerary building was wealthy enough, and possibly also important enough, in terms of the public life of the city. Another elite family, living at the beginning of the 4th century in Aquae Balissae, owned a lavishly decorated underground funerary building whose walls were provided with marble and limestone slabs inscribed with liturgical hymns. It held at least two sarcophagi, one of which was a marble chest of a married couple (AB[t] II.1). Unfortunately, their names were not preserved either on the sarcophagus, which reached us as inscriptionless, or on a separate inscribed slab. Nevertheless, the fact that the man portrayed on the sarcophagus is wearing a *pallium* (*hymation*) and is holding a scroll in his hands, clearly suggests that the family's cultural profile was intellectual.[320] This is further supported by the religious symbolism of the scenes depicted on the sarcophagus in tandem with the liturgical hexameter-hymn cut into a stone slab and probably fitted on a wall of the chamber. While the hymn is difficult to fully comprehend on account of its many allegories and metaphors, it certainly conveys the dogma of original sin and redemption through Christ's suffering. The scenes on the short sides, featuring small trees, vines, grapes, and two heraldic panthers next to a vase holding fruit and vines, at the first sight suggest a Dionysian religious background. When, however, viewed in tandem with the hymn, this iconography seems to be echoing the religious-philosophical atmosphere characteristic of the higher classes of Roman citizens at the beginning of Late Antiquity, still close to Dionysian paganism but adumbrating a syncretistic Christianity in its incipient phase.[321] On balance, the intellectual and material characteristics thus evidenced certainly classified the married couple from Aquae Balissae among the highest rank of the city's inhabitants. The cost of their sarcophagus must have been considerable, and is perhaps comparable to north Italian or Salonitan sarcophagi imported from Athens, which were more expensive than those made of Prokonnesian marble.[322] In the example from Aquae Balissae, Pohorje marble can be perceived as perhaps less costly than Gummern marble, but still in the uppermost range of luxury stone in Pannonia; to this, the sarcophagus' measurements and rich sculptural repertoire should be added, as well as a separately carved base. A comparison can be made between the social background of some Salonitan and south-west Pannonian monuments. It has been established that sarcophagi made of Prokonessian marble in Salona were used by the equestrian class, while senators purchased imported Attic sarcophagi.[323] This data can be juxtaposed with three examples from the area of research. The only equestrian family had a marble stele (S I.8), while a single senatorial person purchased a marble sarcophagus (S II.10 – see below), as did the family whose social rank remains unknown, but must have been high and accompanied by considerable wealth (AB[t] II.1).

The third of the civilian elite families is represented on a missing sarcophagus, whose epitaphs and depictions luckily survive as drawings (S II.9). The stone in which it was carved remains unknown, but marble can be presumed on account of the sophistication of both the epitaph, composed as a dialogue between the deceased and her husband, and the depictions on the lateral and short sides. These present mythological scenes most probably interpretable as a life snatched away, but juxtaposed with an eternal love that is able to survive in the spheres of the Underworld's mythical security.[324] A further member of the civilian elite is represented on sarcophagus S II.10; its owner was a senatorial lady from Siscia, as the only example of her social class in this city, as well as in the entire evidence. Despite some hesitance about the details of her clothes, her servant girl represented on the front is most probably wearing a toga (see Chapter II.3.1), thus exemplifying a particular case of a woman acting in accordance not with her sex, but her social class.[325] The remainder of the elite, or at least wealthy owners, belong to the military.

The first representative of the military to own a marble tombstone (A[t] I.3) is a local man, *peregrinus*, who joined the army and finished his career by acquiring civil rights. This was not in itself an extraordinary accomplishment, but the man evidently distinguished himself during service and advanced to the position of adjutant to the praetorian prefect, ultimately amassing enough property to be able to purchase a luxurious marble funerary monument. Another military man, Heliodorus, whose resting place was

[318] B. Bergmann 2010: 59, *passim*.

[319] B. Bergmann 2010: 211, *passim*. On examples of funerary monuments proving this presumption, see *lupa* 3100 (a legionary veteran), 3288 (a married couple).

[320] On the scroll, see Migotti 2007b: 71; Birk 2013: 76. On the social symbolism of the *pallium*, see Gabelmann 1973: 70; Fejfer 2008: 200; Cleland, Davies and Llewellyn-Jones 2012: 137; Birk 2013: 76, fn. 322.

[321] Migotti 2017b: 507–509.

[322] Cf. Gabelmann 1973: 37; Cambi 2010: 45/280. On the relatively low cost of Prokonessian marble in the Diocletian's Price Edict, see Russell 2013: 34.

[323] Cambi 2010: 45/280.

[324] Migotti 2017a: 44–49; Migotti 2017b: 504–506.

[325] Migotti 2007b; Migotti 2017a: 40–43; Migotti 2017b: 506–507.

probably in a Late Roman funerary building (S IV.1), was mentioned in his epitaph as having an excellent military career, with no other details specified. The third instance was a praetorian (AB[t] I.7) who possibly shared his tomb with a veteran, apparently from his own family. A family whose equestrian social position was probably rooted in the military service, but whose male representatives presumably won positions in the higher civil administration, purchased a fairly luxurious marble stele (S I.8). They did it at the turn of the 3rd and 4th centuries, thus showing allegiance to traditional values, as by that time the portrait stele became an outdated form of tombstone.[326] Significantly, the family (or the dedicator) brought attention to their intellectual profile through representing one of the daughters as holding writing implements.[327]A similar social background can be presumed for the owners of stele AB I.1, but only on account of the insignia held by the men in the portrait niche (rather a sword or a centurion's stick than a scroll), as the epitaph does not survive.[328] The same can be hypothesized for the family commemorated on stele S(ptSW) I.3, judging by a scroll held by one man and writing implements by another, although in this case no military background is attested.

The remainder of the owners of marble tombs were Roman citizens, be they immigrants from north Italy or the western provinces (A[t] I.2), or local people awarded civil rights (A[t] I.4, 8). In one case (A[t] VI.1), Roman citizenship can be presumed on account of the *sagum* fastened with a round brooch, as worn by the adult man in the portrait niche, since this item of clothing fastened with a metal brooch was generally a mark of citizenship, either from a civilian or military background.[329] Therefore, a boy represented on the same monument as wearing only a tunic should not be judged as lacking citizen rights or any other social position; the lack of *sagum* in this case can be rather attributed to his age. On the other hand, the only adult man depicted as wearing only a tunic (A[t] I.4) was, judging by the name, a local freedman, ultimately a Roman citizen. This single example, however, cannot be helpful in fathoming the symbolism of the tunic when worn as the only item of clothing worn by adult men, servants excluded. Admittedly, such examples are not customary in Norico-Pannonian funerary art.[330]

Although the literary sources and monuments of art are not specific enough on the topic of slaves' clothing, and, in spite of the fact that some authors believe that the difference from the clothing of citizens was in the quality rather than the type, the artistic evidence still suggests that the clothes acceptable for slaves did not include metal brooches and belt buckles, and that slaves strove towards presenting themselves in the guise of citizens.[331] One such case of a respected slave presented in the guise of a citizen is stele A[t] I.6, owned by a slave family from Andautonia. In other words, men wearing *saga* fastened with a brooch on the right shoulder should be perceived as representing Roman citizens even with the epitaph missing, but this still cannot be taken as an unquestionable fact. In sum, the material funerary culture of the wealthy, and probably more-or-less prominent people in terms of social rank and position, was marked by the use of marble masonry buildings (S VII.1; S[ptSE] I. IV.1), sarcophagi (S II.9(?), 10; AB[t] II.1), *stelai* (S I.8; A[t] I.2, 3, 4, 8; AB[t] I.7), and possibly funerary medallions (A[t] VI.1).

12 monuments made of local stone pertained to the military persons of various units and ranks: legionaries (S[ptSW] I.4; S[ptSE] I.4; A[ptNW] I.1 – two soldiers mentioned; AB[t] I.5), veterans (S I.3; S[t] I.1; A I.1; AB[t]) I.1), and one each praetorian (A[ptNW] I.1), legionary *cornicularius* (S III.1), a *beneficiarius consularis* (S III.2), an auxiliary soldier (AB[t]) I.6), and a military physician (S I.2). Several of those monuments were owned by the families who can be tentatively labelled as the military elite or sub-elite, or possessing such aspirations, simply on account of their funerary monuments. Such was possibly the family who produced as many as three soldiers: one each legionary centurion and standard-bearer, and a praetorian (A[ptNW] I.1). Their stele is of local stone and of plain formal structure, and is lacking in symbolic depictions and decoration, but the people depicted on them seem to be showing their social pride through a unique family picture, representing full standing figures of the two deceased soldiers (with the third mentioned as a dedicator) and their mother.[332] The same can perhaps be said of the owner of stele AB(t) I.5, as the monument was made of local stone, but on the model of the most sumptuous marble prototypes.[333] One tombstone (S[t] I.6) was owned by a family whose background can be tentatively defined as military-civilian (see above). The stele in question was elaborately carved with portraits of two men flanking a woman and with additional motifs and scenes of both military and civilian associations. The men are dressed in a *sagum* and a belted tunic fastened with a round brooch and a round buckle, respectively. Both have a

[326] Cf. Pflug 1989: 28–29; Cambi 2005a: 88/225; Hope 2001: 48, 93; Weber-Hiden 2014: 72; Migotti 2016a: 186.
[327] Representations of children in a learned sphere are a phenomenon known from the 'City-of-Rome' sarcophagi (Birk 2013: 167).
[328] On insignia, see fn. 162.
[329] See fn.78. On the social symbolism of clothes, see Bonfante 1994; O. Harl 2003: 342; Fejfer 2008: 40, *passim*; Cleland, Davies and Llewellyn-Jones 2012: viii; Rothe 2012: 150, *passim*.
[330] E. Pochmarski (2011: 18, 65, no. 31) does not comment upon this phenomenon, while O. Harl (2003: 342) ascribes it tentatively to local civilians (*peregrini*). The example from Pochmarski (no. 31) further testifies that citizens could also have been portrayed as wearing only a tunic. According to Ursula Rothe (2012: 144), the tunic worn on its own marked craftsmen and tradespeople.

[331] Pflug 1989: 103, fn. 632; Croom 2001: 23–24; Joshel 2013: 132–136.
[332] In the 3rd century, this type of *stelai* was almost exclusively owned by individual soldiers and their families. Cf. Braemer 2005: 39. See also fn. 106.
[333] See fn. 331.

scroll in their left hand, but the one on the right-hand side has a sword tucked under his left arm, while a bundle of scrolls and a codex is depicted to the left of the figure on the left, symbolizing some administrative role, either military or civilian.[334] Furthermore, a shield shown behind the mounted rider in the lower register, with a lance leaning on it, possibly hints at the military occupation of the owners of the stele, even though the relief represents a hunting scene. In funerary contexts hunting is associative of battle, and not only in terms of iconography but also of symbolism; therefore, it is frequently depicted on soldiers' funerary stones.[335] Thus, stele S(t) I.6 is the most telling of the several above-mentioned examples of the social background created in south Pannonia in later Roman times and marked by a class of citizens who can tentatively be labelled as members of the military-civilian elite.

The remainder of the monuments made of local stone belonged to civilian families. The highest on the social scale in this group is the owner of funerary altar AB V.1, who performed multiple duties in the civil administration of Aquae Balissae. Yet his funerary monument was quite unassuming and made of poor-quality stone, testifying either that the family's wealth was not in correlation with their social standing, or that they did not care for an ostentatious funerary display.[336] The only other civilian official (a low-ranking one, in charge of public records) is recorded on sarcophagus S II.2; he was a bearer of the function of *Augustalis*, and as such probably a freedman. Curiously, his compatriot buried in a sarcophagus in the vicinity (S II.3), who shared with him the office of the *Augustalitas*, was not a freedman but a Roman citizen, as proved by his father's name.[337] The only expressly evidenced freedman was a native man from Andautonia (A[t] I.1), and the only slave family, whose tomb was marked by a stele more sumptuous than any other made of stone other than marble in the whole territory under research (A[t] I.6), also originate from Andautonia.[338] The remainder of the monuments' civilian owners and bearers of Roman citizenship were probably not officials, as no such clues appear in their epitaphs. The first of them in terms of chronology (S I.1) was presumably an Italian immigrant who came to Siscia in pursuit of business as a merchant or contractor. A Roman citizen from Andautonia (A[t] I.5), most probably of Italian origin, proudly exercised his citizenship through his portrait as a *togatus*, which is the only example of the toga in the entire evidence relative to men. Another Roman citizen probably of Italian or Western origin was commemorated on stele S(ptSE) I.1. Civilians of native stock can be established on several monuments on account of the epitaph, or both the epitaph and depictions, showing women in native clothes (S II.1; S[t] I.2, 3, 4; S[ptSE] I.2, 3, 4); the latter's social status (Roman citizens, freedmen, *peregrini* or slaves) understandably remains unfixed. A *peregrinus* is indicated on stele A(t) I.11 through a single name, while people on the funerary medallion AB(t) VI.1 can only be presumed to have been Roman citizens on account of the *sagum* for the man, and the lack of any recognizable local traces in the woman's clothes. Similarly, the social background of two men and a woman portrayed on stele S(ptSW) I.3 remains inconclusive, as one man certainly holds a scroll, while the other has writing implements. Still, the family's social background could be the same as that depicted in the marble stele S I.8, and a limestone example S(t) I.6, i.e. containing civilian and (former) military-career components. The latest and most singular example of a civilian tomb is a large and fairly sumptuous Christian sarcophagus, whose significance lies in the fact that Christian sarcophagi are generally rare in Pannonia.[339] Apart from the inscribed slab AB VII.1, a possibly crypto-Christian sarcophagus from the same context (AB[t] II.1), and another probably Christian sarcophagus (AB[t] II.2), no other Christian monuments are from Aquae Balissae, while none whatsoever has been recovered in Andautonia. On the other hand, in addition to the undoubtedly Christian sarcophagus S II.22, several other pieces, probably or possibly affiliated with the Christian religion, are known from Siscia (S IV.1, 2; S VIII.15, 18). Nevertheless, Christian finds are unexpectedly few considering that Siscia was a see from at least the second half of the 3rd century; this certainly results from the lack of research and loss of monuments.[340]

As many as twelve of a total of sixteen military tombstones are *stelai* (S I. 2, 3; S[t] I.1; S[ptSE] I.4; S[ptSW] I.1; A I.1; A[t] I.3; A[ptNW] I.1; AB[t] I.1, 5, 6, 7), which makes this kind of funerary monument the most popular among the soldiery. As a matter of fact, military *stelai* surpass the ratio of *stelai* and other kinds of monuments, above all sarcophagi and ash-chests, in the total evidence under discussion (Figure 10). This is indirectly in accordance with the fact that all of the sarcophagi revealing the social status of the deceased

[334] The same phenomenon is observed on a tombstone from Enns (Lauriacum, *lupa* 495), on which one of the two men has only a scroll in his hands, while the other is holding a scroll, a sword, and writing implements (*lupa* 495). Such families correspond to the phenomenon labelled by S. Birk (2013: 91–92) as 'learned figures outside the intellectual sphere'.

[335] Mercando and Paci 1998: 209–210; Braemer 2005: 34; Birk 2013: 107; Migotti 2013c: 266. On the claim that in funerary contexts the hunt primarily represents an economic activity, see Kaszab-Olschewski 2007: 174.

[336] See fn. 173.

[337] *Augustales* were mostly recruited from the class of freedmen, but there were exceptions. On the office of *Augustalitas*, see Šašel Kos 1999; Carroll 2006: 138; Mouritsen 2011: 249–260; Himelrijk 2015: 15–17.

[338] Wealthy slaves who could even afford marble monuments were not such a rarity in the Roman Empire (Hope 2001: 28–29; Mouritsen 2011: 168; Pochmarski and Weber-Hiden 2016: 41–42).

[339] Cf. Migotti 1997; Gáspár 2002.

[340] On the Siscian episcopacy, see Jarak 1994:31/171, *passim*.

were employed by civilians, mostly elite or sub-elite citizens (S II. 1, 2, 3, 9, 10, 22; AB[t] II.1), which can hardly be a mere coincidence.

As adduced at the start, of a total of 40 monuments furnishing some data on social background, 20 belonged to civilians, 16 to soldiers, alone or with their families, and three or four to the class labelled conditionally as a military-civilian elite of the later Roman period. Thus, the society seems to have been balanced if viewed in terms of the whole of the chronological span. Nevertheless, the majority of the monuments are from the 2nd and 3rd centuries, while 1st-century *stelai* are rare, although they must have been more numerous. In the 2nd and 3rd centuries the society of southern Pannonia was no longer strongly militarized, while in the second half of the 3rd century it witnessed the creation of a class of citizens whose careers originated in the military, but who did not represent the military class in the narrow sense of the word, defined here as a military-civilian elite (see above). Otherwise, two social classes in the Roman Empire are best represented in the epitaphs: freedmen and soldiers.[341] While the number of the latter in the present evidence corresponds with such statistics, the number of freedmen remains inconclusive, as of the 20 civilian monuments only one was explicitly owned by a freedman (A[t] I.1), and two probably (S II.1, 2), on account of their service as *Augustales*.

The ratio between the monuments set up for families versus individuals is 30:17. Such prevalence is in accordance with the chronology, as the majority of the monuments fall into the 2nd and 3rd centuries. It was the time when the increased significance of the family as the most important unit of society became strongly reflected in the funerary evidence, contrary to the 1st century, which was strongly marked by individual, mostly military commemorations.[342] The stress on the importance of familial attachments is further marked by gestures of *dextrarum iunctio* between the couples, and hugging between the members of the family, as an overall social phenomenon well reflected in funerary art.[343] Unexpectedly, depictions of married couples represented alone (S I.3, S III.1, 2; S[t] I.2, 3, 6; S[ptSW] I.3; S[ptSE] I.4; A[t] I.1, 4; AB II.1; AB III.1; AB[t] I.5) are slightly more numerous than those accompanied by their children (S I.7, 8; S[ptSW] I.1, S[ptSE] I.3, 6; A[t] I.6, 8; A[t] VI.1, AB[t] VII.1). Such a ratio is even more in favour of the former when the mention of children (or the lack of it) is counted by including monuments furnished by epitaphs but lacking portraits. At first sight, such evidence suggests that in the three cities of south-west Pannonia children were less cherished as an essential component of the family than was the case in northern Pannonia. This, however, does not seem likely, and should be rather ascribed to the limited and therefore insufficiently representative evidence. Otherwise, the importance of children as family members transpires from their treatment as 'small adults' in Roman funerary iconography. This means that they were combed and dressed on the model of their parents, and were not treated as an individual social category unless they were very small.[344] A slight wavering in such approach is seen on stele S I.8, on which a four-year-old girl is dressed as her mother and her elder sister, but was still estimated to be too young to wear a necklace like her mother and sister.

While the social status was greatly expressed through clothes, it was not customary in Pannonian funerary iconography to illustrate the deceased's profession or occupation.[345] In this aspect of social communication within funerary cultures, Pannonia was closer to the city of Rome than to the northern provinces of the Empire.[346] In other words, in funerary self-representation, the social aspect in the narrow meaning of the word seems to have been more important than the material aspect in terms of earning and occupation. A slight exception to this rule is represented by the military, whose depictions with weapons can be perceived as a sign of their social status and occupation at the same time. In any case, such depictions are quite rare in the present evidence (S[t] I.6; A[t] I.3; A[ptNW] I.1; AB I.1(?); AB[t] I.5). On only one soldier tombstone (S[t] I.6) is an occupation suggested through the depiction of a diptych and a bundle of scrolls, probably indicating a military *scriba* or *librarius*. As for the civilians, an occupation is illustrated in only two cases. What we have on *aedicula* A(ptNW) VII.5, featuring a woman milking a goat, is not an illustration of an occupation in the narrow sense, but of the work done within the household.[347] Therefore, its symbolism should be rather in the sphere of social status reflected in the representation of servants (see above) than in the occupation of the deceased, who

341 Pflug 1989: 78; Weber 2007; Mouritsen 2011: 127–128; Edmondson 2015: 561; Mitthof and Cenati 2016.

342 Cf. Hope 2001: 62–73; Carroll 2006; 132, 223; Jaeger 2007: 30–32, 35; Rose 2007: 207–208; Carroll 2015; Migotti 2016a: 185. On the importance of the family in Roman society, see Mouritsen 2011: 285–286; Osgood 2014; Edmondson 2015.

343 Gabelmann 1973: 51; Jaeger 2007: 32; Facsády 2014: 197; Carroll 2015: 505; Pochmarski and Weber-Hiden 2015: 24.

344 Pflug 1989: 73; Jaeger 2003: 477; Jaeger 2007: 32; Birk 2013: 157; Facsády 2014: 198; Carroll 2015: 504–505.

345 Significantly, the same is true for the grave goods (Migotti 2010b: 30).

346 Depictions representing occupation or household activities are found in funerary iconography of the city of Rome and all of Italy to a certain measure but are particularly widespread in the provinces of Gallia and Germania. Cf. Braemer 1959: 118; Zimmer 1982: 1; Pflug 1989: 96; Mercando and Paci 1998: 97–122, nos 43–59; Hope 2001: 52–57; Langner 2003; Braemer 2005: 32; Kremer 2006: 38; Kaszab-Olschewski 2007: 179; Rose 2007: 208; Facsády 2014: 197; F. Harl 2016: 25. Overall, occupation is rarely mentioned in the epitaphs (Chioffi 2015: 634).

347 Cf. Mercando and Paci 1998: 99–101; Kaszab-Olschewski 2007; Birk 2013: 13.

probably owned a farm and servants working on it; this transpires from the kind of monument used. On balance, an illustration of the owner's occupation is depicted only once, remaining, however, inconclusive. The first association about the whip in the hand of the slave owner of stele A(t) I.6 is of a *vilicus* (a foreman of an estate). However, the epitaph supplies some prosopographic indications that the slave in question was travelling in performing his duties; thus, the whip possibly connotes his travels.[348]

As was posited in Chapter II.3.1, each and every motif employed on funerary monuments had some inherent symbolic connotation. Some of this symbolism was connected with death in its basic meaning, such as perishing (animals chasing or devouring each other), diverting of evil (*gorgoneia*), grieving (mourning Erotes), transporting the dead to the Underworld (dolphins and fabulous beasts) or watching over the graves (lions), while others can be tentatively connected with specific religious affiliations. Such is, for instance, the motif of lions suppressing rams, represented on *stelai* of type 3a and 5a and some other monuments (A[t] I.2, 3, 5, 6, 8, 9, 10; S I.8; S VIII.16; S[t] I.2, 3; S V.2; AB I.2; AB[t] I.5), which possibly stems from the iconology of Cybele.[349] In a similar way, ivy leaves, as well as vines and grapes, are presumably affiliated with Dionysius (S I.8; S II.22; S[t] VIII.1; S [ptSE] I.3; A[t] I.4, 7; A[t] VIII.3; A[ptNW] I.1; A[ptNW] VII.4; AB[t] II.1),[350] *lupa Romana* (AB[t] I.7) with the idea of *romanitas*,[351] and eagle (S[ptSE] I.6) with the worship of Jupiter.[352] The motif of *gorgoneion* (a Medusa head) in the *stelai* pediments (S I.8; A[t] I.3, 4, 5, 6, 8; AB I.1) should not be associated with any structured religious conceptions on one hand or mere death symbolism on the other, but rather with the idea of the protection of the grave, while rosettes in the same place (S I.1, 2, 4; S[t] I.3) possibly bear the same association, thus surpassing a decorative role exactly on account of its central place on monuments. On the other hand, society in south Pannonian was receptive to mythological themes, which is true of the Norico-Pannonian cultural circle in general. If from this circle we exclude Erotes, which turned into a sort of a generic symbol in funerary iconography, mythological motifs and episodes include subjects such as Iphigenia in Tauris (AB[t] I.4), the legend of the Sabine women (S[ptSE] I.3), a centaur (S I.6), Icarus (A[t] VIII.2; AB VIII.1), Helena and Eros (S II.9), Apollo chasing Daphne (S II.9), Hercules rescuing Hesione (S II.9), Hercules alone (S[t] V.1), maenad or/and Venus (A[ptNW] I.7, VII.3, 4; S[ptSW] VIII.2), sphinx (S VIII.17), and possibly Endymion (AB[t] I.8). The appearance of myths in Norico-Pannonian funerary art, and especially its origin, has been widely discussed in the literature, among others, in view of the scarcity of this phenomenon in the funerary art of north Italy.[353] Regardless of whether the impetus came directly from Attica and the wider regions of the Greek East, or via Rome, or still via Aquileia, it is most important that mythological motifs and episodes took root in Norico-Pannonian art, suggesting a cultural ambiance open to their messages.[354] Significantly, it has been established that imported marbles probably reached Aquileia directly from the East and not via Rome.[355] As no direct artistic influences from the city of Rome could be securely established in Aquileia either, mythological themes presumably reached this town's stone-carving workshops from the same (eastern) direction, but they were apparently more effective in spreading them towards Noricum and Pannonia than in its immediate surroundings.[356] Transmitters of such art and backgrounds for its receptivity in Pannonia have been contemplated, as well as directions of its spreading, with its bearers ultimately recognized in the class of veterans and oriental traders on the Danube *limes*, whose political and cultural affiliation was with the Roman state and not with the indigenous society.[357] However, what could have been true for northern Pannonian boundary regions, need not have been relevant for the province's south-western part, as its social background was different. Unfortunately, not much can be said about this topic, as epitaphs survive that show only two (S II.9; S[ptSE] I.3) of the mythological scenes, allowing a very limited insight into the social background of the tombs' owners. Curiously, those backgrounds are completely opposing in the two examples, despite the fact that both relate to civilian families. Judging from the flamboyant epitaph and a sophisticated repertoire of themes on sarcophagus S II.9, its owners must have been highly intellectual. On the contrary, a mythological scene unparalleled elsewhere was somewhat unexpectedly used on stele S(ptSE) I.3, whose owners were local inhabitants cherishing native customs in clothing, and

[348] Migotti 2008a: 458–459.
[349] Gregl and Migotti 2000: 151–152.
[350] Migotti 1997: 43–44; Migotti 2017a: 50–56.
[351] Cf. Szirmai 1991; Mercando and Paci 1998: 201–207, nos 129–133; Gabler 2007: 26; Pochmarski and Weber-Hiden 2015: 25.
[352] The motif of the eagle as the bird of Jupiter can be perceived as a funerary metaphor standing for immortality (Pochmarski and Weber-Hide 2015: 24), but also as representative of the Imperial cult (Perdrizet 1900).

[353] Schober 1923: 204; Gabelmann 1973: 88–89; Jaeger 2003: 479; Koch 1993: 135; Verzár-Bass 1996: 258–260; Mercando and Paci 1998: 165–201, nos 97–128; Kremer 2006; Gabler 2007: 23; Ertel 2010: 105–121; F. Harl 2016: 25; Djurić, in press.
[354] The reasoning of S. Priester (1998), that such receptivity had less to do with cultural sophistication and more with everyday life, seems to be a simplification.
[355] Previato and Mareso 2015: 306.
[356] Gabelmann 1973: 25, 68, *passim*; Gabelmann 1977: 13, *passim*; Koch 1993: 135; Cambi 2003: 19–22; Kremer 2006: 46; Verzár-Bass 1996. Whether the northern Italian region of Piedmont had a role in the transmitting of influences in both structural typology of the monuments and iconography, mythological themes included, remains open. Cf. Verzár-Bass 1996: 262, *passim*; Hagenweiler 2001: 80; Weber-Hiden 2014: 71–72. See also fns 62 and 63 in Chapter II.1.
[357] Verzár-Bass 1996: 262–265; Gabler 2007: 24.

evidently taking no special place or public function in the community.

In contemplating a psychological background to the choice of mythological motifs, various possibilities have been considered, two of which come to the fore. First, it is a correlation of the deceased's characteristics with the virtues of gods and heroes as mythical protagonists, and second, it is the conception of being snatched as a metaphor for dying or being rescued as an allegory of conquering death.[358] Ultimately, it boils down to a juxtaposition of an optimistic versus pessimistic approach to death. In four of the six examples the idea of conquering death prevails: Endymion in an eternal sleep and not dead (AB[t] I.8); the rescue of Iphigenia (AB[t] I.4); the suppression of a centaur (S I.6); and preventing the Romans and Sabines from clashing (S[ptSE] I.3). In two of them, both featuring Icarus, tragedy prevails. A 'narrative' from sarcophagus S II.9 is the most intricate and ambiguous at the same time. While the episode of Heracles rescuing Hesione evidently signifies victory, the remainder of the scenes (Helena and Eros and Apollo chasing Daphne) are allegories of love, but love with ominous connotations. In sum, the prevailing atmosphere of the sarcophagus in question is gloomy and thus in accordance with the epitaph brimming with grief and despair. On balance, an attempt at fathoming mythical episodes should be ventured only when they are accompanied with an epitaph, because even episodes with an optimistic message can figure on monuments whose surviving owner remained hopeless and inconsolable. Also, no religious-psychological background detected on individual monuments should be perceived as marking a common cultural-religious attitude of the provincial society, even when the evidence is larger and more representative than in the present case; the choice of themes must have been individual and deeply personal. Thus, one can only wonder whether mourning Erotes should be understood as representing particular, individual grievance (S II.2, 4, 7; A[t] I.3), while those holding garlands or other objects (S II.8; S III.2; S[t] I.2, 5, 6; AB[t] I.6) should indicate a more hopeful experience of death. If the latter is true, the evidence, meagre as it is, tips the scales in favour of a more positive attitude, which, on the other hand, contradicts the messages conveyed through the mythological episodes. According to one theory, optimistic themes and attitudes expressed through mythological episodes in Roman art were introduced in the Severan period, suppressing those gloomier scenes that had prevailed in the 2nd century.[359] Unfortunately, the evidence at our disposal here is too meagre to prove conclusively the relationship between the mythological persons and episodes used and the chronology of the respective monuments. Much has been written on various nuances of the phenomenon of the deceased's 'divinization', tentatively hidden behind their depiction in the guise of gods, goddesses, heroes, and mythological mortals, or else, in the depiction of divine characters wearing fashionable hairdos.[360] Only one monument in the entire evidence, a sarcophagus from Siscia (S II.9), possibly reflects such an attitude, but on two conditions: that the hairdo of the four mythological female figures depicted is understood as representing a wigged hairstyle of the empress Iulia Domna; and that those figures indeed impersonate the deceased Maximiana Aemilia.[361]

Two individual motifs from everyday life need to be commented on. The first relates to the depiction of servants, usually a boy and a girl (S II.10; S[t] I.6; A[ptNW] VII.2), or a boy alone (S I.3). Their interpretations as either epitomizing the banquet or generally illustrating the role of servants in a family of the owners, has been shifting from the idea of a symbolic funeral banquet towards the conception of the owner's idea of well-being and social status.[362] In Norican funerary art the iconography of servants is well established, while there are still some curious examples of girls wearing unusual clothes.[363] A curiosity in the present evidence is the two examples apparently depicting servant girls in the guise of their mistresses (S II.10; S[t] I.6). This seems to be more certain on sarcophagus S II.10, on whose front a servant girl wearing a toga and holding a huge scroll is depicted, than on tombstone S(t) I.6, as the reliefs of the servants on its short sides are very badly damaged. Another frequently employed motif with social implications is the scroll in the hands of the deceased (S I.8; S II.10; S III.2; S[t] I. 2(?), 6; S[ptSW] I.1, 3; S[ptSE] I.4, 5; A[ptNW] I.1; AB I.1[?]; AB[t] II.1; AB[t] III.1[?]), also widely commented on in the literature and generally perceived as implicating various conceptions of symbolism in the spheres of law and culture.[364] These two conceptions are discernible through different backgrounds of the monuments' owners, such as military, civilian, as well as culturally irrelevant people versus those with evident cultural aspirations. Two symbolical aspects of the scroll in funerary iconography need to be addressed. The first is the phenomenon of women holding a scroll, which has been variously treated in the literature. Some commentators claimed its rarity, with others holding it as a common feature of funerary iconography, especially from the 3rd century when the social

[358] Koch and Sichtermann 1982: 615–617; Gabler 2007: 22; Fejfer 2008: 133–134; Birk 2013: 94–95.

[359] Newby 2007: 247; Birk 2013: 96.

[360] Gabelmann 1973: 119, *passim*; Koch 1993: 59; Carroll 2006: 41–42; Fejfer 2008: 124–127; Birk 2013: 99, *passim*.

[361] See fn. 59.

[362] Mercando and Paci 1998: 123–124, nos 60–67; Rose 2007: 2017; Škrgulja and Migotti 2015: 30; F. Harl 2016: 24–25.

[363] Pochmarski 2005: 361–363, fig. 6.

[364] Migotti 2007b: 71 (with literature).

position of women and their role in society increased.[365] There are two such examples in the present evidence. On a sarcophagus from Siscia (S II.10), a servant girl is most probably depicted in the guise of her mistress, a senatorial tomb owner, holding a scroll in her left hand, whose symbolism should be related to the sarcophagus' owner. Another case is represented on tombstone S(t) I.6, on which two men and a woman are portrayed with scrolls in their hands, with a presumed stress on the social equality of all three of them. Curiously, this monument also holds servants on its short sides, with the girl servant probably depicted in the guise of her mistress represented on the front, that is, wearing the same clothes and holding a scroll; unfortunately, such interpretation remains uncertain due to the damage and wear of the stone. The other aspect concerns the fact that the gesture of holding a scroll takes two forms equally widespread on tombstones in Noricum and Pannonia, and sometimes depicted on one and the same piece. In the first one the person is holding a scroll firmly in one of his or her hands (S I.8; S II.10; S[t] I.2, 6; S[ptSW] I.1; S[ptSE] I.4, 5; A[ptNW] I.1), or in both hands, while in the other they are pointing at it with the opposite hand (S I.8; S[t] I.6; S III.2; S[ptSW] I.1, 3; A[ptNW] I.1; AB I.1; AB[t] II.1; AB[t] III.1). Depictions of the two gestures at the same monument (S I.8; S[t] I. 6) suggests different symbolisms, but such that are probably not generally defined but should be figured out for each case individually. The presumption of different symbolisms in terms of the two gestures is based on the importance of symbolical meanings of gestures as culturally ritualized indicators of social relationships among the Romans.[366]

The funeral rite is a complex phenomenon which reflects social and religious attitudes of the deceased and their families, comprising also a chronological component that further complicates its understanding.[367] The main argument for the shift from incineration to inhumation should be the ratio between sarcophagi on one hand and all other types of monuments on the other, which regularly held the deceased's ashes. Unfortunately, this method has proved unworkable for two main reasons. First, the number of sarcophagi is too small in the present evidence, except in Siscia, where they are more numerous, but quite a lot of them cannot be dated precisely. Although it is probable that the majority of them were used in the 3rd and 4th centuries, this remains inconclusive. At the same time, we cannot be sure that the many ash-chests from the Kordun (S[ptSW] III.1–50) all date from the 2nd century, with the possibility of a continued use in the first half of the 3rd century.[368] This is on account of ash-chest S III.2, dated to the second half of the 3rd century. The same has been presumed as valid for the entire area of Noricum (and by extension to Pannonia), as it was established that inhumations started later and incinerations lasted longer than previously presumed, meaning that the two rites were practiced contemporarily during the entire 3rd century.[369] This is further indicated by the number of *stelai* in the total evidence (Figure 10), the majority dating from the 2nd and 3rd centuries. On the other hand, it should be borne in mind that the stele does not automatically denote incineration, as witnessed by S(ptSW) I. 4, which was found as related to a skeletal grave.[370] If the evidence under research were not so lacking in epitaphs, a valuable insight into social relations would probably be available from these two kinds of funerary monuments. In other words, it would be possible to establish which social classes stuck to a more traditional rite (incineration) in the 3rd century. As we do not know how many of the inscriptionless sarcophagi actually *were* used, while more ash-chests from the Kordun were found at the quarry workshops (unused) than as *spolia* (possibly having been used previously), such insight remains unattainable.

### *II.3.3. Material and workshops*

Much work has been done in recent decades in the field of archaeometric studies of Roman stone, revealing the quarry origin of material and the manner of their use in reference to the categories and kinds of monuments, with an ultimate goal to widen the knowledge of the economic aspect not only of stone trade, but of the Roman economy generally.[371] However, the Roman economy was not an administratively structured and planned system at the state level.[372] Therefore, the most important and potentially rewarding aspects of this research should concern local and regional patterns of production and trade, with the stress on the interregional connections. The first and ultimate rule in the use of Roman stone had been established as *acquiring from the closest possible sources*, be it marble or other stone.[373] As every rule, this one is liable to many exceptions for various reasons; at this point the issue of trade routes comes to the focus. A well-established and generally accepted axiom is the one known from literary sources and archaeological research: the huge advantage of waterways over land routes. At one point, or with some commentators, this advantage tended to be taken as absolute, while a more balanced approach

[365] Migotti 2007b: 71; Fejfer 2008: 134–135; Birk 2013: 93.

[366] Cf. Kleiner 1977: 30–46; Peachin 2014:18, *passim*.

[367] Cf. Carroll 2006: 3; Fejfer 2008: 106.

[368] Perkić 2012: 298.

[369] O. Harl 1991b: 19–20; Pochmarski 1996: 139. In north Italy an example is known of a sarcophagus containing remains of two people, the skeleton of one and the ashes of another (Carroll 2006: 7).

[370] Migotti 2010a: 95.

[371] The bulk of this research has been concentrated around the 'Association for the Study of Marble and Other Stones in Antiquity' (ASMOSIA). Cf. Migotti 2018: 19, 33, with a list of all ASMOSIA conferences and conference volumes so far.

[372] Greene 1990: 9–15; Russell 2013: 3–5.

[373] Djurić 2004: 162; Hudeczek 2008: 11; Djurić and Müller 2009: 115; Russell 2013: 12.

is needed. In other words, waterways (both by seas and rivers) certainly had precedence in transportation of heavy cargoes such as stone, but they also had their disadvantages and in many cases land routes were used either exclusively or in tandem with waterways.[374] When speaking of the latter in the present context, only river traffic comes into consideration, with its many obstacles and weaknesses when compared to overseas sailing.[375] Therefore, in some cases an overland transport could have been either more suitable or the only one available.

Although it has generally become a must to include scientific analyses of materials when discussing any kind of Roman stone monuments, such examinations have never been conducted systematically on Roman funerary stones from the area of the present research; this is true of both marble and other stones. Some of the marble pieces from all three towns, kept in the Archaeological Museum in Zagreb, were characterized in 2008 by Bojan Djurić and Harald Müller (The University of Natural Resources and Applied Life Sciences, Vienna), but the results remained unpublished for the reasons that will be explained below.[376] Hence, the evidence of Siscia, Andautonia and Aqauae Balissae remained an 'island' among quite advanced research of other south-Pannonian towns, notably Poetovio, Sirmium and Mursa.[377]

A study of the production of Roman marble monuments and trading in them in the Norico-Pannonian region rests on the fact that Pannonia does not have marble sources in its territory, and therefore depended on eastern Alpine (Norican) quarries.[378] Such a statement is true only in a relativising archaeological discourse, because one of the main producers of monuments, and mediators in their trading, was Poetovio, administratively a Pannonian town on the border between Noricum and Pannonia, which sourced from the quarries on Mt Pohorje, also presumed to have belonged to the province of Pannonia.[379] Therefore, when positing the lack of marble sources in Pannonia and mentioning the export of Norican marble monuments into Pannonia throughout this discussion, the inclusion of Poetovio in so defined scheme is justified by the fact that in terms of geology Mt Pohorje belongs to the complex of the eastern Alpine marbles.[380] After roughly two decades of intensive research on the Alpine marbles and the trade in marble monuments in Pannonia, mostly by Bojan Djurić and his collaborators, the picture emerges as clearer in many details on one side, but also as more complicated than previously believed on the other. The above-mentioned rule of the *closest possible source* remains valid and is mostly reflected in the fact that the nearer the marble source, the larger the number of marble monuments in Norican towns. The same is true of Pannonian cities, although modified in terms not of the vicinity of sources, but the most effective transport routes. Notwithstanding the rule of closeness, exceptions are quite conspicuous, revealing unexpected practices. Thus, some Norican towns with their own marble sources still imported other marbles or products made of such material, and even traded in them, with many aspects of this production and trade remaining open. The most perplexing of them is probably the knowledge that the (characterized) material of a monument does not automatically reveal its workshop origin.[381] One of the most important but still contested issues is the location of those workshops that participated in the trade, that is whether they were parts of quarries or located in their near vicinity, as archaeologically proved, or whether they were established in the nearest towns, as tentatively presumed. Thus, the issue of how exactly the traffic in prefabricates was organized still needs further investigation.[382] For this reason the term workshop(s) has been used somewhat loosely throughout this discussion, possibly referring to the quarry workshops or the workshops in the respective cities, as well as wavering between the conceptions of *a workshop*, *workshops* and *the workshop(s)*. Whatever the case, at this stage it has been securely established that two quarries or two workshop centres took leading parts in the exportation of eastern Alpine marbles to Pannonia: Gummern in the vicinity of Villach (originally supplying Virunum, Zollfeld near Klagenfurt, AU) and Šmartno on Pohorje (originally supplying Poetovio, Ptuj, SI), both on account of the vicinity of the Drava River, and the former also on account of superior and more abundant reserves of marble.[383]

As mentioned above, the characterizations carried out by Djurić and Müller in 2008 in the Archaeological Museum in Zagreb remained partly inconclusive because at that time the methods of distinguishing between various Alpine marbles, notably between

[374] Gabelmann 1973: 169; Russell 2013: 95–114; Djurić 1991: 39; Djurić 1997: 80; Djurić 2004: 161–162; Djurić 2005: 76; Tomasevic Buck 2007: 94; Đurić, Davidović, Maver and Müller 2008: 116; Djurić 2010: 201; De Stefano 2015: 517.

[375] The most important river routes in Norico-Pannonian stone transport were the rivers Drava, Mura, Danube, and Sava. Cf. Djurić, in works from the previous footnote. On the difficulties of river transport, see Russell 2013: 105–106.

[376] Djurić 2013.

[377] Djurić 2004; Djurić 2005; Djurić and Müller 2009; Djurić, Müller and Filipović 2010; Đurić, Davidović, Maver and Müller 2008.

[378] Djurić 2004: 166; Djurić and Müller 2009: 116.

[379] Djurić 2004: 150; Djurić and Müller 2009: 113.

[380] Djurić and Müller 2009: 111, *passim*. See also Migotti 2018: 27, fn. 24.

[381] Djurić, Hebert *et al.* 2005; Hudeczek 2008: 11; Djurić and Müller 2009.

[382] Kane and Carrier 1992: 124; Djurić 1991: 46, 99; Djurić 2004: 157–160; Djurić 2005: 77; Hudeczek 2008: 11; Djurić, in press.

[383] Djurić 2004: 163; Djurić 2005: 76; Steiner 2007: 396–397; Đurić, Davidović, Maver and Müller 2008: 116; Djurić and Müller 2009: 113.

those from Gummern and Pohorje in our example, were not refined enough. All of the samples stemmed from the southern part of the province of Pannonia. The method used comprised oxygen ($\delta^{18}O$) and carbon ($\delta^{13}C$) isotope analyses for stable isotopes, microscopic (x-ray diffraction) analyses of thin-sections for minerals, and geochemical analyses for trace elements. Ultimately, in this case as in characterizations made before and after 2008, the most serious obstacle to gaining dependable results turned out to be overlapping isotopic fields of Gummern and Pohorje marbles.[384] While only two of the samples from the Archaeological Museum in Zagreb were determined as stones other than marble, of the 64 marble examples, six could not be determined more precisely than just as marble. Of the remaining 58 samples, five were established as originating from Carrara, 23 from Gummern, and 18 from Pohorje, while 12 samples remained inconclusive in terms of origin either from Gummern or Pohorje.[385]

Arguably, the biggest overall obstacle in the present study on stone material was its limitation to funerary monuments alone. Thus, one of the important goals inherent in such researches generally, that is, the use of particular kinds of stone for different categories of monuments, remained beyond the scope of the present work. Theoretically, this line of investigation can be applied to various kinds and types of monuments within the same category.[386] However, such research could not have been conducted on the present material either, because of a too small database; only some hints in this regard could have been attempted (see below). Another obstacle to ideally conducted characterizations is the fact that not the best and most usable sample in terms of the dimensions and position within the stone can be taken from delicate archaeological monuments, but one that can be obtained in the least destructive way. Notwithstanding all that, the archaeometric research on the funerary monuments from Siscia, Andautonia and Aquae Balissae produced some new and relevant findings.

In the last decade much work has been dedicated to refining the methods of characterizing white marbles in order to differentiate between those with overlapping isotopic fields. Recently, Walter Prochaska from the Department of Applied Geological Sciences of the University of Leoben distinguished himself as one of the leading researchers in this field; he conducted the majority of marble characterizations within the present project in 2017. While also using oxygen and carbon isotope ratio analyses and trace element chemistry, he refined the whole procedure by introducing another method: fluid inclusion investigations, that is analyses of extractable total dissolved solids (TDS) from marble and carbonate rocks. The evaluation of the data was carried out by a customary procedure comprising a multivariate discrimination analysis, displayed graphically through statistical ellipses which represent quarries or marble-producing sites.[387] As mentioned above, of the material characterized by Harald Müller in 2008, 12 samples remained inconclusive; only three of them fall under the present evidence, because they have been sampled on funerary monuments from Siscia (S I.6) and Andautonia (A[t] VIII.2; A[t] I.2 ) (Figure 20: 19, 24 and 25). When characterized anew, A(t) I.2 turned out to be Pohorje marble, while the samples of S I.6 and A(t) VIII.2 revealed that the respective monuments were not made of either Pohorje or Gummern marble, but of that from Salla, a quarry supplying Flavia Solva (Leibnitz, AU).[388] Apart from those three, some new marble samples were characterized in 2017, with final results producing a couple of novel insights into the researched subject matter. Given that the characterizations in 2008 came with an approximately equal number of Gummern and Pohorje origins for the analysed south-Pannonian finds, the same was expected in relation to the three south-western Pannonian cities discussed here. However, it emerged that apart from three monuments made of Salla marble (S I.6; A[ptNE] I.1 and A[t] VIII.2) and ten pieces that for various reasons could not be determined more precisely than just as marble (S IV.4; S VIII.5, 10, 13, 14, 17; S[t]) V.2; A[t] VIII.1; AB[t] I.4; AB[t] III.1), the remainder revealed a conspicuous domination of Pohorje over Gummern marble. As many as 22 monuments were certainly or very probably made of Pohorje marble (S I.4; S II.10; S IV.2, 3; S VII.1, 2, 3; S [ptSE] IV.1; A[t] I.2, 4, 7, 8, 9, 10; A[t] VI.1; A[ptNW] VII.1, 7; AB I.1; AB[t] II.1; AB[t] VII.1: 1–3) and only three of Gummern marble (S I.8; A[t] I.3; AB[t] I.7). The presence of marble monuments in the three cities is as follows: 18 pieces in Siscia, nine in Andautonia, and six in Aquae Balissae, which is roughly in accordance with the total number of finds from these cities and their territories (S 143, A 30, and AB 27). The ratio between Pohorje and Gummern material is slightly unexpected in view of a different picture in some other Pannonian cities, where this ratio is either levelled or even showing a prevalence of Gummern marble.[389] Given that the products from both Virunum and Poetovio could reach south-western Pannonian cities only by overland routes, or by a quite impracticable waterway route via the Drava

[384] Cf. Müller 2002; Müller 2007; Djurić and Müller 2009: 113.

[385] Djurić 2013.

[386] This is based on the fact that the choice of the rock depended on its suitability for a particular usage purpose. Cf. Djurić 1991: 49, *passim*; Bozzo, Merlanti and Rocca 1992; Tomasevic Buck 2007: 94–95; Djurić 2008: 160; Đurić, Davidović, Maver and Müller 2008; Russell 2013: 9; Kremer and Kitz 2016: 77–80; Djurić, in press.

[387] Cf. Prochaska and Attanasio 2012; Onuk and Prochaska 2015; Prochaska 2015; Prochaska, Attanasio and Bruno 2015.

[388] See fn. 16.

[389] The ratio between Gummern and Pohorje marbles in Sirmium is 52:48 and in Mursa 62.5:37.5 (Đurić, Davidović, Maver and Müller 2008: 116; Djurić and Müller 2009: 116; Djurić, Müller and Filipović 2010: 10).

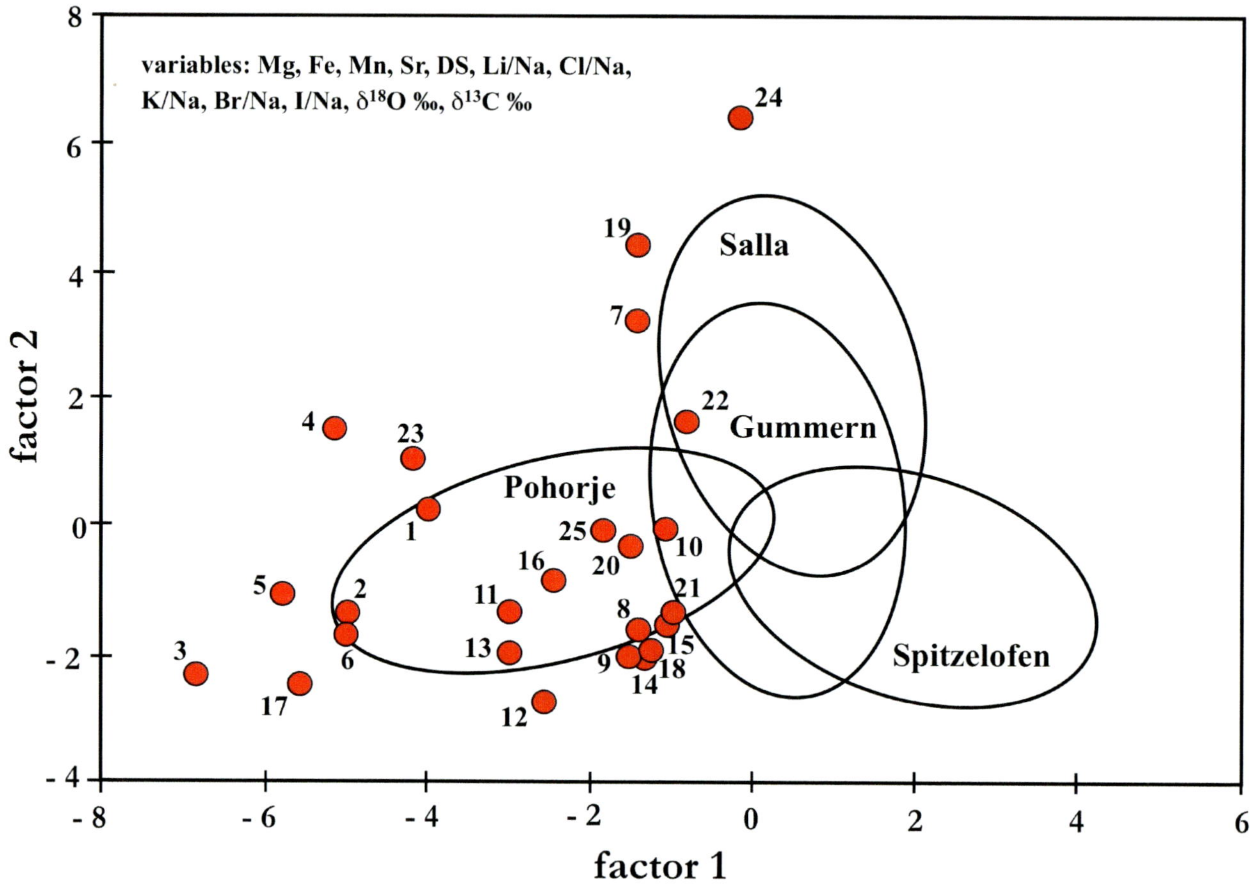


Figure 20. Bivariate diagram of the two most powerful factors of the multivariate discriminant analysis of the investigated samples and the compositional fields (90% ellipses) of the Alpine marble sites (made by: W. Prochaska)

and Danube, and back all the way westwards via the Sava, there is no straightforward explanation for such results.[390] Therefore, the possibility that some of the Pohorje material could have reached Andautonia and Siscia from Celeia (Celje, SI), by way of the Savinja and Sava, should be considered.[391] It is true that the Roman-period working of the quarry situated at the village of Hudinja on the southern slopes of Pohorje Mountain and looking towards Celeia has not been confirmed archaeologically, although some indications do exist. Still, the hypothesis about trading in stone monuments between Celeia on one side and Andautonia and Siscia on the other remains plausible, as Celeia workshop(s) certainly used Pohorje marble. Significantly, a stonemason workshop has been presumed at the point where the Poetovio–Celeia road branched off towards Flavia Solva.[392] Furthermore, several monuments from Siscia and Andautonia (S I.4; S VII.3; A[t] VI.1; A[ptNW] VII.1) characterized as most probably carved in Pohorje marble (Figure 20, nos 3, 4, 5 and 17), were interpreted as outlayers in the isotopic field of Pohorje and tentatively explained as stemming from sites on Pohorje not yet included in the databases.[393] On balance, such a presumption provides an acceptable clue for an explanation of the absolute prevalence of Pohorje over Gummern marble in south-western Pannonia, but this issue needs further enquiry.

The other of the two unexpected results of the 2017 characterizations was the establishment of three monuments made of Salla marble (S I.6; A[ptNE] I.1; A[t] VIII.2), which did not emerge in the 2008 analyses (Figure 20, nos 19, 7 and 24). They equal in number those made of Gummern marble, which is slightly surprising in its own right, but also because, so far, monuments made of Salla marble have not been established in other Pannonian towns except in Savaria (Szombathely, HU), where they could have come only by overland transport.[394] One of the three monuments in question here (lion top A[ptNE] I.1) was found near the right

[390] On the Drava-Danube route, see Đurić, Davidović, Maver and Müller 2008: 116.
[391] Such a possibility was mentioned in passing by B. Djurić on several occasions (Djurić 1997: 80; Djurić 2010: 206), but was never considered engagingly.
[392] Djurić 1991: 36; Djurić 1997: 80; Djurić 2004: 159; Djurić and Müller 2009: 113.
[393] Prochaska 2017: 6.
[394] Hudeczek 2008: 11; Djurić and Müller 2009: 116.

bank of the Drava River, and was thus possibly lost during transportation towards one of the Pannonian or Moesian towns further east. Another possibility, the one favoured here, is that this piece was intended for an unknown site in the territory of Andautonia, since another marble lion top originates from the same area (A[t] I.9). The lack of Salla marble monuments in south-eastern Pannonia is an additional argument in favour of Andautonian affiliation of lion top A(t) I.2. In any case, monuments made of Salla marble could have arrived to southern Pannonia only by the Mura and Drava Rivers, and via a combination of the river and overland routes. On the basis of finds made of Gummern and Pohorje marble in Flavia Solva, E. Hudeczek presumed that the local marble from Salla mostly satisfied the needs of the city's wider territory to the north, as well as that of the eastern Styria along the middle stretches of the Mura River, lacking in marble.[395] As noted above, the exportation from Flavia Solva went further east, reaching Savaria, but also towards the south and south-east to Andautonia and Siscia, as transpires from the present research. There should be no doubt that the monuments made of Salla marble were carved in Flavia Solva, although in the case of a stele base from Siscia (S I.6), which cannot be dated more precisely, no structural or iconographic traits reveal such workshop affiliation. While mythological motifs were quite popular in Noricum and Pannonia, centaurs were not depicted often, and no close comparison could be found for the Siscian fragment. Curiously, with the lion top from the presumed territory of Andautonia (A[ptNE] I.1), the Salla workshop emerges as a new centre for the exportation of lion tops, following in the steps of the well-known workshops of such pieces, Virunum and Poetovio.[396] In doing so, the Salla workshop faced market competition from the Pohorje workshop(s), as evidenced by pieces made of Pohorje marble from Andautonia (A[t] I.9, 10). Finally, the last piece (statue of Icarus A[t] VIII.2) needs no justification in stone material to witness its affiliation to the Flavia Solva workshop, as this is manifestly obvious in the bulk of the production of such pieces in Flavia Solva.[397] A further indication for the influence of Flavia Solva stone-carving workshops in the territory of research is the fact that one of the sandstone *stelai* from Topusko has a parallel in a marble tombstone from Voitsberg in the territory of Flavia Solva (see below).

One of the three monuments made of Gummern marble is the most attractive and skilfully carved Roman marble stele from north-western Croatia (A[t] I.3), coming from the territory of Andautonia. Structurally, its type has been established by Bojan Djurić as a sub-group of a very widespread architectural stele type, comprising a base, an inscription field flanked by columns, a frieze (possibly also an architrave), and either a lion pediment or a rectangular upper part, with variously relief-decorated spandrels (Figure 11, types 3a and 3b).[398] As the upper parts of such *stelai* were presumably carved only in a workshop at the place of its final use, while the quarry workshop left open both possibilities of finishing a monument, the area of the research for the origin of individual pieces cannot be securely limited by the shape of the upper parts of these *stelai*.[399] Nevertheless, since the material itself cannot be decisive, limiting the area of distribution and establishing iconographic comparisons remain the only available tools for determining the workshop origin. So far, *stelai* with lion pediments were found only in Celeia and Flavia Solva and their territories; significantly, one of the findspots is on the southern slopes of Pohorje looking towards Celeia.[400] The whole group, that is, both types 3a and 3b here, were made of Gummern, Pohorje and Salla marbles, with the quarry workshop of Gummern established as the first and most important producer of such *stelai* as roughed-out products; still, no examples with lion pediments were found in Virunum.[401] No comparisons for specific iconographic traits of stele A(t) I.3 (an imposingly ornamented tripartite base and an elaborate architrave superposed by a frieze) could be found in the evidence from Celeia and Flavia Solva, or anywhere else in Noricum and Pannonia.[402] The motif of mourning Erotes with legs crossed and holding a reversed torch pulled through a wreath, as employed on the base, is absent from Norican marble *stelai*. Of the four pieces (apart from Andautonia) known to the present author, three originate from Pannonia and one from Moesia Superior, but none is marble.[403] On the other hand, a very similar tripartite scheme is found on a century or so later *aediculae* in Šempeter, in the Celeian territory.[404] Since the Celeia workshops used Gummern marble,[405] as did those in Poetovio and Flavia Solva, a Celeian origin for stele A(t) I.3 emerges as probable, which is further substantiated by a commercial logic, that is, the availability of the transport route via the Sava. The other of the three monuments made of Gummern marble is stele S I. 8. It testifies to a continued production of the Gummern quarry workshop for the Pannonian market as late as the turn of the 3rd and 4th centuries, although in a simplified version. This type seems to have developed

[395] Hudeczek 2008: 11.

[396] On the workshops of Virunum and Poetovio as producers of lion tops, see Djurić 2005: 79; Đurić, Davidović, Maver and Müller 2008: 116; on Flavia Solva's workshop of the same products, see Hudeczek 2008: 49–55, nos 54–66.

[397] Hudeczek 2008: 25–28, nos 21–28, especially no. 21.

[398] Djurić 2005: 78; Djurić 2008: 162–163; Djurić 2010.

[399] Djurić 2010: 205.

[400] Djurić 2010: 204, fig. 3.

[401] Djurić 2010: fig. 3; Djurić 2016: 178–179, 180, fig. 2.

[402] The only comparison, but only for the base, is found on a slightly later and differently structured marble stele from Tác (Gorsium, Pannonia Inferior, *lupa* 3586).

[403] Migotti 2016b: 228, fig. 15.

[404] Cf. *lupa* 13256, 13263, 13271, 13272, 13333.

[405] Bojan Djurić, personal communication.

from *stelai* comprising a portrait niche provided with a frieze above and a decorative band below, such as A(t) I.8 (Figure 11, type 5a; see below at the discussion of the monuments made of Pohorje marble). The last monument made of Gummern marble stems from the territory of Aquae Balissae (AB[t] I.7). As its motif of the *lupa Romana* is not indicative of the workshop affiliation, the stele could have arrived there either via the Drava or Sava, and could have been produced in Virunum, Flavia Solva or Poetovio. Norico-Pannonian trading in marble monuments after the early 2nd century, which until recently had been less well researched than the previous period, has now been considerably upgraded. Nevertheless, not all debatable issues have been cleared. Thus, the production of the Pohorje/Poetovian workshops, juxtaposed with their role as mediators in trading products made of Gummern marble, especially *stelai* and lion tops, requires further research.[406]

As the majority of the marble monuments were made of Pohorje marble, in some cases their workshop origin remains obscured, because it could have been any of the workshops using this marble: Poetovio, Celeia, or Flavia Solva. In other cases, Poetovian origin appears to be more clearly demonstrated, but final conclusions remain debatable for several reasons: the fragmentary state of some of them, obscuring their full shape and/or a workshop affiliation; and the proverbial difficulties in dating stone monuments, even when they are well preserved, let alone those fragmentary ones. The importance of the chronological aspect is related to the fact that the Poetovian workshop started exporting stone monuments only after the town had become *colonia Ulpia Poetovio* at the beginning of the 2nd century.[407] Presumably the earliest of the *stelai* made of Pohorje marble (S I.4) probably dates from the Flavian or Trajanic periods. It could have been modelled on the scheme of similar Poetovian *stelai* from the Flavian period, whose pediments were mostly decorated with a central *gorgoneion* flanked by dolphins, as by that time the rosette had become outdated, although still appearing in tandem with dolphins.[408] The type of stele A(t) I.4 cannot be restored with certainty. Depending on the shape of its upper part, it could have been of type 5a or 5b, figuring as a step forward in the structural development of the *stelai* of type 3a and 3b. In spite of its tentative determination in the Catalogue as type 5, its pedimental section could have also been pentagonal, thus forming narrow, elongated spandrels. Therefore, this stele possibly represents a continued line of production of similar mid 1st-century examples from Poetovio, which were there carved exclusively in Gummern marble.[409] All in all, the stele from Andautonia suggests that either there was some mistake in the characterization procedures, or else, that *stelai* containing a pentagonal upper part and a portrait niche were in Poetovio made of both Pohorje and Gummern marble. If, however, the stele in question had the upper part rectangular, as suggested at the start, it can be perceived as belonging to the type whose presence in Noricum and Pannonia in the 2nd and 3rd centuries results mainly from the large output of Gummern workshops, but which could have also been carved in Pohorje marble.[410] Another example testifying that some uncertainties concerning such reconstruction of distribution patterns remain, is stele A(t) I.8 (type 5a), the only entirely preserved marble piece from Andautonia in addition to A(t) I.3. As noted above, despite being carved in Pohorje marble, its appearance in Andautonia should be put in the context of a continued trade in marble monuments carved under the influence of Gummern production. Curiously, while the majority of *stelai* with the portrait niche fall under the subgroup with the rectangular upper part (type 5b), a simplified version from Siscia (S I.8), made of Gummern marble, has a lion pediment (type 5a). Since *stelai* of both types are almost absent from Noricum, the presumption that they were initially produced by the Gummern quarry workshops to satisfy the needs of Pannonian customers is best confirmed through monuments made of Gummern marble, such as stele S I.8.[411] Stele A(t) I.8 from Andautonia was, however, made of Pohorje marble, which leaves open its ultimate workshop origin; it could have been carved in any of the workshops using Pohorje marble, notably Poetovio, Celeia and Flavia Solva. The same is true of *stelai* A(t) I.2 and A(t) I.7, whose structural classification and workshop origin is even more uncertain due to their fragmentary state. The former could have belonged to types 3(a or b) or 5(a or b), while the latter was probably of type 5a or 5b. Similar uncertainty about a structural typology and workshop origin refers to stele AB I.1. Irrespective of whether it had a lion pediment or a rectangular upper part (types 5a or 5b), it can be perceived as a simplified version of A(t) I.8, on account of the lack of a frieze; the same is valid for stele S I.8.

Contrary to lion pediments, lion tops were typical Poetovian products, possibly differing from the same pieces made of Gummern marble in terms of the inferior quality of their production.[412] A presumption of their Poetovian origin in Andautonia seems to be justified through the stone material (Pohorje marble), as well as on account of abundant production of such monuments

[406] Cf. Djurić 2005: 78–79; Djurić 2008: 164; Djurić and Müller 2009: 116; Djurić, in press.

[407] On Poetovio's exportation of *stelai* made of Pohorje marble, see Djurić 2004: 163; Djurić 2008; Djurić, in press.

[408] Djurić 2008: 162, 165, figs 1: 7–8.

[409] Djurić 2008: 161, fig. 2: 3.

[410] Djurić, in press, fig. 23.

[411] Djurić, in press, fig. 17.

[412] Djurić 2005: 79; Đurić, Davidović, Maver and Müller 2008: 116.

in Poetovio and their lack in Celeia. However, caution is needed in formulating a final conclusion, because one lion top, presumably found in the Andautonian territory (A[ptNE] I.1), was made of Salla marble. At this point it is necessary to remember that Flavia Solva was a leading Norican centre for the production and exportation of lion tops.[413] Stone characterization of a marble lion top from the southernmost part of Pannonian territory at its border with Dalmatia (Catalogue, Peripheral areas III.1 – Mahovljani, BA) produced too wide a scope of possible origins. Even though, on account of the common knowledge of lion tops and the presumed Poetovian affiliation for the examples of lower quality than those made of Gummern marble, this piece was probably imported from a Poetovian workshop as a half-product and finished locally. Norico-Pannonian lion tops did reach the province of Dalmatia, but to this author's knowledge, no marble examples have yet been found in this province.[414] Therefore, the lion top from Mahovljani should be the southernmost marble specimen found so far, provided that visual determinations of other pieces from Bosnia were correct.

Another intricacy of the Norico-Pannonian trade in stone monuments is represented by a funerary medallion from the territory of Andautonia (A[t] VI.1). As it was made of Pohorje marble, its Poetovian (or Celeian) origin should be expected, but the established production and market procedures testify otherwise. Not only were the workshops of Virunum and Flavia Solva the most important centres for the production of funerary medallions, but the Poetovian workshop produced funerary medallions for Flavia Solva without using them in its own territory.[415] This means that the medallion from Andautonia should probably be added to a group of monuments imported from Flavia Solva (see above). A Poetovian affiliation of two sarcophagi, one each from Siscia (S II.10) and Aquae Balissae (AB[t] II.1), need not be doubted as both are of a well-studied Poetovian type (Figure 12, type 3a and 3b, respectively).[416] What remains inconclusive is the question whether the sculptural parts were executed in the workshop of origin or at the place of use; in both cases a highly individual choice of themes testifies to works to order and not to a purchase from stock.[417] As for sarcophagus AB(t) II.1, its detached ornamental base, otherwise rare in the corpus of sarcophagi generally, points to an oriental prototype.[418] The only marble ash-chest (AB[t] III.1) could not have been sampled and characterized because it is walled and mostly covered in paint. Nevertheless, its typical Poetovian structural typology (type 3) and the fact that it evidently arrived via the Drava, speak for this piece's origin in Poetovio. All of the marble *aediculae* most probably date from the Severan period or slightly later, and all were made of Pohorje marble (S VII.1, 2, 3; A[ptNE] VII.1, 7). One of those from Siscia, a corner of a cornice with architectural decoration and masks (S VII.2), has a close parallel in Poetovio, while fragment S VII.3 gives the impression of coming from the same workshop, as indicated by the choice and execution of motifs.[419] One of *aediculae* fragments from Andautonia (A[ptNW] VII.7) counts among the funerary architecture well known in Celeia and its territory.[420] Several monuments of various kinds, but mostly *tituli* (S IV.2, 3; S[ptSE] IV.1) testify to a continued use of Pohorje marble in the Late Roman period, without pointing to a specific workshop origin; this could possibly be classified only through detailed comparative epigraphic and paleographic analyses. Because Pohorje marble was established for all the Siscian *tituli* available to be sampled and characterized, the same can be tentatively presumed for another Late Roman *titulus*, whose characterization failed (S IV.4). The monument (S VIII.17) which remained structurally undetermined due to its unusual shape, otherwise unknown in Norico-Pannonian funerary evidence, was probably made of Pohorje marble. Even if this is true, its origin should be associated with its owner's probable oriental/Greek origin rather than with any workshop repertoire. Only one funerary marble altar (S[t] V.2) has been identified among the entire evidence, and this was determined as carved in marble *breccia* from the central eastern Alps. From the point of structure and iconography, the piece could have been produced anywhere in north Italy and Noricum, as an iconographic conception similar to the one from Siscia is found at various places within northern Italy, Noricum and Pannonia.[421]

As was mentioned at the beginning of this chapter, the database was too limited to be applied in the research of relations between the kinds of monuments and kinds of stone used. Nevertheless, what strikes us immediately is an aberration in the ratio between marble *stelai* (13) and marble sarcophagi (two), as compared with their ratio in the entire evidence overall (52:39). At this stage it is not possible to figure out whether such a result is purely accidental, or whether there is some reason behind it, for instance a longer time during which the stele was used in tandem with an overall decline of

[413] Hudeczek 2008: 49.

[414] For the examples roughly executed and made of local limestone from the interior of the province of Dalmatia, see Paškvalin 2012: 146–147, nos 83–86. E. Hudeczek (2008: 49) mentioned the export of funerary medallions in Dalmatia, but without any details, such as findspots or material.

[415] Piccottini 1972; Djurić 1991: 99; Djurić 2004: 165; Pochmarski 2011: 22–26.

[416] Djurić 2001a: 48 (sarcophagi of type 3); Djurić 2001b (ash-chests).

[417] Koch 1993: 45; Cambi 2010: 19/262; Birk 2013: 31–32. The mass production of sarcophagi as stock has been questioned in recent scholarship (Russell 2013: 3, 294–301).

[418] Gabelmann 1973: 18; Koch 1993: 42.

[419] For the Poetovian parallel see *lupa* 9442.

[420] See fn. 113.

[421] See fn. 37.

marble monuments in late Roman Pannonia, which should have affected the production of sarcophagi more. A total lack of marble ash-chests to as many as 54 limestone pieces in the total evidence is of no concern, because their large number is the result of local production in the area of Kordun.

In conclusion, it can be postulated that marble characterizations within the present research produced two new and unexpected pieces of information: the considerable scale of imports from Flavia Solva; and the complete prevalence of Pohorje marble in the circumstances otherwise testifying to the leading role played by products from the Gummern quarry workshop in the Pannonian market. However, the ratio between Gummern and Pohorje marble for the whole of Pannonia relates to all categories of monuments, while the present research comprised only funerary. This suggests that Pohorje marble could have been preferably used for funerary monuments overall, but this, understandably, is only a working hypothesis. Finally, the present study has demonstrated that monuments made of Pohorje marble should not be automatically affiliated with Poetovio, and that the role of Celeia in trade with towns in south-western Pannonia needs to be considered instead.

As far as archaeometric study of stones other than marble is concerned, before the present research (very few) characterizations were made by macroscopy and exceptionally by archaeometric methods, with the majority of the remaining evidence being assessed by the naked eye by non-specialists. Thus, quite a few monuments were published with broad and quite often incorrect determinations.[422] Understandably, the limitation in establishing relations between categories of monuments and kinds of stone used, or the same relations between various kinds in the same (funerary) category, is valid for the pieces made of limestone and sandstones as well as marble, and for the same reasons. Although the number of monuments made of limestone and sandstone is around four times greater (or even six times if 66 monuments from the quarry and stone-cutting area of the Kordun are included) than those made of marble, it was still insufficient for research into the above-mentioned relationship for each of the three cities individually. A further two postulates of archaeometric studies on stones could not be carried out within the present research: the comparison of samples from the monumental material with those from the quarries; and identification of layers of the relative rocks in their primary position to establish the quarry origin of the monuments under research. This was because very few Roman quarries in north-western Croatia have been verified archaeologically, while such potential sites are either inaccessible or blurred by modern use. Only a project targeted exclusively at finding traces of Roman quarries in north-western Croatia could possibly result in a better insight into this subject matter. In such circumstances, the guesses about a precise origin of the respective monuments' material remained conjectural, with two levels of results offered: possible and probable areas of (potential) quarrying in antiquity (see Chapter III).

While it is clear that marble monuments could have reached Siscia, Andautonia and Aquae Balissae only from Poetovio, Noricum or even further-away parts of the Roman Empire, the main goal of the archaeometric research on the stone other than marble was to establish the (presumably) local sources of rock available for carving. Within this frame, the starting point was checking whether there were exceptions to the general rule of the preference for the closest possible source, which mostly relates to marble, but not exclusively.[423] Scrutiny was necessary because this rule could have been obstructed by some quirk of economics or logistics.[424] As a matter of fact, the possibility for long-distance trade was suggested for several examples through archaeological indications. Three of them will be illustrated here, two of them certainly, and one allegedly, originating from Siscia. The first example is stele S I.1. When first published, it was reported as having been made of limestone from the quarry of Bijele stijene in the village of Hrastovica, near Sisak (see Chapter III), allegedly ascertained by a geologist.[425] This suggestion was suspicious for various archaeological reasons, such as the findspot in terms of micro-topography and the structural typology of the stele.[426] When the characterization conducted in Zagreb suggested that the stone did not originate from northern Croatia, the first thought was that it came from Aurisina on the north-east Adriatic coast (Italy), well known for its Roman quarrying and currently under research by a multidisciplinary team of Italian geologists and archaeologists.[427] This reasoning was based on the fact that the owner of the stele was an individual from Roman Tergeste (Trieste, IT), 14 km to the south-east of Aurisina. A new characterization was conducted in Padua (IT) by Claudio Mazzoli of the Department of Geosciences of the University of Padua (IT), and through the mediation of Caterina Previato from the same university. The characterization proved a northern Adriatic origin for the stone, with the Roman quarry in Aurisina as a very probable

[422] Migotti 2018: 23–25.

[423] See fn. 236. On limestone and long-distance trade, see Wenner and Herz 1992: 199; Djurić, Kele and Rižnar 2015.

[424] Pearson 2006: 86; Russell 2013: 169, 178; Kronberger, Mosser and Pülz 2016: 93.

[425] Migotti 2018: 31–32; Migotti, in press (*ASMOSIA* XI).

[426] The structural typology of the stele is Adriatic rather than Pannonian, while its findspot on the bank of the Sava excludes its arrival from the south of Sisak, i.e. from the quarry of Hrastovica.

[427] Bonetto and Previato 2013; Djurić and Rižnar 2016: 138–140.

immediate source. This is further substantiated by typology, as parallels for the stele from Sisak are lacking in Noricum and Pannonia, but are known from north Italy, Istria and central Dalmatia.[428] Therefore, the stele could have been carved in the workshops of Tergeste or Aquileia, or, even more probably, the quarry workshop in Aurisina. Admittedly, Aquileia has always been pinpointed as the source of workshop influences, on account of the fact that it was indeed the main trade centre of the area, especially towards Noricum and Pannonia, despite the fact that no stone-cutting activity has been archaeologically ascertained there, or in Tergeste for that matter.[429] Otherwise, the Siscian stele seems to represent the easternmost appearance of Aurisina limestone, which so far was ascertained in Emona (Ljubljana, SI), but not further eastwards.[430] The second example is another stele from Siscia (S I.3). From the point of formal structure and iconography, which in this particular case cannot be divorced, the stele in question represents an isolated easternmost appearance of its type. Such *stelai* were otherwise produced in the army workshops on the Rhine *limes*, above all in *Mogontiacum* and *Colonia Agrippina*, and are found almost exclusively in those regions.[431] Several indications suggested the possibility that the Siscian stele arrived all the way from the province of Germania: i.e. the fact that this piece is an isolated case in Pannonia; and it was made of a yellowish stone very similar at first sight to the sandstone used for some of the same type of *stelai* from *Germaniae*.[432] The characterization made in Zagreb in 2017 came up with the result that the stele was carved in Baden biocalcarenite, which, theoretically, could come from Austria and southern Germany. Nevertheless, since it has been established that such stone could also have been quarried in the surroundings of Sisak, it would be beside the point to claim its supra-regional origin. As a matter of fact, the stone in question is at sight very similar to that used for carving the monuments from Topusko (S[t] I.1–5; III.1, 2). The last example is the lidded sarcophagus S II.11, decorated with the *peltae* motif. Again, it is the only such example in western Pannonia, as its area of distribution is the region around Aquincum (Budapest) and Brigetio (Szőny) on the Danube *limes* in north-east Hungary. Since the characterization proved that it was made of north-Pannonian (Budakalász) travertine, for a moment there seemed to be no doubt that this piece was an exception to the *nearest source* rule. Despite the fact that Budakalász travertine was widely exported along the Danube,[433] transporting one (fairly unassuming) monument all the way down the Danube and Sava, with a possible reloading along the smaller tributaries of these two rivers, seems illogical. On the other hand, nowhere in the archives of the Archaeological Museum in Zagreb are there any clues as to the original findspot of this sarcophagus. Later, it turned out that its provenance from Sisak was not ascertained, and that its publication was accompanied by the remark 'probably from Sisak'.[434] Given that monuments made of Budakalász travertine, as the nearest in quality to the eastern Alpine marbles available in Pannonia, could be found in Mursa (Osijek, HR),[435] the possibility cannot be discarded that sarcophagus S II.11 was transported to Zagreb from Osijek at an unknown date, without being properly recorded or noticed by museum staff. To conclude: the only certain case of a monument that did not come to its final destination from the vicinity is stele S I.1.

The methods used in characterizations conducted in the Croatian Geological Institute in 2017 and 2018 by Mirko Belak and his collaborators were those customary for this type of analyses. It included macroscopy, polarized light microscopy and electron microscopy, X-ray diffractions and chemical analyses, and identification of rock sources by using geological maps; each analyzed sample was provided by a microscopic image.

Not all monuments made of stones other than marble and included in the Catalogues were available for characterizations. Of a total of 128 samples that were available, 34 were found in Siscia and its presumed territory, with a further five stemming from its south-eastern areas and 61 from its south-west hypothetical territories, amounting to a total of 100 pieces. Only six of the 61 pieces from the city's hypothetical territories were characterized (S[ptSW] I.2, 4; S[ptSW] III.9, 12; S[ptSW] VIII.2, 3), but a local origin for all of them is still indisputable (see Chapter III). Andautonia furnished seven samples from the city itself and its presumed territory, as well as a further eight from its hypothetical territory, totalling 15 pieces, while Aquae Balissae and its presumed territory provided 13 samples. Apart from one each imported stone (S I.1; S II.11), the remainder were sourced from the immediate or slightly more distant surroundings (Figures 21 and 22). The ratios of marble and other funerary monuments are as follows: Siscia 18:127, Andautonia 9:30, and Aquae Balissae 6:26. A more realistic result for Siscia (18:61) emerges when 66 monuments from the Kordun, a quarry area for the immediate production of stone monuments, and only hypothetically included in Siscian territory, are excluded from this calculation. Overall, the percentage

[428] Starac 2002: 65–66 (type Ic, karta III); Cambi 2005a: 11, fig. 5; *lupa* 14483 (Aquileia).
[429] Caterina Previato, personal communication. On the trading role of Aquileia, see fns 62 and 63 in Chapter II.1.
[430] Djurić and Rižnar 2016: 139.
[431] Škrgulja and Migotti 2015.
[432] *Lupa* 7071, 7073, 15876.
[433] Djurić, Kele and Rižnar 2015; Djurić, in press.
[434] Migotti 2018: 30–31.
[435] Bojan Djurić, personal communication.

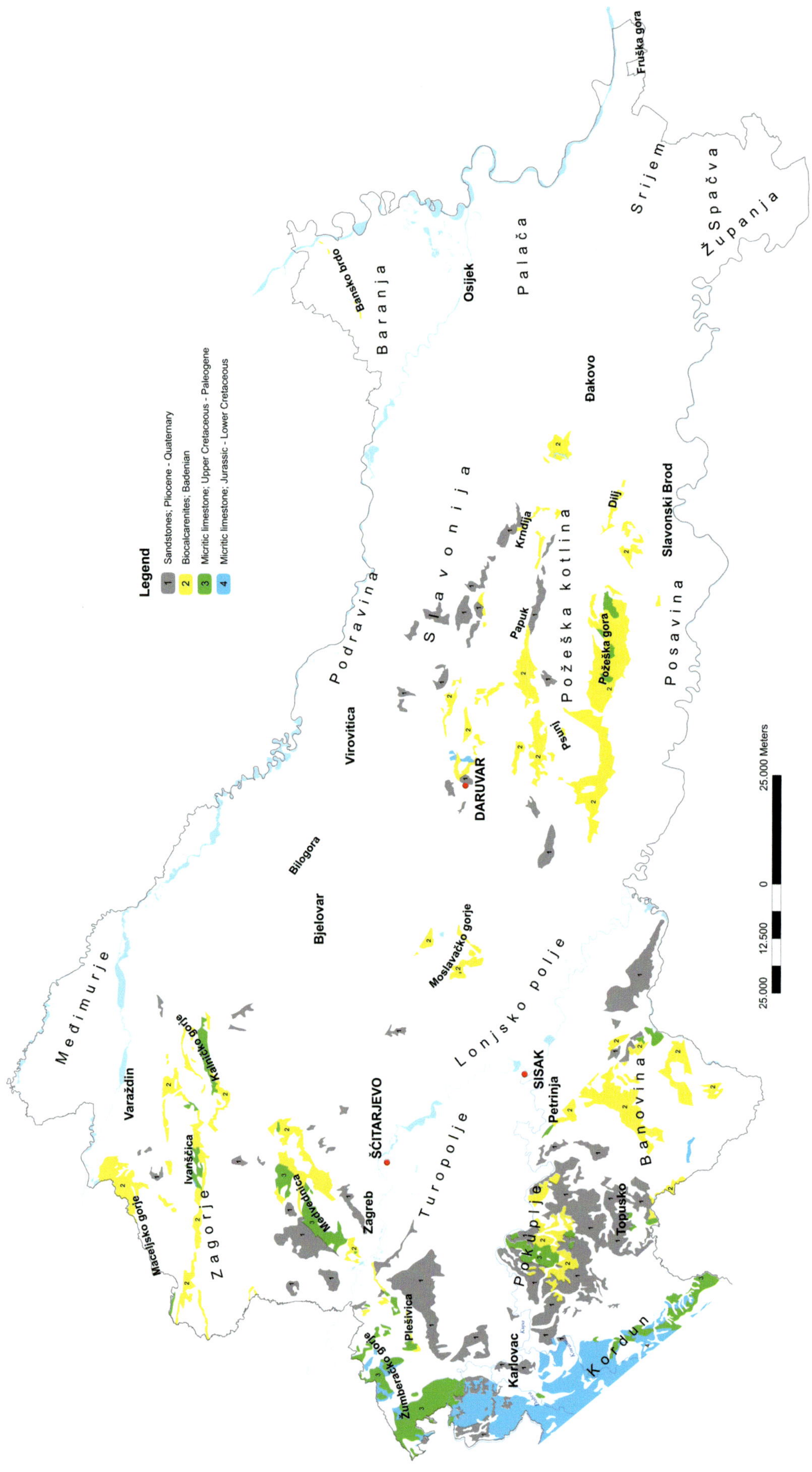


Figure 21. Geological map of the sources of stone for the Roman funerary monuments of Siscia, Andautonia and Aquae Balissae (drawn by: M. Belak, L. Fuček and P. Ferić)

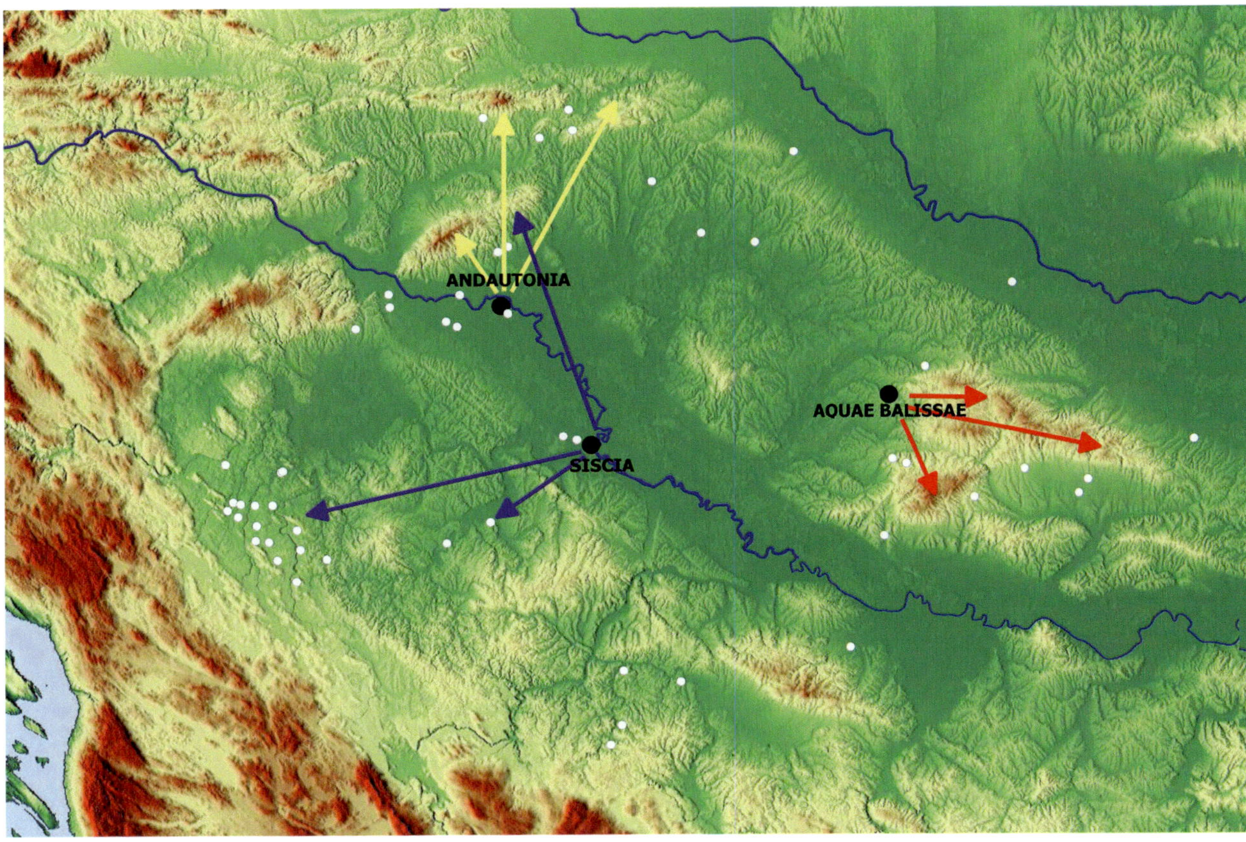


Figure 22. Schematic map of the stone sourcing areas for Andautonia, Siscia, and Aquae Balissae (drawn by: T. Leleković)

of marble in the total evidence for all three cities is not very different: 29.5 in Siscia, 30 in Andautonia, and 23 in Aquae Balissae. As a higher number for Siscia should be expected, it remains inconclusive whether the above numbers are realistic, or whether they are entirely accidental resulting from the enormous loss of monuments, possibly greater in Siscia, and the fact that only funerary stones have been considered in this research.

Stone for the city of Siscia was possibly quarried in the mountains of north (I.7; II.1, 2, 3, 7; VIII.16) or north-west Croatia (S II.4; S III.4; S IV.1), the Banovina (I. 2, 3, 5; S II. 6, 12, 13, 14, 15, 16, 17, 19; S VIII.8), and the Kordun (S II. 8, 18, 20, 21, 22; S III.2). Only one site (Topusko) in its presumed territory yielded data for the stone origin; all of its monuments were carved from the quarry situated on the town's periphery (S[t] I.2, 3, 4, 5; S[t] II.1; S[t] III.1, 2). The sites in Siscia's hypothetical territories on the south-east also seem to have been sourcing from the vicinity (S[ptSE] I.1, 2?, 3, 4, 5), while those in the southwest in the area of the Kordun (S[ptSW] I. 2, 4; S[ptSW] II.1–8; S[ptSW] III. 1–50; S[ptSW] V.1; S[ptSW] VIII. 2, 3) certainly used stone from local surface quarries. We cannot be sure whether the monuments that could be only broadly characterized as originating from the mountains in north Croatia should actually be classified with those from north-west Croatia. This hesitancy comes from the fact that the term *north* in this case actually comprises mountains in the north-west and north-east, with the latter denoting the Slavonian mountains, some 100 km to the east of Sisak, while those in the north-west are some 50 or 60 km away. As both are approachable by the Sava and its tributaries, and the rocks in both are of approximately the same (quite low) quality, the possibility should be left open that Siscia's stonemason workshops sourced from the quarries in both the city's north and north-west surroundings. In two cases were north-west quarries indicated with more certainty through archaeometry, although the geographical circumstances are concordant with such characterization only in the first case (tombstone S VIII.2), because Mt Medvednica was only 50 km away from Siscia and was directly accessible by the Sava. In the other example (S II.4), the presumed quarry origin was on Mts Ivanščica or Kalnik, which are some 100 km away from Sisak and do not lie on the Sava. Admittedly, the most important areas of quarrying for Siscia were in its immediate southern vicinity on the Banovina, and in a slightly more distant region of the Kordun in the south-west, in the middle course of the Korana River. What strikes us about the latter is the fact that, with very few exceptions, only funerary monuments were found there. Of a total of 120 stone pieces, there are 110 sarcophagi and ash-chests, two *stelai*, only one each altar (?) and a fragmented column, three cylindrical receptacles interpreted as utilitarian objects, and four undetermined fragments.[436] At first glance, it looks like this reflects the choice of stone according to the use for which it was intended, but it can hardly be true, because this rural area must have used its own (abundant) sources of the same limestone for all the needs of its several settlements. Therefore, the scarcity of stone artefacts other than funerary in the Kordun quarry area remains an enigma.

The Andautonia stonemason workshops sourced locally, in the narrow meaning of the word (A[t] II.1), from the city's nearest vicinity on Mt Medvednica (A II.1; A[t] I.1, 6; A[t] VIII.3), as well as from more distant mountains of northern Croatia (A[t] I.11; A[ptNW] I.1; A[ptNW VIII.2]). One of the sites in its presumed territory possibly used stone from the Banovina (A[t] I.5), while the site hypothetically belonging to its territory in the north-west used stone from its immediate vicinity on Mt Ivanščica, which is more probable, or the somewhat more distant Mt Kalnik (A[ptNW] I.2; A[ptNW] VII. 2–6). As with Siscia, we cannot be sure whether the monuments from Andautonia, which could be only broadly characterized as originating from the mountains in north Croatia, indeed come from north-west Croatia. For Siscia this remains debatable on account of roughly equal distances from both quarry areas, while it is hardly likely that Andautonia sourced from the Slavonian mountains, as the mountains in northwestern Croatia are significantly closer. Thus, the characterizations coming with a wide determination as *north-Croatian mountains* presumably result from haphazard sampling and inadequate potential of the samples for characterization. Aquae Balissae and its presumed territory seem to have used nearly exclusively the stone from the Slavonian mountains, notably Psunj and Papuk, on whose western fringes the city was located (AB I.2, 3; AB V.1; AB VII.1; AB VIII.1; AB[t] I.5, 6, 8; AB[t] II.2; AB[t] VI.1; AB[t] VII.1, and probably AB V.1; AB[t] I.3, and AB[t] II.3).

As was mentioned in the Introduction, stonemasonry was introduced to southern Pannonia with the Roman conquest, and was normally related to the army and its workshops.[437] The nearest to our region and the most revealing example is Poetovio, whose army quarry and stonemason workshop(s) were activated in the beginning of the 1st century AD.[438] The same should be expected for Siscia, because of the presence of the army from the early 1st century AD to about the years 42 or 45.[439] The evidence, however, is still missing, and the earliest stele from Siscia was set up for a civilian (S I.1) and was possibly carved in one of the two stonemason workshops on the north-eastern Adriatic coast, either Tergeste as the owner's domicile, or Aquileia as the leading production centre in north Italy at that time. The third, and even more probable option, is the Aurisina quarry workshop. There should be no doubt that an army workshop for stone monuments existed in Siscia in the first half of the 1st century, but early military tombstones have not been discovered so far; the earliest one stems from the mid 1st century or slightly later (S I.2).[440] It was made of local limestone and thus most probably in the Siscian workshop. Since north Italy has been established as the main source of artistic influences for the Norico-Pannonian region, at least until the early 2nd century, a north-Italian background can be presumed for this stele as well. Close parallels for architectural *stelai* with a pentagonal upper part from north Italy are rare, but this element of formal structure was in the 1st century sometimes applied to pieces of different overall formal characteristics.[441] Significantly, non-architectural *stelai* with a pentagonal upper part are found among military monuments from Noricum (Lauriacum) and Pannonia (Carnuntum), and can therefore be brought into connection with north-Italian artistic influences on army workshops in the Danube and Rhine provinces, which is otherwise a well-established fact.[442] Admittedly, the army workshop of Poetovio played a role in transmitting models for funerary *stelai* towards Pannonia. It exported those made of Gummern marble during the whole of the 1st century, with some of them showing individual structural characteristics corresponding to that of the Siscian piece.[443] In some other cases, imitations of marble models, whether those are available or presumed although not surviving, are clearer and the phenomenon as such understandable, because local imitations are a common place of the transmission of arts in Classical Antiquity.[444] Stele S I.3 only seemingly represents a case of imitation, although not of a marble model. Its design was most probably the result of a private initiative of the owner, who had served in the unit stationed in the area of production of such *stelai* and probably brought a draft of his funerary stone as he settled in Siscia to spend his veteran career there.[445] On the other hand, it seems clear enough that three *stelai* from Topusko (S[t] I. 2, 3, 5), all made of local sandstone, are imitations of

436 Perkić 2012: 5.

437 See fn. 1 in the Introduction.

438 Djurić 2008.

439 Radman-Livaja 2012: 169; Radman-Livaja and Vukelić 2015: 399.

440 Škrgulja and Migotti 2015: 46.

441 Architectural *stelai*: Merkando and Paci 1998: 106, no. 45; *lupa* 18928. Non-architectural *stelai*: Merkando and Paci 1998: 196, no. 125; *lupa* 19554.

442 Cf. Gabelmann 1972; Weber-Hiden 2015 (type D); *lupa* 470 (Lauriacum, Noricum).

443 One of them was supposedly shaped as the Siscian example (Djurić 2008: 161, fig. 2.2).

444 Gabelmann 1973: 11, *passim*; Rebecchi 1978: 208, *passim*; Djurić 1991: 100; Braemer 2005; Djurić 2005: 77; Carroll 2006: 8–9; Cambi 2010: 64; Migotti 2013b: 205–208; Weber-Hiden 2014; Palmentieri 2015: 283; Djurić, in press.

445 Škrgulja and Migotti 2015.

Norican marble prototypes. The first two comprise a lion pediment, a portrait niche, an inscription field and a base, imitating the above-mentioned examples made of various east-Alpine marbles (type 5a). What remains less evident is how and where the local workshop got insights into marble models, as the only stele of this type (S I.8) was of a much later date and of a slightly simplified structure. One detail is significant, although not straightforwardly explainable: the only marble stele known to the present author, approximately from the same period as S(t) I.2, which features the same type of high bases of the columns flanking the portrait niche, comes from Voitsberg, some 50 km from Leibnitz (Flavia Solva).[446] There is no knowing whether such (marble) *stelai* existed also in Siscia, or whether the design in question reached the stonemason's workshop in Topusko by way of intricate and as yet poorly known routes, along which Norico-Pannonian funerary stone sculpture circulated on the market. Whatever the case, the workshops of Flavia Solva should be taken into consideration. Similarly, a pediment flanked by floating Erotes (S[t] I.5) cannot be found in the surviving Siscian evidence, but it is a well-known motif on various kinds of funerary monuments in north Italy and the Norico-Pannonian region.[447]

The 22 sarcophagi and ash-chests of Siscia make a complex whole comprising seven types (Figure 12, types 1–8), with types 3a and 7 represented by a small number of pieces.[448] Two of these were carved in Pohorje marble (S II.10; AB[t] II.1); they clearly exhibit Poetovian typology, that is, a tripartite front side comprising a central inscription panel and lateral niches topped by a Norico-Pannonian volute, and so testify to importation from Poetovio, while others presumably disclose structural traits from various sources.[449] However, two distinct and well-defined groups of those sarcophagi come to the fore, both made of local stones. Unfortunately, the characterizations were not precise enough to conclusively determine the possible use of a particular stone material for any of the two groups, but indications still exist. The first one comprises two sarcophagi (S II.4, 7), with the possibility of adding one sarcophagus (S II.8) and one ash-chest (S III.2), all made on the Poetovian structural model (type 3a). It is difficult to hypothesize on the workshop origin within the Siscian stone-carving production, because no archaeological data exists on the organization of such work. In other words, it remains unknown whether there was one workshop, possibly with branches, or whether there were more of them. In any case, sarcophagi S II.4 and 7 are part of a series carved to the same structural typology and also iconography, with the lateral niches topped by simply stylized, slightly pointed volutes, holding extremely roughly executed mourning Erotes of the same appearance, and with blank short sides. The third piece (S II.8) can tentatively be put into the same group, as it is also of type 3a, but with differently shaped, more rounded volutes and with the Erotes executed equally roughly, and in a different guise, that is, as season Erotes. The stone of sarcophagi S II.4 and II.7 was characterized as originating from somewhere in the mountains of north Croatia, but most probably in the Banovina area, in the immediate vicinity of Sisak in a southerly direction. On the other hand, sarcophagus S II.8 was made of stone quarried in the Kordun. The fourth piece (S III.2), which can be tentatively attributed to a somewhat later stage of the same production, differs from the rest in some aspects. Its short sides are sculpted, while the frontal portrait figures are executed more skilfully than the Erotes on the short sides, with the latter different from the rest of the Erotes in this group in terms of iconography; finally, it was made of stone quarried in the Kordun, but still different visibly from pieces S. II.4, 7 and 8. On balance, these four pieces indicate that the same workshop (or its branch) evidently used various stones to produce sub-variants of the same sarcophagus type, resulting in pieces different in material, the quality of execution, and iconography. The other group of Siscian sarcophagi representing a serial production comprises five pieces (S II.13–17) of a basic type defined by plain chests and flat edges (Figure 12, type 7). All were made of the same stone, most probably quarried in the Banovina, Siscia's nearest quarrying area. Significantly, all were found in the south-east cemetery, indicating the workshop location.[450] Indicatively, the last in the series of plain sarcophagi (S II.18) from the same cemetery is different from the previous pieces in terms of structure, in that it has cuts around the edges for taking the lid, and in its stone material, which most probably came from a different source (the Kordun). Finally, two Siscian sarcophagi (S II.2 and S II.22) are of unique structure and iconography, thus defying any sustainable guess about their structural and, partly, their iconographic prototypes; as such, they add to the variegated picture of Siscian production.[451] As sarcophagus S II.2, otherwise the only one of an architectural type (Figure 12, type 2), shares the motif of mourning Erotes with the sarcophagi of the first group, its slightly more competent execution should be noted. However, neither the same motif applied, nor the fact that all of these sarcophagi were probably carved from the same

[446] *Lupa* 1609.
[447] Gabelmann 1977: 69–70, pl. 12: 1; *lupa* 18508 (Aquileia); 674 (Celeia); 3106 (Poetovio); 4338 (Sirmium).
[448] Migotti 2013b.
[449] Migotti 2013b: 205–208.
[450] On cemeteries as a customary location for stone-carving workshops, see Boschung 1987: 42; Cambi 2010: 19/262.
[451] Cf. Migotti 2013b: 205–208.

stone, are sufficient arguments for (or against) any presumption about the same workshop origin.

Overall, if the stone fabric and the quality of execution is considered, it seems at first sight that the sarcophagi and ash-chests made of stone quarried in the Kordun (S II.18, 22; S III.2) were of a somewhat better quality of execution than those carved in the stone quarried most probably in the Banovina region (S II. 1–4, 6–8, 13–17, 19). This, however, is not an absolute rule, as, for instance, sarcophagus S II.2 was more competently carved than the others from the same group, all probably carved in stone from the Banovina. Therefore, it cannot be posited that the Siscian workshop(s) used stone from the Kordun to produce sarcophagi and ash-chests of higher quality than those carved in stone quarried in the Inner Dinarides, most probably the Banovina. On the other hand, it seems that the form of the chest edge is related to the stone used, as the pieces with flat edges are mostly carved in limestone from the Inner Dinarides (probably the Banovina), while those provided with the cut to hold a lid are mostly made from stone quarried in the Outer Dinarides (probably the Kordun).[452] Curiously, the workshops in the Kordun carved chests with flat edges for local use. As for Siscia, it remains unclear whether this structural detail was used by only one city workshop, while other(s) would have produced chests with cuts to hold a lid. The issue of Siscian workshop(s) for sarcophagi gets even more complex when its variegated production is juxtaposed with the claim that the sheer number of Salontian sarcophagi testify to many workshops operating at the same time, despite the fact that the sarcophagi were fairly uniform.[453] In sum, although the elements suggesting a stone-carving workshop (structural, formal, iconographical, stylistic and technical traits) are well known, the evidence for the three cities under research is too limited to allow a systematic insight into this aspect of stone carving.[454] This, however, is truer for Andautonia and Aquae Balissase than for Siscia.

Finally, the production of monuments in the quarry area of the Kordun should be mentioned for one curious detail. The workshops were established there beyond doubt on the basis of half-finished and discarded pieces found at several places, while exports to Siscia were ascertained through shipwrecks along the Korana, a tributary of the Kupa and joining the Sava at Sisak,[455] and through archaeometric analyses within the present research. The only specific iconographic characteristic of the local production is the motif of a circular inscription panel, applied on one stele and six ash-chests (S[ptSW] I.2; III. 9, 12, 14, 27, 35, 49). With two structurally singular *stelai* added (S[ptSW] I.2 and 4), the area has produced a total of three individual monument types (Figure 11, types 8 and 9; Figure 12, type 8). Curiously, so far not a single such example of those typologies was found in Sisak, suggesting that the Kordun quarry workshops supplied Siscia with blank rough-outs. The monuments found in the open in the Kordun differ between themselves considerably in terms of the visual appearance of stone, and the same is true of the pieces made of the stone from the Kordun, now in the Archaeological Museum in Zagreb. Thus, archaeometric analyses of the monuments *in situ* in the Kordun would produce important results, in that it would probably reveal various lithotypes. It has been presumed that the circular motif suggests north-Italian or/and Norico-Pannonian influences.[456] Another north-Italian workshop influence is suggested by a structurally unique stele S(ptSW) I.4, whose tent-shaped pediment is strongly associative of funerary *aediculae* and altars with pyramidal roofing, typical of northern Italy from as early as the 1st century AD (Figure 11, type 9).[457] The evidence from the south-eastern hypothetical Siscian territory enables a significant insight into the use of stone and trans-regional workshop influences in the production of funerary monuments. *Stelai* S(ptSE) I.3, 4 and 5 were made of local stone from their immediate northern surroundings, while the one found somewhat more to the south (S[ptSE] I.1) was made of a different kind of stone, quarried somewhere in the Outer Dinarides, probably in the closest vicinity. Now, while the former reveal Pannonian structural and iconographical characteristics, the latter falls under a typical Dalmatian structural typology and iconography, typified by *stelai* containing a moulded inscription panel flanked by columns.[458] It would be tempting to affiliate the first three *stelai* to the territory of Siscia and the fourth one to Dalmatia, although no final conclusions are sustainable on the basis of such limited evidence. In any case, it seems more probable that the monuments from the possible south-eastern *ager* of Siscia are exactly comparable to those from the bordering areas, which have invariably revealed a blend of Pannonian and Dalmatian influence (see Chapter II.1).

Some workshop influences can be fairly clearly traced in the production of Andautonia, notably as concerns

[452] The first type was typical of workshops in the city of Rome, while the second was employed in north-Italian and Oriental production (Gabelmann 1973: 7; Koch 1993: 26, 42).

[453] Cambi 2010: 19/262.

[454] Here is a random list of works dealing with, or touching on, the problem: Rinaldi Tufi 1983: 27, 44; Koch 1993; Pflug 1989: 5, 32, fn. 198; Hope 2001; Carroll 2006: 114–117; Fejfer 2008: 308–327; Gaggadis, Hermary, Reddé and Sintès 2009; Birk 2013: 161; Weber-Hiden 2014: 71–72.

[455] Perkić 2012: 157, *passim* (on production at quarry sites); 181 (on shipwrecks).

[456] See fns 35 and 36 in Chapter II.1.

[457] Migotti 2010.

[458] Such *stelai* are found in the central and southern coastal part of the eastern Adriatic, notably in the city of Narona, modern-day Vid in the vicinity of Metković (*lupa* 24331, 24374), as well as in the military camp of Tilurium (modern-day Gardun) in the hinterland of the provincial capital of Salona (Sanader 2003: 507, Typentafel 3: XX).

*stelai*. One of the earliest (A[t] I.1) was in the immediate vicinity of the Roman quarry on the slopes of Mt Medvednica, above Zagreb (see Chapter III). The stele was found as broken, and because of its findspot and unfinished state, there is the possibility that it never made it to its destination, despite holding the epitaph. While the use of unfinished Roman funerary monuments was a common phenomenon,[459] the findspot, which has yielded no traces of a settlement, indeed indicates the possibility of a failed and discarded stone. If this presumption is correct, stele A(t) I.1 is an important proof that the quarry workshop in question supplied customers with monuments ready to be used, even in the unfinished state, as transpires from the epitaph. As for the possible origin of the structural prototype of the stele, north-Italian and Norican influences seem to be evident.[460] The same is true of stele (A[t] I.5), whose structural typology points to the Adriatic area, with some details suggesting the north Adriatic rather than Dalmatia, and ultimately resulting in a blend of Adriatic and Pannonian traits.[461] On balance, this can be interpreted as resulting from a synergy between the local workshop and the owner's Italian origins. Otherwise, this stele and another one (A[t] I.6) are the only examples which point to the same workshop, not only on account of the stone material and the finding place, but possibly also on account of the iconography. Both feature the motif of scrolled rosettes on the decorative band dividing the portrait niche and the inscription field. Although the motif in question is not completely identical in design or execution on the two *stelai*, the same workshop should still be considered. Decoration of this type is rarely found on Norico-Pannonian funerary monuments, and is mostly executed in a more 'baroque' manner, as whirled rosettes.[462] However, of the rare geometrically stylized examples, one is close in appearance to the motifs on the two Andautonian *stelai*; significantly, it was applied to a stele from Mokrice castle in the vicinity of Brežice (SI), and thus from the territory of the neighbouring Pannonian city of Neviodunum, connected with Andautonia by the Sava.[463] Imitations can also be traced in some *stelai* made on the marble models used in the *ager* of Andautonia, as witnessed by stele (A[t] I.6) and its marble prototype (A[t] I.8). The same is true of the local production at the site of Lobor, presumably in the farthest north-west corner of Andautonia's territory. Theoretically, the site could have been supplied by the production of the city, but a local workshop is much more probable on account of plentiful stone sources in the immediate vicinity on Mt Ivanščica (see Chapter III) and possibly also on account of structural and iconographic traits of Norican descent, which can be traced among the material, while they have not been witnessed in Andautonia so far.[464] The settlement associated with the finds of funerary monuments in Lobor has not been traced yet. Because of the ample use of (Pohorje) marble, a fairly luxurious rural villa can be presumed. On the other hand, a lot of local limestone has also been used for the monuments, so a settlement associated with quarrying of local limestone could also have existed in the vicinity, comparable to several minor settlements established in the quarrying area in the Kordun.[465] Finally, because Siscia and Andautonia lay on the Sava, there is the possibility that their workshops traded in stone monuments, or at least exercised mutual influences. The only two monuments that can raise such a question are S II.2 and A(t) I.3, as both hold a rare motif of mourning Erotes with legs crossed and holding a reversed torch pulled through a wreath; the former is earlier and marble, and could have provided a model for the latter.[466] Another possibility is that both sourced from a pattern book individually.[467]

Local production on the basis of marble models is evidenced also in Aquae Balissae and is best illustrated by the statue of Icarus AB I.2, lion top AB I.2, and stele AB(t) I.5, as the last-one is a local simplified reworking of the known marble prototypes with lion tops (Figure 11, type 6). No such marble stele is found in the area under study, but they must have existed judging by marble lion tops found individually in the territory of Andautonia. Otherwise, marble lion tops were produced in the Gummern and Pohorje quarry workshops and in Flavia Solva (see above). A possible model for the stele from Aquae Balissae should be looked for in examples containing a frieze and a base, but they are not readily found because separately made lion tops were usually found detached. Nevertheless, a broad prototype for stele AB(t) I.5 should be looked for in variously structured *stelai* with a lion top either in one piece or in two pieces fixed together.[468] The local workshop in Aquae Balissae presumably supplied also its territory, as nowhere in the wider surroundings of the city was any concentration of finds established to indicate otherwise. The local production seems to be characterized by two main traits: fairly to extremely unskilled work, and quite a few singular monuments unparalleled elsewhere either in terms of structural typology, such as AB(t) I.6 (Figure 11, type 7), or iconography (AB V.1). The only funerary medallion (AB[t] VI.1) should not be perceived as a direct local

[459] Rebecchi 1977: 114, fn. 32; Djurić 1991: 85; Koch 1993: 45–47; Carroll 2006: 112; Cambi 2010: 25/265; Russell 2013: 265, 273.
[460] See fn. 109.
[461] See fns 110–111.
[462] *Lupa* 3105, 3401, 34419.
[463] *Lupa* 9165. On Neviodunum see Lovenjak 2003.

[464] See at fns 113–116.
[465] Perkić 2012: 236–242.
[466] See fn. 266.
[467] On pattern books, see Koch 1993: 54; Verzár-Bass 1996: 263; Gabler 2007: 23.
[468] *Lupa* 3106, 3779 (Poetovio), 3622 (Maribor), 3633 (Spodnji Jakobski Dol). Examples comprising detached lion tops found together with the shaft are rare (cf. *lupa* 1379, Flavia Solva).

imitation of a Norican model, but rather as one of the many Pannonian examples, whose production spread from Noricum to develop along the own lines of Pannonian workshops.[469]

### *II.3.4. Summary*

Despite many tools in tentative fixing of the territories of Roman autonomous towns, such as boundary stones, milestones, literary sources, epigraphic evidence of city officials and soldiers, onomastics, small finds, economic potential, geographical features, the logic of 'spatial geometry', medieval administrative borders, and so on, the *ager* of the majority of Roman provincial cities remains notoriously conjectural. The sorest point of the controversy is two sharply contrasted theories, of which one states that all of the land was assigned to the *agri* of respective autonomous towns, while the other claims that a lot of land remained outside such division, resulting in much smaller town territories than customarily presumed. The latter theory is in accordance with the fact that indeed the Romans clearly distinguished between the administered and unadministered land. Such a stance has been exercised in the present research, resulting in the division of the three south-Pannonian cities' territories into presumed, that is most probably administered *agri* and possible territories (probably unadministered), oriented towards the respective cities. All in all, it was not aimed here at fixing the border of the three cities more precisely than had been done in previous attempts, but to delineate the territories for practical reasons. The present fixing of the cities' borders was based on the combination of the grouping of findspots and, in a smaller measure, landscape features. Such an agenda, however, left quite a few findspots only tentatively affiliated with one of the cities, producing at the same time apparently too large territories, particularly as concerns Aquae Balissae (Figure 1). In any case, the aim of the territorial division so devised was to enable the study of the monuments of the probable or hypothetical individual town territories in order to check overall relations and potential differences between them. Such expectations were based on the differences between historical, political, administrative and cultural roles and importance between *colonia* Siscia (present-day Sisak), and two *municipia*: Andautonia (present-day Ščitarjevo) and Aquae Balissae (present-day Daruvar), all three of them in north-western Croatia. However, expectations towards establishing particular artistic traits in the evidence in terms of suggested territorial-administrative affiliations have failed in the present research. Instead, it has been substituted with the knowledge that material culture in the bordering areas between the cities, especially when these corresponded with provincial boundaries, produced typical cultural blends; this was especially clear in the case of Siscia, in parts of its south and south-eastern territory bordering on the province of Dalmatia.

If the three cities are compared by sites and finds, the differences both in numbers and nature of the evidence can be observed (Figure 10). Of the 70 funerary stones from Siscia (the hypothetical ager excluded), as many as 56 pieces stem from the cemeteries and only 14 from its presumed ager; the latter are distributed among four sites. To this, the region of Kordun should be added as a hypothetical Siscian territory, or at least the area gravitating towards Siscia and furnishing as many as 66 finds distributed among several smaller and larger sites, with only seven monuments stemming from the hypothetical territory in the south-east. Owing to the particular stone-cutting nature of the Kordun, the total amount of finds from the hypothetical Siscian *ager* is, so to say, artificially augmented, and this should be kept in mind in the comparative analysis of the entire area of study. Eight of a total of 27 funerary stones from Aquae Balisssae stem from the urban cemeteries, while 18 were recovered from the presumed *ager*, with one of the sites (Donji Podgradci) only tentatively belonging to or just gravitating towards Aquae Balissae. Of a total of 30 finds from Andautonia, only two (three, if a missing sarcophagus, not included in the Catalogue, is counted) stem from the urban cemeteries, while its presumed *ager* furnished 16 finds distributed across 11 sites. The hypothetical *ager* of Andautonia has furnished an additional 12 funerary monuments, with as many as nine of them stemming from the site of Lobor, which actually holds many more finds.

A distribution by sites reveals various relationships at two levels: between each city and its territory individually, and between the three cities in comparison. The large number of finds from the Siscian cemeteries is realistic, while the extremely small number of them from its presumed territory undoubtedly results from an inadequate state of research. The circumstances in Andautonia are comparable, but in the opposite direction, yielding an extremely small number of monuments from the urban cemeteries, which can only be explained as accidental, and producing an abundance of rural finds. The latter may be pointing to the fact that the majority of the population of Andautonia able to afford stone monuments, and especially marble ones, lived in their country estates in the city's *ager*. The ratio of urban and rural finds from Aquae Balissae seems to be slightly more balanced than in the previous two cities, but the original number of urban finds should have been much larger in proportion to those from the *ager*. Otherwise, the larger number of sites in the territory of Andautonia as compared to that of Aquae Balissae is realistic if the geography of the areas is concerned, with a fertile valley taking a considerable part of the *ager* of

[469] See fn. 112.

Andautonia and a mountainous landscape prevailing in Aquae Balissae's surroundings.

If the number of finds is juxtaposed with the hypothetical territories, Siscia expectedly emerges as the site with the largest number of monuments (143), followed by Andautonia (30) and Aquae Balissae (27). It seems that such a huge discrepancy is realistic, not only on account of the historical importance of Siscia, but because the loss of Siscian monuments has been well substantiated. In other words, the possibility that the discrepancy is due to a heavier loss of monuments in the remaining two cities is practically negligible. If now the approximate surfaces of the city territories are juxtaposed with the respective number of finds, we see that the Siscian *ager* is not four of five times larger than the other two, as the number of finds would suggest. As a matter of fact, the territories of the three cities seem to be surprisingly levelled in extention (Figure 1). However, more balanced spatial relationships would probably be gained by the knowledge of the territory of Iovia, situated between Andautonia and Aquae Balissae, as well as the location and territory of *municipium Faustinianum*, which must have been somewhere to the east of Siscia. The former would have probably reduced the territories of both Andautonia and Aquae Balissae, while the latter would have presumably further reduced the *ager* of Aquae Balissae, and would have also affected the area of the Siscian *ager* in the south-east. With these calculations in mind, the predominance of the evidence from Siscia still seems to result from the importance of this city rather than from its territorial scope.

Given a huge discrepancy in the number of finds between Siscia on one side and the remaining two cities on the other, there was no point in comparing structural typology and iconography in chronological contexts statistically between the cities. On the other hand, such comparisons could be more realistically conducted through an observation of conspicuous differences (or similarities) between the cities. Still, due to the scanty and fragmentary evidence often lacking any archaeological context, the results gained mostly remained hypothetical, but should nevertheless be taken as directions toward a commendable future research which would include all categories of monuments, and not only funerary. The sorest point in this research was chronology; because of the so many stones dated tentatively or within the span of two or even three centuries (such as plain sarcophagi among others, dated to the 2nd–4th centuries), only a broad chronological orientation can be given at this stage. Siscia is the only city furnishing (scanty) evidence before the Flavian period, followed by a fairly steady distribution from the 2nd century to the Late Roman period, but still with the stress on the 2nd and 3rd centuries. Quite a lot of monuments stem from the Severan period, but many of those dating from the 2nd *or* the 3rd century, together with those dating very probably from the 3rd century, suggest that no dramatic decline in their production occurred during the so-called '3rd-century crisis'. The concentration of the monuments in Andautonia is on the 2nd and 3rd centuries, with the Severan period presumably dominating, and with no single monument dating from the 1st century or the Late Roman period. The circumstances in Aquae Balissae seem to have been similar, with one difference: the city produced a luxurious Late Roman underground chamber yielding several exceptional finds. While the evidence from Andautonia failed to produce any find remotely connected with Christianity, several such monuments stem from Siscia and Aquae Balissae, but their total number is still extremely moderate. As for the monument kinds, what straightforwardly catches one's eye is the difference between the distribution of sarcophagi and *stelai* in the three cities. While *stelai* are numerous in all three cities, figuring as a leading kind of monument in Andautonia and Aquae Balissae, sarcophagi are more numerous in Siscia. Surprisingly, and with no obvious explanation, only two ash-chests were found in Siscia and one in the hypothetical territory of Aquae Balissae, while none stem from Andautonia. The issue becomes even more perplexing if we remember that a huge number of ash-chests were produced in the hypothetical territory of Siscia in the Kordun. A further curiosity is an extremely small number of kinds of funerary monuments that were very popular in Noricum, such as *aediculae*, other funerary buildings and enclosures, as well as altars and medallions. None of these kinds of monuments were particularly widespread in Pannonia outside the area of the present research, but they still come in respectable numbers, especially funerary buildings and medallions. Therefore, while a small number of altars is perhaps reciprocal to their general lesser popularity in Pannonia, more funerary *aediculae* should be expected. As a matter of fact, the present research has just brought attention to *aediculae*, altars and medallions, which were completely ignored in previous studies of funerary stones from south-western Pannonia. On the other hand, a large number of types of *stelai*, and especially those that occur only once, suggests two conclusions. While the four types more frequently represented (3a, 3b, 5a and 5b) testify to a standard stock production, those many examples of a *hapax* nature suggest that in the area of study the original number of *stelai* was much larger, and that they were probably the most popular kind of funerary monument from the later 1st to the end of the 3rd century. In terms of iconography, notably portraits, the three cities share several prevailing characteristics, with a negligible number of exceptions. Thus, men are mostly dressed in a tunic and *sagum*, and are rarely depicted with military equipment. Women feature a blend of native and Roman portrait characteristics and fashions, with the former predominating, and especially so in the territory of Andautonia, where a clearer affiliation with neighbouring Norican styles is detectable. The evidence for all three cities is in accordance with Ursula Rothe's claim that there is no such thing as a 'Norico-Pannonian national costume', but a lot of local styles. Contrary to Noricum, where a native headwear was customarily blended with a local hairdo, it was typical for Pannonia in the 3rd century to merge native headwear (as the only residue of local fashions) and a Roman hairdo by pushing back the bonnet and exposing the hair, at least partly, as substantiated through several examples from Siscia and Andautonia. This phenomenon is especially curious in the case of Andautonia, otherwise under a stronger influence of the Norican ethnic-cultural circle than Siscia, the more so as the stele in question (A[t] I.8) dates from the mid 2nd century.

Since Pannonia had no marble sources, except for Mt Pohorje on the border of Noricun and Pannonia, the use of marble monuments in the latter province was an expensive luxury, generally indicative of the owners' well-being. Unfortunately, the overall database, both by individual cities and as a whole, is too limited to allow insights into the preferences of various social classes towards specific kinds of marble monuments in statistical terms. While marble monuments vary between 23% and 30% in the cities of the present study, the number of those that yield information on both the kind of monument and its owners is quite small (40 pieces). Such information is preferably gained through an interaction of the epitaph and picture, or the epitaph alone, while in some cases it can be hypothetically, and in a limited manner, restored through the depiction alone. The database is somewhat richer if the monuments made of other kinds of stone are added, but still inadequate for a statistical overview of the relations between the kinds of monuments, material used, and social classes.

Of the above 40 pieces, 16 were owned by the military, 20 by civilians, and three or four by a class of citizens who can tentatively be labelled as a military-civilian elite. Despite the above-mentioned hindrances, an attempt has been made to get an insight into the kinds of monument preferably employed by various classes of people, looking at the same time into the relation between the material used and the place of the monuments' owner in social hierarchy. Several of quite a few aspects of these relationships come to the fore. Presumably, the civilian elite families favoured (marble) sarcophagi over other kinds of monuments, but unfortunately data are missing on their (or other social classes') possible preference for marble *aediculae*, as another generally popular monument with higher classes in Noricum and Pannonia. On the other hand, military *stelai* surpass the ratio of the *stelai* and other kinds of monuments in the total evidence, suggesting that the stele was the most popular tombstone with soldiers. The number of military *stelai* would be even larger if those from the early 1st century were not missing from the preserved record, although they must have existed. On the other hand, the ratio between military and civilian tombstones (16:20) does not conclusively agree with the general claim that soldiers and freedmen were the best represented in epitaphs. Firstly, the number of civilians is larger than the number of soldiers, and secondly, the number of freedmen among the civilians remains unknown. The ratio between the monuments set up for families versus individuals is 30:17. Such prevalence is in accordance with the chronology, as the majority of the monuments fall into the 2nd and 3rd centuries, the time when the increased significance of the family became strongly reflected in the funerary evidence. Contrarily, the 1st century was strongly marked by individual, prevalently military commemorations. While social status was very much expressed through clothes, it was not customary in Pannonian funerary iconography to illustrate the deceased's profession or occupation. This phenomenon transpires in the present evidence, with only one each depiction of the occupation and genre scene representing a household work (goat milking). Thus, in funerary self-representation, the social aspect in the narrow meaning of the word seems to have been more important than the material aspect recording occupation and earning. A slight exception to this rule is represented by the military, whose depictions with weapons can be perceived as a sign of their social status and occupation at the same time. On the other hand, mythological motifs and episodes are quite numerous, featuring Iphigenia in Tauris, the legend of the Sabine women, a centaur, Icarus, Helena and Eros, Apollo chasing Daphne, Hercules rescuing Hesione, Hercules alone, and possibly Endymion; they testify to a cultural ambience open to their messages. Early Christian monuments, all dating from the 4th century, are extremely few in view of the fact that Siscia was a bishopric. The main point concerning the social and religious aspect of the funerary ritual, the diachronic relation between inhumation and incineration could not be conclusively established from the limited evidence but needs to be viewed against the background of the knowledge that inhumations started later and incinerations lasted longer than previously presumed. This means that that the two rites were practised contemporarily during the entire 3rd century.

Two lines of research have been conducted regarding the subject matter of stone used and workshop origin of the monuments, one concerning marble and the other local stone. The investigation was based on several premises: the dependence of Pannonia on Eastern-Alpine quarries for marble, traffic routes (preferably waterways) between the marble quarries and Pannonian cities, and the rule of acquiring from the closest possible source, which is valid for all kinds of stone.

Archaeometric analyses as the main methodological condition of such study had never been conducted systematically on Roman funerary stones from Siscia, Andautonia and Aquae Balissae before; therefore, petrographic characterizations were the focal point of the present research. Marble samples were analyzed by Walter Prochaska from the Department of Applied Geological Sciences of the University of Leoben, and those of other stones were processed in the Croatian Geological Institute in Zagreb. In sum, the workshop origin of monuments was tentatively established on the basis of stone used in tandem with their structural and iconographic characteristics. Research on marble monuments in Pannonian cities, conducted by Bojan Djurić and Harald Müller in 2004–2008, has established a slight prevalence of Gummern over Pohorje marble. However, the results of the new research, based on refined methods whose aim was to overcome the obstacle of overlapping isotopic fields of similar marbles, such as Gummern and Pohorje, were different. Ten of a total of 38 samples, including those from the previous analyses, could not be determined more precisely than simply as marble. Of the remaining 28, as many as 22 turned out to be Pohorje marble, while the quarries of Salla and Gummern each produced three samples (Figure 20). The so-far established frame of the production and trade in marble monuments in Pannonia is represented by two quarries taking the leading role in the exportation of eastern Alpine marbles to Pannonia: Gummern in the vicinity of Villach (originally supplying Roman Virunum, Noricun) and Šmartno on Pohorje (originally supplying Roman Poetovio in Pannonia and Celeia in Noricum), both on account of the vicinity of the Drava River. Marble reserves in the Gummern quarry were more abundant and of a superior quality, and the workshop there was established as leading a more important role in comparison with the Pohorje or/and Poetovio workshops. Therefore, the results from Siscia, Andautonia and Aquae Balissae have considerably changed this picture, bringing into focus a prevalence of Pohorje marble in the area of research. Presumably, this was realized through the role of Celeia, whose workshops also used Pohorje marble, and which was connected with Andautonia and Siscia via the Savinja and Sava Rivers. Another piece of new information on the trade in marble monuments in Pannonia concerns products made of Salla marble (the quarry supplying Roman Flavia Solva, Noricum), which have so far not been established in other Pannonian towns except in Savaria (Szombathely). With three monuments made of Salla marble, and some other indications for the influence of Flavia Solva workshops on monuments from Siscia and Andautonia, a new dimension of the role of Flavia Solva has emerged from the present research. As far as archaeometric study of stones other than marble is concerned, before the present research (very few) characterizations were made by macroscopy and exceptionally by archaeometric methods, with the majority of the remaining evidence being classified with the naked eye by non-specialists. Therefore, quite a few monuments were published with broad and quite often incorrect determinations. The present research has established that generally there was no exception to the rule of sourcing from the closest quarries of usable stone, except in the example of one stele owned by a civilian from Roman Tergeste (S I.1). His stele was made of stone from the Adriatic area, most probably sourced from the quarry of Aurisina near Trieste, and can therefore be perceived as a single, private initiative. This source of stone was not used more in the area of research, at least for the funerary monuments, as the remainder of them were all carved in local stone. At this point it was not possible to determine the samples more precisely than as potentially originating from relatively broad areas, nor was it possible to relate the characterized samples with potential Roman quarries, as none of them but possibly one (*Bijele stijene* in the village of Hrastovica near Sisak) reveals traces of Roman quarrying. On the other hand, the conducted analyses do not allow for the presence of stone from outside continental Croatia in any of the three cities, actually indicating that each of them probably sourced from the nearest vicinity (Figure 22). Siscia can be considered as an exception in this regard, as its workshops used stone from the Kordun area on the border between the provinces of Pannonia and Dalmatia, some 80 km to the south-west of Sisak.

# Chapter III
# Geology and quarries

Branka Migotti

## III.1. INTRODUCTION

No archaeological research had been done on the topic of Roman quarries in the area of study before 1997 (see below), although some such sites were recognized by archaeologists and described in the literature as early as the end of the 19th century. In the course of the preparations for the present research, and through subsequent surveys, it has been established that there were two kinds of potential places of Roman quarrying, both practically lost for study. The first are those that were used in the modern period, thus obliterating any traces of Roman quarrying, regardless of whether they are still in operation or abandoned for decades. If they are still in use, this is mostly occasionally and/or at a small scale, and because of the exhaustion of the dimension stone deposits, stone is quarried for construction purposes. Another type of site are those which were possibly used in antiquity, but are completely overgrown and inaccessible except by using professional equipment and expertise. The research team visited several accessible quarry sites on the mountains of Medvednica and Ivanščica in the territory of Andautonia, several quarry sites in the village of Hrastovica and its surroundings, as well as some surface sites in the Kordun, both in the territory of Siscia. Finally, several sites were surveyed on Dilj, one of the mountains in the territory of Aquae Balissae. None of these surveys resulted in any sound traces of Roman quarrying, except possibly the site of *Bijele stijene* (White Rocks) in Hrastovica. Nowhere could typical extraction trenches and chisel and point marks on vertical quarry faces be spotted, such as can still be seen on many sites across the former Roman Empire, including the province of Dalmatia and its border stretches with Pannonia in the region of Lika.[1]

## III.2. SISCIA

While publishing stele S I.1 in 1935, Josip Klemenc stated that the stone had been characterized by geologists from the Museum of Mineralogy and Petrography in Zagreb (the modern-day department of the Croatian Natural History Museum) as lithotamnion limestone, with the remark that such stone was available in Hrastovica, 11 km to the south-west of Sisak.[2] From then on, *the quarry of Hrastovica* figured in the archaeological literature as the main source for stone monuments of Siscia, with no further enquiry about the precise site where this stone could have been quarried.[3] As a matter of fact, there are several smaller and somewhat larger quarries in Hrastovica and in neighbouring villages in the vicinity (especially Sibić and Župić), some inaccessible and some still in use, the latter revealing no traces of Roman quarrying. In the quarry in Župić, vertical groove-traces of modern exploitation of sand are clearly visible, as a typical feature of this whole quarry area exposed to long-term usage and crumbling ('sugaring') of stone surfaces during which rocky hills turned into vertical sandy planes (Figure 1). The largest and most significant of all of them is the site of Bijele stijene in the village of Hrastovica. Its present appearance suggests long-term usage in the past, although the quarry went out of use three or four decades ago. Even the older villagers do not know the place as a proper stone-quarrying site, remembering only that it was used by the Yugoslav Army after the Second World War for quarrying sand. Indeed, the site today has the appearance of a huge amphitheatre some 300 m long, surrounded at places with crumbled and washed-out upstanding rock faces turned into sandy surfaces with no traces of quarrying whatsoever, but with some larger stone blocks scattered in clusters at their foot (Figures 2 and 3). Among those stones there is a fragmented coffin measuring approximately 1 x 1.5 m and seeming to be a roughed-out or an unfinished and discarded piece which should have been carved into a

Figure 1. A detail from the quarry at Župić (photo: Lj. Perinić)

[1] Šarić 1980; Djurić 2004: 155, fig. 9; Rižnar and Jovanović 2008: 145, figs 11 and 12; Scardozzi 2012; Zachos and Leka 2012; De Stefano 2015.
[2] Klemenc 1935: 67.
[3] Perkić 2012: 159.

Figure 2. The quarry at Hrastovica
(photo: Lj. Perinić)

Figure 3. The quarry at Hrastovica, a detail
(photo: Lj. Perinić)

Figure 4. A discarded sarcophagus chest (?) (photo: Lj. Perinić)

sarcophagus or an ash-chest (Figure 4). Apparently, this is the only trace, but a quite probable one, of Roman quarrying at the site. Otherwise, a characterized sample from this quarry turned to be Baden biocalcarenite, the same as in several Siscian monuments (S II.4, 16; S III.1; S[t] I.6 S[ptSE] I.3, 4), while some others were characterized as Baden siliciclastic biocalcarenite (S I.5; S II.13, 14, 15, 17). The presumption of Roman quarrying at the site of Bijele stijene is further substantiated by scattered remains of Roman building materials, including ashlars, in the village of Hrastovica and its surroundings.[4]

Topusko (presumably Roman Quadrata) is one of the sites in the *ager* of Siscia whose source of stone is clear without any doubt, as it used rocks from the quarries on the very edge of the town. Two quarry sites (*Draškovićevi podrumi* [the Drašković Family's Cellars] and *Žuti kamen* [Yellow Stone]) are actually recognizable on the spot, situated on the slopes of two hills divided between themselves only by a modern road, but without leaving traces of Roman quarrying.[5] Roman traces on the site of *Draškovićevi podrumi* have been obliterated by the building of 18th-century cellars cut into the slope of a hill. Across this site is another one – *Žuti kamen*, whose name reveals the appearance of yellowish sandstone quarried there; the only remains of a former quarry cut into a hill slope are two lumpy rocks at a small distance to each other (Figure 5). The stones from the two sites differ in their outward appearance, as that from *Draškovićevi podrumi* is slightly reddish-brownish in colour, while that quarried at the *Žuti kamen* is yellowish. Curiously, when viewed with the naked eye, only one monument from Topusko reveals the outward appearance of the former site (S[t] I.1), while the remainder point to the latter (S[t] I.2, 3, 4, 5; S[t] III.1, 2). Even more curious, at least to a non-specialist in geology, is the fact that the archaeometric picture of all the monuments, as well as a sample taken from the *Žuti kamen* site, happen to be identical, that is, Upper-Miocene sandstone. As the two sites must have been parts of a single quarry complex comprising two adjoining hills, these differences are a further illustration of the fact that everywhere the same quarries contain visually and compositionally various layers of stone.[6] There is another quarry in Topusko, that is, in the village of Hrvatsko Selo, 3 km to the east of the town. It is still in use today but was obviously not exploited in Antiquity, as it supplies a greenish rock (diabase) usable as construction material but unsuitable for carving. This needs to be mentioned because it was exactly the greenish stone of a sarcophagus lid (Figure 6) now in the Archaeological Museum in Zagreb (inv. KS-480) that prompted Josip Brunšmid to presume Topusko as its findspot.[7] However, the green stone it was made of is actually a green tuff rock widely used in Roman Sirmium, which is certainly the findspot of this piece.[8]

The surface quarries in the karst area of Kordun, abounding in outcrops of rocks suitable for the carving

[4] Aleksandar Durman (a long-term researcher of the area, whose findings unfortunately remained unpublished), personal communication.

[5] Čučković 2007: 50–51.

[6] Wenner and Herz 1992; Rižnar and Jovanović 2008; Beltrán, Ontiveros, Loza Azuaga and Romero 2012: 506; Girardi Jurkić 2012: 643.

[7] Brunšmid 1909: 216, no. 480.

[8] Djurić, in press.

Figure 5. The Žuta stijena site in Topusko
(photo: B. Djurić)

Figure 6. The sarcophagus lid, AMZ inv. KS-480
(photo: O. Harl; *lupa* 4286)

of monuments, were first described in the literature at the end of the 19th century, but proper researches were conducted through field surveying, trenching and even some small-scale test excavations, and systematically documented in the 1997–2001 research by Domagoj Perkić.[9] This research has established that the whole of the area in the middle course of the Korana River was occupied by surface quarries with a concentration of scattered pieces, some unfinished and/or discarded, testifying to a lively stone-cutting activity. As many as 29 smaller surface quarry sites were registered, which can be grouped into 11 clusters, formed either by digging into karst slopes or valleys (Figure 7).[10] The monuments found in the open in the Kordun differ between themselves considerably in terms of the appearance of stone when viewed by the naked eye, meaning that various kinds of limestone can be found there. The exact nature of this area in terms of Roman settlement remains obscure. Possibly, several smaller settlements documented through remains of building materials, funerary monuments and small finds,[11] were created and maintained just for the sake of quarrying and stone-cutting, as can be tentatively presumed for (an) unknown settlement(s) at the site of Lobor (see below). Clustering of quarries around the towns and within their territories is a widely evidenced phenomenon, as is the settlement of quarry areas related exactly to this activity.[12] In the case of the presumed relation between Siscia and the Kordun, the distance (50–70 km) is more than double than that taken to be customary (20–30 km). This, however, is remedied by the fact that the quarries in the Kordun practically had a direct access to the waterways leading to Siscia, and that the stone quarried there, although not of supreme quality, was still better than the sandstones available in the immediate vicinity of the city in the region of Banovina. On balance, the city of Siscia could have owned and managed the quarries in the Kordun, which further supports this region's hypothetical administrative affiliation with the city (see Chapter II.1).[13] In theory, the quarries could have been supervised by the army, as it was a bordering region and the presence of the military, which is self-explanatory in boundary areas, has also been widely identified in relation to quarrying in other parts of the Roman Empire.[14] However, the inscribed stone monuments from the Kordun are extremely rare; only one of them, dating from the early 4th century, was certainly owned by a soldier (S[ptSW] I.4), while another (S[ptSW] I.3) possibly belonged to a civilian-military elite family.[15]

[9] Perkić 2012.

[10] Perkić 2012: 5, *passim*.

[11] Perkić 2012: 236–240.

[12] Cf. Russell 2013: 65–81. On quarry settlements in the rural landscape, see Koželj and Wurch-Koželj 1992; Nervi 2015.

[13] On the (municipal) ownership of quarries, see Djurić 2004: 153; Russell 2013: 53–77.

[14] Cf. Koželj and Wurch-Koželj 1992; Djurić 2008; Perkić 2012: 160; Russell 2013: 41–42, *passim*; Migotti 2016a: 186.

[15] See Chapter II.3.2.

### III.3. ANDAUTONIA

Several of the many now abandoned quarries on the southern slopes of Medvednica Mountain above Zagreb have been mentioned in the literature as sites of Roman quarrying, all now situated within various districts of Zagreb: Podsused (Bizek), Gornje Vrapče, Crna voda (Markuševec), and Čučerje (Dubrava).[16] Three of them have yielded artefacts which conclusively testify to their use in the Roman period. In 1896, in the then private quarry of *Crna voda* (Black Water), an altar to Jupiter was found together with undressed stones.[17] In the quarry of Vrapče, an altar to Silvanus was found in 1894, at the site that had previously yielded an unfinished Roman column and several iron quarry tools, all subsequently lost.[18] The quarry in Podsused (Figure 8) is the findspot of stele A(t) I.1, which was found there in 1884 during the building of a footway above the quarry. Because of its fragmentary state, it was presumed that it had broken during transportation from the quarry and left there unused.[19] In the subsequent digging at the spot, no archaeological finds were recovered that should have helped in recovering the archaeological context of this find. On the other hand, it was established that the stele was lying in the ground consisting of crumbly soil, cultivated in the past. This means that there could have been a settlement in the vicinity of the quarry in Roman times, possibly belonging to the owner of the quarry and the land around it, who could have been buried there.[20] Various kinds of rocks are available in the quarries on Medvednica (limestones, sandstones, schists, basalts, dolomites, marble, and marmorized limestone).[21] So far, only limestone and sandstone Roman monuments were mentioned in the archaeological literature, the former presumably made of stone quarried in the Podsused quarry, and the latter made of stone from the Gornje Vrapče quarry.[22] When first published, sarcophagus A(t) II.1 was determined as probably carved in sandstone from the quarry of Vrapče on Medvednica, but the authors rightly observed that archaeometric analyses would be necessary to prove this beyond doubt.[23] Curiously, the analysis conducted within the present research came with the result that the stone for the sarcophagus in question could have been found locally on any surface elevation, which unfortunately leaves its quarry origin undetermined, that is, without excluding or conclusively corroborating a quarry on Medvednica. Although the presence of marble on Medvednica is a well-known fact (Figure 9),[24] so far it

[16] Gregl 1991: 37, 49; Knezović 2012: 240, map on p. 229.

[17] Brunšmid 1907: 119, no. 213.

[18] Brunšmid 1907: 125, no. 253.

[19] Brunšmid 1909: 164, no. 359; *AIJ*: 216–217, no. 482.

[20] For villa and other settlements related to quarries, see Nervi 2015.

[21] Knezović 2012: 240; http://www.stari.pp-medvednica.hr/Medvednica_hr/Medvednica_priroda_geologija2.htm (last accessed on 27 May 2018).

[22] Galić and Radman-Livaja 2002: 241–242 (with literature in fn. 5).

[23] Galić and Radman-Livaja 2002: 241.

[24] Knezović 2012: 226. See also fn. 21.

Figure 7. A detail from the Kordun quarry area (after Perkić 2012: pl. X)

Figure 8. The quarry in Podsused
(photo: M. Jambrović)

Figure 9. An outcrop of marble on Mt Medvednica (photo: M. Belak)

was not thought that it was used for Roman sculpture, mostly because of the lack of archaeometric research of the monuments. Such an hypothesis can also be substantiated by the fact that the surface layers of marble on Medvednica, at least those accessible today, are tectonized and do not seem appropriate for cutting blocks large enough for carving funerary monuments. Nevertheless, one of the tombstones from Siscia (S VIII.2) was determined as carved in marmorized limestone from Medvednica. Thus, while it is quite probable that in Roman times marble from Medvednca was used as a construction aggregate material, its use for monuments remains open, but cannot be completely discarded.

One of the most remarkable sites possibly belonging to the *ager* of Andautonia is Lobor, situated at the

Figure 10. The quarry at the site of Zazide (Lobor) (photo: K. Filipec)

south-western foot of Ivanšcica Mountain. Apart from importing monuments made of Pohorje marble (see Chapter II.3.3), the site was self-sufficient in terms of stone supply, as Ivanšcica abounds in limestone quarry sites, none of which are active now, except for small-scale private or temporary use, mostly for the quarrying of building stone.[25] In none of them can traces of Roman use be spotted, for the reasons stated at the beginning of this chapter. One of the largest such quarries is situated on the site of Zazide within the town of Lobor. The vertical face of the quarry site is inaccessible, but broken-off small stones can be found at its foot (Figure 10). A sample taken from such material produced the same result as all the limestone monuments from Lobor, that is, that they were carved from bioaccumulated limestone that can be found on two mountains close to Lobor: Ivanšcica in its immediate vicinity, and Kalnik, some 50 km to the east, with Ivanšcica evidently representing the most likely source of stone for the site in question.

### III.4. AQUAE BALISSAE

As far as this present author is aware, none of the researchers who discussed the Roman monuments of Aquae Balissae ever mentioned the origin of stone used for their carving. According to the characterizations conducted within the present research, all the monuments (apart from marble) were most probably made of stone quarried from the Slavonian Mountains (Papuk, Psunj, Krndija, Požeška gora and Dilj) surrounding the Požega Valley to the east of Daruvar (Aquae Balissae). Experimentally, several potential sites of Roman quarrying were surveyed on Dilj Mountain, with the negative result due to the fact that all of them were found as either inaccessible or ruined through modern activity. One such site is the quarry in the village of Brodski Zdenci. It is still in use, with its vertical face featuring a solid rock at places, but with the majority turned into a crumbled, sandy surface, produced by the working of modern machinery (Figure 11). The southern slopes of Dilj are composed of alternating canyons and ridges, with quite a lot of caves and potential Roman quarry sites, most of them inaccessible except by using special equipment (Figure 12). A characteristic feature of this area is natural cleavage of the rock, which breaks along the vertical and horizontal axis, creating forms that have anthropogenic appearance. Archaeologists would most probably be inclined to proclaim one such piece (Figure 13) as a pedimental fragment of a funerary stone.

In conclusion, only a project targeted exclusively on the research of Roman quarries in northern Croatia could possibly bring more concrete results on this subject matter, but the majority of data seems to be irretrievably lost.

[25] Šimunić 1986; Knezović 2012: 226, 241; Filipec 2017: 118–119.

Figure 11. The quarry at Veliki Zdenci
(photo: B. Migotti)

Figure 12. An outcrop
on Mt Dilj
(photo: B. Migotti)

Figure 13. A rock fragment from Mt Dilj
(photo: Lj. Perinić)

# Chapter IV

# Population of south-western Pannonia: Siscia, Andautonia, and Aquae Balissae

Marjeta Šašel Kos

## IV.1. BRIEF HISTORICAL INTRODUCTION

South-western Pannonia was attached to Illyricum after the Roman conquest, when this province had been probably founded early under Augustus. It is difficult to imagine that Illyricum could have been established as an administratively governed territory before the Illyrian War, fought by Octavian (Augustus since 27 BC) in 35–33 BC, when major parts of the future Pannonia had not yet been conquered, including the regions of these three towns.[1] Hence, it seems less likely that Illyricum would have existed as a province already under Caesar;[2] at that time it must still have been 'provincia' in the original sense of the word (a task assigned to a magistrate). After the great Pannonian-Dalmatian revolt Illyricum was probably divided into an Upper and Lower province, Pannonia having been called Lower Illyricum at that time.[3]

The first wave of municipalisation in Pannonia occurred under Vespasian, when five indigenous settlements became autonomous towns: Neviodunum, Andautonia, and Scarbantia acquired the status of *municipia Flavia*, while Sirmium and Siscia became *coloniae Flaviae*. The only town in Pannonia, founded earlier, was Savaria, a *colonia Claudia*.[4] This happened during the first phase of urbanisation of Noricum, which had begun earlier, under Claudius, when five important Celtic *oppida* in southern Noricum became *municipia Claudia*: Celeia, Virunum, Teurnia, Aguntum, and Iuvavum (Pliny, *N. h.* 3. 146). It may therefore be presumed that Savaria was under the strong influence of the Norican kingdom at the time of Roman conquest; Pliny's formulation that Savaria and Scarbantia 'adjoined the Norici' seems to confirm this thesis (*N. h.* 3. 146).[5] Two other important autonomous towns in south-western Pannonia were Neviodunum (*municipium Flavium Latobicorum*, modern Drnovo near Krško) and Poetovio (*colonia Ulpia Poetovio*, present-day Ptuj). Neviodunum was situated relatively near the border with the 10th Region of Italy (later called Venetia and Histria),[6] the border settlement having been Praetorium Latobicorum (present-day Trebnje), both located along the main route leading towards the eastern part of the Empire.[7] Poetovio had first been a legionary fortress of the Legion *VIII Augusta* under Augustus and became a Roman colony under Trajan; it was the second largest city in Upper Pannonia next to Carnuntum, situated along the Amber Route.[8] Aquae Balissae achieved the status of a *municipium* under Hadrian and from then was officially called *Municipium Iasorum*, in a similar way as was Neviodunum: *Municipium Latobicorum*.[9]

More or less all the available epigraphic evidence (partly combined with archaeological, historical, and philological data) has been taken into account in the attempt at assessing the population of Siscia, Andautonia, and Aquae Balissae in the Roman period. The tombstones included in the catalogue were analysed, cited with the catalogue numbers, as well as other tombstones, honorific inscriptions, votive monuments, including inscriptions of the inhabitants of the three towns under investigation that had been found outside the boundaries of these communities and their territories. It should be emphasized that the evidence is far too limited to allow for any kind of meaningful evaluation of the population. It illuminates some aspects of the past life in the towns and the countryside, permitting us to get acquainted with certain individuals or families, while it is not possible to draw any far-reaching conclusions.

## IV.2. COLONIA FLAVIA SISCIA

### *IV.2.1. A short historical outline and available evidence*

Siscia, also Segesta (thus called by Appian, *Illyr.* 23. 67) or Segestica (as in Strabo, 7. 5. 2 C 314), was one of the most significant Pannonian/Celtic trading centres and harbours, since it was very conveniently located at the confluence of the rivers Sava (Savus) and Kupa (Colapis), close to the stream of Odra. The settlement was the centre of the region of the Segestani, also known as Sisciani, and closely related to the Colapiani, who must have extended west of Siscia.[10] Siscia and Segesta/ Segestica (the latter name being the adjective derived

[1] Freber 1993: 125–127 and *passim*; Šašel Kos 2000; cf. Santangelo 2016.
[2] Čače and Milivojević 2017.
[3] Kovács 2008; Šašel Kos 2010, for a slightly different opinion.
[4] Scherrer 2003.
[5] Cf. Scherrer 2017: 433; Kovács 2018.
[6] Lovenjak 1998; Lovenjak 2003.
[7] See recently Fodorean 2017.
[8] Slapšak 2001; Horvat *et al.* 2003; Šašel Kos 2014a.
[9] Mócsy 1968a; Schejbal 2004.
[10] Mócsy 1959: 24-26; Šašel 1974: 702 ff. (1992: 600 ff.).

from Segesta) must have been two settlements close to each other, one of them possibly, but not necessarily, older than the other, which eventually developed into one large town. The name Segestica has recently been defined as possibly Celtic, etymologically related to the word for 'rush' or 'reeds'.[11] The toponym Segesta was explained as having been of Indo-European origin, meaning a settlement near the fields with crops.[12] Earlier sources mainly refer to Segesta/Segestica, an Iron Age Pannonian (Celtic) settlement at Pogorelec on the right bank of the river Colapis. Siscia was a different site within the same broad area, located on the left bank, closer to the river Sava, where it developed as a settlement favoured by the ancient river traffic along this river.[13] Recent excavations have shed new light on both prehistoric settlements, at Pogorelec and on the opposite bank, as well as on the early Roman settlement and the location of a military camp (camps?).[14]

Strabo writes that 'near Segestica there are the fortresses Siscia and Sirmium, located along the way to Italy' (7. 5. 2 C 314). However, his text can be regarded as slightly misleading since it may be implied that there were two settlements, Segestica and Siscia, placed at a certain distance from each other. In the Roman period the name Siscia prevailed and was given to the Roman colony, while Segestica remained the name of one of the town's districts.[15]

Siscia/Segest(ic)a had been conquered by the Romans in 35 BC,[16] two and a half legions having been stationed in it immediately after the conquest. Due to its excellent strategic position, the town remained a significant legionary fortress even afterwards.[17] Obviously it must have supported Vespasian during his ascent to imperial power, the emperor then rewarding several communities in Pannonia for their loyalty.[18] Siscia became a *colonia Flavia*, in which fleet veterans were settled. The voting district of the inhabitants of Siscia, who were Roman citizens, was *Quirina*; *tribus Flavia*, which appears in some of the inscriptions, was a *pseudo-tribus*.[19] A station of the *beneficiarii* is documented in the town,[20] as well as the provincial archives (*tabularium provinciae*); Siscia soon developed into a most prosperous Pannonian town, becoming one of the principal administrative centres in Pannonia and later Upper Pannonia.[21] This was made possible due to a local and international trade and, consequently, lively traffic, based both on land and river transport. The Ljubija iron mines and other natural resources added to the prosperity of the city.[22]

It is therefore not surprising that residents of Siscia were composed not only of indigenous population and colonists who arrived in the town under Vespasian, but also of many other foreigners, attracted by various possibilities offered by the city. Such was, for example, the case of Titus Tullius Tertius from Tergeste, perhaps a merchant, who died at Siscia, when he was fifty (S I.1). Many more immigrants, who were clients of fullers and dyers at Siscia, from Orientals and Iberians to Africans and others, are documented on the *tesserae* (see below),[23] but it is difficult to know whether they were long-term residents in the city, or whether they only stayed at Siscia for a short time.

Marriages between indigenous inhabitants and newcomers contributed to the integration of the local population and to the creation of new identities; this process was unavoidable in view of cultural and ethnic diversities. It can be claimed that each town had its own specific identity, while all towns in the province, together with the countryside, which developed in a different way, contributed to the creation of a diversified provincial society. Immigration from Italy, particularly from northern Italy, but also from those western provinces that had earlier been Romanized, must have obscured the autochthonous population, which can only be sporadically identified.[24] At a later date, after the Marcomannic Wars, immigration from the East had an impact on the population in Pannonia generally. On the other hand, several Housed Sisciani have been attested in Italy, most notably in Rome, mainly soldiers, but also in various other parts of the Roman Empire, even in the distant regions.[25]

Any population study is made difficult due to the very small percentage of Roman inscribed monuments preserved to date, notably tombstones, a few honorific public inscriptions, and altars. The evidence thus obtained is not only very limited but also random, as is the case more or less everywhere in the Roman Empire.[26] Conclusions suggesting any synthetic reconstruction of the society living at Siscia under the Empire should therefore be avoided, but some results based on the available data nonetheless shed light on certain aspects

11 Anreiter 2001: 179–182.
12 Anreiter 2001: 122–124 (and 189).
13 Šašel 1974: 702–707 (1992: 600–603); Durman 1992: 120.
14 Drnić and Miletić Čakširan 2014; Miletić Čakširan 2014; Škrgulja and Tomaš Barišić 2015: 45–64; Radman-Livaja and Vukelić 2018.
15 Radman-Livaja 2007.
16 Šašel Kos 2005: 437–442; Radman-Livaja 2014: 13–14; both with earlier literature; Škrgulja and Tomaš Barišić (eds) 2015.
17 Possibly of the legions *IX Hispana, XX* or even *XV Apollinaris*, see Radman-Livaja 2010: 179–201; Radman-Livaja 2012: 162–165; 169–170, with earlier citations; Radman-Livaja and Vukelić 2015: 400–401; Radman-Livaja 2016: 185–186; cf. Mócsy 1974: 43, 48; Wilkes 2000: 116; Bishop 2012: 101–102.
18 Šašel 1983 (1992).
19 Forni 1985: 83, cf. also 20, 56, 66.
20 Radman-Livaja and Vukelić 2016.
21 Šašel 1974 (1992); Hoti 1992; Buzov 1996; Lolić 2003; cf. Mrozewicz 2015: 152.
22 Hirt 2010: 138, 162, 239, 245.
23 Radman-Livaja 2016.
24 See a detailed discussion of the population of Siscia in Šašel Kos 2018.
25 Šašel Kos 2017b; Radman-Livaja 2016: 169–191.
26 Eck 2007.

of the population of the town and its countryside. However, the case of Siscia is different in so far as 1123 *tesserae*, lead tags used in the textile industry of the town, and mainly referring to the customers of fullers and dyers in Siscia, were rescued from the Kupa River near the Roman harbour on the right bank of the river. They have been published in an exemplary way by Ivan Radman-Livaja.[27]

Since almost everyone could afford to pay for the maintenance of their garments,[28] the *tesserae* do allow us also a glimpse at the lower classes of Siscia. The tags are mainly dated between the reigns of Claudius and Hadrian, and none is later than the Severan period. 949 individuals are documented on them, bearing 743 names, of which more than 110 occur for the first time; almost 40% of the names are female. *Gentilicia* are not very frequent (presumably they must have often been omitted for the sake of brevity): several Flavii, Iulii, Claudii, and Ulpii are documented, but only one Aelius and one Aurelia.[29] The majority of the names (386) are Latin, some 50 may have been 'Illyrian' (referring to Roman Illyricum, i.e. Dalmatia and Pannonia)[30] or Pannonian, while 163 could tentatively be identified as Celtic, of which a certain number already had to be re-interpreted.[31] Non-Latin, possibly indigenous, names are much more frequent on the *tesserae* than on stone monuments. However, the impact of Romanization is confirmed by the Latin names which prevail, both on the inscribed stone monuments and on the lead tags. Two factors should not be forgotten: first, names very often shed light on current fashions and not on the ethnic background; and, second, indigenous names do not always mean local personal names characteristic of Siscia. Celtic names, which could theoretically be regarded as names of autochthonous Siscian inhabitants, could also have been borne by immigrants from northern Italy or Noricum.

### *IV.2.2. Members of the upper class*

As in every Roman autonomous town, members of families which belonged to the local upper class, performed high functions in the city's administration or were members of the municipal council. Some were rich, possessing a large number of freedmen and freedwomen; such was the family of the Idimii, attested in a damaged epitaph, now lost and only preserved in a drawing (S[t] V.1). They certainly belonged to the well-to-do inhabitants of Siscia, since their tombstone was also intended for their *liberti* and *libertae*. It is questionable, however, whether Quintus Idimius, whose cognomen is damaged and cannot be read, would have been *v(ir?) l(audabilis?)* or not. The reading is most doubtful: if indeed he held this title, he would have belonged to *ordo equester* and would have probably been a municipal official. However, what is problematic is the combination of this title (rarely used before the 4th century AD) with the formula *libertis libertabusque* and the interdiction to inherit the tombstone, which were usually earlier.[32] Idimius is the only non-Latin *gentilicium* preserved on the stone monuments in Siscia, probably derived from an indigenous name, possibly Celtic.[33]

A costly marble stele, probably from Poetovio, on which 'eternal home and lasting security' are invoked at the beginning of the epitaph, was erected by Cenius [...] ianus, in his lifetime, for a family of probably four or even five persons (S I. 8). The stele – its lower part is missing – is most likely from the 3rd or the first half of the 4th century AD. A couple and two daughters are depicted in the niche, with another man on their left. Cenius is styled *vir egregius*, a member of the equestrian order; he may have bought the tombstone in his lifetime for the family of one Flavius Tiberinus or Tiberianus (the cognomen is partly damaged), who had perhaps died at thirty (the number of years may have continued in the next, fragmentary, line and he could have been older). He, too, may have been *vir egregius*. His wife [---]nia Ursa was most probably not Cenia, as has been suggested earlier, because the missing space is too large; hence she would not have been a relative of Cenius [...]ianus. Their daughter was four years old, of her name only *[---]n(a)e* has been preserved; the older daughter is depicted next to her, but neither her name nor her age are known.

The explanation of the epitaph is not easy due to the uncertain identity of the first man on the stele, depicted somewhat in the background. If not Cenius, as suggested by Branka Migotti (see Catalogue), possibly, but less likely, the brother of the deceased [Ce?]nia Ursa,[34] he could have been the adult son of the deceased couple. He would have been mentioned in the missing part of the inscription; perhaps he was still alive when the stele was commissioned, thus it would be understandable that his name was mentioned at the end of the epitaph. It is not clear what would in this case be the relationship of Cenius to the family: *socer?*, *pater?*, *avus?* (father-in-law, father, or grandfather) have been suggested in the *lupa*. On the other hand, it is not necessary to explain Cenius as a relative of the family. All these questions must remain without a reliable

[27] Radman-Livaja 2014.
[28] Radman-Livaja 2011; Radman-Livaja 2013.
[29] Radman-Livaja 2014: 145; on chronology: 144–147; on female names: 94.
[30] Radman-Livaja 2014, 291; Falileyev 2014: 111, 133; Radman-Livaja 2017.
[31] Falileyev 2014; Falileyev and Radman-Livaja 2016.

[32] *Diz. ep.*, s.v. *vir laudabilis*; according to Mihailescu-Bîrliba 2006: 74–75, the formula *libertis libertabusque* would have been characteristic of the 1st century AD inscriptions in Dalmatia, but it also occurs in the 2nd century AD.
[33] Prof. Alexander Falileyev, personal communication.
[34] Gregl and Migotti 2001: 141.

answer, but it is clear that both families mentioned in this epitaph, the Flavii and Cenii, belonged to the town's upper class.

The most illustrious person documented to date in Siscia was undoubtedly Romania N(a)evia, known from a marble sarcophagus, which had been in all probability erected in the second half of the 3rd century AD (S II.10). This is suggested by the discovery of the coins of Valerian and Aurelian in the proximity of her sarcophagus, dated to the years 253 to 275.[35] Nevia was a *clarissima femina*, and as such belonged to the highest class of Roman society generally; she undoubtedly enjoyed a most prestigious position among members of the upper class in Siscia. Nevia had a *memoria* set up in her lifetime for herself and for two deceased men, Cletius Romulianus and Aurelius Calemerus. Although their identity is not certain, they must have most probably been her husbands who had died before her. Since neither of the men seems to have belonged to the senatorial order, clearly Nevia's father must have been a senator.[36]

A *duumvir* of Siscia and also a member of the city council, Gaius D(omitius?) Victorinus, is known from an altar dedicated to Iuppiter Optimus Maximus Nundinarius (patron of the *nundinae*), discovered at Degoj on the Kupa River. The altar was set up in AD 238 (the mentioned consuls were Pontius Proculus Pontianus and Fulvius Pius) for the prosperity of the emperor Gordian.[37] Gaius D(omitius?) Victorinus was an eminent personality at Siscia, since he had also performed the significant function of the high priest of Upper Pannonia (*sacerdos p(rovinciae) P(annoniae) Sup(erioris)*); his wife, Lucilia Lucilla, was called *coniux sacerdot(alis)*, referring to her husband's post as a former priest of the province.[38] Their son Gaius D(omitius?) Victorinianus was also a member of the city council; the family belonged to the equestrian order.

Another *duumvir* is known from a fragmentary inscription discovered at Siscia: Titus A[-] Iulianus; nothing more has been preserved of his career, but his high social status is confirmed by the title of his wife (her name is broken off), who was *usu stolae ornata*.[39] She was distinguished by wearing the *stola*, a long upper garment, which was a sign of socially outstanding and wealthy married women, referring to the equestrian rank in the late 2nd and 3rd centuries AD.[40]

[35] Migotti 2013: 190.

[36] Barbieri 1952: 384 no. 2198.

[37] *CIL* III 3936 = 10820 = *ILS* 7116 = *AIJ* 500 = *HD*071792 = *lupa* 20218; The only *gentilicium* beginning in *D* at Siscia and broader region is Domitius; cf. Radman-Livaja 2014: 200. On the consuls: *PIR*² P 818 and F 553.

[38] Fishwick 2002: 274; Szabó 2006: 244 P 37.

[39] *Lupa* 22444 = *EDCS*-62500014 (no publication cited). *Usu* is required by the genitive, *stolae*.

[40] Holtheide 1980.

Gaius Ingenuius Rufinianus was a *decurio* at Siscia at a later date, under the Severi or later; he dedicated a marble altar to Hercules. He was in charge of a municipal treasury (*quaestor*) and a prefect of the college of craftsmen of cloth and clothing (*praefectus collegii centonariorum*), an association, which also performed the tasks of municipal firemen.[41] His wife was Ingenuia Rufina, bearing the same, relatively rare, *gentilicium*, indicating that they had either been of freedman descent or belonged to two branches of the Ingenuii.

Another *decurio* and *duumvir* of Siscia, Lucius Valerius Valerianus, is known from Savaria, where his dedication to Nemesis has been discovered.[42] He must have spent some time at Savaria, where the seat of the provincial council was located, since he held the high posts of the *flamen divi Claudi* and of the high priest of Upper Pannonia. Valerii were frequent all over the Empire;[43] they may have come from northern Italy to Pannonia, since many are documented at Aquileia.[44] Due to the frequency of the *gentilicium*, it can safely be asserted that members of the *gens* belonged to various social classes. Certainly they were an eminent family both at Siscia and Savaria, as they were also elsewhere; Poetovio, for example, was the hometown of the first known Pannonian senator, Marcus Valerius Maximianus.[45] Other Valerii were settled at Siscia: Valeria Saturnina is known from an altar to Liber Pater.[46] A veteran of the Legion *XV Apollinaris*, Lucius Valerius Verecundus moved from Siscia to Topusko, where his tombstone was discovered (S[t] I.1). He was a member of an association of veterans located at Siscia, which contributed a sum of money to erect his tombstone. *Collegia veteranorum* have only been documented so far at Aquileia, Ateste, Carnuntum, and Ancyra.[47] Valerius Saturninus served in an unknown legion; his tombstone was discovered in the countryside of Siscia, in the village of Svojić, 24.5 km south-west of Karlovac (S[ptSW] I.4). One Gaius Valerius Spectatus from Siscia, a soldier of the 8th Praetorian Cohort, died at the age of 24 in Rome, where his funerary monument was found.[48]

Members of the *Titia gens* most probably also came from northern Italy to settle at Siscia,[49] where they were an

[41] *CIL* III 10836 = *AIJ* 528 = *HD*056618 = *lupa* 5731; Sanader 1994: 105–106 no. 35; Liu 2009: 382 no. 228; cf. Barkóczi 1964: 301; 29/38 (p. 330). On textile guilds and their hypothesized role as firemen: Perry 2006: 7–18; Liu 2009: 57–96.

[42] Balla 1971: 89 no. 38 = *AE* 1972, 389 = *RIU* 20 = *HD*009523 = *lupa* 7957; Fishwick 2002: 273; Szabó 2006: 221 P 70; 244 P 70; *Šašel Kos* 2017b.

[43] *OPEL* IV: 143–146.

[44] Indexes of *Inscr. Aquil.*; cf. Mácsy 1959: 160; Šegvić 1988: 60.

[45] *Šašel* 1982: 567–568 (1992: 176–177); Filippini and Gregori 2014.

[46] *CIL* III 3956 = 10834 = *AIJ* 540 = *HD*071824 = *lupa* 11578.

[47] Bennett 2006: 93–99 (Ancyra); Aquileia: *CIL* V 784 = *EDR*116867; Ateste: *CIL* V 2475 = *EDR*130463; Carnuntum: *CIL* III 11189 = *HD*073138; *CIL* III 11097 = *HD*000911.

[48] *CIL* VI 2689 = Dobó 1975: no. 63 = *EDR*102983; Ricci 1993: 186 no. P. 69; Šašel Kos 2017b.

[49] Mócsy 1959: 57/18 (p. 212), also suggested Dalmatia as a possibility; Harding and Jacobsen 1988: 160, 195; cf. Solin and Salomies 1994: 187.

upper class family. Lucius Titius Proculus is recorded on a base for a statue of Hadrian, which he had ordered to be erected at Siscia in AD 124.[50] He is very likely also a benefactor mentioned in a fragmentary building inscription referring to a portico, a provision-market (*macellum*), and a basilica. These public buildings were either donated to the city of Siscia by Proculus, or they were renovated at his expense. His generosity vis-à-vis his native town was partly prompted by his desire to perpetuate the memory of his father Gaius Titius [-] and his brother Gaius Titius Materninus. Proculus had a distinguished municipal career as a *pontifex* and a priest of the imperial cult (*flamen*), as well as probably a *duumvir* and *duumvir quinquennalis*. After having held all these high municipal posts, he was granted a public horse and was enrolled in the five panels of judges (*adlectus in quinque decurias*).[51] Such *cursus honorum* was characteristic of the reigns of Trajan and Hadrian. The family must have had several freedmen; one of them was perhaps Gaius Titius Agathopus, who was *Augustalis* at Siscia and Sarmizegetusa, where he is attested in a dedication to Aesculapius and Hygia.[52]

Two state officials active at Siscia should also be mentioned; however, many more must have been employed to carry out supervision or perform other tasks in this important administrative provincial centre. Most unusually, there is no other evidence of their presence in the town so far. One Flavius Severus supervised building of a presumably large structure in the Tetrarchic period, of which nothing else is known because the inscription is fragmentary.[53] Severus' rank is not known, he could well have been a provincial procurator, but also only a centurion or a *beneficiarius consularis*.[54] On the other hand, Flavius Verus Metrobalanus was indeed an imperial procurator, perhaps under Commodus or under the Severi. He is known from a dedication of an altar to Jupiter, Best and Greatest, Fulminator and Fulgurator, which was erected by one of his subordinate clerks, Asclepiades, a treasurer of the office in Siscia (*arkarius stationis Siscianae*).[55] Metrobalanus was also in charge of the revenue from the iron mines (*praepositus splendissimi vectigalis ferrarium*) and therefore perhaps *procurator vectigalis ferrariarum Pannoniae Superioris* (?).[56] In view of his two titles it seems more reasonable to postulate that he performed two different offices: as an imperial *procurator* he was in charge of four provinces (Noricum, both Pannonias, and Dalmatia),[57] also having been appointed to supervise the revenue from the Pannonian iron mines. It seems less likely that he would have held these two assignments one after the other.[58] Siscia was almost certainly the headquarters of the Ljubija iron mining district, and perhaps also the seat of the financial procurator responsible for Pannonia Superior.[59]

### IV.2.3. Soldiers and veterans

Siscia became a Flavian colony, colonized by fleet veterans from the Ravenna navy, which can be inferred from a military diploma issued under Vespasian on 5 April AD 71, in which it is declared that the veterans who had served in the navy of Ravenna under Sextus Iulius Bassus were settled in Pannonia.[60] Afterwards, military presence in the city other than the station of the *beneficiarii* was either temporary or exceptional. The *beneficiarii* are well documented at Siscia, both *beneficiarii procuratoris* and *beneficiarii consularis*.[61]

Several soldiers are attested as serving at Siscia. One of the earliest was a centurion of the Legion *XV Apollinaris*, Tiberius Claudius Pontius, from the first half of the 1st century AD, who died at the age of 25 and was commemorated by his heir (S VIII.1). The legion was stationed in Carnuntum from the early Claudian period onwards; shortly before that it was briefly garrisoned at Vindobona.[62] However, during the revolt of the three Pannonian legions in AD 14, after the death of Augustus, the *XV Apollinaris* perhaps had its camp for several years at Siscia. The rank of *centurio*, which Pontius attained very young, could imply that his family had been of local origin,[63] belonging to the indigenous elite. His father must have acquired Roman citizenship under Claudius or even under Tiberius (when he commanded the army in Illyricum under Augustus), and had perhaps himself served in the Roman army. Pontius was either sent on a mission to Siscia, or his legion was stationed there at the time of his death. Roughly from the same time is the tombstone of Marcus Mucius Hegetor, a physician of the Cohort *XXXII voluntariorum* (S I.2). This cohort was

[50] *CIL* III 3968a = *AIJ* 559 = *HD*072112 = *lupa* 5734.

[51] *CIL* III 6476 = 10856a-c. e-h = *AIJ* 561 = *HD*012614 = *lupa* 6021; restored and linked to the already known L. Titius Proculus by Christol and Demougin 1986 (*AE* 1987: 820). Cf. Vukelić 2011: 238–239. On the five panels of judges: Demougin 1975.

[52] *AE* 1914: 109 = Dobó 1975: no. 556 = *IDR* III 2, 165 = *HD*021011 = *lupa* 15172; cf. Šašel Kos 2017b: 177.

[53] *CIL* III 10851 = *AIJ* 564 = *HD*072118 = *lupa* 22359.

[54] Fitz 1994: 1157 no. 837; cf. 1133 and 1995, 1421. Hoffiller (*AIJ*) suggested a supplement [*v(iro) e(gregio)*?], but with no arguments.

[55] *CIL* III 3953 (p. 1742) = *ILS* 03049 = *EDCS*-28701037 (*ILS* and *EDCS* have 'Myrobalanus', not plausible); cf. *PIR*² F 396 (as M[y]robalanus); Winkler 1970; Fitz 1993: 738–739 no. 422; cf. 405; Sinobad 2010: 154, 156, 208–209 no. 172.

[56] Pflaum 1961: 1064.

[57] Or even more provinces: Mócsy 1962: 594; Hirt 2010: 138, 239.

[58] Fitz 1993: 738–739 no. 422; Hirt 2010: 245.

[59] Haensch 1997: 352–353: procurator's seat in Upper Pannonia may have either been at Poetovio or at Siscia; Hirt 2010: 162.

[60] *CIL* XVI 14 = *ILS* 1991 = *EDCS*-12300218 (Mesihović 2011: 306; Grbić 2014: 125–126 no. 57); Chiriac, Mihailescu-Bîrliba, and Matei 2004.

[61] Radman-Livaja and Vukelić 2015: 403–404; Radman-Livaja and Vukelić 2016; on the role of the station: Nelis-Clément 2000: 49–50, 183–186, 194–196, 255, 515.

[62] Šašel Kos 1995: 237; Mosser 2002; Mosser 2003: 141, 261 (Siscia); 144–145 (Vindobona); Radman-Livaja 2012: 162–165, 169–170, with earlier citations; Šašel Kos 2014b: 80; Radman-Livaja and Vukelić 2015: 401.

[63] Suggested by Mócsy 1959: 57/1 (p. 211).

among the first auxiliary units garrisoned in Pannonia, having been most probably stationed at Siscia in the first half of the 1st century AD, until the reign of Vespasian, when it was transferred to Germania.[64] As is indicated by his Greek cognomen, Hegetor must have been recruited in one of the eastern provinces. The name of his freedwoman is Greek, Corinthia, and indeed this name was frequently borne by slaves.[65]

Two soldiers from the Legion *XIV Gemina*, stationed at Carnuntum, are documented at a later time at Siscia: Publius A() Ci[t]us, an *optio* known from an altar to the Genius of the province,[66] and a *cornicularius* (adjutant) Gaius Sempronius Severus, whose wife Iulia Florentina had an ash-chest erected for her husband (S III.1). The first probably came to Siscia on a special mission, while the second very likely originated from the town and had family there.[67]

Several veterans are attested as residents of Siscia; perhaps their family lived in the town or they married local women, or else they decided to settle at Siscia because this prosperous Pannonian city could offer them a good economic and/or social position to start a new career or a business. An interesting case is that of Marcus Minicius Saturninus, veteran and former *optio* of the praetorian navy from Ravenna, who may have belonged to the first colonization wave and to the veterans from the Ravenna fleet, who, as has been noted, had been settled at Siscia under Vespasian. There is indeed nothing that would contradict the date of his tombstone in the late 1st century AD. Saturninus had it erected in his lifetime for himself and his numerous family, first of all his wife Crispia Celerina and his sons Celer, Certus, and Gratus, but also for his parents Minicii Calvus and Prisca (S VIII.3). The fact that both his father and his mother bear the same *gentilicium* probably betrays their freedman origin. Due to the fact that Saturninus belonged to the Ravenna fleet, it was presumed that the family had originated from Dalmatia, but there is no other indication for this assumption.[68] On the contrary, the *gentilicium* Minicius, which is particularly frequent in Italy and Hispania, is very rare in Dalmatia.[69] However, Crispii, which are elsewhere rare, do occur three times at Siscia and it may well be suggested that Crispia Celerina had been a native of Siscia.[70] The cognomina Saturninus (most popular in Italy and in the western part of the Empire), Celer, Gratus, Calvus, and Prisca also appear on the *tesserae*.[71]

Gaius Antonius Sentinus was a veteran of the Legion *XIV Gemina Martia Victrix*, who died at Siscia most probably towards the very end of the 1st century AD, when he was 50 years old (S I.3). The legion took part in the great Pannonian-Dalmatian Revolt in Illyricum, but was presumably posted as early as AD 9 to Mogontiacum in Upper Germania, where it stayed until AD 43. The legion then participated in the conquest of Britannia, earning honorary titles *Martia Victrix* and remaining there until AD 70. However, two years before that, it was briefly posted either in Pannonia or Dalmatia. Between AD 70 and 92, it had been in Mogontiacum again, from where it went to Carnuntum, to take part in Domitian's war in Pannonia against the Marcomanni and Quadi in AD 96, which ended under Nerva.[72] There, the legion had briefly stayed at Mursella near Mursa, not far from Siscia, and it can be hypothesized that Sentinus came to know the city at that time.[73] The remains of the once rich decoration of Sentinus' funerary monument, commissioned by his heir, indicate that he was depicted as reclining on a *klinē*, with a servant and a tripod in front of him. Such *stelai* were made for legionaries in Upper and Lower Germania under the Flavian emperors and Trajan, particularly at Mogontiacum (Mainz) and Colonia Agrippina (Köln).[74] Antonii are documented both in the western and eastern provinces of the Empire, while his Latin cognomen Sentinus may have been a Celtic assonance name.[75] It seems very likely that he had been recruited around the mid 1st century AD in northern Italy, or an early Romanized western province, where a Celtic language was spoken. Another veteran of this legion, Marcus Aurelius Glabrio, is known from a much later time, the 3rd century AD, from an ash-chest which he had bought for himself and his wife Septimia Matrona (S III.2). He had formerly been a *beneficiarius consularis*, possibly stationed at Siscia,[76] where he could have met his wife. She seems to have been from a local family, such as Aelia Matrona, known from a sarcophagus (S II.1).[77]

Two auxiliary veterans are known from military diplomas found at Sisak. Sapia, son of Sarmosus, from Anazarbus in Cilicia, received a military diploma in AD 100, after having served in the Cohort *I Antiochensium*

---

[64] Lőrinz 2001: 44, 57–58 (cf. Spaul 2000: 47–48, not plausible); Radman-Livaja 2012: 171.

[65] Pape and Benseler 1911: 697; Alföldy 1969: 180; *OPEL* II: 75; Solin 2003a: 632.

[66] *CIL* III 3943 = *EDCS*-28701031.

[67] See also Radman-Livaja and Vukelić 2015: 402; on the whole, relatively few military inscriptions have been found at Siscia: Zaninović 1981.

[68] Mócsy 1959: 57/2 (p. 211).

[69] *OPEL* III: 82; *Epigraphik-Datenbank Clauss/Slaby*; Alföldy 1969: 100.

[70] Mócsy 1959: 57/2 (p. 211); *OPEL* II: 85; Crispus and Crispinus occur on several *tesserae*, Radman-Livaja 2014: 192.

[71] Saturninus: *OPEL* IV: 51–53; Radman-Livaja 2014: 259; id. 2014: 184 (Celer); 213 (Gratus); 178 (Calvus); 248 (Prisca).

[72] Franke 2000: 197–198; Strobel 2010: 155 (the third Pannonian war); Radman-Livaja 2012: 170–171.

[73] Radman-Livaja 2012: 170, 175; Dušanić 1983: 20–21 (2010: 958–959); Hoti 1992: 143; Škrgulja and Migotti 2015: 39.

[74] Škrgulja and Migotti 2015.

[75] Solin and Salomies 1994: 400; Delamarre 2007: 166; on this phenomenon: Raepsaet-Charlier 2012.

[76] See also Radman-Livaja and Vukelić 2015: 399; Radman-Livaja and Vukelić 2016: 211.

[77] Mócsy 1959: 57/6 (p. 211).

in Moesia Superior, not in Pannonia.[78] He could have come to know Siscia during some unrecorded detached service.[79] The other veteran, attested in a fragmentary diploma from the first half of the 2nd century AD, remains nameless, merely his father's name Saraba is known.[80] Perhaps his unit, whose title is damaged, was stationed somewhere in southern Pannonia and he decided to settle at Siscia due to various advantages offered by this prosperous city.

### *IV.2.4. Sisciani serving in the Roman army abroad*

Two soldiers from Siscia are attested at Brigetio: the first is Lucius Lucanius Festus, who served in the Legion *XIV Gemina*. Marcus Ulpius Victor, a trumpeter from the same legion and his heir, who may have himself been from Siscia, had the funerary monument commissioned for him, presumably at some time in the 2nd century AD.[81] The legion had its permanent camp at Carnuntum after Trajan's Dacian Wars, while at Brigetio the Legion *I Adiutrix* was stationed.[82] The second soldier from Siscia, one Iulius Nigellio, was on service at Brigetio at the beginning of the 3rd century AD, as a *beneficiarius* of the legionary tribune of Legion *I Adiutrix Antoniniana*; his heir Aelius Paternianus arranged for the tombstone to be erected for him. They were both serving as *regionarii*, having been in charge of policing a district (*regio*) that had been assigned to them.[83] *Regionarii* are seldom documented in the Roman Empire, except in Lower Moesia, where they are relatively well attested.[84] The *gentilicia* of these soldiers and the cognomen Nigellio also occur on the *tesserae*.

During the first years of Trajan's reign the Legion *XI Claudia* was also stationed for some years at Brigetio, when perhaps Publius Crescentinius Flavius Saturninus from Siscia had been recruited in this legion. After Trajan's Dacian Wars, the legion was based at Durostorum in Moesia Inferior.[85] However, Saturninus' tombstone, commissioned for him by his daughter Flavia Ianuaria, was discovered at Portus Magnus in Mauretania Caesariensis (modern Arzew in Algeria),[86] to where a detachment of his legion must have been sent on some mission around the mid 2nd century AD. Both Saturninus and Ianuarius/Ianuaria were most popular names among the Sisciani, particularly the former, which has been attested 16 times, also on the *tesserae*.[87] Ianuarius/Ianuaria was an assonance name in the Celtic-speaking areas.

A *beneficiarius consularis* from Siscia, Titus Aelius Victorinus, appears on a list of the members of the *officium* of the legate of Numidia, from the second half of the 2nd century AD. It was discovered at Lambaesis (Lambèse near Tazoult in Algeria), where the Legion *III Augusta* was stationed.[88] At Apulum in Dacia (Alba Iulia) an unknown soldier of the Legion *XIII Gemina* from Siscia has been attested.[89] At Tyana in Cappadocia a tombstone from the 3rd century AD came to light, commissioned by a centurion of this legion from Siscia, Marcus Aurelius Lucius, for his former slave (*verna et libertus*), Marcus Aurelius Saturninus.[90] Not only Saturninus, as has been noted, but also Lucius were extremely popular names at Siscia, the latter occurring on 14 *tesserae*.[91]

An inhabitant from Siscia, Gaius Publicius Priscilianus, is known from Bonna (Bonn) in Lower Germania, from the Legion *I Minervia*, in which he was a commander of the first cohort (*primus pilus*). Soon after the accession of Severus Alexander on 11 March 222, Priscilianus had an altar to Victoria Augusta erected on 1 May 222 AD, in the name of the governor of the province, Titus Flavius Aper Commodianus, and the legionary legate of his legion, Aufidius Coresnius Marcellus, by origin from Sagalassus in Pamphylia, the future governor of Lower Germania.[92] Two soldiers from Siscia are registered in a badly preserved list of soldiers from the 3rd century AD, from Albanum in *Regio I* (Latium and Campania). Only their cognomina are known: Valentinus and Marcellinus; very likely they belonged to the Legion *II Parthica*, enrolled by Septimius Severus.[93]

### *IV.2.5.* Augustales *and other members of the middle class*

The previously mentioned Gaius Titius Agathopus was not only *Augustalis* at Siscia, but also at Sarmizegetusa. At Siscia only *Augustales* have been attested to date, no *seviri* or *seviri Augustales*.[94] Pontius Lupus *was Augustalis* at Siscia, but carried out the supervision of public records (*scriba*) at *municipium Faustinianum* (S II.2), an Upper Pannonian town that has not yet been located. It has been suggested that in small Pannonian towns such as Faustinianum, the *scriba* had perhaps been the

78 *CIL* XVI 46 (*AE* 1912: 128) = *ILS* 9054 = *HD*028179; for the name Sapia, see Zgusta 196: 456 § 1372a (from Lycia); Mócsy 1962: 621; Lőrinz 2001: 46.
79 Cf. Radman-Livaja and Vukelić 2015: 399–400, 404.
80 *CIL* XVI 103 = *HD*027802 (with earlier literature); cf. *RMD* V, p. 702; for the name: Mócsy 1959: 57/23 (p. 212).
81 *CIL* III 11029 = *RIU* 566 = *HD*039484.
82 Franke 2000; Lőrincz 2000.
83 Borhy 2010: 70–73.
84 M. P. Speidel 1984 (1992), for the Montana region; Sarnowski 1988.
85 Fellmann 2000.
86 *CIL* VIII 9761 = Dobó 1975: no. 180 = *EDCS*-25500576.
87 Radman-Livaja 2014: 259, 215.
88 *CIL* VIII 2586, v. 22 = *ILS* 2381 = *CBI* 783 = *EDCS*-58800037 (cf. *AE* 2010: 1828); cf. Radman-Livaja and Vukelić 2015: 404.
89 Forni 1982: 190 no. 2 (= *AE* 1982: 826b) = *HD*001774 = *lupa* 11992.
90 *AE* 1941: 161 = Dobó 1975: no. 263 = *HD*021468.
91 Radman-Livaja 2014: 224.
92 *CIL* XIII 8035 = Dobó 1975: no. 157 = *EDCS*-11100259 (*EDH*: FO18705); cf. Radman-Livaja and Vukelić 2015: 404. Commodianus: *PIR*[2] F 210; Thomasson 1984: 59 no. 100; Eck 1985: 207–208 no. 53; Marcellus: *PIR*[2] A 1383; cf. *AE* 2002: 1772.
93 *AE* 1964: 14 = Dobó 1975: no. 161 = *HD*015583 = *EDR*074321; M. P. Speidel 1983: 60–61 no. 33 (1992: 38–39 no. 33); Faure 2013: 60–61 no. A 56.
94 Duthoy 1976: 187.

chief official in charge of municipal affairs;[95] however, this may be true of settlements without autonomous status, but not of *municipia* and *coloniae*. As *Augustalis*, Lupus did not belong to the municipal aristocracy; he obviously performed the function of *scriba* as a member of the middle class and this must rather be regarded as a lower municipal post, subordinate to a *quattuorvir* or *aedilis*. His sarcophagus from the second half of the 2nd century AD was also intended for his sister Pontia Victorina and Domitius Crescens, who may have been her husband. The Dioscuri on the two side panels hold a standard similar to a *vexillum*, perhaps referring to the association of the *Augustales* in Siscia, to which Lupus belonged.[96] The *gentilicium* Pontius is sporadically attested almost in all western provinces, but was frequent in Italy,[97] from where the family had probably originally emigrated. At Siscia it certainly was of some significance, since the name occurs on two commercial tags.[98] As can be inferred from the sister's cognomen, Victorina, very popular in Pannonia particularly after the Marcomannic Wars, she and her brother Lupus probably belonged to the indigenous Siscian community.

Another *Augustalis* from Siscia, Titus Aurelius Anicetus, is known from an altar erected to Silvanus Augustus.[99] The last to be mentioned is Marcus Mulvius Narcissianus, *Augustalis* of the *colonia Septimia Siscia*, recorded on a sarcophagus, set up for him and for his father Marcus Mulvius Narcissus by his mother Mulvia Furnia; her husband was sixty when he died, her son only twenty-six. Mulvii came most probably from Italy to settle at Siscia. (S II.3).

Publius Aelius Iulianus, a member of one of the units of the association of timber workers (*ex numero collegii dendrophorum*), commissioned a sarcophagus for Aurelia Veneria, his 'most devoted girl, snatched by a sudden death' (S II.6). The relationship of Iulianus to the girl seems enigmatic; he was not her father, but might have been her (future?) husband. *Collegia dendrophorum* are usually associated to the cult of Cybele;[100] it can therefore be concluded that the goddess was worshipped at Siscia.[101] *Collegium dendrophorum* counted among three principal colleges in each Roman city in the western provinces, the other two being the *collegium fabrum* and *collegium centonariorum*.[102] These were typically Roman institutions, open particularly to members of the middle class, where they could display their wealth and enjoy social respect as craftsmen, artisans, and builders, as well as members of fire-brigades.[103]

### *IV.2.6. Inhabitants of Siscia: some additional interesting cases*

As much as from their tombstones, the residents of Siscia are also known from altars and other dedications to various gods and goddesses.[104] Titus Aetius, with a damaged cognomen, erected an altar to Jupiter.[105] His Latin *gentilicium* has not been attested to date in the western provinces and northern Italy,[106] but it is recorded a few times in Rome and possibly once in Dalmatia.[107] Links between Siscia and Dalmatia are also documented in two other cases. A young man, Marcus Plusius Ammianus, who had been born in Siscia, died at the age of twenty-three at Dalmatian Nedinum (Nadin), where his funerary monument has been discovered. The tombstone was set up to him by his friend Titius Genialis.[108] The family name, Plusius, has not been attested to date elsewhere, but it does occur as a cognomen and a personal, mainly feminine, name.[109] Ammianus, too, only occurs a few times;[110] it could possibly be regarded as an indigenous name. The *gentilicium* Titius, however, is quite frequent, particularly in northern Italy, Dalmatia, Pannonia, and Noricum.[111] At Siscia it occurs five times, having been borne both by members of the upper and middle classes: perhaps Genialis had also come to Dalmatia from Siscia. One Aurelius Victorinus, a citizen of Siscia and a *librarius* (a secretary or a bookseller), has further been documented in Dalmatia, the find-spot of his marble ash-chest (now in Vienna) not being known. Victorinus' two colleagues, Macrinus and Maturinus, who had provided for his ash-chest to be erected, were also from Siscia, calling themselves *c(ives?) et collegae*.[112]

Some more families from Siscia are documented on funerary monuments. Cassia Candida had a sarcophagus commissioned for her dearest daughter Septumia Marcella, who was only twenty years old (S II.7). The *gentilicium* Cassius is frequently documented in the Roman Empire, but has only been attested at Siscia once more on a *tessera*, on which one Cassia Ignastia is mentioned, Ignastia being elsewhere unknown.[113]

---

[95] Mócsy 1974: 222; on the complex public role of the *Augustales*: Abramenko 1993: 43–82 and *passim*; Mouritsen 2011: 248–261.
[96] Migotti 2005a.
[97] *Ibid.*; Solin and Salomies 1994: 147; Mócsy 1959: 158; Alföldy 1969: 112; *OPEL* III: 153.
[98] Radman-Livaja 2014: 246; once possibly a personal name.
[99] *CIL* III 3961 = *EDCS*-28701043.
[100] Cf. Van Haeperen 2012: 61.
[101] As has already been noted by Zaninović 1981: 204.
[102] Thus called, e.g., in *CIL* XI 5749 = *EDR*016319 from Sentinum.
[103] Liu 2008: 65; Haeussler 2013: 286.
[104] All are collected, including statuettes and other depictions, by Buzov 2012; for Egyptian cults at Siscia see Selem 1997; for the oriental deities in general: Selem 1980: 6–10, 182–186, 201–207, 217–223, 270–271 and *passim*.
[105] *CIL* III 3946 = *EDCS*-28701033; Sinobad 2010: 207 no. 165.
[106] Solin and Salomies 1994: 8; *OPEL* I does not record it.
[107] *Epigraphik-Datenbank Clauss/Slaby*; cf. *ILJug* 2568,3161.
[108] *CIL* III 9962 = Dobó 1975: no. 261 = *HD*058156 = *lupa* 23021.
[109] *OPEL* III: 147; *Epigraphik-Datenbank Clauss/Slaby*.
[110] *OPEL* I: 49; Radman-Livaja and Ivezić 2012: 142; cf. Alföldy 1969: 148.
[111] *OPEL* IV: 124.
[112] *CIL* III 3166 = Dobó 1975: no. 262 = Noll 1986: 69 no. 275 = *HD*056762 = *lupa* 9650.
[113] Radman-Livaja 2014: 182–183.

However, the name Candidus/a has been documented 16 times at Siscia, also occurring on *tesserae*.[114]

Other family names at Siscia or its administrative territory include the Numisii and Pomponii, Ravonii, Sempronii (S III.1), Valentinii (S[t] I.4), Vitalii (S[t] II.1), and Urbicii (S VIII.7), a rare *gentilicium*.[115] Only the name Numisius also occurs on a *tessera*, recording one Numisia Tripena. Tripena could have been an indigenous Pannonian name, she was perhaps of a freedman descent. A funerary altar found rather far from Sisak, in the village of Mrežnički Varoš, was erected for the family of Gaius Numisius Ursinus and Pomponia Iuventilla (S[ptSW] V.1). The Numisii may have arrived in Siscia as colonists from Italy or from Hispania.[116] The *gentilicium* Pomponius was frequent in Italy and Hispania, but also in Dalmatia, while in Pannonia it only occurs sporadically.[117]

Several Siscian families are known from votive inscriptions. An individual with an interesting name, Marcus Iulius Vesnis, is attested in a dedication to Jupiter and to the Genius of the *colonia Siscia*.[118] The linguistic explanation of the cognomen Vesnis is not certain, but it has been suggested that he may have arrived in Siscia from Italy.[119] Gaius Heius Capito and his wife Caecilia Rufina are known from an altar dedicated to Jupiter and the *Genius loci* by an unknown dedicator for the well-being (*pro salute*) of the couple.[120] Since both Heius and Capito are names characteristic of Hispania, it can be presumed that the couple came from Hispania. Moreover, Heius is a very rare *gentilicium*, which has not been attested to date in Pannonia.

One or more sanctuaries of Mithras existed at Siscia, which is first of all documented by an altar dedicated to the god by an imperial slave Iucundus, a treasurer of Upper Pannonia (*dispensator provinciae Pannoniae Superioris*) in the mid 2nd century AD or later. He paid for a portico and a room where preparations for a sacrifice (*apparatorium*) took place. A relief with a depiction of Cautes was dedicated by a municipal slave Urbicus.[121] One Aurelius Eutyches, perhaps from one of the eastern provinces, dedicated an altar to Sol invictus Mithras for the well-being of either Caracalla or Elagabalus.[122] His Greek cognomen may indicate his origin in the Greek East, which would be in accordance with the eastern origin of the cult of Mithras. Another dedication to Mithras by the 'brothers' Aurelius Heraclides and Aurelius Agathopus is on a bronze *tabula ansata*.[123] An Imperial freedman (*Augg. nn. libertus*) Augustinus, employed in the archives of Upper Pannonia (*tabularius provinciae*), is attested on an altar to Venus.[124] The seat of the provincial archives (*tabularium*) was at Poetovio, while at Siscia a subordinate office must have been located.

Several Sisciani worshipped the Italian god Silvanus, Silvanus Domesticus having been one of the most popular gods in Pannonia, and very prominent also in Dalmatia.[125] Nine dedications are known so far from Siscia, such as an altar erected by one Octavius Cornelianus.[126] Octavii are documented almost everywhere and the *gentilicium* also occurs on three *tesserae*; like so many other residents at Siscia, the Octavii, too, may have come to settle in the town from northern Italy.[127] Silvanus and the Celtic goddess Magla ('Great Goddess') were worshipped by Longinia Ingenua;[128] it should be noted that Magla was not a Celtic epithet of Silvanus. Silvanae were worshipped by one Marcus Iulius Ingenuus: an altar dedicated to them was discovered in the nearby village of Topolovac.[129] The Ulpii are relatively well attested at Siscia, also appearing on six lead tags.[130] One Ulpius Priscianus erected an altar to all the gods and goddesses.[131] Two religious associations are documented in the city: Iuno was worshipped with Jupiter, Best and Greatest, by two *magistri* (the chief officers of the *collegium*), Aelius Valerius and Aelius Secundinus.[132] Two other *magistri*, responsible for the cult of Mars Marmogius, Iunius Philocrates and Iulius Crispinus, dedicated to the god a statue on a semicircular couch (*signum cum stibadio*).[133]

A most interesting and almost unique funerary monument from the Late Roman period was erected to Leburna, master of mime actors, who lived to a great age, and on whom see more in the Chapter on 'Palaeographic and epigraphic characteristics'.

[114] Radman-Livaja 2014: 179.
[115] *OPEL* IV: 185.
[116] Radman-Livaja 2014: 239, 274.
[117] *OPEL* III: 151–152.
[118] *CIL* III 3951 = *EDCS*-28701036; Sinobad 2010: 208 no. 170.
[119] Meid 2005: 306; cf. Mócsy 1959: 57/13 (p. 212), who presumed that Vesnis had come from a western province.
[120] *CIL* III 3952 = *AIJ* 538 = *EDCS*-26600422; Sinobad 2010: 208 no. 171.
[121] *Dispensator*: *CIL* III 3960 = *CIMRM* 1478 = *HD*071853; cf. Weaver 1972: 251; Selem 1980: 82 no. 11; Urbicus: *CIL* III 3977 = *AIJ* 556 = *CIMRM* 1473 = *lupa* 6028 = *HD*071852; Selem 1980: 83–84 no. 14.
[122] *CIL* III 3958 = *AIJ* 543 = *CIMRM* 1476 = *HD*071830; Selem 1980: 81–82 no. 10.
[123] 'Brothers': *CIL* III 3959 = *AIJ* 544 = *CIMRM* 1477 = *HD*071837; Selem 1980: 83 no. 13.
[124] *CIL* III 3964 = *EDCS*-28701045; Šašel 1974: 736 (1992: 617); Fitz 1993: 392.
[125] Dorcey 1992: 72, 130 and *passim*; Map 6; Perinić Muratović and Vulić 2009; cf. Matijašić and Tassaux 2000; Dészpa 2012: 165–168 referring to Siscia.
[126] *CIL* III 3962 = *EDCS*-28701044.
[127] On the *tesserae*: Radman-Livaja 2014: 239; from northern Italy: Mócsy 1959: 57/15 (p. 212).
[128] *CIL* III 3963 = *AIJ* 548 = *HD*071838; Perinić 2016: 97 no. 11.
[129] *CIL* III 10847 = *AIJ* 550 = *HD*071847 = *lupa* 22539.
[130] Radman-Livaja 2014: 276.
[131] *CIL* III 10835 = *EDCS*-29900780.
[132] *CIL* III 10845 = 15179 = *AIJ* 539 = *HD*071823 = *lupa* 11531; cf. *EDCS*-29900783.
[133] *CIL* III 10844 = *ILS* 4574 = *AIJ* 542 = *HD*067309 = *lupa* 11579; Hainzmann 2011: 213 no. 6.

### *IV.2.7. Persons known from two curse tablets and an inscribed fragment of a bowl*

The curse tablet, in which the river god Savus and Muta Tagita (most likely the goddess of silence Tacita Muta) were asked to harm the adversaries of the authors of the curse, was found in 1913 at Sisak in the Kupa River (the Colapis in antiquity).[134] The authors and their opponents must have been involved in some lawsuit; the names of six adversaries appear on the inner side of the lead tablet. They were not local people, except perhaps one: Lucilius Valens, mentioned without an *origo*. And indeed, the Lucilii occur frequently at Siscia, both on stone monuments and on *tesserae*,[135] on which also the cognomen Valens is well attested.[136] Three individuals were most probably from Cibalae, another Pannonian town (modern Vinkovci): Gaius Domitius Secundus, Lucius Larcius, and Valerius Secundus. Domitii, a frequent *gentilicium*, are also documented at Siscia, while Larcii are very well attested in Rome and Italy, Africa and eastern provinces, and relatively well also elsewhere.[137] Three individuals came from far: Publius Caetronius and Gaius Corellius were from Narbo in Gallia, Lucius Licinius Sura from Hispania. The family name Caetronius occurs frequently in Italy, but elsewhere it is very rare,[138] while Corellius is well documented in Italy, but only occurs sporadically in the provinces, except at Salonae in Dalmatia.[139] The two men from Narbo were obviously of Italian descent. The whole group may have been merchants, although not necessarily; in any case, however, they travelled on boats and Savus was invoked to cause them to drown. Despite a recent attempt at identifying Lucius Licinius Sura with the Roman senator from Tarraco, who, as a close friend and adviser of Trajan was three times consul (*ordinarius* in AD 102 and 107),[140] this does not seem plausible. It can hardly be imagined that he would have personally been involved in a lawsuit at Siscia, appearing in a company of quite insignificant individuals. Hispanic Licinii are not rare and one of them obviously chose the name Sura for their son, since giving famous names to one's own children was a well-known phenomenon in antiquity, recorded in all social classes.[141]

Some years earlier, in 1898, another curse tablet was found in the river bed of the Kupa, inscribed with Greek letters. It was horizontally folded three times, displaying five nail-holes. 29 names are recorded, with five magical *characteres* below the list, the letter-forms suggesting a date in the 3rd century AD.[142] 16 names are classified as Latin, several among them being quite common, such as Festus, Pusilla (written as Posilla), Vitalis, Optatus, Aprio, Laberius, and Gratus. 12 names are Greek: Eupor, Epiteuxsis, Kosmos, Philetos, possibly Karpe, Eutychas (!), Heraklas (!), Euplous, Kallisttos, Hermes, Sosis, and Doryphoros. These names cannot shed light on the social status of their bearers, but they well illustrate the heterogeneous society of such a significant Pannonian town as was Siscia.

A fragment of a ceramic bowl or a plate with a list of ten or 11 names was found in 2001 during archaeological excavations in the centre of Sisak, in front of the Holy Cross Church on Ban Jelačić Square, near the remains of the south-western city walls. The names, except the first one, seem to be in the vocative, each line, except perhaps the fifth, containing one name: *[F] irmianus / Rufin[a?] / Victorine / Valeri /*[5] *[C?]larevinator / Sextilla / Kardo / Sasura / Salviane / Silvane*.[143] The fifth line is explained as two names by the editors: (C)larus and Vinator, i.e. Venator, a name and not the man's occupation, a hunter.[144] Some of the names, such as Firmianus (derived from the common name Firmus), or Victorinus, are characteristic of the 3rd century; Rufina, Victorinus, Valerius, Salvianus, and Silvanus are all Latin names, which are rather frequent in Pannonia.[145] Kardo has not been attested to date and should be regarded as a name belonging in the group of names derived from parts of a building, *cardo* meaning a pivot forming the hinge of a door.[146] Sasura, not known elsewhere, may have been a Celtic or perhaps Pannonian (?) name.[147] The editors dated the list to the Severan period, interpreting it tentatively as a record of guests invited to a feast, or of persons participating in a joke or a game; their social status cannot be determined.

### *IV.2.8. Territory under the possible influence of Siscia*

Six *stelai* were discovered as chance finds in the region distant between 51 and 57 km south-east of Siscia, in the territory of Bosnia. The first was found at Japra. Although the inscription is fragmentary, it is a testimony of a person with Roman citizenship, since remains of his *tria nomina* are preserved, as well as his filiation: Marcus M[-], son of Gaius, T[-] ... (S[ptSE] I.1). Of the two *stelai* found in the village of Blagaj one is fragmentary with only a portrait niche preserved, while the other mainly consists of an inscription field bearing an interesting epitaph (S[ptSE] I.2). The monument

[134] *AIJ* 557 = *HD*027805; Marco Simón and Rodà de Llanza 2008 (*AE* 2008: 1080); cf. *AE* 2010: 109; emended reading: Barta 2017.
[135] Radman-Livaja 2014: 224.
[136] *OPEL* IV: 139–140; Radman-Livaja 2014: 277.
[137] *OPEL* II: 105–106 (Domitii); *OPEL* III: 19; *Epigraphik-Datenbank Clauss/Slaby*; Barta 2017: 34–35 (Larcii).
[138] *OPEL* II: 21; Alföldy 1969: 70.
[139] Alföldy 1969: 78; Barta 2017: 35.
[140] Rodà 2014.
[141] Solin 2001; 2015b: 31.
[142] *AIJ* 526 = Curbera and Jordan 1996 = Kovács 2007: no. 69 = *HD*035604.
[143] Demicheli, Radman-Livaja, and Wiewegh 2012.
[144] Demicheli, Radman-Livaja, and Wiewegh 2012: 188–189; the name Venator occurs rarely, but is twice attested in Pannonia: *OPEL* IV: 153.
[145] *OPEL*, s. vv.
[146] Solin 2015a: 198; Kajanto 1965: 347.
[147] Demicheli, Radman-Livaja, and Wiewegh 2012: 189–190.

was erected for Ulp() Nonnatio, who had died at forty, by Campanius Marcianus and Iulia Marciana, pious parents. However, it is neither clear whether Nonnatio was male or a female, nor what was the relationship of the three mentioned individuals, since each of them bears a different *gentilicium*. If Campanius Marcianus was Marciana's husband, Nonnatio had most probably not been his child. The name, perhaps Celtic, was carved in two lines and could also have been Nonntio: either form has not been attested elsewhere.[148] Marcianus/a was a popular cognomen or a personal name and may have been an assonance name in the Celtic speaking provinces.[149]

Three *stelai* are known from Cikote near Prijedor, one is fragmentary, with only a portrait niche preserved. The other is a richly decorated stele, erected by Aelius Quartus for his wife Ulpia Matera; this was obviously an indigenous family (S[ptSE] I.3). The two *gentilicia* shed light on the granting of Roman citizenship under Trajan and Hadrian. The third stele, on which two men and a woman between them are depicted in the portrait niche, belonged to a rich family of a veteran of the 10th Twin legion (*legio X Gemina*) Marcus Aurelius Surus, a former *beneficiarius* (soldier granted special privileges). He erected the tombstone to his wife Tatonia Procula, who died at thirty-three, together with their two sons Marci Aurelii Optatus and Firminianus, and their freedmen Aurelii Baosus, Sumpa, Zipandus, and Quintus. One of their freedmen, Caro, had already been dead by the time the stele was commissioned (S[ptSE] I.4). The family name Tatonius has only twice been attested to date,[150] including the occurrence on this stele, in both cases in Pannonia. Although regarded as a Latin name,[151] it may have been a *pseudo-gentilicium*, derived from an indigenous 'Illyrian'-Pannonian name, such as Tato. The veteran probably returned to his home village, where his family lived. Most interesting are the 'foreign' names of the first three veteran's freedmen, which are not documented elsewhere,[152] hence it is not possible to draw any certain conclusions concerning their origins.

### *IV.2.9. Late Roman funerary monuments*

The name Messilla, known from a lost Late Roman stele, on which she is mentioned together with her godfather Romanus, is rare, occurring once in Italy, a few times in Dalmatia, and once in Noricum, and may possibly be regarded as an indigenous name.[153] Severilla and her husband Marcellianus are known from a Christian sarcophagus from the early 4th century AD; Severilla is called a servant of Christ, indicating that the secular social status of her family was no longer important (S II.22). Both names are characteristic of the Late Roman anthroponymy, although Severilla as a feminine name occurs rarely.[154] Another tombstone, long lost, may have also been Christian; however, the reading is uncertain and the explanation of the text is perhaps not correct (S VIII.15). It seems that the names of a couple are mentioned, Felicissima and dominus Victorinus. The latter name is typical of the Late Roman period, while the first name occurs particularly often among freedmen.[155] A decorated sarcophagus belonging to another deceased couple has also long been missing. The text is composed poetically in the form of alternating passages by Aemilia Maximiana (written in the inverted word order as Maximiana Aemilia) and her husband Flavius Ursinus (S II.9). Their names occur frequently, the cognomina being typical of the Late Roman period.[156]

## IV.3. MUNICIPIUM FLAVIUM ANDAUTONIA

### *IV.3.1. Some short historical and topographical notes*

As could be inferred from the name, Andautonia (present-day Ščitarjevo near Velika Gorica, 12 km south-east of Zagreb) must have been an ancient settlement, of much lesser importance than Siscia, presumably located at a more defendable site in the vicinity, since prehistoric finds are almost entirely lacking at Ščitarjevo. It must have undoubtedly been at least of a limited significance in the prehistoric period, as is testified by its name and location. A prehistoric Andautonia, a centre of the Varciani,[157] was probably less exposed to inundations; however, no such site has so far been identified with certainty. The settlement was a harbour, reloading and raft-station on the right bank of the Sava River (part of the ancient trade route related to the legend of the Argonauts), situated on the road linking Siscia with Poetovio and not on the land route Sirmium–Emona. Its location was strategic enough to have been of logistic significance during all phases of Roman occupation, even during the Republican period, most notably during Octavian's Illyrian war in 35 and 34 BC, when Siscia and Metulum were conquered, as well as later, during the Pannonian war in 13 to 9 BC, and the great Pannonian-Dalmatian rebellion AD 6–9. It remained an important harbour, which is not least indicated by an altar from the second half of the 1st, or from the 2nd century AD, dedicated to the river god Savus by Marcus Iuventius

[148] Alföldy 1969: 254.
[149] Radman-Livaja 2014: 227.
[150] *OPEL* IV: 109.
[151] Solin and Salomies 1994: 182.
[152] *Epigraphik-Datenbank Clauss/Slaby*; see, however, the name Sipandus on one of the *tesserae*: Radman-Livaja 2014: 266, possibly 'Illyrian'.
[153] Alföldy 1969: 247.
[154] *OPEL* IV: 75; *Epigraphik-Datenbank Clauss/Slaby*; cf. Barkóczi 1964: 324 (Severilla); *OPEL* III: 53; cf. Barkóczi 1964: 317 (Marcellianus).
[155] Kajanto 1965: 68; Barkóczi 1964: 328; Lochner von Hüttenbach 1989: 182; Alföldy 1969: 201–202.
[156] Alföldy 1969: s. vv.
[157] Mócsy 1959: 21–24; Šašel 1970: 71–75 (1992: 580–582).

Primigenius with his associates.[158] It was discovered at the Savišće oxbow near Ščitarjevo, where the Roman-period port of Andautonia on a meander of the Sava had been located. Iuventius Primigenius, or his immediate ancestors, most probably originated from northern Italy, probably Aquileia, where the Iuventii are well represented, while they are not so frequent in Pannonia and Noricum.[159]

The name of the settlement is noted in the *Antonine Itinerary* (266. 2) as Dautonia, but the given distances well correspond to Ščitarjevo and it is highly unlikely that two settlements would have existed one opposite the other on the two banks of the Sava.[160] Varciani had been recruited in the Roman army at an early date, as has been confirmed by the tombstone of Titus Flavius Bonio (see *infra*). They are further mentioned on an altar found in Samaria (Sebaste) in Judaea, dedicated to Jupiter by soldiers of a detachment (*vexillatio*) of the Pannonian cohorts, as *cives Varciani*, together with the *cives Sisciani* and *Latobici*.[161] These *vexillarii* most probably took part in one of the two bloody wars caused by the insurrection of the Jews in the first half of the 2nd century AD, either under Trajan (AD 115–117) or under Hadrian (AD 132–135). In the distant land their identity awareness manifested itself differently and more intensively than would have been the case had they stayed in Pannonia.

The location of Andautonia has been confirmed by a milestone from the reign of Maximinus Thrax, found *in situ* in the village of Jelkovec, 30 miles from Siscia, corresponding to 3 miles between Ščitarjevo–Jelkovec, and 27 miles between Ščitarjevo–Sisak.[162] The name of the settlement is Pannonian, pre-Celtic (note, for example, the Pannonian tribe of Andizetes, and the toponym Anderva): settlement at or near the river.[163] The way of life of the indigenous population can also be reflected in the La Tène-derived shapes of pottery and the ritual of burials under *tumuli*, which have been discovered in the entire territory of the town.[164] Part of the town has been excavated and preserved, including several (monumental) buildings, *insulae*-type constructions, sewage system, cemetery, as well as baths and remains of the Late Roman town walls. The archaeological remains are presented as an open-air museum.[165]

Andautonia was most probably a *municipium Flavium*, as is indicated by the voting-tribe *Quirina*, mentioned on a tombstone from the 1st century AD, from Aquincum (Buda; now in the National Museum at Budapest), of a rider of the *ala Tungrorum Frontoniana*, who was by origin from Andautonia, Titus F(lavius) Bonio. Bonio can be regarded as a Celtic name.[166] The monument was commissioned for him by his heir Campanus (his family name is not mentioned), who served in the same ala and who may or may not have also been from Andautonia.[167] The ala was transferred at the beginning of the reign of Vespasian from Dalmatia to Aquincum, and under Titus to Carnuntum. Subsequently it was briefly garrisoned at Campona (the outskirts south of Budapest) and was *c.* 105 relocated to Intercisa.[168] Andautonia was also mentioned on a tombstone from the 3rd century AD of a rider of the *cohors I Thracum*, found at Brigetio, Aurelius Vindex.[169] The cohort (*I Thracum equitata civium Romanorum*) had been transferred between AD 113 and 126 from Germania (Upper or Lower) to Upper Pannonia, where it was stationed at least until 250 at Ad Statuas (Ács-Vaspuszta).[170]

Andautonia may have attained autonomous status thanks to the support of its patron, the provincial governor Lucius Funisulanus Vettonianus, who was honoured by a decree of a city councillors (*d(ecreto) d(ecurionum)*), probably in AD 86 or a few years later, after he had been awarded *dona militaria*, earned in the Dacian Wars (85–88) under Domitian. He governed Dalmatia between AD 79 and 84 and Pannonia between September 84 and September 85; until 88, he was the first governor of Upper Moesia after the division of the province.[171] On the other, reverse, side of the same marble slab, a standing Nemesis is depicted; beneath the relief, the dedication to Nemesis Regina Augusta was engraved by Iulius Victorinus (his *praenomen* being damaged), who had almost certainly been veteran and very probably a city councillor (*decurio*) of Andautonia.[172] The marble monument of Funisulanus Vettonianus must have been reused for the votive monument towards the end of the 2nd century AD or even later; this is not contradicted by the name Victorinus, which was especially characteristic in Pannonia after the Marcomannic Wars.[173] Andautonia has been noted as *municipium* only in the Nemesis dedication.[174] The

[158] *AIJ* 475 = *HD*057335; Šašel Kos 1999: 100 no. 12; Knezović 2010: 188–189 (= *AE* 2010: 1239): carved *Iuentius*.

[159] *Inscr. Aquil.*, index; *OPEL* III: 211.

[160] Šašel 1970: 73 (1992: 581); for the obsolete opinion cf. Graf 1936: 62.

[161] *AE* 1909: 235 = *AE* 1938: 13 = Dobó 1975: 219 = *HD*021921.

[162] *AIJ* 602 = *HD*072139 = *lupa* 22334.

[163] Anreiter 2001: 29–30; it was earlier defined as Illyrian by Mayer 1957: 43; 1959: 36, or Celtic by Holder 1904: I 139; It has even been suggested that the toponym might be Venetic, since indeed personal names with the element *and-* are actually attested in the Venetic regions: Šašel 1970: 72 (1992: 580).

[164] Nemeth-Ehrlich and Kušan Špalj 2003: 113.

[165] Nemeth-Ehrlich and Kušan Špalj 2003.

[166] Meid 2005: 258.

[167] *CIL* III 3679 = *Tit. Aq.* II 616 = Lőrinz 2001: no. 216 = *HD*068590 = *lupa* 2726.

[168] Lőrinz 2001: 26.

[169] *CIL* III 4316 = *HD*039438 = Lőrinz 2001: no. 461.

[170] Lőrinz 2001: 42–43.

[171] *CIL* III 4008 = *HD*030742 = *lupa* 23285; Knezović 2010: 193–194 (= *AE* 2010: 1239). On Funisulanus: *PIR*² F 570; Fitz 1993: 157–158 no. 61; his support was suggested by Šašel 1970: 74 (1992: 581).

[172] *CIL* III 4013 = *HD*030742 = *lupa* 5739; Knezović 2010: 193–194 (= *AE* 2010: 1239). 'Veteran' was abbreviated as *ve()*, while *decurio* has been supplemented.

[173] *OPEL* IV: 168–169; Barkóczi 1964: 328.

[174] The town must have legally functioned as most of the *municipia Flavia*, see a useful survey in Karlović, Milotić and Petrak 2015.

monument also sheds light on the amphitheatre games in the town; topographical investigations have indeed confirmed the existence of an amphitheatre in the southern area of Andautonia.

Most significant for the mid-3rd-century Roman history of Andautonia are two honorific inscriptions on marble slabs from AD 250, erected by the *res publica Andautoniensium* for the emperor Decius and his wife Herennia Etruscilla.[175] The emperor, born at Budalia near Sirmium (Martinci near Sremska Mitrovica), may well have stopped at Andautonia on his way through Illyricum to the Danube, to fight the Goths who had crossed the river to invade Moesia and Thrace, or was perhaps expected to stay in the town.

### *IV.3.2. Population*

One of the first known colonists may have been Sextus Valerius Secundus, a veteran of the 13th Twin legion (*XIII Gemina*; A I.1). The Valerii are frequently documented over the broad area: Valerius Lucilianus, a praetorian, has been attested at Lobor, that may or may not have belonged to the *ager* of Andautonia; on this stele also Marcus Cocceius Superianus, centurion of the 10th Twin Legion is mentioned, as well as their mother Septimia Lucilla (A(ptNW) I.1). Another Cocceius, whose cognomen is damaged, is recorded next to the mother Lucilla; he was *signifer* in the same legion as Cocceius Superianus and very likely Superianus' father, while Valerius Lucilianus must have been Lucilla's son from her previous marriage or relationship.

A few other family names documented on the monuments from Andautonia and its territory, included in the catalogue, are Iulii, Aconii, Flavii, Ulpii, Egnatulei, Caesernii, and Marcii. Gaius Iulius Adietumarus, who was a hundred years old, was a freedman of Maximus (A[t] I.1); both may have belonged to the indigenous population, as Maximus could have been an assonance name; in any case, it was quite popular also among local inhabitants. Adietumarus is a Celtic name ('great in zeal'),[176] while the name of his wife, Sporilla, has only been attested once in Noricum; their tombstone could well have been from the 1st century AD. Perhaps Adietumarus' master Gaius Iulius Maximus came from northern Italy to settle at Andautonia, possibly from Emona, where Maximi were particularly frequent.[177] Another Iulius was Gaius Iulius Paternus, who had a funerary stele commissioned for his wife Aconia Salvia; the stele may be dated to the end of the 1st, or to the 2nd century AD (A[t] I.2). Aconius is a rare Latin *gentilicium*, only sporadically attested outside Rome and Italy.[178] However, it is interesting that it occurs three times at Siscia, also on two *tesserae*,[179] hence it may be suggested that Aconia Salvia was a local woman, while her ancestors had perhaps come from northern Italy to Pannonia.

An interesting case is that of Titus Flavius Ateboduus, a 75-year-old veteran from the *cohors II Varcianorum equitata*, who had also been adjutant of the cohort's prefect. The cohort was stationed in Lower Germania. Ateboduus is known from a stele from the early 2nd century AD (A[t] I.3). His father was called Aterix; clearly the veteran belonged to an indigenous Varciani family. Having been recruited under the Flavian emperors, he returned to his hometown. His Celtic name has been attested three times in Noricum, but not elsewhere.[180] His father's name, Aterix, is also Celtic and has only been documented at Mursa to date.[181] On the other hand, the Latin name of his wife, Crispina, occurs very frequently and is also documented six times on the *tesserae* from Siscia,[182] indicating that it was favoured among the local population. Ateboduus' five-year-old grandson Flavius Augurinus, also mentioned on the stele, bears a Latin name, too: if he alone appeared on the tombstone, nothing would betray his indigenous origin.

Ulpii and Aelii belonged to the second phase of Romanization and urbanisation in Pannonia. The stele of Marcus Ulpius Geminus was found in the territory of Andautonia (A[t] I.4); his name is Roman, both his imperial *gentilicium* and his cognomen are Latin, but the name of his father, Savus, indicates the local origin of his family. It does occur occasionally that parents gave their children names after river divinities.[183] Geminus' wife was Volusia, a frequent *gentilicium* in Italy, but rare elsewhere.[184] Publius Aelius Verus is known from a dedication to Hercules.[185]

The Caesernii were an illustrious family at Aquileia, from where they migrated at an early date to Emona, which became their secondary centre. From Emona they spread partly to Noricum, but most of all to Pannonia.[186] The stele with the depiction of the family of Valens and Melania, slaves of Caesernius Avitus, and their two children, Valentina, twenty years old, and Donicus, two years old, was discovered in the village of Donji Čehi in the territory of Andautonia (A[t] I.6). As is indicated by the name Avitus, an assonance name in the Celtic speaking provinces and very popular, their patron and his family was most probably of a

[175] *CIL* III 4010 = *AIJ* 477 = *HD071779*; *CIL* III 4011 = *AIJ* 478 = *HD071780*.
[176] Meid 2005: 94; Delamarre 2007: 12.
[177] Šašel Kos 2017a.
[178] Alföldy 1969: 54; *OPEL* I: 18 and 20.
[179] Radman-Livaja 2014: 257.
[180] Delamarre 2007: 28; Lochner von Hüttenbach 1989: 19; Kakoschke 2012: 267–268.
[181] *ILJug* 292: *Urbana Aterigis liberta*; *OPEL* I: 84; Delamarre 2007: 29.
[182] Mócsy 1959: 171; *OPEL* II: 85; *tesserae*: Radman-Livaja 2014: 192.
[183] Šašel Kos 2017c: 445–446.
[184] Alföldy 1969: 139; *OPEL* IV: 183.
[185] *CIL* III 4007 = *HD057296* = *EDCS-28800764*.
[186] Šašel 1960 and 1981 (1992); Zaccaria 2006.

local background.[187] Donicus is a rarely attested Celtic name.[188]

The Latin *gentilicium* Marcius occurs on a stele from Lekneno with a very damaged inscription (A[t] I.8). The Latin family name, possibly a *pseudo-gentilicium*, derived from the cognomen Marcus, is frequent, particularly in Italy and Pannonia,[189] and is also attested on a tombstone from Siscia in the present Catalogue (S VIII.4). One Vepon(ius/a) Panun() is known from an altar dedicated to Silvanus, carved with letters that very much resemble the cursive script. The *gentilicium* was published as Veron(ius?) in *CIL* (following Brunšmid), in *AIJ*, and *EDCS*, but on the photograph *P* is quite clear, therefore I see no reason not to accept the reading Veponius. This is a *pseudo-gentilicium*, documented several times in Noricum and once in Dacia, derived from the Celtic name Vepo, known from Celeia and Neviodunum.[190]

Nothing is known of the life at Andautonia in the Late Roman period nor during the migration period. However, since the not far-distant Siscia still functioned as a town in the 6th century AD and was not destroyed even later, as is indicated by the continuity of the name, it may be presumed that also at Andautonia some population, much reduced and mixed with various migrants, lived on into Late Antiquity.

### IV.4. AQUAE BALISSAE/MUNICIPIUM IASORUM

Aquae Balissae were an important native settlement in Pannonia Superior, in the territory of the Iasi, which developed by the thermal springs at present-day Daruvar (the region of Slavonia), on the road Siscia–Sirmium, as is indicated in the Antonine Itinerary (265. 7).[191] The settlement of the Iasi is located in the Daruvar basin on the hill of Stari Slavik, where an early Roman military camp was constructed.[192] Aquae Balissae were noted as *origo* of an *eques singularis* from Rome: *Ulpius Cocceius ... ex Pan(nonia) Sup(eriore) natus ad aquas Balizas pago Iovista vic(o) Coc[co]netibus ....*[193] This inscription is a most interesting testimony of the administrative organization of the town's territory, divided into *pagi* and *vici*; *Coc[co]netibus* is undoubtedly identical with the *mutatio Cocconis*, mentioned in the *Itinerarium Burdigalense* (562. 4) or *Cucconis* in Anonymus from Ravenna (4. 19).[194] During the reign of Hadrian the town presumably received municipal status (*IIIIviri* are mainly known in the *municipia* founded by Hadrian),[195] its official name having been *Municipium Iasorum*. In a similar way, another of the southern Upper Pannonian towns was named *Municipium Latobicorum*, generally known by its epichoric name Neviodunum (Drnovo near Krško). Municipium *Iasorum* as the name of the town has not been known before the discovery of the funerary altar for Publius Aelius Aelianus, secretary in charge of public records, member of the town council, and *quattuorvir* of *municipium Iasorum* (AB V.1). *Res publica Iasorum*, however, was mentioned earlier: most interesting is an honorific monument erected in honour of Divus Commodus, perhaps soon after AD 195, when Septimius Severus had the Senate deify the assassinated emperor, who had suffered *damnatio memoriae*.[196] In 202, a base for a statue of the consort of Caracalla, Fulvia Plautilla was set up by the *res publica Iasorum* at Aquae Balissae;[197] Plautilla was also honoured with a statue at Siscia,[198] indicating close connections of both towns with the Severan dynasty. Gordian III and his wife Furia Sabinia Tranquillina were also honoured by the *res publica Iasorum*.[199]

#### *IV.4.1. Aquae Iasae – a note*

The name *Municipium Iasorum* would indicate that the entire *civitas* of the Iasi would have been included in the municipal territory, but Aquae Iasae (present-day Varaždinske Toplice) did not belong to the *ager* of *Municipium Iasorum*, although the settlement had been inhabited by the Iasi in pre-Roman times and later. Even in the 2nd or 3rd centuries AD, in addition to Nymphae Iasae *Fortuna Iasoniana* was also worshipped at the spa.[200] Aquae Iasae were a settlement with a prehistoric tradition, which developed at the site of the curative sulphur thermal springs. A base for the statue of Quintus Gavius Fronto was discovered in 2011, the most high-ranking centurion (*primus pilus*) of the 13th Twin legion (*XIII Gemina*), which had been stationed at Poetovio since around 45 AD. The base is dated to the second half of the 1st century AD and was dedicated to him by his freedman Bachylus. Fronto acted as a prefect of the communities (*civitates*) of the Scordisci, Breuci, and Iasi.[201] There is hardly any doubt that this significant spa site belonged to the *ager* of

[187] Lochner von Hüttenbach 1989: 26–27; Meid 2005: 254–255; Delamarre 2007: 34; Kakoschke 2012: 288–289; Radman-Livaja 2014: 171.

[188] *OPEL* II: 107; Alföldy 1969: 191; Delamarre 2007: 88.

[189] Alföldy 1969: 97–98; *OPEL* III: 56–57; Kakoschke 2012: 130–131.

[190] *CIL* III 14354,24 = *AIJ* 476 = *lupa* 11600 = *HD*032886 = *EDCS*-30100718; *OPEL* IV: 155; Kakoschke 2012: 208. In *lupa* and *EDH* the letter *P* is correctly noted, but emended in *R*.

[191] Mócsy 1968; Schejbal 2004.

[192] Schejbal 2004: 106, 114.

[193] *CIL* VI 3297 = *EDR*152501 = Dobó 1975: 38–39 no. 115; M. P. Speidel 1994: 355 no. 657; cf. Tarpin 2002: 326.

[194] Mayer 1957: 194; see Schejbal 2004: 102; Iovista may or may not be one of the *Ioviae*.

[195] Schejbal 2004: 108; Migotti 2017: 86–87, 117; Grbić 2014: 180–181; any later date can likely be dismissed, cf. Mócsy 1959, 27–28

[196] *CIL* III 4000 = *HD*074407.

[197] *AE* 2007: 1144 = *HD*065289 = *lupa* 26272.

[198] *CIL* III 3968 = 10850 = *AIJ* 560 = Buzov 2008: 476–478, 488 (= *AE* 2008: 1079) = *HD*066898.

[199] *AIJ* 586 = Schejbal 2004: 108 fig. 2 on p. 105 = *HD*028185 = *lupa* 15443; *AIJ* 587 = Schejbal 2004: 108, fig. 3 on p. 106 = *HD*028188 = *lupa* 15444.

[200] Kušan Špalj 2014: 161 no. 76 = *HD*075021.

[201] Kušan Špalj 2014: 152 no. 64 = *HD*075016.

Poetovio (Ptuj in Slovenia),[202] as indicated most of all by a marble slab dedicated to the Nymphs by the *res publica Poetovionensis*: the city must have restored or donated, just before the Marcomannic Wars (between AD 162 and 166), some large building to the spa community, under the supervision of the procurator of Upper Pannonia Titus Geminius Rufinus, who acted under the order of the governor Lucius Dasumius Tullius Tuscus.[203] Most significant is also a statue base, dedicated in the 1st century AD to Iuno Regina and Fortuna by the inhabitants of Poeovio (*Poet(ovionenses) publ(ice))*.[204]

Further light on the close links with Poetovio is shed by two altars erected by two *Augustales* (priests responsible for organizing imperial cult) from *colonia Ulpia Poetovionensis*. The first, found in 2011, was dedicated to the Nymphs by Gaius Valerius Posphorus; the second, discovered a year later, was erected to Isis and Serapis by Gaius Valerius Priscus.[205] It is noteworthy that both *Augustales* belonged to the most illustrious family at Poetovio, the Valerii; the two priests or their immediate ancestors must have been freedmen, as *Augustales* usually were, and as is also confirmed by the Greek name of the first mentioned *Augustalis*. Under Hadrian, Poetovio became the seat of the central administration of the Illyrian customs, *publicum portorii Illyrici*;[206] it is therefore not surprising to find customs employees at Aquae Iasae.[207] Two dedications were erected to the local Nymphs by high officers of the Legion *XIII Gemina*, stationed at Poetovio between *c.* 45 AD and the accession of Trajan.[208] Outstanding is also a dedication to Pollux for the health of Lucius Fabius Cilo, at that time (197–202) governor of Upper Pannonia; the altar was erected by his freedman Menander.[209] Some other individuals of the highest rank stayed at Aquae Iasae for a shorter or longer time, perhaps due to some illness, such as Lucius Alfenus Avitianus, the governor of Arabia and Lower Pannonia. In AD 213, he dedicated a marble slab with one of the most beautiful reliefs at the spa (depicting Hygia, Telesphorus, and Asclepius), for the health and victories of Caracalla, who may have himself stopped at Aquae Iasae when visiting Pannonia, seeking recovery from illness.[210] The spa-site, destroyed by fire, was restored under the Emperor Constantine.[211] Baths and basilica have been excavated and preserved, as well as the spring pool, part of the forum and sanctuaries (*capitolium*).[212]

A recently discovered marble monument from the 3rd century AD, found in the secondary use during the excavations of the northern wall of the spring pool (2011 and 2012), deserves attention due to its specific character. The relief of a couple is in a deepened rectangular upper field, below which the inscription was carved. The couple is depicted in a posture that usually appears on funerary *stelai*: the wife is holding her left hand on her husband's shoulder, while with her right hand she points towards her husband's right arm. He is holding a scroll in his left hand, pointing to it with his right hand. Their names are mentioned in the inscription field below: Aurelius Aquilinus and Aurelia Maximilla. Aquilinus' name is followed by two abbreviations: *DEC* and *DVM*.[213] While the first one could be explained in a usual way as *dec(urio)*, the second is not clear, possibly it would mean *du(u)m(vir)*? At first glance the monument should be interpreted as a funerary stele, however, the inscription only contains the names of the couple and no sepulchral formula. Moreover, the plaque is rather thin (6 cm), rather corresponding to a donor's label; it must have either been immured or firmly set into a pedestal. The most decisive argument speaking against the monument being a tombstone is the fact that no funerary monuments have so far been unearthed at or near the spa.

### *IV.4.2. The population of Aquae Balissae/Municipium Iasorum*

Less than 30 Roman inscriptions have come to light at Daruvar and its environs, therefore it is not possible to draw any far-reaching conclusions regarding the population of the (pre-)Roman spa town, where life pulsated at least for five centuries.[214] The Iasi were recruited in the Roman army at an early date. In a military diploma from AD 85 (5 May), under the reign of Vespasian, one Fronto Sceni f(ilius) Iasus was dismissed from military service in the 1st Lusitanian cohort (*cohors I Lusitanorum*, perhaps *equitata*),[215] which had been stationed in Pannonia as early as about 50 AD.[216] An officer in charge of the Iasi (*princeps Iasorum*), Titus Flavius Serenus, is known from a military diploma, issued under Vespasian on 5 April 71 AD for the soldiers serving in the Ravenna fleet, for their bravery in the civil war. They had been dismissed before their time (*ante emerita stipendia*), to settle in Pannonia. Serenus was probably given citizenship slightly earlier.[217]

[202] See recently Kušan Špalj 2014: 54.
[203] *AIJ* 461 = *ILJug* 357 = *HD*069301 = *lupa* 5346 = Lučić 2014: 220–221 no. 12; Fitz 1993: 487–489 no. 289 (Tuscus); 727 no. 404 (Rufinus).
[204] *ILJug* 1168 = *HD*034992 = *lupa* 5324 = Lučić 2014: 230 no. 23.
[205] Kušan Špalj 2014: 164 no. 79 = *HD*075024; 159 no. 73 = *HD*074972.
[206] Vittinghoff 1953: 358–361.
[207] Kušan Špalj 2014: 165 no. 80; Lučić 2014: 216–217 no. 6.
[208] *CIL* III 4118 = *AIJ* 463 = *HD*069304 = *lupa* 5338 = Lučić 2014: 215 no. 5; Fitz 1993: 187 no. 78 (M. Fabius Fabullus); *CIL* III 10893 = *AIJ* 462 = *HD*069303 = *lupa* 5330 = Lučić 2014: 218 no. 9; Fitz 1993: 186–187 no. 77 (M. Rutilius Lupus).
[209] *AIJ* 466 = *HD*053171 = *lupa* 5342 = Lučić 2014: 232–233 no. 25; Fitz 1993: 506–509 no. 299.
[210] Kušan Špalj 2014: 70–74, 96–99, 156 no. 69: 156 = *HD*075019; Okoń 2017: 27; on the cults at Aquae Iasae see Kušan Špalj 2017.
[211] *ILS* 704 = *AIJ* 469 = *HD*064415 = *lupa* 5339 = Lučić 2014: 236–237 no. 30.
[212] Vikić-Belančić 1978: 165–170; Kušan Špalj 2014.
[213] Kušan Špalj 2014: 80–81, 188 no. 146 = *HD*075030.
[214] On the population of the town, see a detailed analysis in Migotti 2017c.
[215] *CIL* XVI 31 = *EDCS*-12300235 = Grbić 2014: 182 no. 84.
[216] Lőrinz 2001: 37–38.
[217] *RMD* 205 = *AE* 2004: 89 = Grbić 2014: 252 no. 170.

### *IV.4.3. Legionaries and auxiliary soldiers*

A stele from the Flavian period, which Marcus Nunnidius Successus, veteran of the 32nd Cohort of Roman citizens (*XXXII voluntariorum civium Romanorum*) had erected for his family when he was 90, was found in the village of Orešac in the territory of the town (AB[t] I.1). His wife Petronia Procula was much younger, she died at the age of 60. While the Latin *gentilicium* Nunnidius has not been documented to date in northern Italy and western provinces, occurring only on stamps from Rome and Ostia, in the context of potters' workshops on estates,[218] the Petronii were a frequent family name, particularly so in northern Italy, Pannonia, and Dalmatia.[219] It is interesting to note that the cognomen of Successus' wife, Procula, a rather common name in the Roman Empire, appears on 20 *tesserae* from Siscia: it was among the most popular names in the city and most probably in other towns of southern Pannonia as well.[220] Two children of the couple, Nunnidia Vitalis and Marcus Nunnidius Saturninus, died young, the daughter was 19 years old, the son 7. They also bore very popular names, which is indicated by the fact that both appear on several *tesserae* from Siscia.[221] It should be noted that Vitalis may have either been a masculine or a feminine name. The Cohort *XXXII voluntariorum civium Romanorum*, in which Successus had served, should very likely be regarded as one of the earliest auxiliary troops attested in Pannonia, with its camp at Siscia. This is mainly inferred from the funerary monument of the physician of this cohort, Marcus Mucius Hegetor (S I.2). It has been suggested that the cohort stayed at Siscia until Vespasian, when it would have been transferred to Upper Germania.[222]

A stele came to light at the site of a *villa rustica* in the village of Brusnik, 17 km south of Aquae Balissae. The monument was erected to Aurelius Naso, soldier of the 4th Flavian Antoninian legion (*IV Flavia Antoniniana*), and to mother Prisca by Proclus, Proclianus, and Provincialis, sons of Maximianus (AB[t] I.5). Prisca was the daughter of Tato, an indigenous, 'Illyrian'-Pannonian name;[223] it may well be that she and her first (?) husband Maximianus, father of the three sons, were a local family. The legion, in which Naso served, was stationed at Singidunum in Upper Moesia; he must have died at some time between 198–222, when the legion bore the honorific title *Antoniniana*.[224]

Approximately from the same time is a stele found in the village of Kusonje, 15 km south of Aquae Balissae, for a soldier of the Cohort of the 500 Mauri (*D Maurorum*), Cassius Sextus, erected by his three brothers Cassii Exsoratus, Surus, and Quintus, who were also his heirs (AB[t] I.6). It is noteworthy that none of the brothers had a *praenomen*, indicating a later date, while on the other hand both Sextus and Quintus were cognomina or personal names derived from *praenomina*, and both very popular in Pannonia, even occurring on several *tessearae* from Siscia.[225] Since also the name Surus, regardless of how it should be explained: as a Celtic, 'Illyrian', Thracian, or Semitic name, was very frequent in Pannonia and most popular at Siscia, appearing on 28 lead tags,[226] it may be suggested that family was almost certainly from Pannonia, where Sextus had probably been recruited into the cohort. This is further confirmed by the name of the first mentioned brother, Exsoratus, which – when not spelled Exoratus – may be considered a Celtic name.[227] In AD 171, during the Marcomannic Wars, this auxiliary troop had been transferred, together with the cohort *milliaria Maurorum equitata*, to Upper Pannonia, where both fought in the first expedition against the Germans.[228] At the end of the wars both cohorts were stationed at Alta Ripa in Lower Pannonia.[229]

A stele from the end of the 2nd, or the 3rd century AD, of another soldier, a praetorian, was discovered in the village of Veliki Bastaji, 10 km north-east of Aquae Balissae. Adatilia A[n?]nia commissioned the tombstone for her two sons, Flavius Val[-], soldier of an unknown praetorian cohort, and to Valerius Dig[-], who may have been a veteran (AB[t] I.7). *VF* has only been preserved, hence the meaning is uncertain. If a veteran, their mother must have been quite old, which, however, would not have been impossible. The right part of the stele is broken off, hence the reading is difficult and the names cannot be restored with certainty. Obviously, however, each son must have had a different father. Both, Flavii and Valerii are very well attested in Pannonia and elsewhere, the latter *gentilicium* was indeed the most wide-spread non-imperial family name generally, and Valerii were one of the most significant known upper-class families at Siscia, whose freedmen were prominent middle-class members, acting as *Augustales*.[230] The *gentilicium* Adatilia, however, has not been attested elsewhere,[231] while the name An[n?]ia could be explained as 'Illyrian' in the broad sense of the word, or even Celtic.[232]

---

218 *OPEL* III: 107; cf. Mócsy 1959: 157; *Epigraphik-Datenbank Clauss/Slaby*.

219 Alföldy 1969: 108–109; *OPEL* III: 135; cf. Mócsy 1959: 157–158; Barkóczi 1964: 302.

220 Radman-Livaja 2014: 249, with literature.

221 Radman-Livaja 2014: 259 and 284, with literature.

222 Lőrinz 2001: 44, with earlier literature; Radman-Livaja 2004: 18–19.

223 Radman-Livaja and Ivezić 2012: 142; cf. Falileyev 2014: 112.

224 See recently Kovács 2017: 263; the exact years are still disputed.

225 Radman-Livaja 2014: 263–264 and 251, with earlier literature; *OPEL*, s. vv.

226 Radman-Livaja 2014: 269, with earlier literature.

227 Delamarre 2007: 100.

228 *CIL* VIII 12066 = *EDCS*-24400196: *expeditio Germanica prima*; Lőrinz 2001: 272 no. 391.

229 Lőrinz 2001: 38–39; cf. Spaul 2000: 468–469.

230 Šašel Kos 2018: 273–274, 297.

231 The name is not listed in Solin and Salomies 1994, and neither in *OPEL* nor in *Epigraphik-Datenbank Clauss/Slaby*.

232 Alföldy 1969: 150–151; Delamarre 2007: 23.

Two soldiers from the legion *I Adiutrix* must have been sent to Aquae Balissae on some special mission, Ser(-) Felix and Valerius Proculus;[233] as noted above, the legion was stationed at Brigetio. At some time in the 3rd century (the soldiers have no *praenomina*) they erected an altar to Nemesis. On the basis of this dedication Berislav Schejbal did not exclude the possibility that a small amphitheatre had existed at Aquae Balissae.[234] In the Severan period, a *beneficiarius consularis* Seius Ingenuus has been attested in the town, known from an altar that he had dedicated in AD 208 to Silvanus Domesticus and Silvana.[235] Two more dedications to Silvanus came to light at Aquae Balissae, as well as an altar with the depiction of Silvanus between two Silvanae, shedding light on the popularity of the god.[236] The worship of Jupiter Dolichenus has also been well documented in the town. In addition to the aforementioned altar dedicated to the god for the health of Lucius Aelius Ver(), an officer from the 1st Legion, two more altars have been discovered at Aquae Balissae, one erected for the well-being of the emperors Septimius Severus and Caracalla by Quintus Carmeus Iulianus, centurion of the Legion *VII Gemina* (stationed in Hispania Tarraconensis, at León), by his wife Iulia Atticilla and their sons Carmaeus Secundus and Atticillianus.[237] Their Latin *gentilicium* is rather rarely attested.[238] The third altar was dedicated to the god by Secundius Restutus, centurion of the Legion *X Gemina* from Vindobona.[239] Lucius Aelius Ver() may have been a high officer in the 1st Legion; an altar was dedicated to Jupiter Dolichenus for his health (*pro salute*).[240] *Municipium Iasorum* may not have been his native town. The most likely 1st Legion in which he had served would have been *I Adiutrix*, stationed at Brigetio in Pannonia.

### *IV.4.4. Civilian residents*

Although more Aelii should be expected in the town, which became a *municipium* under Hadrian, only two have so far been documented. In addition to the just discussed officer of the 1st Legion, the aforementioned Publius Aelius Aelianus must be noted. He was a member of the city council (*decurio*) and was active as an official in charge of public records (*scriba*), later also performing the function of the highest town magistrate (*quattuorvir*). He is known from a funerary altar, erected to him by his son Aelius Laelianus (AB V.1). Neither Aelianus nor Laelianus were rare cognomina, although Laelianus has been less frequently attested.[241]

A stele from the second half of the 2nd or from the 3rd century AD was found in the village of Donji Podgradci, which perhaps belonged to the territory of Aquae Balissae. It was erected by Potamilia Proba to her brother Ursio, her sister Sirmia and her mother Oginia (AB[t] I.3). The first letter in the name Oginia, *O*, is not certain, but seems a plausible supplement. If correctly restored, the name would be Celtic, and indeed a typical Gaulish name.[242] Ursio appears on two *tesserae* at Siscia, so it may be due to chance that the name has rarely been attested in Pannonia.[243] Sirmia and Potamilia do not occur elsewhere, although the feminine name Potamilla occurs a few times in Italy.[244] The family may have belonged to the Romanized indigenous population, living rather far from any urban centre.

In the village of Vetovo, 48 km north-east of Daruvar, a stele was discovered among Roman ruins in the 19th century, but which has long been missing (AB[t] I.9). Busts of a man and a woman, probably a couple, were depicted in a niche above the epitaph, in which the name Gr(a)ecus is mentioned, undoubtedly that of the father, followed by a few seemingly meaningless letters and Max(), most likely the name of the mother, perhaps Maxima, a generally popular name among indigenous population,[245] or Maximilla. Graecus is not a particularly frequent name, except in Italy, and must have also been a name characteristic of slaves, since it was often borne by freedmen.[246] Their son was called Mumapius, a name elsewhere not documented to date. However, the name cannot be regarded as certain, since it may not have been correctly transcribed.

### *IV.4.5. Epilogue: Christian monuments*

Little is known about the spread of Christianity in the town and still less about Late Antiquity. However, in the village of Veliki Bastaji, 10 km north-east of Daruvar, a sarcophagus of Pohorje marble should be mentioned, adorned with early Christian reliefs (AB[t] II.1), as well as an inscription in hexameters, carved in elegant Late Roman rustic capitals, found in a funerary context. The translation of the hymn should be regarded as not entirely certain in places, but clearly the text is concerned with a dogma of original sin, sacrifice of Jesus Christ, son of God, for mankind, and redemption (AB[t] VII.1).

[233] *AIJ* 585 = *HD*072133 = *lupa* 15442; Selem 1980: 266 no. 4; Schejbal 2004: 111 fig. 14.

[234] Schejbal 2004: 111.

[235] *CBI* 277 = *HD*004711 = *lupa* 26271; Schejbal 2004: 111 fig. 8 on p. 109; cf. Rendić-Miočević 1980: 114–115, 123 (1989: 508–509, 514).

[236] Schejbal 2004: 110–111.

[237] *CIL* III 3998 = *AIJ* 583 = *HD*072131; Selem 1980: 227 no. 3; Schejbal 2004: 111 fig. 12 on p. 110.

[238] Solin and Salomies 1994: 47; *OPEL* II: 37.

[239] *CIL* III 3999 = *AIJ* 584 = *HD*072132 = *lupa* 15441; Selem 1980: 227 no. 4; Schejbal 2004: 111 fig. 13 on p. 110.

[240] Selem 1980: 228 no. 5 = *HD*074650 = *EDCS*-29200177 (*I(ovi) O(ptimo) M(aximo) Dol(icheno) pro salut(e) L(uci) Ael(i) / Ver() Pan(noni?) co(n)s(ularis) I leg(ionis)*); most probably he was not a *benficiarius consularis*, as suggested in *EDH*.

[241] Mócsy 1959: 177; *OPEL* III: 16.

[242] Meid 2005: 238; Falileyev 2014: 112.

[243] Radman-Livaja 2014: 277; *OPEL* IV: 187; cf. Mócsy 1959: 194.

[244] *OPEL* IV: 85; Solin and Salomies 1994: 148; *OPEL* III: 155; *Epigraphik-Datenbank Clauss/Slaby*).

[245] Šašel Kos 2017a: 441.

[246] Alföldy 1969: 212; *OPEL* II: 169–170; Kakoschke 2012: 441.

# Chapter V
# The social impact of Roman soldiers in Siscia, Andautonia and Aquae Balissae

Ivan Radman-Livaja

## V.1. INTRODUCTION

The title of this chapter could hardly be described as short but it remains nonetheless rather vague and does not convey precisely what was in this author's mind when deciding to tackle this issue. Pannonia, being a frontier province, or rather two frontier provinces after it had been divided into *Pannonia Superior* and *Pannonia Inferior* around AD 102, was certainly not lacking garrisons. By the late 1st century AD and afterwards, soldiers must have been a fairly common sight to people inhabiting the border area that followed the long stretch of the Danube from Vindobona to Taurunum. It is approximately at that moment that the last large garrison within the province, the one in Poetovio, left. From that moment on, all the permanent military bases were concentrated in the *limes* area, with legionary garrisons quartered in Vindobona, Carnuntum, Brigetio and Aquincum, remaining there until Late Antiquity. Permanent auxiliary garrisons were, just as well, all stationed along the Danube during the 2nd and most of the 3rd centuries AD. On certain occasions, in case of war or imminent threat, reinforcements were temporarily rushed to certain urban centres in the interior, but most of the time the military presence within *Pannonia Superior* and *Pannonia Inferior*, away from the Danube, was far from being noticeable by civilians, and was likely limited to *beneficiarii* and soldiers in transit. Presumably, an inhabitant of Andautonia, a *municipium* far away from the *limes*, seldom interacted with army personnel, especially during longer peaceful periods, such as the reigns of Hadrian or Antoninus Pius. In all probability, it would not be too far-fetched to assume that during longer periods of the 1st and 2nd centuries AD some civilians living in south-western Pannonia almost never had occasion to meet an active duty soldier. While the army likely did not have much of an influence on everyday life in the serene Pannonian urban and rural communities far away from the *limes*, its social impact could still be felt in several ways. Without explicit references in written sources and in the epigraphic record, it is not easy to assess the impact it could have had on local economies, although one might reasonably presume that army units met many, if not most, of their everyday needs in the local or regional markets, be it food, fodder, clothing or equipment.[1] Besides standard military necessities, soldiers had their own requirements, although one might expect that they met their personal needs, at least the basic ones, in the very vicinity of their garrisons, at first in *canabae* around their forts and later in the settlements which flourished in all the garrison places in Pannonia. Thus, one may hardly claim that the army directly influenced economic growth in the interior of the province, although it is a fair presumption. For instance, it is plausible as far as some industrial activities are concerned, for instance metallurgy in Siscia,[2] although one should keep in mind that production in those workshops must have started while there was still a legionary garrison in Siscia. Siscia, therefore, may not truly be described as a provincial industrial centre that did not owe its economic development to army presence. With metallurgical production facilities firmly established in Siscia by the mid 1st century AD, one could expect Pannonian garrisons to continue acquiring their metal implements and equipment from a city which was, after all, not too distant. Lacking reliable data, such economic aspects will not be explored further here and the discussion can turn instead to the impact the Roman army in general and soldiers in particular had – or could have had – on Pannonian communities situated further away from the *limes*, in places without permanent garrisons (most of the time, in any case), i.e. in areas where the military presence was certainly not felt as much as on the Danube. This being a vast topic, only three communities will be focused on, in *Pannonia Superior*, one colony and two *municipia*, each with a more or less relevant epigraphic record. It was not a random choice, since monuments from Siscia, Andautonia and Aquae Balissae were extensively studied within the framework of the current project and were thus obvious candidates for this analysis. Those three case studies, despite sharing certain similar patterns, present nonetheless enough particularities enabling us to provide a nuanced picture of the soldiers' social impact in the south of Upper Pannonia. It is worth emphasizing the soldiers' impact because the Roman army as an institution

[1] See Elton 1996: 81–83; Roth 1999: 165–169; Whittaker 2004: 88–110; Kehne 2007: 325–329; Dross-Krüpe 2012: 13–18; Liu 2012: 19–28.
[2] Koščević 2013: 4–9.

appears to have been far less involved than individual soldiers, either active duty military personnel or veterans.

When discussing the social issues in the Roman world, one necessarily has to rely on the epigraphic record. While the sources, i.e. the writings of Roman authors, remain a useful resource as far as the higher strata of the society are concerned, the lower one goes on the social ladder, the less data there is. Common people and lower ranked soldiery were hardly in the focus of Roman writers and when one wants to research different aspects of everyday life in Roman provincial communities, ancient sources will not provide much valuable data after all. Fortunately, in most provinces, especially in urban centres, the epigraphic record is usually far from scant and provides quite enough useful data about social life. While Pannonia does not benefit from the papyrological record, epigraphic data is sufficiently rich for assessing the importance and impact of certain social groups, the military certainly being one of them. Funerary inscriptions, being in a way a *curriculum vitae* of the deceased, are particularly enlightening in this regard, offering us important information about the soldiers' and veterans' origins, their age, service record, family ties, etc., but they are definitely not the only helpful source of information about them. One may discover valuable evidence by checking the votive inscriptions erected by soldiers, the military diplomas and even the graffiti: any study of the Roman military man's social impact in a given community has to make good use of such data as well. Thus, besides relying on the funerary inscriptions from the Catalogues, all the other available epigraphic evidence was used.

## V.2. ANDAUTONIA

J. Fitz hypothesised, with rather sound arguments, that an auxiliary garrison could have been stationed in Andautonia during the rule of the Julio-Claudian dynasty, or at least during Augustan and Tiberian times, but this claim has not been corroborated by archaeological evidence.[3] While one may only conjecture about a military garrison in Andautonia for the time being, this *municipium*, whose territory covered what used to be the tribal area of the *Varciani*,[4] has nevertheless yielded several epigraphic finds related to the Roman military. None could be associated to the presence of an auxiliary unit or a legionary outfit, but they, nonetheless, illustrate links existing between this community and the Roman army. Almost all of them belong to the same category of monuments, i.e. they are funerary monuments. One stele (A I.1.), whose discovery in Ščitarjevo was recorded in 1958 and unfortunately lost since, was erected by his son to a deceased veteran of the *XIII Gemina*, a certain *Sextus Valerius Secundus*, from the Quirina tribe, a native of Mediolanum.[5] The stele could be dated to the early 2nd century AD at the latest – perhaps even several decades earlier, but certainly within the second half of the 1st century AD – and we may presume that *Sextus Valerius Secundus* served in the *XIII Gemina* while this legion was stationed in Poetovio, from the mid 1st century AD till the very beginning of the 2nd century AD, i.e. from Claudius' reign to the time of Trajan.[6] The deceased legionary veteran had an extremely common *gentilicium*[7] and the same can be said for his cognomen.[8] Fortunately, his *origo* is clearly stated and we know that he was from northern Italy, a traditional recruiting ground for legions in Illyricum during the 1st century AD.[9] Since his heir was also his son, he must have founded a family which resided in Andautonia or nearby.

A fragmentary inscription (A[t] VIII.1), built in the parish house in the village of Čučerje, is what remains of a funerary monument belonging to a *primus pilus*[10] who served, among other units, in the *IV Flavia*.[11] Due to the bad state of preservation, his name is mostly missing (his cognomen might have been *Severus* or *Verus*) and we do not know much about his career, except for his rank. It would seem that the inscription mentions his wife *Proculina* as well, who died at the age of fifty. Considering her age, we may reasonably suppose that her husband died as a veteran and not as an active duty soldier. Since she had a rather widespread name, one may only conjecture about her origin, but it has to be pointed out that the highest number of occurrences was actually recorded in Pannonia, although most of them should be dated to a later period, i.e. from the end of the 2nd century AD

---

[3] Fitz 2003: 50; Šašel Kos 2010: 220.

[4] Mócsy 1959: 22–24; Mócsy 1974: 136; Fitz 1980b: 143; Nemeth-Erlich and Kušan Špalj 2003: 107–112; Šašel Kos 1997: 36; Domić Kunić 2006: 80–81; Radman-Livaja and Ivezić 2012: 139 ; Grbić 2014: 272–274; Kovács 2014: 8, 12, 14, 28, 75, 96, 102.

[5] *ILJug* 1133; Reali 2000: 661, cat. 39.

[6] Ritterling 1925: 1713–15; Mócsy 1962: 613; Mócsy 1974: 48; Fitz 1980a: 131; Wolff 2000: 203–204; Ferjančić 2002: 29–36; Radman-Livaja 2012: 170.

[7] Mócsy 1959: 160; Barkóczi 1964: 295, 303; Alföldy 1969: 131–133, *s.v. Valerius*; Mócsy 1983: 300, s.v. Valerius; Mócsy 1984b: 47; Mócsy 1985a: 80–87; Abascal Palazón 1994: 232–244, *s.v. Valerius*; Solin and Salomies 1994: 197, *s.v. Valerius*; Minkova 2000: 93–96, *s.v. Valerius*; *OPEL* IV: 143–146, 199, *s.v. Valerivs*; Tataki 2006: 424–431; Radman-Livaja 2014: 278.

[8] Dean 1916: 49–51; Frank 1916: 692; Gordon 1924: 106; Mócsy 1959: 17, 21, 189–190; Barkóczi 1964: 295, 323; Lochner-Hüttenbach 1965: 35; Kajanto 1965: 30, 74–77, 292; Alföldy 1969: 291–292, *s.v. Secundus*; Ben Abdallah and Ladjimi Sebai 1983: 43–44, *s.v. Secundus*; Mócsy 1983: 258, *s.v. Secundus*; Pflaum *et al.* 1983: 87, s.v. Secundus; Mócsy 1984a: 209, 216; Mócsy 1984b: 47; Mócsy 1985a: 92–98; Abascal Palazón 1994: 499–501, *s.v. Secunda, Secundus*; Solin and Salomies 1994: 399, *s.v. Secundus*; Minkova 2000: 249–250, *s.v. Secundus*; Dondin-Payre 2001b: 543–548, 571–576, 594–595; *OPEL* IV: 59–61, 195, *s.v. Secvndvs*; Tataki 2006: 513–515; Radman-Livaja 2014: 261.

[9] Mann 1983: 30–32; Keppie 1984: 152; for Mediolanenses in particular, see Reali 2000: 655–661.

[10] Domaszewski and Dobson 1967: XXIX–XXX, 112–119.

[11] Bekić and Lovenjak 2002: 197–201; *AE* 2002: 1126.

onwards.[12] This inscription was dated to the late 1st or early 2nd century AD at the latest and we may assume that this *centurio* served in Dalmatia and/or Moesia with the *legio IV Flavia* during the Flavian period, but part of his service certainly took place elsewhere, presumably in the last years of his career. He may have attained the ultimate rank of *primus pilus* while serving in the *IV Flavia*, but it could have happened towards the end of his career as well, perhaps in one of the Pannonian legions.

The funerary stele of an auxiliary veteran *Titus Flavius Ateboduus* (A[t] I.3) was discovered in 1992 in the village of Odra, close to the Roman road connecting *Emona*, *Neviodunum* and *Siscia*, with one branch going to the neighbouring town of *Andautonia*.[13] A Roman necropolis was discovered a few hundred meters away from the stele findspot and everything points to the fact that the whole area was covered with smaller settlements and *villae rusticae*, established in close proximity to the urban centre of the *municipium*. The deceased, *Titus Flavius Ateboduus*,[14] son of *Aterix*,[15] was clearly a veteran, who died at the age of seventy-five. He served in the *cohors II Varcianorum equitata* as a *cornicularius praefecti*, basically a clerk and a non-commissioned officer in the cohort's staff, a task which implies that he must have been totally proficient in spoken and written Latin.[16] The funerary stele was also erected to his wife, *Crispina*, aged 45, and their grandson, the boy *Flavius Augurinus*, who died at the age of five. *Ateboduus* was clearly discharged under one of the emperors of the Flavian dynasty between AD 69 and 96. Considering his age, he obviously died many years afterwards, probably two or three decades later, depending on the age he enlisted. He could also have stayed somewhat longer in the army. *Ateboduus* was certainly an old man when he passed away and he was incontestably much older than his wife. Since her cognomen was quite widespread,[17] we may only conjecture about her origin but it would not be too far-fetched to assume that *Ateboduus* married her after returning to his homeland, although one cannot entirely dismiss the possibility that he met *Crispina* while still on active service in *Germania Inferior*. The relative frequency of this Latin name in areas inhabited by the Celtic people suggests that it may have been perceived as related to Celtic anthroponomy (*nom de traduction, nom d'assonance*).[18] Thus, *Crispina*'s western Pannonian origin is not to be excluded. The Latin cognomen of their grandson, *Augurinus*, while not extremely frequent, is nevertheless far from being uncommon and has already been encountered in Pannonia.[19] The early 2nd century AD is accepted as the most likely dating by all the scholars who studied this monument. Since the *cohors II Varcianorum equitata* was probably raised at the very beginning of the Flavian period, perhaps during the Civil War of AD 69,[20] *Ateboduus* likely joined the unit from the very start. Therefore, he could have been discharged towards the very end of Domitian's reign. If so, he likely lived for another two or three decades, which means that he passed away in the late Trajanic or early Hadrianic period. The absence of the acronym *D(iis) M(anibus)* definitely favours an earlier dating, meaning that the stele could not have been erected much after Trajan's reign at the latest.[21] *Ateboduus*, son of *Aterix*, could have been born towards the end of Tiberius's reign at the earliest, perhaps during Caligula's short rule or more likely in the first decade of Claudius', presuming that he joined the *cohors II Varcianorum equitata* as a young man around AD 69.

One interesting stele was found in the 19th century in the village of Lobor (A[ptNW]) I.1],[22] perhaps at the very border of the territory of Andautonia. While fragmented, most of the inscription is still preserved, as well as the portraits of the two deceased soldiers with their mother, who erected the monument with another soldier, presumably a family member, either her husband or perhaps her third son. Interestingly, the two dead brothers do not share the same *gentilicium*, which would imply that they had different fathers. The older brother, aged forty, was named *Valerius Lucilianus*[23] and served in the Praetorian

[12] Mócsy 1959: 186; Barkóczi 1964: 321; Kajanto 1965: 177; Solin and Salomies 1994: 385, *s.v. Proculinus*; *OPEL* III: 166, *s.v. Procvlinvs*.

[13] Rendić-Miočević 1993: 28–31; Rendić-Miočević 1994: 104, cat. 147; Gregl 1996: 12; Šegvić 1996: 132, cat. 2; Gregl 1997: 21, 57, 69; Gregl and Migotti 2000: 130, 132–136, 146; Cambi 2002: 153–154; Cambi 2005b: 11; Bulat 2007: 9–12; Heger 2007: 185–187; Nikoloska 2010: 62; Radman-Livaja and Ivezić 2012: 141, 154; Rendić-Miočević 2013: 343–378; *AE* 2013: 1206; Falileyev and Radman-Livaja 2016, 52–61; Migotti 2016a: 177, cat. 10.

[14] Schmidt 1957: 136–141, 152; Evans 1967: 142–145, 151; Mócsy 1983: 32, *s.v. Ateboduus*; Lőrincz and Redő 1994: 191, *s.v. Atebodvvs*; Delamarre 2003: 57, 81; Meid 2005: 165; Delamarre 2007: 28, *s.v. Ateboduus*, 211–213.

[15] Schmidt 1957: 136–141, 260–261; Evans 1967: 142–145, 243–249; Mócsy 1983: 33, *s.v. Aterix*; Lőrincz and Redő 1994: 195, *s.v. Aterix*; Delamarre 2003: 57, 259–260; Meid 2005: 78–79; Delamarre 2007: 29, *s.v. Aterix*, 211–212, 230.

[16] Domaszewski and Dobson 1967: 58–59; Clauss 1973: 17–45; Clauss 1999: 32–35.

[17] *TLL*, Vol. Onom. 2: 718–722, *s.v. Crispinus*; Mócsy 1959: 171; Barkóczi 1964: 310; Kajanto 1965: 223; Alföldy 1969: 183, *s.v. Crispinus*; Mócsy 1983: 93, *s.v. Crispinus*; Mócsy 1984a: 216; Abascal Palazón 1994: 339, *s.v. Crispina/-us*; Solin and Salomies 1994: 320, *s.v. Crispinus*; *OPEL* II: 85, 215, *s.v. Crispinvs*; Minkova 2000: 144, *s.v. Crispina, Crispinus*; Rémy 2001: 118–119, 157; Tataki 2006: 483; Radman-Livaja 2014: 192.

[18] Alföldy 1977: 257–258; Alföldy 1993: 11; Delamarre 2003: 130, *s.v. crixsos*.

[19] Mócsy 1959: 165; Barkóczi 1964: 306; Kajanto 1965: 83, 318; Alföldy 1969: 159, *s.v. Augurinus*; Mócsy 1983: 38, *s.v. Augurinus*; Lőrincz and Redő 1994: 225–226, *s. v. Avgvrinvs*; Solin and Salomies 1994: 298, *s.v. Augurinus*; *OPEL* I: 95, *s.v. Avgvrinvs*.

[20] Cichorius 1900: 347–348; Alföldi 1948: 13–18; Kraft 1951: 191; Alföldy 1968: 74; Spaul 2000: 329; Radman-Livaja 2012: 168.

[21] Thylander 1952: 50–51.

[22] *CIL* III 4114 =10888; *AIJ* 455; Brunšmid 1909: 165–166, cat. 360; Schober 1923: 71–72, cat. 154; Migotti 2011a: 155–167; Migotti 2013c: 265–266; Migotti 2016a: 180, cat. 17.

[23] For the *gentilicium Valerius*, *vide supra*; Barkóczi 1964: 316; Kajanto 1965: 149, 173; Solin and Salomies 1994: 353–354, *s.v. Lucil(l)ianus*; *OPEL* III: 35, *s.v. Lvcilianvs*.

Guard, while his younger brother, *Marcus Cocceius Superianus*,[24] was a thirty-year-old centurion who served in the *X Gemina* when he died. The mother's name was *Septimia Lucilla*,[25] while the name of the last man to be mentioned in the inscription is not entirely preserved. We do know, however, that his *gentilicium* was *Cocceius* as well, and that he was a *signifer* in the same legion as the centurion *Marcus Cocceius Superianus*. If the signifer was the latter's father, we could presume that he was a veteran, but it is not explicitly stated in what survives of the inscription. Thus, he may have been an active duty soldier as well, possibly even the brother and half-brother of *Marcus Cocceius Superianus* and *Valerius Lucilianus*. The *nomina* and cognomina of these four individuals have no significant regional connotations – except for *Superianus*, which is only known in Pannonia and Dacia – and may thus hardly provide clues about their origin. Nonetheless, since the mother must have lived in the area, we may assume that they were of local origin. As Pannonians, they chose to serve in their province, in the *legio X Gemina* garrisoned in Vindobona since the beginning of the 2nd century AD.[26] *Valerius Lucilianus* was likely also a Pannonian legionary before joining the Praetorian Guard, presumably after Septimius Severus's accession to the purple.[27] The stele is usually dated to the first half of the 3rd century AD, but this dating could probably be narrowed to the first decades of the 3rd century, i.e. to the Severan period.

Speaking of veterans in Andautonia, a famous votive monument to Nemesis was erected by a certain *Iulius Victorinus*, seemingly the *decurio municipii Andautoniensium* who also stated that he was a veteran, without, however, mentioning the name of his unit.[28] The inscription is not entirely preserved, but it seems quite likely that the dedicator was a former soldier who became a magistrate in the late 2nd or early 3rd century AD.

The number of preserved monuments in the territory of Andautonia is too small to enable us to distinguish clear patterns. Two of the funerary monuments certainly belonged to veterans: one stele was erected to the memory of two deceased soldiers on active duty, while the last fragmentary inscription, considering the age of the deceased wife, likely belonged to a veteran as well. As such, they provide evidence for natives joining the army and them also demonstrate that veterans were willing to settle there. Nonetheless, we may assume that the reasons that enticed veterans to settle in the territory of Andautonia were manifold and cannot be solely attributed to nostalgia, i.e. to the desire of returning to their homeland. Indeed, not all the veterans recorded on those tombstones were native *Varciani*. As a matter of fact, only one must have considered himself as a *Varcianus*. One veteran was clearly an Italian (A I.1). This is far from being exceptional: although quite a few Italian veterans who served in Pannonia during the 1st century AD clearly preferred returning to Italy after their discharge (not necessarily to their home towns), a large number of legionary veterans settled in Pannonia. However, the epigraphic record suggests that they usually chose to settle in their former garrison places or in veteran colonies.[29] In this regard, choosing to settle in Andautonia does not seem to follow the typical pattern, but one should keep in mind the fact that the *municipium* of Andautonia is actually quite close to Poetovio and *Sextus Valerius Secundus* did not have to move far away. What enticed him to move to Andautonia remains an open question. There are no sources suggesting that veterans were ever deducted to Andautonia but considering two legionary veterans' tombstones from the late Flavian or early Trajanic period at the latest (A I.1; A[t] VIII.1), one may wonder if a *missio agraria* took place there. Not all legionary veterans were settled in colonies, some veterans appear to have been deducted in Pannonian settlements which were not given the status of a colony. This certainly happened in Scarbantia in Tiberius' time, where this practice may have continued afterwards, during Claudius' reign, under whose rule such deductions may have occurred in the Balaton lake area as well.[30] It might have happened also in the territory of the Eravisci in the 2nd half of the 1st century AD.[31] However, although Andautonia became a *municipium* under the Flavians, one may not claim that this had anything to do with deductions. Be it as it may, veterans who were given a *missio nummaria* were free to go wherever they wanted, and a retired *primus pilus*, being eligible for equestrian rank, could certainly afford buying real estate anywhere in Pannonia. A retired legionary such as *Sextus Valerius Secundus* likely had the financial means to consider himself as middle class and he could easily have been

[24] *TLL*, Vol. Onom. 2: 516–518, *s.v. Cocceius*; Schulze 1904: 426; Mócsy 1959: 149; Alföldy 1969: 41–42, *s.v. Cocceius*; Mócsy 1983: 83, *s.v. Cocceius*; Solin and Salomies 1994: 58, *s.v. Cocceius*; *OPEL* II: 67, *s.v. Cocceivs*; Minkova 2000: 43, *s.v. Cocceius*; Tataki 2006: 189; Barkóczi 1964: 325; Kajanto 1965: 277; Solin and Salomies 1994: 409, *s.v. Superianus*; *OPEL* IV: 100, *s.v. Svperianvs*.

[25] Schulze 1904: 229; Mócsy 1959: 159; Barkóczi 1964: 294, 299; Alföldy 1969: 53, *s.v. Septimius*; Mócsy 1983: 262, *s.v. Septimius*; Mócsy 1985a: 86; Mócsy 1985b: 403–410; Abascal Palazón 1994: 218–219, *s.v. Septimia, Septimius, Septumia/-ius*; Solin and Salomies 1994: 168, *s.v. Septimius*; *OPEL* IV: 69–70, 195, *s.v. Septimius*; Minkova 2000: 85, *s.v. Septimius*; Tataki 2006: 387–389; Mócsy 1959: 179; Barkóczi 1964: 316; Kajanto 1965: 173; Alföldy 1969: 41–42, *s.v. Lucilla*; Solin and Salomies 1994: 354, *s.v. Lucillus*; *OPEL* III: 35, *s.v. Lvcillvs*.

[26] Ritterling 1925: 1683–1687; Gómez-Pantoja 2000: 188–190.

[27] Dury 1968: 247–249; Birley 1969: 64–65.

[28] *CIL* III 4008; Brunšmid 1905: 65, cat. 125; *AIJ* 474; Knezović 2010: 193–195.

[29] Mócsy 1959: 36–37, 43–44, 89–92; Mann 1983: 31–35, 56–59, 123–124, T. 18.1; Todisco 2000: 670–671, 673; Ferjančić 2002: 21–55, 129–136, 140.

[30] Mócsy 1959: 43–44, 89–91; Ferjančić 2002: 119–125.

[31] Mócsy 1959: 62.

a prosperous person anywhere in late 1st-century Pannonia.

The origin of the *primus pilus* from Čučerje (A[t] VIII.1) may only be conjectured about, although one is inclined not to consider him as a local man, while *Titus Flavius Ateboduus*, as already pointed out, clearly appears to have been a native (A[t] I.3). The nomenclature of the brothers from Lobor (A[ptNW] I.1) does not really have recognizable regional features – although the cognomen *Superianus* appears to be confined to Danubian provinces – but they were presumably of local origin since their mother evidently dwelled there. However, this stele belongs to a later period when regional, i.e. Pannonian, identity likely mostly replaced the original ethnic affiliation of their forefathers.[32] Interestingly, most of these men had distinguished military careers: two were legionary centurions (one even reached the rank of a *primus pilus*), one was an auxiliary *cornicularius*, while one enlisted man was a praetorian, quite a prestigious achievement in itself. The legionary from Mediolanum is the only one who does not appear to have held any rank or prominent position while on active service. It should also be pointed out that the two dead brothers – the centurion of the *X Gemina* and the praetorian – were commemorated, besides their mother, by another soldier, presumably their father (or perhaps their brother), a signifer in the *X Gemina* as well. The whole family was seemingly following military careers and they appear to have been rather successful in their profession.

Taking into account only these funerary monuments, we could hardly claim that the *civitas* of the Varciani, later on the *municipium* of Andautonia, was attracting large numbers of veterans or that it was a major recruiting ground for the Roman army. There is however sufficient evidence confirming that not a few Varciani were actually interested in becoming Roman soldiers. The sources for the *cohors I Latobicorum and Varcianorum equitata* suggest that it was likely raised sometime before the *cohors II Varcianorum equitata*, as it appears to have been active during Nero's reign, but it may have been raised earlier.[33] The earliest evidence for the second cavalry cohort of the Varciani (*cohors II Varcianorum equitata*) – the unit of the cornicularius *Titus Flavius Ateboduus* – belongs to Vespasian's time and it does not seem likely that this cohort was raised much before AD 69.[34] Besides these two units, Varciani were certainly joining auxiliary units on an individual basis as early as Tiberius's reign. The Varcianus *Iantumarus Andedunis filius* joined the *cohors II Hispanorum* around AD 35, a unit garrisoned in Pannonia during the 1st century AD.[35] A certain *Vercaius*,[36] *Me[di*[37] *f(ilius)]*, trooper in the *ala Claudia nova*,[38] was buried in Dalmatia towards the middle of the 1st century AD, and his epitaph clearly states *domo Varcianus*.[39] Another anonymous *Varcianus* is mentioned on a fragmentary funerary monument from another fort in Dalmatia, Bigeste,[40] where his unit, the *cohors III Alpinorum* was stationed during most of the 1st century AD.[41]

*T. Flavius Bonio*[42] from Andautonia, who died at the age of thirty-four in the late Flavian period, towards the end of the 1st century AD, after having served 16 years as a trooper in the *ala Frontoniana*, also joined a unit stationed in his native province.[43] At the time of his death, his ala was likely garrisoned in Campona.[44] Although serving in an auxiliary unit, he was a Roman citizen enrolled in the Quirina tribe but his local origin is quite certain.

A votive altar devoted to Jupiter was erected, probably in Hadrianic times (or perhaps rather towards the very end of that century, depending on interpretations), by *milites vexillationis cohortium Pannoniae superioris, cives Sisciani et Varciani et Latobici* in Samaria, in Judea.[45]

Somewhat later, in the 1st half of the 3rd century AD, more or less contemporaneously with the soldiers mentioned on the stele from Lobor, *Flavius Respectus* served in the *equites singulares*.[46] Interestingly, his epitaph points out that he was a Varcianus, illustrating that Pannonian regional identity still had not completely replaced the

[32] Cf. *CIL* VI 3257, the *eques singularis Flavius Respectus* is still called a Varcianus!

[33] *CIL* XVI 158; Kraft 1951: 27, 36; Alföldy 1968: 63–65; Spaul 2000: 328.

[34] Cichorius 1900: 347–348; Kraft 1951: 191; Alföldy 1968: 74–75; Spaul 2000: 329.

[35] *CIL* XVI 4; Cichorius 1900: 299; Wagner 1938: 152–153; Kraft 1951: 51, 178; Mócsy 1959: 210, cat. 52/1; Domić Kunić 1988: 89, cat. 18; Lőrincz 2001: 36, 153, cat. 2; Raybould and Sims-Williams 2007: PAN 045; Grbić 2014: 274–275, cat. 187. Although numerous scholars believe that cohorts manned by Pannonians were raised shortly after AD 9 (Mócsy 1959: 75, 120–121; Mócsy 1962: 645; Alföldy 1968: 48; Bogaers 1969: 36–46; Dušanić 1978: 467), considering the available epigraphic evidence, I am more inclined to believe that the first Breuci, Pannonian and Varciani auxiliary units (with the exception of the *ala Pannoniorum*) were not raised before the very end of Caligula's and/or the beginning of Claudius' reign (Radman-Livaja 2012: 167–169).

[36] A. Holder III: 182; Mócsy 1959: 195; Alföldy 1969: 324; Mócsy 1983: 306, *s.v. Vercaius*; *OPEL* IV: 156–157, *s.v. Vercaivs*; Delamarre 2007: 195, *s.v. Vercaius*.

[37] Cf. Mócsy 1983: 184–185; *OPEL* III: 72–73; Delamarre 2003: 221–222; Meid 2005: 235; Delamarre 2007: 130–131, *s.v. Medius, Medo, Medos, Medus*, 226.

[38] Cichorius 1894: 1237–1238; Alföldy 1962: 261–262; Wilkes 1969: 141, 470–471; Bojanovski 1988: 355; Spaul 1994: 89–91.

[39] *CIL* III 9796; Mócsy 1959: 210, cat. 52/2; Dobó 1975: 53, cat. 221; Grbić 2014: 275, cat. 188.

[40] *AE* 1950: 110; *ILJug* 116; Dobó 1975: 50, cat. 195; Grbić 2014: 275–276, cat. 189.

[41] Alföldy 1962: 263–265; Wilkes 1969: 141, 470–472; Bojanovski 1988: 356; Spaul 2000: 266–268; Tončinić 2009: 1454.

[42] Mócsy 1959: 167; Mócsy 1983: 52, *s.v. Bonio*; Lőrincz and Redő 1994: 310, *s.v. Bonio*; Matasović 2003: 14; Meid 2005: 258–259; Delamarre 2007: 45, *s.v. Bonio*, 213–214.

[43] *CIL* III 3679; Mócsy 1959: 210, cat. 51/3; Lőrincz 2001: 223, cat. 216.

[44] Cichorius 1894: 1267–1268; Wagner 1938: 76–79; Spaul 1994: 117–119; Lőrincz 2001: 26.

[45] *AE* 1938: 13; Saxer 1967: 60–61; Dobó 1975: 53, cat. 219.

[46] *CIL* VI 3257; Dobó 1975: 39–40, cat. 121; Grbić 2014: 276, cat. 190.

ethnic affiliation. A fragmented funerary inscription from Cherchell, Caesarea in Mauritania, most likely mentions a Varcianus who served in the navy. It has been dated quite broadly, to the 2nd or 3rd century AD, and while his name and rank remain unknown, it surely proves that Varciani, like other Pannonians, were also serving in the navy.[47] A trooper from the cohors *I Thracum equitata civium Romanorum*, *Aurelius Vindex*, was also born in Andautonia.[48] He probably died towards the end of his service, at the age of forty-five, and his stele, dated to the 1st half of the 3rd century AD, was found in Brigetio. His unit was stationed in the nearby auxiliary camp at Ad Statuas (Ács-Vaspuszta).[49]

Considering the aforementioned, it is obvious that many Varciani, i.e. people from what was to become the *municipium Andautoniensium* in Flavian times, were quite eager to join the ranks of the Roman army from Tiberius' reign onwards. Since it was not exactly a very large community, we may assume that many families had kinsfolk in the Roman army and that *Titus Flavius Ateboduus* was probably not the only one who returned home after discharge. Actually, over the generations, serving in the army must have become a tradition in more than one family. As far as legionaries of Italian descent are concerned, settling in a place where many locals had family ties with their comrades from auxiliary units and where some auxiliary veterans must have been present as well was probably not an unpleasant prospect. While the number of funerary inscriptions mentioning veterans and soldiers in the territory of *municipium Andautoniensium* is far from being impressive, one should take into account the relatively small total number of funerary monuments discovered in that area (30 more or less well-preserved monuments and fragments). From this perspective, we cannot dismiss the veterans or soldiers' relatives as an insignificant social group in Andautonia. Furthermore, their monuments are all of decent or even high quality, and confirm that those men were part of an affluent class, many were likely landowners, and we may even have evidence for them engaging in local politics and occupying municipal charges, if the inscription *CIL* III 4008 has been correctly interpreted. Thus, although there was no permanent garrison in Andautonia, we may presume that native soldiers, through their families, and veterans played a certain role in the local social, economic and cultural life. Besides being well-off and able to contribute to the local trade and economy in general – probably not only as customers but also as producers or craftsmen – their presence must have played a determining role quite early in the spread of Latin as a language of everyday communication and the acceptance of the Roman way of life, especially as far as its cultural and religious aspects are concerned. When Andautonia became a municipium in the early Flavian period,[50] this community must have been composed of thoroughly romanised Varciani – and there were among them quite a few who served in the *auxilia* – as well as a fair number of settlers who likely counted some army veterans in their ranks.

[47] Agusta-Boularot and Bousbaa 1993: 122–125; *AE* 1995: 1798; Grbić 2014: 276, cat. 191.
[48] *CIL* III 4316; Lőrincz 2001: 292, cat. 461.
[49] Lőrincz 2001: 42–43.

### V.3. AQUAE BALISSAE

Also known as the *res publica Iasorum* and the *municipium Iasorum*, Aquae Balissae was a Roman *municipium* created in the ancient tribal territory of the Iasi, approximately in what is now western Slavonia, i.e. more precisely the larger area around Daruvar.[51] It most likely reached municipal status during Hadrian's reign, when it must have already been a rather important urban centre.[52] While the hypothesis that Aquae Balissae may have had a military garrison during the 1st century AD has been put forward in the recent past,[53] indisputable evidence is lacking.[54] Just as in the case of Andautonia, it is not impossible that a military outfit had been stationed there during Augustan and/or Tiberian times, perhaps during and/or after the quelling of Bato's revolt, but this has to remain mere conjecture for the time being.

For a town which has not yielded much as far as architectural remains and archaeological finds are concerned,[55] quite a few – 20 altogether – inscribed stone monuments have come to light, and, lacking much information from written sources, it is precisely this epigraphic record that provides most of the data about this Roman *municipium* and thermal station. Of a total of 20 monuments nine are funerary, seven are votive, while the remaining four are honorary.[56] Considering the scope of this chapter, we are mostly interested in the five funerary monuments associated with military personnel. Not all of these monuments are from Daruvar, a few were discovered further away, but they are nonetheless considered to have been found in the territory of the *res publica Iasorum*. Such is one relatively early stele, unfortunately lost, said to have been found in the village of Orešac, 40 km north-east of Daruvar (AB[t] I.1).[57] According to its inscription, it has been dated to the Flavian period and it commemorates *Marcus Nunnidius Successus*, a veteran of the *XXXII cohors voluntariorum*, which was stationed in Siscia during the

[50] Šašel Kos 2010: 219–220.
[51] Mócsy 1959: 27; Schejbal 2003: 393–408; Schejbal 2004: 99–122; Migotti 2017c: 83–120; Grbić 2014: 177–182; Kovács 2014: 6, 11, 19, 22, 94, 96, 101.
[52] Schejbal 2004: 108; Migotti 2017c: 86–87, 117; Grbić 2014: 180–181; a late 2nd- or early 3rd-century date can likely be dismissed, cf. Mócsy 1959: 27–28.
[53] Schejbal 2004: 99–100, 106.
[54] Migotti 2017c: 86, 111.
[55] Schejbal 2004: 100.
[56] Schejbal 2004: 108–114; Migotti 2017c: 88–110.
[57] *CIL* III 4006; Kraft 1951: 199, cat. 3141; Lőrincz 2001: 297, no. 480; Migotti 2016a: 175, cat. 6; Migotti 2017c: 88–91.

Julio-Claudian period. The veteran lived a long and likely prosperous life, passing away at the age of ninety. His wife, *Petronia Procula*, also a Roman citizen, died aged sixty, but their children were less fortunate. His daughter *Nunnidia Vitalis* died as a teenager, at the age of nineteen, while his son *Marcus Nunnidius Saturninus* died as as a small boy, aged seven. The man was most likely not a native, and the same appears to be true for his wife.[58] We do not know why he chose to settle in the territory of *Aquae Balissae* after his discharge – presumably in the late Julio-Claudian or early Flavian period – and why he did not settle in his old garrison place, Siscia. Nonetheless, we may presume that he managed to live as a wealthy man there, perhaps after being granted land. Since there is no other reliable evidence, we may only conjecture that he was not the only veteran who was given this opportunity. A. Mócsy surmised once that veterans may have been deducted in the territory of the *civitas Iasorum* in the 1st century AD, without entering into details and, as a matter of fact, likely basing his conjecture on *Nunnidius Successus*' lost tombstone.[59]

Another missing monument not too far away, near the town of Našice, almost 70 km to the southeast of Daruvar, dated to the 2nd century, has been believed to mention a deceased commanding officer of the *ala I Pannoniorum* (AB[t] VIII.1).[60] The inscription was recorded as fragmentary and we may doubt the transcription, but contrary to the veteran *Marcus Nunnidius Successus*, the anonymous prefect of the ala mentioned in the inscription appears to have been a native from *Incerum* or *Inicerum* (according to this dubious lost inscription), a settlement situated on the Siscia–Mursa road, and variously located by different authors in Treštanovačka gradina near Tekić,[61] Velika[62] or Vetovo.[63] Since the monument has been lost and there is no data whatsoever about its typological and iconographical features, one has to rely on the transcription of what was left of the inscription at the moment of discovery. The name of the officer belonging to the *tribus Claudia* was not recorded, but if the epitaph was really mentioning a *praef(ectus) alae I Panno(niorum)*, several questions without straight answers come to mind. Considering that the inscription started with the formula *D(iis) M(anibus)*, the tombstone was quite certainly not anterior to the 2nd century AD, probably not even anterior to Hadrian's reign. The prefect obviously must have been a Roman citizen and there is definitely nothing odd about that: by the 2nd century AD Pannonia must have been inhabited by many Roman citizens and a few were certainly members of the *ordo equester*,[64] but a *vir egregius* from *Incerum* raises an eyebrow. Presuming that he really was an equestrian born in Pannonia, which *ala I Pannoniorum* could he have commanded and when? As a matter of fact, it is not a simple issue. The oldest ala raised among Pannonians was most likely in existence already in Augustus' time and was commonly referred to as just the *ala Pannoniorum*, without the numeral.[65] That unit is assumed to have been transferred from Pannonia to Moesia after AD 69 and only came back to Pannonia Inferior after the Marcomannic wars, presumably to *Cusum*. This *ala Pannoniorum* should perhaps not be equated with several other units which bore that title and a numeral, i.e. *ala I Pannoniorum*,[66] but to make matters more complicated, it is not entirely excluded that this very unit may have been occasionally referred to as the *ala I Pannoniorum*, at least on some diplomas and stamped tiles, while it was in Moesia and Dacia, unless it was actually a different *ala I Pannoniorum*.[67] One *ala I Pannoniorum* was raised during the Julio-Claudian period, maybe already by Augustus, and was sent to Africa where it stayed for the next three centuries. Spaul argued that this could have been the original *ala Pannoniorum* which was transferred from the Balkans to Numidia by Trajan.[68] Which, then, was the *ala Pannoniorum* recorded in *Pannonia Inferior* in the early 3rd century AD?[69] In any case, Le Bohec's arguments for the parallel existence of one *ala I Pannoniorum* in Africa and another one in Moesia are strong, and several of the African inscriptions belong quite certainly to the 1st century AD. The *ala I Pannoniorum Tampiana Victrix* was founded in the 1st century AD, not unlikely during the Civil War in 69, and moved between Pannonia and Britain a few times from the Flavian period onwards, until it was finally stationed in Noricum around AD 125–130.[70] The *ala I Pannoniorum Sabiniana* was also raised in the 1st century AD, perhaps already during Tiberius' reign, although this remains conjectural. However, it was soon sent away from Pannonia, first probably to Spain and

[58] Mócsy 1959: 212, cat. 60.
[59] Mócsy 1959: 90–91.
[60] Pinterović 1975: 129; Domić Kunić 1988: 89, cat. 17; Migotti 2017c: 99.
[61] Bulat 1977: 82–83.
[62] Schejbal 2004: 102–103.
[63] Fodorean 2017: 343–345.

[64] The available data suggest that Pannonian equestrian families were becoming more common towards the end of the 2nd century AD and more so during the 3rd century AD. They usually belonged to the elite of the largest urban centres, see for instance *PME*: A-27, A-55, A-151, A-232, C-211, F-57, F-67, F-105, L-22, U-2, V-23, V-36, V-39, Incerti-59, Incerti-61; Birley 1969: 82; Mócsy 1974: 144, 245; Zaninović 1981: 205–206; Gregl and Migotti 2000: 141–150. Nonetheless, some members of the *ordo equester* could have been residing in Pannonia already in the 1st century AD (presumably rather towards the end of that century), see Mócsy 1974: 114–115. The above-mentioned *primus pilus* from Čučerje (A[t] VIII.1) was also eligible for the equestrian rank.
[65] Cichorius 1894: 1255; Wagner 1938: 56–58; Kraft 1951: 25; Alföldy 1962: 262–263; Lőrincz 2001: 22; Ferjančić 2015: 37–43.
[66] Spaul 1994: 167–172; Spaul 1995: 63–73.
[67] *CIL* III 6242; *RMD* 222, note 2.; *RMD* 225, note 3; *RMD* 338, note 2; Kraft 1951: 25, 155. Was this unit amalgamated with an *ala Gallorum* to become the *ala I Gallorum et Pannoniorum*?
[68] Cichorius 1894: 1253; Le Bohec 1989: 33–63; Spaul 1994: 169–171; Spaul 1995: 63–64, 66–68.
[69] *CIL* III 3252 = 10241; CIL III $15138^3$; Lőrincz 2001: cat. 157, 158.
[70] Cichorius 1894: 1254; Wagner 1938: 59–60; Spaul 1994: 215–216; Lőrincz 2001, 22.

then, in the early 2nd century AD to Britain, where it remained throughout the 2nd and 3rd centuries.[71] This necessarily brief overview shows that during the 2nd century AD the anonymous prefect from *Incerum* may have served in Africa, Noricum, Britain, Moesia or even in Pannonia, if the lost inscription is to be dated rather towards the end of the century. Nonetheless, without the autopsy of the monument, these are mere conjectures and this monument may thus hardly contribute to any scholarly study of Pannonian *alae* and equestrians from Pannonia.

Presuming that this lost tombstone really belonged to a man of local origin whose remains were brought back home, the burials of other soldiers in the territory of Aquae Balissae could correspond to the same pattern. Although it is never explicitly stated, they could all have been natives whose last resting place was in their homeland. Since their funerary monuments belong to a later period, one cannot rely on onomastics to prove a local origin, but the absence of military garrisons in that area likely excludes the possibility that they died while serving in Aquae Balissae, even more so since their units are known to have been garrisoned elsewhere.

*Beneficiarii* are actually the only soldiers who might have been serving in *Aquae Balissae*, but the only evidence for the possible existence of a *statio* is one votive altar, erected in 208 AD to *Silvanus Domesticus* and *Silvanae* by the beneficiarius *Seius Ingenuus*.[72] The presence of a station in Aquae Balissae is not to be excluded but *Seius Ingenuus* may simply have been visiting the spa there. The altar not being devoted to Jupiter, it was clearly an act of private worship.[73] Other active soldiers were also likely visiting the thermal baths and some left votive altars. The centurion *Secundius Restutus* of the *X Gemina* had an altar devoted to Jupiter Dolichenus approximately at the same period, i.e. in the late 2nd or early 3rd century AD.[74] Another centurion, this one from the *VII Gemina*, *Quintus Carmeus Iulianus*, erected an altar to Jupiter Dolichenus as well, with his wife and sons, also in the early 3rd century AD, i.e. between 209 and 211 AD.[75] Interestingly, the garrison of *Carmeus Iulianus* was far away from Pannonia, since the legio *VII Gemina* was in *Hispania Tarraconensis* at that time.[76] Assuming that he just visited the spa with his family seems a bit far-fetched, but conjecturing about an official reason for his presence in Pannonia at that time is even more puzzling.[77] The Marcomannic wars were long over by then and one may hardly find a good explanation for the presence of a *VII Gemina* vexillation in Pannonia, even less in a small *municipium* not quite close to the Empire's borders. Finally, and again towards the same period, in the late 2nd or early 3rd century AD two legionaries from another Pannonian legion, the *legio I Adiutrix* stationed in Brigetio, *Serg(ius) Felix* and *Valerius Proculus* erected an altar to Nemesis.[78] Except for *Carmeus Iulianus*, all of these soldiers were stationed in Pannonia and while we may presume that they came to enjoy the spa, we may not entirely exclude the possibility that they spent some time in Aquae Balissae on official duty.[79] Be that as it may, their votive altars do not reveal much about the personal ties they may have had with the *municipium Iasorum*. Their presence attests at least that inhabitants of *Aquae Balissae* must have been pretty accustomed to seeing soldiers in their town during the Severan period, even if only as tourists and spa visitors. However, the men who erected those altars were probably not of local origin and whatever the reason of their stay might have been, they presumably did not reside there over a longer period of time.

Soldiers' funerary monuments from the Severan period found in the territory of the *municipium Iasorum* probably imply a more personal relationship of the deceased with their last resting place. A stele of a legionary of the *IV Flavia Antoniniana*,[80] *Aurelius Naso*, was found in the village of Brusnik, less than 20 km away from Daruvar (AB[t] I.5).[81] His wife, *Prisca Tatonis*, was buried with him and she was likely a native.[82] The three men who erected the monument appear to have been Prisca's sons, but they are perhaps not related to *Naso*, unless the fourth name in genitive – *Maximiani* – is not a stonecutter's mistake. If so, *Maximianus* would be their fourth son. In any event, *Naso*, who must have served in Moesia, appears to have been buried at home after dying on active service during the Severan period. A stele roughly from the same period (AB[t] I.6), also found in the vicinity of Daruvar, in the village of Kusonje, was erected by the four brothers of the deceased, a certain *Cassius Sextus*, who died while serving in the *cohors quingenaria Maurorum*.[83] The unit was stationed in Alta Ripa after taking part in the Marcomannic wars. Presumably, *Cassius Sextus*,

[71] Cichorius 1894: 1259; Spaul 1994: 189–190.

[72] *CBI* 277; Fitz 1993: 385–387; Ott 1995: 92; Nelis-Clément 2000: 184, 186; Schejbal 2004: 111; Radman-Livaja 2012: 180; Migotti 2017c: 106–107.

[73] The remains of a Silvanus temple may have been found during construction works in 1967 (Schejbal 2004: 110).

[74] *CIL* III 3999; *AIJ* 584; Schejbal 2004: 110–111, fig. 13; Migotti 2017c: 103–104.

[75] *CIL* III 3998; *AIJ* 583; Schejbal 2004, 110–111, fig. 12; Migotti 2017c: 101–103.

[76] Ritterling 1925: 1632–1633; Le Roux 2000: 383–396.

[77] Ritterling 1925: 1636–1637; Le Roux 2000: 387.

[78] *AIJ* 585; Pinterović 1975: 142; Schejbal 2004:111, fig. 14; Migotti 2017c: 105–106.

[79] Migotti 2017c: 110–113.

[80] Ritterling 1925: 1540–1549; Le Bohec and Wolff 2000: 239–242.

[81] *AIJ* 590; Pinterović 1975: 126–127; Ubl 2013: XXIII no. 7, pl. 3.12 (figs 180 and 243); Migotti 2013c: 265; Migotti 2016a: 179–180, cat. 16; Migotti 2017c: 96–98.

[82] A. Holder II: 1751; Krahe 1929: 111; Mayer 1957: 330; Alföldy 1969: 305; Barkóczi 1964: 325; Mócsy 1983: 283, *s.v. Tato, Tatto*; *OPEL* IV: 109, *s.v. Tato, Tatto*; Meid 2005: 291–292; Delamarre 2007: 179, *s.v. Tato, Tatto*; Radman-Livaja and Ivezić 2012: 156.

[83] Cichorius 1900: 315; Wagner 1938: 167–168; Kraft 1951: 180, cat. 1640; Spaul 2000: 468–469; Lőrincz 2001: 39; Migotti 2017c: 94–96.

a Roman citizen and presumably a Pannonian from Aquae Balissae, joined the unit in the late 2nd century (the unit must have arrived in Pannonia in AD 171) and died in service towards the end of the 2nd or in the early 3rd century AD. His brothers, who do not appear to have been soldiers, were able to bury him at his home. A rather high-quality, albeit fragmentary marble stele was found even closer to Daruvar, in the village of Veliki Bastaji (AB[t] I.7).[84] While the inscription is incomplete, there is no doubt that a mother erected the monument to her two sons, a praetorian soldier named *Flavius Val[erius]* and *Valerius Dig[_ _ _]*, possibly a veteran of an unknown unit. It has been dated to the 2nd half of the 2nd century AD, and while it is not unlikely that *Flavius Valerius* became a praetorian only after Septimius Severus took the Imperial throne, like many other Pannonians,[85] it is not entirely excluded that he may have joined this prestigious elite corps even before that. Interestingly, the brothers do not share the same *gentilicium*, which implies that they had different fathers, although both were Roman citizens. *Valerius Dig[_ _ _]* was thus likely a legionary, if he indeed was a veteran. Another fragmentary stele from the Severan period, whose inscription is completely missing, was also found in Aquae Balissae. One of the two men depicted in the portrait niche is likely a soldier and we may assume that this funerary monument also belonged to a family with relatives in the military.[86]

If we compare their respective epigraphic records, the *Iasi* appear to have been less interested in joining the *auxilia* than their western neighbours, the Varciani. Only one native of the Iasi tribe is recorded on a 1st-century diploma,[87] *Fronto Sceni filius*,[88] who was honourably discharged in AD 85 from the *cohors I Lusitanorum*,[89] stationed at that time somewhere in Pannonia. He must have joined the army during Nero's reign and he chose to serve not too far from home, in his own province. He is the only known auxiliary soldier from the Iasi tribe in the 1st century AD but an extremely interesting diploma from AD 71, issued to an Eraviscan sailor, mentions a *princeps Iasiorum* among the witnesses, a certain *Titus Flavius Serenus*.[90] This gentleman was obviously granted Roman citizenship shortly before that. It would thus seem that even members of the elite among the Iasi did not become Roman citizens during the Julio-Claudian period. Having him as a witness (as well as his colleagues, the *principes Breucorum, Boiorum* and *Antizitium*) was a clever political move, a propagandistic way of emphasising the importance of the support their respective communities must have given to Vespasian during the Civil War of AD 69. Since only his function of *princeps Iasiorum* is pointed out, we may only conjecture about *Titus Flavius Serenus*' service in the Roman army, but that he was an active soldier or perhaps a veteran is almost certain.[91] He must have been in Rome when this diploma was issued and it does not seem unlikely that he came there when the first Flavian emperor occupied the throne. He was presumably rewarded with citizenship on that occasion and we may assume that he was serving in an auxiliary unit in Vespasian's army. In any event, there is not much evidence for the Iasi in the Roman army before the 2nd century AD. Incidentally, no *cohors* or *ala Iasorum* had ever been raised. Despite the honour given to *Titus Flavius Serenus*, his *civitas* did not become a *municipium* for another 50 years. It may not be unreasonable to assume that the *civitas Iasiorum* was still not perceived as deserving municipal status during the Flavian period. The fact that the neighbouring *civitas Varcianorum* became a *municipium* under the Flavians, likely already under Vespasian, is probably not a coincidence. There were certainly many political, administrative, economic and even cultural factors in play, but a high number of native soldiers and veterans, a fact which directly influenced the level of romanisation and loyalty to the Empire, must have played a role. Even after the *civitas Iasiorum* became a *municipium*, its inhabitants do not seem to have started particularly favouring the military profession. It is only in the second half of the 2nd century AD that joining the army seems to have become more popular. Besides the tombstones mentioned above, one has to mention an *eques singularis* who died in Rome in the early 3rd century AD. The deceased *Ulpius Cocceius* is explicitly said to be from Aquae Balissae, more precisely '*natus ad Aquas Balizas pago Iovista vic(o) Coc[co]netibus*'.[92]

As far as the territory of Aquae Balissae is concerned, military tombstones represent a fair share of funerary monuments, even taking into account the fact that the overall number is not exactly impressive. Since most of those monuments are not to be dated before the late 2nd century AD, soldiers and veterans presumably started having more of a social impact in the local community towards the end of the Antonine period and especially during the Severan dynasty. The funerary monuments suggest that soldiers buried in that *municipium* were usually locals who joined the

[84] *CIL* III 4001 = 10865; *AIJ* 588; Schejbal 2004: 104, Map IV/7, 111–112, fig. 15; Djurić 2013: 8, no. 32; Migotti 2016a: 178, cat. 12; Migotti 2017c: 93–94.
[85] Dury 1968: 247–249; Birley 1969: 64–65.
[86] Migotti 2013c: 266; Migotti 2016a: 180, cat. 18.
[87] *CIL* XVI 31; Mócsy 1959: 216, cat. 71; Radman-Livaja and Ivezić 2012: 156; Grbić 2014: 182–183, cat. 84.
[88] Krahe 1929: 101; Krahe 1955: 65; Mayer 1957: 313; Mócsy 1959: 189; Katičić 1963: 274–275; Alföldy 1969: 289, *s.v. Scenus*; Mócsy 1983: 256, *s.v. Scenus*; Križman 1991: 130; *OPEL* IV 2002: 55, *s.v. Scenvs*; Delamarre 2007: 162, *s.v. Scenus*; Radman-Livaja 2014: 260.
[89] Cichorius 1900: 312; Wagner 1938: 162; Kraft 1951: 180; Radnóti and Barkóczi 1951: 199, 214; Fitz 1962: 65; Mócsy 1962: 623; Visy 1986: 491, 494–496, 507, 509–510, 512–514, 516; Spaul 2000: 61–62; Lőrincz 2001: 37–38.
[90] *RMD* 205; *AE* 2002: 1771; *AE* 2004: 89; Grbić 2014: 182, cat. 83 = 170; Kovács 2014: 68.
[91] Morris and Roxan 1977: 300, 326.
[92] *CIL* VI 3297; Dobó 1975: 38–39, cat. 115; Radman-Livaja and Ivezić 2012: 156; Grbić 2014: 183, cat. 85.

army, served mostly in Pannonia, and were buried by relatives in their homeland. Since cremation was still the most widespread burial custom, transporting their ashes for a few hundred miles at most was hardly a big issue for their relatives. Their service in the army was not necessarily a family tradition but it certainly insured sufficient financial means for a decent burial and good-quality funerary monument.

V.4. SISCIA

Contrary to the two previously discussed settlements, there is absolutely no doubt that Siscia had a military garrison over rather long periods of time, from the very beginning of Augustus' reign until the end of the Julio-Claudian period or even into the beginning of the Flavian era, as well as later on, from Gallienus' reign onwards.[93] It is not the only major difference. Besides being a much larger urban centre, Siscia was also a colony, albeit a relatively later one, i.e. a Flavian colony, although the settlement must have had an urban structure far earlier.[94] Many civilians presumably chose to settle in the immediate vicinity of the legionary camp already during the Augustan period, when the garrison was at the peak of its strength. As a matter of fact, Siscia takes its roots from the earlier protohistoric settlement. However, considering the circumstances of 35 and 34 BC, as well as the likely very high number of casualties among the inhabitants of Segestica, not to mention that many survivors must have been sold into slavery, one may wonder to what extent native people may have influenced the development of Roman Siscia during the Julio-Claudian period.[95]

That Siscia was a legionary camp from the time of Augustus to the reign of Claudius is quite certain.[96] Written sources, i.e. Appianus, Cassius Dio and Velleius Paterculus, provide more than enough evidence for the presence of legions in Siscia, admittedly not throughout the entire Julio-Claudian period, but we may reasonably assume that this was the case. The definitive identification of the legions that made up the garrison is however more problematic, more so since there is currently no unquestionable evidence about the location of Roman military camps in Sisak itself, Pogorelac, or anywhere else in the immediate vicinity of the town. After analyzing the sources, it is possible to establish with a rather high level of certainty which ten legions stayed in *Siscia* in AD 7 – *IV Scythica, V Macedonica, VII, VIII, IX, XI, XIII, XIV, XV* and *XX* – but this really was an exceptional situation which did not last for very long as a matter of fact.[97] Which legions were garrisoned in *Siscia* over a longer period of time is not an easy question to answer, but there are enough credible arguments in this respect for at least three legions. We may presume that those were the IX, XV and XX legions, but we still lack epigraphic finds which would corroborate this claim. Indeed, the early epigraphic finds from Siscia are confined to Roman military equipment with incised inscriptions, but no military units are explicitly mentioned. The Hagenau-type helmet found in the Sava River probably belonged to a soldier who was stationed in *Siscia*, but nothing else is known about the unit that *Varro* and his centurion *Luccius* belonged to.[98] The *gentilicium Luccius* is not very common, and most recorded examples are known from Italy and southern Gaul.[99] The cognomen or idionym *Varro* appears far more often, at least in Italy and Dalmatia, but remains less common in other provinces.[100] The helmet may have belonged to a legionary, i.e. a Roman citizen, but the owner could have been an auxiliary soldier as well. The horse bit found in the Kupa River can undoubtedly be ascribed to a Roman citizen, *Veturius Vivus*.[101] His *nomen gentile, Vet(t)urius*, is not very widespread, but is still not an uncommon *gentilicium* among Italians.[102] The cognomen *Vivus* appears far less frequently, and is very uncommon outside Italy.[103] The man was almost certainly of Italian origin, possibly a legionary cavalryman, maybe even an officer, i.e. a commander of an auxiliary unit,[104] but it should be mentioned that this type of bit can be dated to a wider chronological span than that of the Julio-Claudian period. While those artefacts definitely provide interesting clues about the Roman garrison in Siscia during the early principate, one has to admit that they do not contribute much to the discussion about Roman soldiers' social impact in that city.

If it were preserved today, the gravestone of the young centurion Tiberius Claudius Pontius (*Ti(berius) Claudius Pontius, centurio legionis XV Apollinaris*) could be the only serious epigraphic trace of legionary presence in *Siscia* during the Julio-Caudian period (S VIII.1).[105] This

[93] Mócsy 1974: 23, 43; Šašel 1974: 732–734; Radman-Livaja 2004: 16–19, 21–22; Radman-Livaja 2007: 161–168; Radman-Livaja 2010: 182–193, 197–200; Radman-Livaja 2015: 30–44.

[94] Mócsy 1962: 597; Mócsy 1974: 112–113; Šašel 1974: 734–736; Hoti 1992: 143.

[95] Radman-Livaja 2007: 161.

[96] Mócsy 1974: 23, 43; Šašel 1974: 732–734; Radman-Livaja 2004: 17–18; Radman-Livaja 2010: 185–190; Radman-Livaja 2012: 163–164, 169–170; Radman-Livaja 2015: 30–42.

[97] Radman-Livaja 2012: 166; Radman-Livaja 2015: 39.

[98] Hoffiller 1937a: 30–31; MacMullen 1960: 35, nr. 33; Radman-Livaja 2004: 68–71, cat. 126; Radman-Livaja and Vukelić 2015: 400.

[99] Schulze 1904: 424; Solin and Salomies 1994: 107; *OPEL* III: 34.

[100] Krahe 1929: 123; Mayer 1957: 354; Kajanto 1965: 69, 118, 264; Alföldy 1969: 321–322; Mócsy 1983: 301; Solin and Salomies 1994: 418; *OPEL* IV: 148.

[101] Radman-Livaja 2004: 102, cat. 388; Radman-Livaja and Vukelić 2015: 400–401.

[102] Schulze 1904: 259, 380, 411, 428, 432, 448; Solin and Salomies 1994: 206; *OPEL* IV: 164.

[103] Kajanto 1965: 274; Solin and Salomies 1994: 425.

[104] At least two people called *Veturius* are known as officers and members of the equestrian order from the Julio-Claudian period, *L. Veturius Homuncio* and *Q. Veturius Pexsus*, *PME* V, 85, 86; Demougin 1992: 170, 327.

[105] *CIL* III 10853; Mócsy 1959: cat. 57/1; Mosser 2003: 261, cat. 188; Radman-Livaja and Vukelić 2015: 401–402; the text allegedly said

lost monument is dated to the 1st century, which is completely acceptable, but it is impossible to provide a more precise dating because only the transcription has been preserved and all other descriptions are non-existent. Considering the youth of the centurion (twenty-five years old), it can be assumed that he came from a wealthier class, enabling him to start his military career as an officer. However, he was clearly not of equestrian rank. His nomenclature clearly points to a peregrine family origin that could have acquired its Roman citizenship during the times of Claudius or Nero, the latter being less likely. His father could have obtained citizenship during Claudius' rule, thereby taking the emperor's *praenomen* and *gentilicium*. Pontius, probably as the eldest son, was given the same *praenomen* as his father. If this assumption is accepted as true, the gravestone could not have been made long before the middle of the 1st century, and certainly not before AD 41. There is no evidence that Tiberius Claudius Pontius became a Roman citizen and officer right at the start of Claudius' rule, and died soon after. Perhaps a more convincing assumption is that he died somewhat later, possibly closer to the end of Claudius' rule or during Nero's time, perhaps even during the Flavian period. Since the monument cannot be dated on the basis of its typological and stylistic characteristics, it is difficult to determine a narrower chronological framework. Mócsy, despite the impossibility of proving the claim, believed the young man was from Pannonia, possibly even a native of *Siscia*. This would certainly explain why he was buried in *Siscia* at the moment when his legion was stationed in *Carnuntum*, or was already waging war in the East.[106] The suggested dating and interpretation seem plausible but what if the monument is significantly older, originating from the time of Tiberius? Could the father of our centurion have obtained citizenship precisely thanks to Tiberius while the latter was not yet the emperor, but only a high ranking commander, and, at the same time, the most important and most influential Roman in Pannonia, i.e. *legatus Augusti pro praetore*?[107] Tiberius undoubtedly spent years waging war in Pannonia, and probably had the opportunity to attract a lot of clients and followers from the native elite. Until Augustus adopted him in AD 4, his name was Tiberius Claudius Nero. The father of our centurion, Tiberius Claudius Pontius, might have served under his command during the Pannonian war and could have distinguished himself enough to obtain Roman citizenship and take over the nomenclature of his patron? This is only an assumption, but if taken as true, young Pontius could have become a centurion much earlier. He could have died in *Siscia* while *legio XV* was stationed there, possibly before AD 14? His death may have occurred when *legio IX* was in Africa, assuming that it was replaced in *Siscia* by a *vexillatio* of another Pannonian legion, in this case *legio XV*? Since the monument is missing, it is impossible to provide more plausible arguments for this hypothesis.

Certain clues for the presence of military units during the 1st century AD – which admittedly cannot be viewed as unquestionable evidence – can be found among other epigraphic material found in Siscia. The lead tags from Sisak, mostly dated to the 1st, as well as to the first half of the 2nd century AD,[108] contain many names that are neither autochthonous nor Latin. The relatively high frequency of Iberian names is particularly interesting.[109] While the multiculturalism of large urban centres in the Roman Empire is a well known fact, we may nonetheless still wonder why all those Iberians came to *Siscia*. Did they come to Pannonia by their own choice or were they somehow forced to do so? At least three legions garrisoned in Siscia at one moment or another – the IX, XV and XX – served in Hispania, from where they were transferred to Pannonia during Augustus' rule. There is no evidence supporting the claim that *IX Hispana* moved directly from the Iberian peninsula to *Siscia* about 16 or 15 BC, as it seems more probable that it spent some time in *Tilurium*, and ended up in *Siscia* after Bato's rebellion.[110] However, this is also just an assumption. In effect, nothing reliable is known about the whereabouts of this legion in Illyricum between 16–15 BC and AD 9, or even later. The same can be said of legion XX, which left the Iberian peninsula around 19 BC, first to participate in the campaigns in the Alps in 16 and 15 BC, only to join Tiberius' army in Pannonia between 13 and 9 BC. It stayed in Illyricum until the end of Bato's rebellion, when it was transferred to the Rhine. It undoubtedly passed through *Siscia*, but it is impossible to determine how long it could have stayed there.[111] Legion *XV Apollinaris* fought against the Cantabrians, Astures and Lusitanians between 27 and 19 BC. After 16 BC, it was almost certainly in Illyricum,

*Ti(berio) Claudio / Pontio / (centurio) leg(ionis) XV Apo(llinaris) / an(norum) XXV / h[e]res / [...].*

[106] Wheeler 2000: 271–282; Mosser 2003: 15–28, 138–157.

[107] Fitz 1993: 51–56.

[108] Radman-Livaja 2014: 144–147.

[109] Radman-Livaja 2014:142–143; *Bardilus Viriatus* (Radman-Livaja 2014: cat. 19.21), *Celtius Exostiti* (Radman-Livaja 2014: cat. 21.28), *Cilia* (Radman-Livaja 2014: cat. 21.20), *Urbani Fulvinus* (Radman-Livaja 2014: cat. 26.70), *Melo Savini* (Radman-Livaja 2014: cat. 08.14), *Oclatia Mela* (Radman-Livaja 2014: cat. 01.35), *Tusculus Congoni* (Radman-Livaja 2014: cat. 26.24), *Tuscul[us?]* (Radman-Livaja 2014: cat. 26.18); somewhat less certain, but possible Iberians could be *Aia* (Radman-Livaja 2014: cat. 08.18, 17.28, 21.100), *Ara* (Radman-Livaja 2014: cat. 13.05), *Astura* (Radman-Livaja 2014: cat. 15.19), *Balausus* (Radman-Livaja 2014: cat. 19.74), *Cebala Callua* (Radman-Livaja 2014: cat. 19.08), *Crespus Flavi* (Radman-Livaja 2014: cat. 19.01), *Fortuna* and *Fortuna Sta..tiis* (Radman-Livaja 2014: cat. 21.86, 26.115), *(H)i(r)suta Cin{e}ae* (Radman-Livaja 2014: cat. 06.15), *(H)ispanus* and *(H)ispanus Felicis* (Radman-Livaja 2014: cat. 07.06, 11.07), *Lovita* (Radman-Livaja 2014: cat. 24.09), *Seppronius* (Radman-Livaja 2014: cat. 20.06), *Simpius* (Radman-Livaja 2014: cat. 23.71), *Spana* (Radman-Livaja 2014: cat. 11.05), *Vera Melvii* (Radman-Livaja 2014: cat. 26.132), *Vitilia (Maxima ?)* (Radman-Livaja 2014: cat. 18.04 ); Radman-Livaja 2016: 179–184.

[110] Ritterling 1925: 1664–1670; Farnum 2005: 21, 34, 42; Radman-Livaja 2012: 163–164, 169; Radman-Livaja 2016: 185.

[111] Ritterling 1925: 1769–1771; Farnum 2005: 24, 65; Malone 2006: 28–31; Radman-Livaja 2012: 163–164; Radman-Livaja 2016: 185.

where it participated in Tiberius' Pannonian war, as well as the crushing of Bato's rebellion, after which it stayed in Pannonia.[112] All of these legions were certainly accompanied by a certain number of civilians when they left the Iberian Peninsula: besides slaves and servants, as well as concubines, there must have been many other followers that always stay close to military garrisons, such as traders, artisans, grocers, as well as shadier characters such as pimps and prostitutes.[113] Since their incomes often exclusively depended on soldiers, it was not unusual for them to follow their military customers when their respective units were transferred elsewhere. *Tesserae* could possibly be the traces of their presence in *Siscia*,[114] and, therefore, evidence of the presence of the legions they had been following to Pannonia. Several Africans, whose names can be found of the *tesserae* from Sisak,[115] could also possibly be connected to the stay of Legion IX in Africa between AD 20 and 24. The legionaries, once they returned to their old garrison, very likely brought along a certain number of civilians, presumably slaves and concubines, although establishing their exact number remains conjectural.[116] Besides names which could point to former garrisons of the afore-mentioned legions, the Siscian lead tags also bear quite a few Celtic anthroponyms typical for Gaul and the Rhine provinces.[117] Some of those Gauls could have been soldiers of auxiliary units or veterans who settled in Pannonia. The same could be assumed for some of the afore-mentioned Spaniards or Africans as well.[118] Finally, one lead tag bears the inscription *Sulpici(i) vet(erani?)*.[119] This man was seemingly explicitly identified by the craftsmen as an army veteran.[120]

However, and despite a large garrison over a relatively long period of time, almost no funerary monument belonging to soldiers who had actually served in Siscia with their units have been preserved. The only notable exception seems to be the funerary stele of *Marcus Mucius Hegetor*, the physician of the *XXXII cohors voluntariorum*, dated to the mid 1st century AD (S I.2).[121] Since he died while on active duty, the *cohors XXXII voluntariorum civium Romanorum* was likely stationed in the city for some time during the 1st century AD. It was perhaps garrisoned in Siscia already during the stay of the legion, presumably the *IX Hispana* and it likely remained there until Vespasian's time, when it was transferred to *Germania Superior*.[122]

A monument (S VIII.3), nowadays lost, was erected in his lifetime by *Marcus Minicius Saturninus*, who died at the age of eighty, a veteran and former optio of the Praetorian fleet from Ravenna, for himself and *Crispia Celerina*, his wife, as well as his sons, *Celer*, *Certus* and *Gratus* but also his parents, *Minicius Calvus* and *Minicia Prisca*.[123] Since the monument is usually dated to the late 1st or early 2nd century AD, *Minicius Saturninus* could likely have been among the first colonists concerned by the deduction of AD 71, even more so since he was a sailor.[124] Presuming that he served till the age of forty-five or fifty, he may have retired around AD 71 and lived a happy life till the beginning of the 2nd century AD. His origin does not appear to be Pannonian, but since his parents were likely freedmen, one cannot exclude that they were natives of Illyricum and that *Saturninus* actually returned to his ancestral country.

The stele of *Caius Antonius Sentinus* (S I.3), a veteran of the *legio XIV Gemina*, who died at the age of fifty, is contemporaneous and could be dated to the late 1st or early 2nd century AD.[125] The man does not appear to have had a local origin and we may presume that he was from northern Italy, like many legionaries in the 1st

[112] Ritterling 1925: 1747–1748 ; Šašel Kos 1995: 236–237; Wheeler 2000: 261–268, 270–272; Mosser 2003: 138–147; Farnum 2005: 23, 64–65; Radman-Livaja 2012: 163–164, 169–170; Radman-Livaja 2016: 185.

[113] Von Petrikovits 1980: 1027–1035; M. P. Speidel 1989: 239–247; Gilliver 1999: 29–31; Roth 1999: 91–115; Feig Vishnia 2002: 265,–272; Radman-Livaja 2007: 166–167; Radman-Livaja 2016: 185–186.

[114] Or perhaps traces of the presence of their direct descendants later during the 1st or in the 2nd century AD.

[115] Radman-Livaja 2014: 142; *Bano Saturi* (Radman-Livaja 2014: cat. 21.18), *Claudia Cnitinia* (Radman-Livaja 2014: cat. 15.21), *G(a)etulus* (Radman-Livaja 2014: 12.08), *Repentinus Afer* (Radman-Livaja 2014: cat. 24.40); Radman-Livaja 2016: 184–186.

[116] It is interesting to note that several *tesserae* from Sisak bear the female name *Fortuna*, which has almost exclusively been recorded on the Iberian peninsula and in northern Africa; Radman-Livaja 2014: cat. 21.86, 26.115; Radman-Livaja 2016: 185.

[117] *Auso* (Radman-Livaja 2014: cat. 01.08), *Saposa* (Radman-Livaja 2014: cat. 24.32), *Satto* (Radman-Livaja 2014: cat. 21.17), *Siarus* (Radman-Livaja 2014: cat. 15.07), *Lucius Toutus* (Radman-Livaja 2014: cat. 01.46), *Touta* (Radman-Livaja 2014: cat. 21.91), *Exoni(i)Verca* (Radman-Livaja 2014: cat. 26.67), *Cabra Terti(i)* (Radman-Livaja 2014: cat. 15.20), *Cappo* (Radman-Livaja 2014: cat. 01.01), *Carisio* or *Carisius* (Radman-Livaja 2014: cat. 23.20), *Mancita A(u)gustia* (Radman-Livaja 2014: cat. 06.16), *Aquilina Lucci* (Radman-Livaja 2014: cat. 19.11), *Adiutor Lucci* (Radman-Livaja 2014: cat. 18.17), *Lucus* (Radman-Livaja 2014: cat. 01.73), *Lecus Liccaius* (Radman-Livaja 2014: cat. 04.14), *Lalos Lani* (Radman-Livaja 2014: cat. 24.31), *Fressa* (Radman-Livaja 2014: cat. 21.24), *Sextus Exomni(i)* (Radman-Livaja 2014: cat. 13.07) and *Gristus Felicis* (Radman-Livaja 2014: cat. 26.37). One may add to this group individuals bearing patronymic *gentilicia*, a rather common practice in Gaul, such as *Exonius Sc(a)ev(i)nus* (Radman-Livaja 2014: cat. 26.112), *Senecius Aper* (Radman-Livaja 2014: cat. 23.40), *Singonius Rufinus* (Radman-Livaja 2014: cat. 19.46), *Vindius Ammamus* (Radman-Livaja 2014: cat. 26.95), perhaps also *Marcus Tescius* (Radman-Livaja 2014: cat. 21.88) and *Nigrinus Cinelius* (Radman-Livaja 2014: cat. 24.19). Interestingly, two lead tags seem to mention inhabitants of Andautonia, presumably clients of Siscian textile craftsmen (Radman-Livaja 2014: 76–78), who also bear patronymic *gentilicia*, *Devesius Nebio* (Radman-Livaja 2014: cat. 07.05) and Melavius Atedunus (Radman-Livaja 2014: cat. 23.63; Radman-Livaja 2016: 174–179).

[118] Radman-Livaja 2016: 186–187.

[119] Radman-Livaja 2014: cat. 23.46.

[120] Radman-Livaja 2014: 89, 119–120, 133.

[121] *CIL* III 10854; Brunšmid 1909: 154, cat. 346; *AIJ* 567; Mócsy 1959: 57/3; Davies 1969: 85, 98, cat. 54; Radman-Livaja and Vukelić 2015: 402–403; Migotti 2016a: 172, cat. 2.

[122] Cichorius 1900: 356; Wagner 1938: 201; Kraft 1951: 199; Mócsy 1962: 624; Mócsy 1974: 81; P. Holder 1980: 156; Lőrincz 2001: 44; Radman-Livaja 2012: 171; Radman-Livaja 2016: 186; J. Spaul disagrees and considers that this unit was not raised before Commodus' reign, see Spaul 2000: 47–48.

[123] Radman-Livaja and Vukelić 2015: 399, 404.

[124] *CIL* XVI 14; Mócsy 1974: 112–114; Ferjančić 2002: 46–47.

[125] Škrgulja and Migotti 2015: 44–45; Migotti 2016a: 175–176, cat. 7.

century AD. Since his heir is not explicitly mentioned, we do not know if he founded a family in Siscia. His unit did serve in Illyricum during the Augustan period but it moved a considerable amount afterwards, having been stationed in Upper Germany and Britain most of the time, with a brief stay in Dalmatia or Pannonia before the Civil War of AD 69. Presumably, *Sentinus* joined the *legio XIV Gemina* during the Flavian period, likely during Vespasian's reign. The legion was transferred from Mogontiacum to Vindobona around AD 92 and might have briefly stayed in Mursella, near Mursa, few years later. After taking part in the Dacian wars, the *XIV Gemina* ended up in Carnuntum 106, where its presence was still recorded in the early 5th century AD. *Caius Antonius Sentinus* was likely discharged by the time his legion moved to Carnuntum, but he could have taken part in Domitian's war against the Marcomanni and Quadi. He may have had the occasion to stay in Siscia while on active duty, if a vexillation of his legion was temporarily stationed at Siscia in Domitian's time, but this is mere conjecture. Nevertheless, he settled in Siscia after his discharge, for reasons we can only guess.

Besides funerary monuments, diplomas may also point to veteran settlement in Siscia. There are only two of them and one is badly fragmented. The first was issued in AD 100 to a Cilician named Sapia, who presumably did not return home after his retirement but settled instead in Siscia.[126] Interestingly, the man quite certainly served neither in Siscia nor even in Pannonia. His unit, the *cohors I Antiochensium*, despite some claims, was most likely transferred to Moesia by AD 75 at the latest and was probably never stationed in Pannonia.[127] The second diploma belonged to another Oriental, an unknown soldier, son of a certain Saraba.[128] It was likely issued in Syria between AD 127 and 136 and one may only conjecture how this veteran could have ended up in Siscia. At best, considering these two diplomas, we may speculate that in the early 2nd century AD the city may have been attractive enough even to veterans who were not natives of the region.

An active duty soldier of the already mentioned Pannonian legion, *XIV Gemina*, was buried in Siscia by his spouse in the late 2nd or rather early 3rd century AD (S III.1). The *cornicularius Caius Sempronius Severus* died at the age of forty-one, but his bereaved wife *Iulia Florentina* provides no clues to his ties with Siscia.[129] While one may imagine a *cornicularius* carrying on certain tasks away from his garrison, his burial in Siscia might rather imply a local origin and not necessarily some official duty there. It has to be pointed out that an altar to the Genius of the province was erected in Siscia during the 2nd century by an *optio* of the same legion, a certain *Tacitus*, or perhaps more likely *T(itus) A( ) Citus*.[130] We may presume that this non-commissioned officer was sent to Siscia for some official reasons and we may thus speculate that men of the Upper Pannonian *legio XIV Gemina* were visiting the city on a regular basis.

Slightly more recent, i.e. closer to the mid 3rd century AD, is the ash-chest of *Marcus Aurelius Glabrio* (S III.2), a veteran from the same legion, *XIV legio Gemina*, but also a *beneficiarius*.[131] The epigraphic record of *beneficiarii* in Siscia is not scant at all, but it consists almost exclusively of votive monuments, this being the only funerary inscription. Nothing in the inscription points to the possibility that he was stationed in Siscia prior to his retirement. He may actually have been a native of Siscia who simply returned home after his retirement. Nonetheless, his legion is known to have provided *beneficiarii* in Pannonia (few of them served in Siscia as a matter of fact), and he may have served in Siscia after all.[132]

A fragmentary stele (S[t] I.6) dated to the mid 3rd century AD was found a bit further away from the city, but definitely within the *ager* of Siscia, in Odra Sisačka. Both the top and the inscription field are missing, but one can see in the portrait niche, besides a man holding a scroll and a woman, a man wearing a sagum fastened with a round brooch and a belted tunic holding a sword tucked under his left arm.[133] Presumably, one of the deceased men was a soldier, likely an active duty soldier, but not much else might be surmised lacking the inscription.

There are several other funerary monuments belonging to veterans and soldiers discovered in the larger area around Siscia. All of them may be considered, with a higher or lesser level of certainty, as being raised by inhabitants of the ager of Siscia and should thus be taken into account in our present discussion. The earliest one is a fragmentary stele found in Topusko (St I.1), 38 km

[126] *CIL* XVI 46; *AE* 1912: 128; Mócsy 1959: cat. 57/4.

[127] Wagner 1938: 86–87; Kraft 1951: 166; Mócsy 1962: 621; Beneš 1978: 16; Spaul 2000: 424; Lőrincz 2001: 46; Radman-Livaja and Vukelić 2015: 399–400.

[128] *CIL* XVI 103; *AE* 1921, 94; Mócsy 1959: cat. 57/23; *RMD* III: 245–246, note 43; *RMD* V: 702, note 18.

[129] *CIL* III 10855 (=3972); Brunšmid 1909: 158, cat. 350; *AIJ* 568; Barkóczi 1964: 29/48; Radman-Livaja and Vukelić 2015: 402–403.

[130] *CIL* III 3943.

[131] *CIL* III 3970; Brunšmid 1909: 151–152, cat. 343; *AIJ* 566; Barkóczi 1964: 29/17; *CBI* 313; Migotti 2005b: 367–381; Radman-Livaja and Vukelić 2015: 404.

[132] *Beneficiarii* were normally spending a relatively short period of time in a given *statio*, six months to a year or two, although returning to the same *statio* after a while was not unheard of. Since they moved a good deal within a province during their service, choosing a place to settle after being discharged probably depended less on bonds of friendship than was usually the case with average soldiers who spent most of their service in the same garrison place. Presumably, family ties must have been the main factor determining the choice of a retirement place, but big cities like Siscia, Sirmium or Poetovio could have been attractive to any Pannonian veteran, whether he lived there previously or not; Rankov 1986: 243–248; Ott 1995: 105–106; Dise 1997: 285–294; Nelis-Clément 2000: 52, 138–140, 144–145, 203–204; Radman-Livaja and Vukelić 2016: 206–207.

[133] Migotti 2013c: 266; Migotti 2016a: 180–181, cat. 19.

from Sisak, erected for the *XV legio Apollinaris* veteran *Lucius Valerius Verecundus*.[134] He died at the age of sixty-five during Trajan's reign and his epitaph clearly states his origin, *domo Siscia*. Interestingly, his funerary monument was partly funded by a *collegium veteranorum*. Considering his age, he could have enlisted already in the late Julio-Claudian period, under Nero, but he must have spent most of his career serving Flavian emperors and was likely discharged before Trajan's accession to the throne. He might have had a fairly eventful military career if he joined the army under Nero. Since the XV legion was sent to the East around AD 63, he could have taken part in the war with Parthia and subsequently in the First Jewish war. If he had joined only after the XV legion came back to Carnuntum from Judea in AD 71, his career may have been less exciting, although towards the end of his career he could have been involved in Domitian's wars. Whether that is true or not, he held no rank and was likely discharged as a simple legionary. This monument is quite significant in several respects. While this is hardly a surprise, it undeniably shows that there were Roman citizens dwelling in Siscia before the town became a colony. We do not know the family background of *Lucius Valerius Verecundus*, except that he was clearly born in Siscia as a Roman citizen, but his parents, or at least his father, could have been originating from a western province or even Italy.[135] We may even conjecture that *Valerius Verecundus* was also the son of a veteran or army follower who stayed in Siscia after his discharge. The mention of the *collegium veteranorum* clearly implies that he was not the only legionary veteran in the area when he passed away. They must have been sufficiently numerous to organise an association and we may thus presume that Siscia and its ager attracted quite a few veterans in the late 1st and early 2nd century AD. *Valerius Verecundus* did not choose to settle in the city itself, but seemingly preferred to retire in Topusko, likely *Ad Fines* or *Quadrata* in Roman times (see at Ch. II.1). Unless his family already had an estate in the area, we may presume that the well-known thermal station may have enticed him, as well as other veterans, to settle there. It has to be pointed out that there is no certain proof that *Ad Fines* was within the ager of Siscia, but this does not seem unlikely considering the distance separating the two settlements.

Further south-east, in present-day Bosnia, a stele (S[ptSE] I.4), dated to the 3rd century AD, was found approximately 60 km from Siscia, in the village of Cikote. The deceased woman was the wife of a veteran of the *X Gemina* and a former beneficiarius, *Marcus Aurelius Surus*.[136] The latter obviously spent his career in Pannonia, and we may conjecture that he was a Pannonian who retired to his homeland. Considering the distance from Siscia, it would be rather far-fetched to claim that the place where he retired (and his presumed place of birth) was in the territory of the colony. Nonetheless, *Marcus Aurelius Surus* may have appreciated the fact that he did not retire too far from the largest urban centre in the area. A slightly later stele (S[ptSW] I.4), erected by his parents to the deceased thirty-year-old legionary *Valerius Saturninus*, was found in Svojić, over 70 km south-west of Siscia.[137] This soldier of an unknown legion may, however, hardly be considered as a native Siscian. At best, we may say that military service in the late 3rd century remained an interesting career for people inhabiting the larger area around Siscia. A similar conclusion might be reached in regard to the contemporaneous fragmentary funerary monument found in Bosnia, in Rakanske Barice, 55 km south of Sisak (S[ptSE] IV.1). The deceased boy *Heliodorus* appears to have just started his military career before he died. As a matter of fact, the eighteen-year-old might have died just when he was joining the army.[138]

The rather small number of soldiers' and veterans' funerary monuments found in Sisak and the surrounding larger area could lead us to the assumption that enlisting into the military was not much of a prospect for Siscians, but this would be quite misleading. There is actually a sizeable epigraphic record from other areas of the Empire, both elsewhere in Pannonia or in other provinces, proving that already from the 1st century onwards quite a few inhabitants of Siscia, both citizens and *peregrini*, chose to follow a military career.[139] Siscians were quite certainly joining the army already during the Julio-Claudian period. Four of the witnesses recorded on an early praetorian diploma issued after the civil war of AD 69, more precisely on 30 May, 73 AD, were Siscians: *Caius Aconius Maximus, Titus Flavius Festus, Marcus Lucilius Saturninus*, and *Marcus Rutilius Hermes*.[140]

[134] *ILJug* 3117; *AE* 1989: 617; Šegvić 1988: 57–62; Mosser 2003: 102, 258, no. 182; Radman-Livaja and Vukelić 2015: 404; Šašel Kos 2017b: 179.

[135] The *gentilicium Valerius* is too widespread to allow conjecture about people's origins, but *Verecundus* appears to be more common in Gaul and Italy, although it is far from being infrequent elsewhere; for *Valerius* as a *gentilicium* see Mócsy 1959: 160; Barkóczi 1964: 295, 303; Alföldy 1969: 131–133, *s.v. Valerius*; Mócsy 1983: 300, s.v. Valerius; Mócsy 1984b: 47; Mócsy 1985a: 80–87; Abascal Palazón 1994: 232–244, *s.v. Valerius*; Solin and Salomies 1994: 197, *s.v. Valerius*; Minkova 2000: 93–96, *s.v. Valerius*; *OPEL* IV: 143–146, 199, *s.v. Valerivs*; Tataki 2006: 424–431; Radman-Livaja 2014: 278; for *Verecundus* see Dean 1916: 57; Mócsy 1959: 21, 195–196; Barkóczi 1964: 327; Kajanto 1965: 68, 264; Evans 1967: 279–280; Alföldy 1969: 324, *s.v. Verecundus*; Mócsy 1983: 307, *s.v. Verecundus*; Solin and Salomies 1994: 420, *s.v. Verecundus*; Degavre 1998: 163, *s.v. condo-*, 441, *s.v. ver-*, 442, *s.v. verco-*; Minkova 2000: 275, *s.v. Verecundus*; Dondin-Payre 2001a: 302, 305; Forier 2001: 505–506; Lefebvre 2001: 597–647; Raepsaet-Charlier 2001: 408, 422; Rémy 2001: 103, 171; *OPEL* IV: 157–158, 199, *s.v. Verecvndvs*; Delamarre 2003: 124, *s.v. condo-*, 314, *s.v. uer(o)-*; Radman-Livaja 2014: 281.

[136] Sergejevski 1957: 116–118, no. 8, pl. III; *AE* 1958: 66; *ILJug* 163; *CBI* 450.

[137] *CIL* III 14333,1; Brunšmid 1907: 183, cat. 339; Migotti 2010a: 93–114.

[138] *CIL* III 8367a = 13242 (+ p. 2328, 159); Patsch 1900: 63–64, Fig. 37; Paškvalin 2012: 36, no. 7.

[139] Radman-Livaja and Vukelić 2015: 403–404; Šašel Kos 2017b: 179–189.

[140] *CIL* XVI 18; Brunšmid 1898b: 144–149; Šašel Kos 2017b: 184–185.

They all happened to be in Rome when the diploma was issued and we may reasonably assume that they were all soldiers.[141] It is impossible to tell in which unit(s) they may have served, but considering their names, three of them were likely born as citizens, and could thus have been enrolled in the legions. *Caius Aconius Maximus* and *Marcus Lucilius Saturninus* could have been Siscians of Italian origin, while *Marcus Rutilius Hermes* could have been the son of a freedman. *Titus Flavius Festus* almost certainly received his citizenship under Vespasian and he must have initially served in an auxiliary unit. After the turmoil of the Civil War, all of those men may have been transferred to Rome from some of the units which distinguished themselves fighting for Vespasian, but some might have been serving there as praetorians or in the urban cohorts, even before Vespasian's accession to the throne.

A late 1st-century AD funerary inscription from Arrabona mentions a deceased trooper of the *ala I Aravacorum*, a certain *Crispus Mac[ _ ]* from Siscia. The tombstone was erected by his *decurio* and his heir, *Ianuarius*, perhaps a countryman as well.[142] The above-mentioned, likely early 2nd-century altar dedicated to Jupiter from Samaria in Judaea, is particularly interesting in this regard. It was erected by the *cives Sisciani et Varciani et Latobici, milites vexillationis cohortium Pannoniae superioris* who were probably suppressing one of the two Jewish revolts, under Trajan or Hadrian.[143] This inscription certainly illustrates well how close those Pannonians from neighbouring communities must have felt in a distant foreign land.

Around the mid 2nd century AD, *Publius Crescentinius Flavius Saturninus*, a soldier aged thirty-seven of the *XI Claudia*, died in Mauretania Caesariensis and was buried in Portus Magnus, rather unusually, by his daughter *Flavia Ianuaria*.[144] His Moesian legion likely sent a vexillation to Mauritania, but we do not know for what reason. The fact that his daughter followed him to Africa is even more puzzling. Somewhat later, in the late 2nd or early 3rd century, another Siscian legionary, this time from the *III Augusta*, served among the *beneficiarii consularis* on the staff of the legate of Numidia.[145] *Titus Aelius Victorinus*' name appears on a list of members of the legate's *officium*. Towards the 2nd half of the 2nd century AD, *Lucius Lucanius Festus*, a native of Siscia and a soldier of the *XIV Gemina*, was buried at Brigetio. His tombstone was erected by a trumpeter of the same legion, a certain *Marcus Ulpius Victor* who may have been a countryman of *L. Lucanius Festus*.[146] Interestingly, their legion was at that time stationed in Carnuntum and we may only wonder what they were doing in Brigetio. A veteran from Siscia, a *duplicarius* of the *cohors I Lusitanorum, [Caius] Art(orius) Satur(ninus)*, was buried in *Tropaeum Traiani* by his children in the second half of the 2nd century AD.[147] Another Siscian, *Iulius Nigellio*, was serving in Brigetio in the early 3rd century AD, as a *beneficiarius* of the legionary tribune of the *legio I Adutrix* and a *regionarius* (i.e. a soldier in charge of policing an assigned district). He died in service and his colleague *Aelius Paternianus* had a tombstone commissioned for him.[148] Two fragments found in Apulum in Dacia in 1961 were believed to belong to a tombstone of a certain *Aelius Propincus*, a *librarius consularis* from the *XIII Gemina*, but it would appear that those fragments did not belong to the same tombstone.[149] In fact, the 3rd-century AD Siscian legionary from the *XIII Gemina* must remain another, unfortunately, anonymous man.[150] In the 3rd century AD, a centurion named *Marcus Aurelius Lucius* erected a tombstone to his former slave, a certain *Marcus Aurelius Saturninus*. One of the two was certainly from Siscia (*domo Sescia(!) ex Pannonia Superiore*), but the inscription is not exactly unequivocal and it is uncertain whether the centurion of the *XIII Gemina* actually originated from that town.[151] An altar to Victoria Augusta, dated to AD 222, is not ambiguous as far as the origin of the *primus pilus* of the *I Minervia* is concerned. *Gaius Publicius Priscilianus* clearly states that he is from Septimia Siscia.[152] We also know of two Siscian legionaries from the *II Parthica*, one *Valentinus* and one *Marcellinus*, whose names appear on a fragmented list of soldiers dated to the 3rd century AD.[153] Two Siscian troopers of the *ala Ulpia contariorum milliaria civium Romanorum*, both of them *librarii*, are mentioned on one funerary stone from Arrabona dated to the 1st half of the 3rd century AD. The deceased man was *Aurelius Saturnio*, while his heir was *Luc(ilius?) Vindex*.[154]

Many Siscians serving in the army are recorded among the praetorians as well as the *equites singulares*. This is undeniably mostly due to Septimius Severus' accession to the throne, when he filled the ranks of the Praetorian Guard with loyal troops from Danubian legions. It also shows that quite a few Siscians served

141 Morris and Roxan 1977: 300, 326.

142 *CIL* III 4373; *RIU* 259; Lőrincz 2001: 197, cat. 130; Šašel Kos 2017b: 184.

143 *AE* 1938, 13; Saxer 1967: 60–61; Dobó 1975: 53, cat. 219; Radman-Livaja and Vukelić 2015: 404; Šašel Kos 2017b: 183–184.

144 *CIL* VIII 9761; Mócsy 1959: cat. 57/9; Dobó 1975: 47–48, cat. 180; Radman-Livaja and Vukelić 2015: 404; Šašel Kos 2017b: 180–181.

145 *CIL* VIII 2586; Barkóczi 1964: cat. 29/77; Dobó 1975: 46, cat. 165; Radman-Livaja and Vukelić 2015: 404; Šašel Kos 2017b: 181.

146 *CIL* III 11029; *RIU* 566; Šašel Kos 2017b: 179.

147 *CIL* III 14214, 9; Mócsy 1959: cat. 57/7; Dobó 1975: 49, cat. 192; Radman-Livaja and Vukelić 2015: 404; Šašel Kos 2017b: 185.

148 *AE* 2008: 1086; Borhy 2010: 70–73; Šašel Kos 2017b: 179–180.

149 *AE* 1965: 35; Dobó 1975: 49, cat. 189; Radman-Livaja and Vukelić 2015: 404.

150 *AE* 1982: 826a; Šašel Kos 2017b: 181–182.

151 *AE* 1941, 161; Dobó 1975: 58, cat. 263; Šašel Kos 2017b: 182.

152 *CIL* XIII 8035; Barkóczi 1964: cat. 29/65; Dobó 1975: 44–45, cat. 157; Radman-Livaja and Vukelić 2015: 404; Šašel Kos 2017b: 182–183.

153 It is less likely for a certain Glauc[us ?]; *AE* 1964: 14; Dobó 1975: 45, cat. 161; Radman-Livaja and Vukelić 2015: 404; Šašel Kos 2017b: 182.

154 The *gentilicium* of the heir might also have been *Lucanius* or *Luc(c)ius*; *CIL* III 13441; *RIU* 282; *AE* 2001: 1641; Lőrincz 2001: 192, cat. 111; Šašel Kos 2017b: 184–185.

in those legions in the last decades of the 2nd century. Pannonia remained an important recruiting ground for the Guard throughout the reign of the Severan dynasty and likely later as well. Thus, not all praetorians from Siscia served under Septimius, but also under his heirs, as shown by the rather high number of *Aurelii* among them.[155] Nonetheless, some probably managed to join the elite guard unit even earlier. It is for instance quite likely that this happened for men such as *Sextus Iulius Augurinus*, a *beneficiarius praefecti* from the 7th Praetorian Cohort,[156] or *Caius Valerius Spectatus*, a soldier from the 8th Praetorian Cohort,[157] both dying in service and buried in Rome. *Marcus Lucilius Proculus* was a *fisci curator* in the 6th Praetorian Cohort and he died in Rome aged thirty-three in the 2nd century AD.[158]

Many more Siscians are mentioned on the fragmentary lists of the members of the Praetorian Guard from the Severan period. One finds a certain *Titus Flavius Provincialis* on a list dated to AD 183–184.[159] We know, from 209 AD, of individuals such as *[ - - - - ]lus*, as well as another anonymous Siscian,[160] as well as *Proculeianus* and *Lucius Septimius Lucanus*.[161] Other praetorians from Siscia are known as well, their names appear on lists which may be dated towards the very end of the 2nd or the early 3rd century AD, such as *Marcus Aurelius Valentinus* and *Marcus Aurelius Firmus*,[162] *Marcus Aurelius Verus* and *Marcus Aurelius Licinius*,[163] *Lucius Marius Candidus* and *Marcus Aurelius Nero*,[164] *Aurelius Tato*,[165] as well as a certain *Restitut[us]*, listed with another anonymous Siscian.[166] Another praetorian from Siscia, *Marcus Aurelius Dasius*, from the 5th Praetorian Cohort, died in Rome at the same period and was buried by his brother who served among the *equites singulares*, *Marcus Aurelius Candidus*.[167] Another anonymous Siscian soldier, a certain *[ - - - - ]nian[u]s*, not unlikely a praetorian, is also mentioned on a 3rd-century AD fragmentary list.[168]

Some of the known *equites singulares* from Siscia were quite certainly recruited before Septimius Severus' reign, following a traditional career path, distinguishing themselves while serving in auxiliary units before being transferred to this elite corps. Nonetheless, it is not unlikely that more joined the *equites singulares* during the Severan period. *Aelius Lucius*, *eques singularis Augusti*, was definitely recruited earlier during the 2nd century AD, since his hometown is still called *Fl(avia) Siscia*.[169] Interestingly, his tombstone was erected by his countrymen, indicating that there were other Sisciani serving with him. Also dated to the 2nd half of the 2nd century would be the tombstone of *Publius Aelius Valentinus*, a young *eques singularis Augusti*, also from *Flavia Siscia*, buried in Reate.[170] His heir, the signifier *Claudius Titianus* may perhaps have been his countryman as well. The already mentioned *eques singularis Marcus Aurelius Candidus*, erected a tombstone to his brother, *M. Aurelius Dasius* from the 5th Praetorian Cohort towards the end of the 2nd century AD, or perhaps rather in the early 3rd century AD.[171] Another Siscian died during the 2nd century AD while serving as the *summus curator equitum singularium consularis*, i.e. the accountant of the bodyguard of the governor of Pannonia Superior in Carnuntum. *Lucius Genucius Exsoratus* was also from *Flavia Siscia* and he must have been transferred to the staff of the governor from a unit garrisoned in Upper Pannonia.[172]

Astonishingly for such a big city that became a colony after a deduction, and which had a large garrison for almost 80 years, the number of military funerary monuments is surprisingly low. Although Siscia has not yielded an impressive number of stone monuments and inscriptions, one would still expect to find more soldiers and veterans among the deceased. Presumably, a large number of funerary monuments have been lost forever, and what is preserved now is hardly representative of what was once on display along the roads leading from Siscia. When the city actually had a large garrison, during the Julio-Claudian period, staying in Pannonia after being discharged was probably not much of a prospect for most veterans and we may speculate that many returned home or settled in the colonies on the Adriatic coast. Nonetheless, one may wonder what happened to the funerary monuments of active duty soldiers. There are quite a few in the Rhine provinces from the very same period but they do not abound in Siscia, except for the lost monument of *Tiberius Claudius Pontius* and *Mucius Hegetor*'s stele.

After the departure of the last garrisoning unit in the Flavian period, the army presence until Gallienus's

[155] Dury 1968: 247–249; Birley1969: 64–65; Šašel Kos 2017b: 186–185.
[156] *CIL* VI 2644; Barkóczi 1964: cat. 29/69; Dobó 1975: 30, cat. 62; Radman-Livaja and Vukelić 2015: 404; Šašel Kos 2017b: 187.
[157] *CIL* VI 2689; Mócsy 1959: cat. 57/20; Dobó 1975: 30, cat. 63; Radman-Livaja and Vukelić 2015: 404; Šašel Kos 2017b: 187.
[158] *AE* 1984: 68; Šašel Kos 2017b: 187.
[159] *EE* IV 891; *CIL* VI 37184 (32523); Barkóczi 1964: cat. 29/70; Dobó 1975: 30, cat. 63a; Radman-Livaja and Vukelić 2015: 404; Šašel Kos 2017b: 187.
[160] *CIL* VI 32533 (2385); Dobó 1975: 30, cat. 63b; Radman-Livaja and Vukelić 2015: 404.
[161] *CIL* VI 32536; Barkóczi 1964: cat. 29/71; Dobó 1975: 31, cat. 31, cat. 63c; Radman-Livaja and Vukelić 2015: 404; Šašel Kos 2017b: 187.
[162] *EE* IV 894; *CIL* VI 32624; Barkóczi 1964: cat. 29/72; Dobó 1975: 31, cat. 63d; Radman-Livaja and Vukelić 2015: 404; Šašel Kos 2017b: 186.
[163] *CIL* VI 32627; Barkóczi 1964: cat. 29/73; Dobó 1975: 31, cat. 63e; Radman-Livaja and Vukelić 2015: 404; Šašel Kos 2017b: 186.
[164] *CIL* VI 32628; Barkóczi 1964: cat. 29/74; Dobó 1975: 31, cat. 63f; Radman-Livaja and Vukelić2015: 404; Šašel Kos 2017b: 186.
[165] *CIL* VI 32640; Barkóczi 1964: cat. 29/75; Dobó 1975: 31, cat. 63g; Radman-Livaja and Vukelić 2015: 404; Šašel Kos 2017b: 186.
[166] *CIL* VI 2388; Barkóczi 1964: cat. 29/76; Dobó 1975: 31, 63h; Radman-Livaja and Vukelić 2015: 404.
[167] *EE* IV 903; *CIL* VI 32680; Barkóczi 1964: cat. 29/68; Dobó 1975: 30, cat. 61; Radman-Livaja and Vukelić 2015: 404; Šašel Kos 2017b: 186.
[168] *CIL* VI 32914; Dobó 1975: 49, cat. 190; Radman-Livaja and Vukelić 2015: 404.
[169] *CIL* VI 3180; Mócsy 1959: cat. 57/5; Dobó 1975: 39, cat. 116; Radman-Livaja and Vukelić 2015: 404; Šašel Kos 2017b: 187.
[170] *AE* 2000: 413; Šašel Kos 2017b: 187–188.
[171] *EE* IV 903; *CIL* VI 32680; Barkóczi 1964: cat. 29/68; Dobó 1975: 30, cat. 61; Šašel Kos 2017b: 186
[172] *CIL* III 4471; Šašel Kos 2017b: 188.

time was likely limited to *beneficiarii*, some vexillations and smaller outfits, which probably never stayed for long, and a small naval base of which we do not know much.[173] According solely to the epigraphic monuments found in Siscia, it would not seem that veterans chose to spend their retirement there on a regular basis. The sailor *Minicius Saturninus* was likely enticed to settle there when the city became a colony and there must have been many other men like him in AD 71, but they have not left epigraphic traces.

Veterans from Pannonian legions were perhaps interested in spending the rest of their lives in a large provincial urban centre such as Siscia during the 2nd and 3rd century AD, and few funerary monuments certainly confirm that, but one may wonder how many actually chose to settle there. It is hard to believe that *Caius Antonius Sentinus* or *Marcus Aurelius Glabrio* were among the few who dwelled in Siscia after their discharge, but the scant epigraphic record does not allow much more than mere conjecture. *Caius Sempronius Severus* died on active duty but was not buried in his garrison city, Carnuntum. He could have been a native of Siscia and we may presume that his wife and family wanted his ashes to be buried in his home town.

Admittedly, one would have expected a stronger presence of veterans, if not active soldiers in Siscia, considering the size, the history and the economic importance of the city. Puzzling or not, this does not appear to be the case according to the rather limited epigraphic record, although one of the preserved monuments (S[t] I.1) undeniably attests to the existence of a *collegium veteranorum* in the early 2nd century AD at the latest. This would likely imply the presence of a significant number of veterans in Siscia. Funerary and votive monuments from other areas of the Empire highlight, however, the relatively high number of soldiers from that town during the 2nd and 3rd centuries AD. As a matter of fact, the available epigraphic record provides evidence for quite a few Siscians serving in the 1st century AD as well. Considering all of the available data, it appears that Siscians were, rather, joining the legions, which is after all quite understandable. While *peregrini* must have been numerous in Siscia, the proportion of citizens in the colony was certainly also important. Even before the Flavian period, Roman citizens must have represented a fair percentage of the population within the settlement itself, but likely a far lower number in the surrounding countryside, which likely provided more recruits for the *auxilia*. Presumably, quite a few citizens were descendants of Italians, among whom many were likely veterans. It would not be too far-fetched to consider that the military profession was a family tradition, both for descendants of legionaries and those natives whose relatives initially served in the *auxilia*. Unlike Andautonia or Aquae Balissae, in Siscia the tombstones of veterans do not represent a major percentage of preserved funerary monuments. In smaller, predominantly rural communities, like those two just mentioned, veterans were likely more visible in the local society, being more affluent than most and thus able to commission higher quality tombstones, which are now basically the only preserved trace of their status in their community. Siscia, as a big town, offered far more opportunities for climbing the social ladder, service in the army being just one of them, besides trade and craft for instance. Also, in a large town where a prosperous middle class, not to mention the really affluent families, was more likely to exist, veterans would probably blend more, not being among the few who could claim a higher social status. However, we should not consider veterans and families of active soldiers as playing a minor role in Siscia's life. As already pointed out, while the number of relevant funerary monuments in Siscia is poor, the available epigraphic evidence clearly demonstrates that many families in Siscia must have had close relatives in the army. Just like in Andautonia and Aquae Balissae, most preferred serving in Pannonia, not too far from home; however serving in Rome was nonetheless an attractive prospect when the opportunity arose. It is impossible to estimate how many of those soldiers eventually returned home as veterans, but even if most preferred retiring in their garrison places, returning veterans, with their families or single, must have been a recurring sight in Siscia. After all, many chose to serve in Pannonia, and returning back to the old home was far from being a daunting task. While campaigning troops were probably not a common sight in 2nd-century Siscia, at least from Trajan's Dacian wars until the Marcomannic wars, soldiers, as sons, brothers and cousins, were present in the mind of many Siscians. Some may even have contributed to the wealth of their relatives back home, if they were sending money or luxury items. For those serving in Pannonia, coming home for leave was fairly easy and quite a few could have been able to preserve close ties with their relatives. Having a cousin serving not too far on the Danube *limes* and enjoying a good life must have enticed many a young man to join as well. Veterans who came back as relatively affluent men likely played an important role in the everyday life of their neighbourhoods. Their sons were also likely to follow their professional path. Although Siscia ceased being a garrison city around the time when it became a colony, it never severed its ties to the army. Those became, incidentally, somehow more personal, one may even say more intimate, since for many Siscians soldiers were not just members of the Imperial armed forces,

[173] *Notitia Dignitatum, OC. XXXII* 56; Klemenc 1961: 9; Mócsy 1962: 625; Šašel 1974: 734; Reddé 1986: 298–299; Hoti 1992: 143; Zaninović 1993: 56; Domić Kunić 1995: 87–91; Radman-Livaja 2004: 19–22; Radman-Livaja 2010: 194–201; Radman-Livaja and Vukelić 2016: 205–211.

but relatives or at least colleagues of their relatives, as well as comrades in arms from the point of view of the veterans. While not easily perceptible, it is reasonable to suggest that soldiers and veterans must have had quite a social impact in Siscia, just as must have been the case in Andautonia and Aquae Balissae, despite the lack of garrisons in the latter cities.

# Chapter VI
# Palaeographic and epigraphic characteristics

Marjeta Šašel Kos

## VI.1. SISCIA

### *VI.1.1. Epitaphs from the 1st and 2nd centuries AD*

Since the sample of funerary inscriptions from Siscia is rather limited, it is very difficult to assess the palaeographic characteristics of Siscian epigraphy in such a way that would enable us to draw any meaningful conclusions. Palaeographic data combined with epigraphic features do shed some light on chronology, but there is very little evidence to identify stone-cutters' workshops that had existed in Siscia during the Roman and Late Roman periods. In this sense, most of the inscribed monuments seem unique. Without the complementary iconographic and onomastic data we would often be hard pressed to determine the date of each monument.

First of all, a supposed tombstone of one Publius Aelius Marcianus should be mentioned, which was most probably invented, or wrongly ascribed to Siscia, since both individuals noted on it, including the age of the first, are mentioned in an epitaph from the territory of Aelium Cetium.[1] Adam Baltazar Krčelić (1770) allegedly copied the text from a tombstone at Sisak as F. P. AEL. MARCIANO FLAM.---PRAEF. AN. XX.---AVVAE MAXIMI---. He interpreted Marcianus as having been *flamen*, since obviously he did not understand the letters he copied. This part of the inscription from S. Leonhard (Aelium Cetium), which is otherwise entirely preserved, reads: *P(ublio) Ael(io) / Marciano f(ilio) / mil(iti) coh(ortis) IIII prae(toriae) / an(norum) XX stip(endiorum) III / et Avvae Maxsimi(!)*. It cannot be *entirely* certain how to explain the hypothesized coincidence of P. Aelius Marcianus and Avva Maximi allegedly appearing on two tombstones,[2] but it seems most plausible to agree with Joseph Wilhelm Kubitschek, who regarded the epitaph from Siscia as misunderstood or invented.[3] It would have been a most unusual coincidence that two different but homonymous individuals (Marcianus even having the same age at death!) would have been mentioned on two tombstones, while an assumption that the same persons appeared on two tombstones in two different towns is *eo ipso* hardly possible. As has been pointed out by Bruna Kuntić-Makvić, falsifications and manuscripts containing confused copies of Roman inscriptions were not an exception at the time of Krčelić.

One of the earliest tombstones from Siscia and the territory of the town is the stele of Titus Tullius Tertius from Tergeste (S I.1), carved with beautiful rectangular capitals, the letters *T* being *litterae longae*, except the second *T* in the third line. In this line, also the letter *L* is *longa*. The letters in the last, fourth line are larger, matching larger letters in the first line. The *impaginatio* of the epitaph is very careful, each second line being shorter and centred. Palaeographic characteristics are in accordance with the fact that Tertius' voting tribe is that of the inhabitants of Tergeste, *Pupinia*, indicating that most probably he became a resident of Siscia before the town became *colonia Flavia*. Only slightly later is the tombstone of Marcus Mucius Hegetor, physician of the 32nd cohort of volunteers (S I.2), displaying entirely different palaeographical features. The decorative rendering and style of the stele, too, is quite dissimilar. Letters, resembling square capitals, are irregular and of uneven height, and also the lines were not carved straight.

However, the majority of funerary monuments are from the 2nd or 3rd centuries AD, often beginning with the formula *D(is) M(anibus)*, the invocation 'to the spirits of the departed'. Such epitaphs can only be dated exceptionally to the end of the 1st century AD, on the whole they are not earlier than the beginning of the 2nd century. The stele of Gaius Antonius Sentinus, veteran of the legion *XIV Gemina*, is slightly later than the aforementioned monuments, probably from the beginning of the 2nd century AD (S I.3); there is no invocation to the *Dii Manes*. Letters are regular, but not so carefully carved as the letters on the stele of Tertius. The stone-cutter, who lacked the needed skill, noticed too late that the space for the letters in the first line will be too short. He therefore carved *N* and *I* in ligature, while the last two letters in the name Sentinus are narrower, with almost no space between them.

Several sarcophagi are from the end of the 2nd or from the 3rd century AD, among them the sarcophagus of

[1] *CIL* III 5663 (and 11806; cf. 3979) = *HD045572* (with extensive literature): *M(arco) Sextio / Vettoniano / aed(ili) m(unicipii) Ael(ii) Cet(iensium) / an(norum) LXX et / Vindae Terti f(iliae) / con(iugi) an(norum) L P(ublio) Ael(io) / Marciano f(ilio) / mil(iti) coh(ortis) IIII prae(toriae) / an(norum) XX stip(endiorum) III / et Avvae Maxsimi(!) / f(iliae) con(iugi) an(norum) LXXX / M(arco) S(extio) Sext(iano) qu(a)estor[i]o [m(unicipii)] / Ael(ii) Ceti(ensium) [a]n(norum) L f(ilio) Potta Croti f(ilia) / e[t] M(arco) S(extio) S[a]m(mo) [---] / [a]n(norum) XLVII f(ilio) s(uo) / et [---]ae an(norum) / e[t] Sex(tio) Sam(m)o / mi[l(iti)] l(egionis) II Sev(erianae) [a]n(norum) XXV.*

[2] Kuntić-Makvić 1993. According to a revised reading, the name of the woman should be Avita Maxsimi (*AE* 2003: 1320).

[3] *CIL* III p. 2187 and no. 3979: 'Dele; est pars tituli n. 5663'.

*Augustalis* Pontius Lupus (S II.2). The epitaph is carved fairly carefully, with regular rectangular capitals. Interesting features are two elongated letters, the first letter in the *gentilicium* and in the cognomen of Augustalis, a rare occurrence in the Pannonian and south-eastern Alpine areas. Another *littera longa* is *I* in *Sisc(iae)* in the third line, while in the fourth the second *O* in *soror* is carved as a small letter, due to the lack of space. Approximately from the same time is a sarcophagus with no decoration and a simply framed epitaph of Aelia Matrona (S II.1). The ductus of the letters is roughly similar; an interesting feature of this inscription are four abbreviations of otherwise current funerary formulas in the first line, instead of the usual two: *D(is) M(anibus)*. In this case the invocation to *Dii Manes*, *DM* is followed by *v(iva) f(ecit)*. In all other epitaphs of the area the formula *v(ivus/a) f(ecit)* appears in other lines, never in the first line together with *DM*. Slightly later is the sarcophagus of Septumia Marcella, adorned with a mourning winged *eros* on each side (S II.7); in terms of rendering the letters the inscription is not unlike the aforementioned sarcophagi.

Very accurately cut square capitals can be observed on a small fragment of a tombstone, which cannot be classified and which may or may not have belonged to an *aedilis* (S VIII.5). Letters are larger in the first preserved line, indicating a careful *impaginatio*; this fact, as well as the letter-forms, would certainly argue in favour of the monument having belonged to a member of the municipal upper class.

### *VI.1.2. Late Roman funerary inscriptions*

As is well known, there was a much greater variety of letter-forms employed simultaneously in the epitaphs of the Late Roman period and Late Antiquity than earlier.[4] The same phenomenon of diversity can be observed in the great variation of funerary monuments generally, ranging from simple funerary slabs to stelae, ash-chests, and sarcophagi. This is particularly true of Siscia, where relatively many Late Roman funerary monuments have been discovered.

A richly decorated stele with five portraits in the niche above the inscription field was erected by a member of the upper class (*vir egregius*) Cenius []ianus for the family depicted in the niche (S I. 8). Letters are longish and carefully carved, and a later date is indicated both by an invocation to eternal home and lasting security (*domus aeterna* and *perpetua securitas*), as well as by several ligatures. Invocation *D(is) M(anibus) et perpetuae securitati* occurs very often in the epitaphs of Noricum, Pannonia, and other provinces in the western part of the Empire, also occurring on three tombstones from Aquileia,[5] as well as on the stele of Tatonia Procula from Prijedor, included in the hypothetical *ager* of Siscia (S[ptSE] I.4). Ligatures in the Cenius inscription are concentrated in the first line: *TE* and *NE* in the word *aetern(a)e*, as well as *ET*. The epitaph of Tatonia Procula, which is carved with long regular letters, contains 16 ligatures. The letter *L* in the third line of the Cenius epitaph has a specific shape, with a small *S* added to the very short lower horizontal *hasta*.

Similar letters *L* can be observed in the epitaph on the sarcophagus of Severilla, *famula Christi*, from the 4th century AD (S II.22); however, other letter-forms in this inscription cannot be compared to those on the Cenius stele. The *impaginatio* of the Severilla inscription is not particularly careful, since the letters in the last three lines are squeezed together because of the lack of space. However, despite the squeezing of the letters, three nonetheless could not fit in the intended last line, hence another line must have been introduced, and only containing three letters that were carved at the very end of the line, not at the beginning. Letters *H* and *A* were not carved with the usual horizontal *hasta*, but with an oblique one. Several letters *T* have a wavy upper *hasta*.

'Eternal dwelling' most probably also figured on the lost stele from the Siscia environs, erected for one Messilla (S[t] I.7). The unusual abbreviations have already been explained by Theodor Mommsen as *Dom(ui) a(eternae)*. The inscription contains a very rare word *compater*, a godfather,[6] written as *conpater*. The end of the epitaph is unusual since the word-order of the last line is inverted, containing two additional errors. Instead of *Hoc est sepulcrum commune cum marito* ('This grave is shared by her husband'), the text in the inscription reads: *comune matito cum*.

The formula *domus aeterna*, written as *domu(i) aetern(a) e* in the Cenius epitaph, partly resembling spoken language, further appears in the epitaph from the 4th century AD for a small boy Gaudentius, whose age at death was described in a most precise manner, which often occurs particularly in the epitaphs for small children. This is another example of Late Roman epigraphy in Siscia, from roughly the same time, i.e. the 4th century AD. Letters are fairly regular, while punctuation marks were carved in a specific, unusual shape, resembling tiny oblique crosses. However, the punctuation mark at the beginning of the second line was engraved in the shape of a small vertical cross, perhaps indicating that the tombstone was Christian.

[4] Salway 2015: 367–369.

[5] *Epigraphik-Datenbank Clauss/Slaby*; *EDR*117175; *EDR*117781; *EDR*117939.

[6] The word is not included in the *Oxford Latin Dictionary*, but see Charles Du Fresne Du Cange *et al.*, *Glossarium mediae et infimae latinitatis*, Niort 1883–1887, s.v. *compater*.

In the Late Roman period, the usual Roman onomastic custom of *duo* or *tria nomina* gradually disappeared and children were only given one personal name. Gaudentius' epitaph reads (S IV.1): 'Eternal dwelling. To Gaudentius, the most devoted son, who lived 6 years, 6 months, 10 days, 8 hours of the night. He fulfilled the duty imposed upon him by fate.' The mention of the *Domus aeterna*, characteristic of Late Roman funerary inscriptions, may refer – although not necessarily – to the fact that the tombstone is Christian, which would be confirmed by the punctuation mark in the shape of a small cross. And, indeed, the formula is often also found in Christian epitaphs.[7] The name Gaudentius, too, was not only popular in the Late Roman period, but also among the Christians.[8] Perhaps the more accurate reading would be *Domus aeterna* (or *aeternalis*) *Gaudentio*: 'Eternal home of Gaudentius', since in vulgar Latin the dative could have been used in place of the genitive.[9]

*Memoria aeterna* occurs in the Late Roman epitaph of the lost tombstone of Aurelius Reginus Alexsius (S VIII.12). This sepulchral formula is more frequently documented in the Late Roman epitaphs of the Latin speaking part of the Empire than *domus aeterna*.[10]

A rustic stele from the 4th century AD with a crude depiction of a head of a young man was found in the village of Svojić, almost 80 km southwest of Sisak. The monument was erected by the parents F. T. and I. C. to their son Valerius Saturninus, a soldier of a legion that cannot be deciphered because of the unusual abbreviations and damaged inscription (S[ptSW] I.4). Abbreviating the names of parents in such a manner is a most unusual phenomenon; neither the father not the mother seem to have borne the *gentilicium* Valerius. The epitaph is very clumsily carved between two columns framing the inscription field. The letters are irregular, partly resembling the cursive script, particularly letters *L* with oblique lower *hasta*; letters *M* are unusually broad. Instead of *titulum posuerunt* the stone-cutter carved *posuerum*; this error, which may indicate the negligent everyday pronunciation, is not unique, since the same mistake occurred at Clusium in Etruria.[11]

### *VI.1.3. The epitaphs of Ianuaria and Leburna and their linguistic features*

Two specific cases deserve closer attention. A most interesting epitaph from the 4th century AD was carved on the marble pedestal of a small column (S VIII.17):

> *Ianuaria // zaesis.*
> Ianuaria (you will) live.

The letters in the first line are elongated, and *N* was carved as a reversed *N* (И), which is a rare occurrence. The correct verb form should be *zeses*, transcribed from Greek, with both letters *e* long. This is an invocation, but also a funerary expression, which occurs several times in Late Roman and early Christian sepulchral inscriptions. The word could also be abbreviated as *z(eses)*. Zaesis in this funerary inscription, however, should not be explained as a name, as has also been suggested.[12] In the western part of the Empire the formula *pie zeses* is frequent both as an acclamation and – more rarely – as a funerary formula;[13] it certainly does occur in epitaphs.[14] As a parallel, a fragmentary tombstone from Salonae from the 3rd century AD can be cited. It seems that in this epitaph only the word *zesaes* was carved, not *pie zeses*, referring to the feminine name Gymnasis.[15] Different transcriptions: *zaesis* and *zesaes* shed light on differing pronunciations of the Greek verb in the predominantly Latin-speaking milieu, as well as an uncertainty of how to render it in Latin. Four Greek Christian funerary inscriptions from Edessa in Macedonia should further be noted, containing the formula 'zéses en theô' (*vivas in Deo*).[16]

The name Ianuarius/a appears several times at Siscia, once on the curse tablet from the Kupa River,[17] but also on six *tesserae*: obviously the name must have been one of the most favourite in the town.[18] Due to the general great popularity of the name in Pannonia, as well as elsewhere in the Celtic speaking provinces,[19] this is not at all surprising. Because of a similarity with certain Celtic names the name can be regarded as an assonance name.[20] However, it should be noted that the name Ianuarius was most frequent both in pagan Late Roman and Christian inscriptions. In Rome this was the second most popular name among the Christian population, preceded only by Leo, Lea.[21] Evoking calendar, the name did not mean that persons bearing it were born in this month.

An unusual case, both in terms of contents and language, is a Late Roman tombstone from the end of the 3rd or the first half of the 4th century, of reddish limestone, erected for a master of the entertainment actors in mimes (S IV.4). He was a centenarian. The text reads:

> *D(is) M(anibus). / Positus est hic Leburna / magister mimariorum / [q]ui vicxit (!) annos plus / [m]inus centum.*

[7] *ILCV* 3650–3680.
[8] Kajanto 1965: 260; Alföldy 1969: 209.
[9] As in *ILCV* 3659a; kindly confirmed by Denis Feissel.
[10] *Epigraphik-Datenbank Clauss/Slaby.*
[11] *CIL* XI 2520 = *EDCS*-22100845.
[12] Cf. *ILJug* p. 6*.
[13] *Epigraphik-Datenbank Clauss/Slaby.*
[14] E.g. *ICUR* 6: 15844 = *EDCS*-20800452; *ILCV* 2216a = *EDCS*-42700514.
[15] *CIL* III 9116 = *HD*063638; see also Alföldy 1969: 213.
[16] Feissel 1983: nos. 8–11. My thanks are due to Nicolay Sharankov for kindly drawing my attention to these monuments, as well as to the tombstone from Salonae.
[17] Curbera and Jordan 1996.
[18] Radman-Livaja 2014: 215–216.
[19] *OPEL* II: 189–190.
[20] 'Deckname': Wedenig 2014: 342–343, 347.
[21] Solin 2003a: 27, 30, cf. 19.

/ *[Al]iquoties mortuus / [sum] set* (!) *sic nunquam. / [Opto v]os ad superos bene / [va]ler{a}e.*

'To the divine spirits of the departed. Here lies Leburna, master of mime actors, who lived more or less a hundred years. I died several times but never in this manner. I hope you are well in the upper world'.

The letters are regular, carefully cut with small interspaces between them and with *hederae* between the words; some letters *T* have oblique upper *hastae*. *Positus est hic* occurs rarely. This may rather be regarded as due to chance: two more instances are noted among the Christian epitaphs in Rome.[22] The name Leburna is a *hapax*, obviously derived from the name of the people of Liburni, who lived along the northern coastal regions of the eastern Adriatic and on the islands in Kvarner (Quarnero). The name Liburnus, too, only occurs once in Dalmatia.[23] Liburna would rather be expected than Leburna. *Vicxit* is carved erroneously instead of *vixit*: in *Epigraphik-Datenbank Clauss / Slaby* such a mistake is not recorded. The more correct form for *aliquoties* is *aliquotiens*, but the former is acceptable; thus it should not necessarily be emended in the transcription. However, *set* is certainly wrong and should be written as *sed*. An obvious error is also *valerae* instead of *valere*, which may perhaps be regarded as a kind of wrongly understood hyper-correctness.

Leburna or his family did not know his exact age; this was a common phenomenon among the inhabitants of the Roman Empire, also illuminated by the frequent rounding up of the years of one's age.[24] The formula *plus minus* occurs often, especially at a later time. As has been seen above this was not true of (small) children, whose ages had usually been recorded very accurately. Leburna's relatives or co-actors, or possibly his heir, clearly wished that even after his death his epitaph should reflect his profession of entertaining the audience.

## VI.2. Andautonia

The sample of the inscribed monuments from Andautonia is much smaller than that at Siscia, which is understandable since the town was smaller and in all respects of much lesser significance.

One of the earliest preserved funerary monuments is the stele of a freedman Gaius Iulius Adietumarus, who was a 100 years old, containing the usual sepulchral formula characteristic of early inscriptions: *h(ic) s(itus) e(st)* and *testamento fieri iussit* (A[t] I.1). The letters are regular and carefully cut, with small *hederae* as punctuation marks. There is a ligature *MA* in the first line, as well as *B* and the second *I* in *sibi*; *T* has a wavy upper *hasta*, letters *V* are specific in as much as their first *hasta* is more bent to the left than the other to the right. From approximately the same time is the stele of Aconia Salvia, the epitaph displaying very different letter-forms, which closely resemble the square capitals. Letters are elegant and carefully cut with rather large spaces between them; *impaginatio*, too, was attentive, as *uxori* in the one but last line is centred, leaving almost equally large uninscribed spaces on both sides.

The elaborately decorated stele from the very beginning of the 2nd century, of Titus Flavius Ateboduus, a 75-year-old veteran of the 2nd mounted cohort of the Varciani, who was adjutant to the prefect, displays perhaps the most elegant inscription among the Andautonian inscribed monuments (A[t] I.3). Letters in the first five lines are elegant and most carefully carved, slightly less so in the sixth line, where interspaces are unequal. *Impaginatio* is also meticulous, which is indicated particularly by the centred seventh, ninth, and tenth lines. The last two lines contain sepulchral formulae typical of the Principate: *ann(orum) V* and *h(eres) f(aciendum) c(uravit)*. Similar formulae appear on the stele of Lucius Egnatuleius from the beginning of the 2nd century; the letters on this monument, however, are less elegant (A[t] I.5). The epitaph of the stele from the territory of Andautonia, in which one Marcus Ulpius Geminus, son of Savus, is mentioned, was also carved with elegant letters: in this case, too, the individuals commemorated on the monument belonged to the Romanized indigenous population, regarding it as prestigious to possess a costly tombstone (A[t] I.4).

A special case is a decorated stele from the mid 2nd century, discovered in the village of Donji Čehi some 13 km southwest of Andutonia, with four portraits above the inscription field, belonging to the slave family of the Caesernii (A[t] I.6). The letters are carved as elegant rectangular capitals, larger in the first line, decreasing in subsequent lines, with carefully carved triangular punctuation marks between the words. *Impaginatio* is careful: lines 1, 3, 5, and the last one are centred. The number of years of the daughter Valentina is carved with two small *X* at the end of the line because of the lack of space, and so is the age of the son Donicus, rendered with two small *I*, perhaps due to symmetry, as the line would be long enough for larger letters. There is a ligature of two *I* in *filiis*, while *karissimis* was written with a stylized *K*.

An entirely different case is represented by a damaged stele found in the village of Kerestinec, 24 km southwest of Andautonia, erected by one Pontius (A[t] I.11). The monument is a very rustic product of an unskilled local stone-cutter. The three portraits: a couple and a child above them, rendered as fully drawn figures, not just

[22] *ICUR* 1: 729 = *EDCS*-38701129 and *ICUR* 1: 2908 = *EDCS*-39900365.
[23] *OPEL* III 25.
[24] Šašel Kos 2006.

as busts, are depicted in an entirely schematic manner, and so is the inscription field, framed with a simple line. This line divides it from the gable, cutting across the portrait of the child. Left and right of the upper part of his body a large letter *M* is carved, instead of the expected *D(is) M(anibus)*. The double *M* should perhaps be understood as *MM(anibus)*. The same abbreviation appears, for example, on a richly adorned sarcophagus from Acumincum (Pannonia Inferior, present-day Slankamen?).[25] The letters are clumsily cut and uneven, and *posit* is carved instead of *posuit*.

## VI.3. AQUAE BALISSAE

At Aquae Balissae even less inscribed sepulchral monuments have been discovered to date than at Andautonia; however, contrary to Andautonia there are more from the Late Roman period.

A stele of Marcus Nunnidius Successus, veteran of the 32nd cohort of Roman citizens, is from the Flavian period (AB[t] I.1). Since the tombstone is unfortunately missing, nothing reliable can be said about the ductus of the epitaph. However, the name of his wife is written as Procla instead of Procula, which may either be a mistake or the record of current pronunciation. There is further an error of syntax: the name of the deceased daughter is in nominative, the name of the deceased son in dative.

There are no other funerary monuments from the 1st century AD, but there are a few from the 2nd century, among them a funerary altar of Publius Aelius Aelianus, the highest magistrate of the town: *quattuorvir* of *municipium Iasorum* (AB V.1). He was a member of the municipal upper class, therefore it may be regarded as unexpected that his epitaph, which is rather damaged, was inscribed with uneven letters and not particularly carefully. Stone-cutter made use of three ligatures: *IR* at the end of the fourth line, *VM* at the end of the fifth line, *IM* in the eighth line. More elegant quadrangular capitals can be observed on a simple fragmentary stele, now lost, from the late 2nd or 3rd century AD, discovered in the (hypothetical) territory of Aquae Balissae and belonging to an indigenous family (AB[t] I.3).

Elegant letters resembling square capitals, carefully carved with equal spaces between the letters, appear in the epitaph of two soldiers (one was praetorian), erected by their mother (AB[t] I.7). A stele from roughly the same time, commissioned for a soldier of the Cohort of the 500 Mauri, was found in the village of Kusonje, 15 km south of Daruvar (AB[t] I.6). The epitaph is very rustic, carved with letters of uneven shape, especially in the last line. An interesting detail is the letter *D* with a horizontal stroke across the vertical *hasta* (Đ), marking the numeral. A stylized star and possibly a moon (?) are carved in the last line, in which only one letter appears, *T* (in *posu(eru)n/t*): the last letter of the inscription. It is not clear how to explain *posunt*: an abbreviation, an error or perhaps a manifestation of vulgar Latin? The role of the star can be regarded as polyvalent: marking the end of the epitaph as a decorative element, filling in the spare empty space in the line, and having some symbolic meaning.

As the last monument, a Late Roman sepulchral slab should be mentioned, broken in six pieces, with a long and rather enigmatic inscription written in hexameters, found in the village of Veliki Bastaji in an underground funerary chamber (AB[t] VII.1). The text was carved very carefully with elegant Late Roman small and neat rustic capitals (*scriptura actuaria*). The letters are slightly elongated and not quite equally broad, with generally small interspaces between them, which, however, are larger in places. There are no punctuation marks between the words and no ligatures. Letters *A*, *E*, *F*, *H*, and *L* have a specific shape. However, the same letters do differ from each other: not all letters *E*, for example, have the same shape. The three *hastae* that should be horizontal, are oblique in this text, but in some letters this feature is more pronounced than in others. Letters *L* have a small *S* attached to the very short lower horizontal hasta, which again is not equal in all letters *L*. Letters *G* also have a specific shape, differing from the letters *C* by way of a large comma attached to the lower end. The inscription gives an impression of an elegant script.

## VI.4. CONCLUDING REMARKS

Letter-forms and other features of the inscriptions on funerary monuments from the three southern Pannonian towns included in the catalogue do not offer us enough data on the basis of which any conclusions concerning stone-cutters' workshops could be drawn. None show such close similarities that would allow for a presumption that they would have been carved in the same workshop, except perhaps the three discussed sarcophagi from Siscia (S II.2, S II.1, S II.7). This is not surprising in view of such a small number of preserved sepulchral monuments in each town. Letter-forms on each of the examined tombstones are more or less different; the differences are even more pronounced in the Late Roman period. Sepulchral formulae, on the other hand, are much more uniform, which is true both for the (early) Principate (e.g. *his situs est / siti sunt*) and for Late Roman times (as, for example, *domus aeterna*). Attention has been drawn to certain specific features, such as the not very frequent Greek sepulchral formula *zaesis* (correctly *zeses*): 'you will live'. Most interesting are also phenomena of haplography and

[25] *CIL* III 10244 = *HD*074756. *MM(anibus)* also occurs several times with *DD(is)*; however, this abbreviation does not figure in the list of abbreviations in Calabi Limentani 1968: 481–502.

other grammatical errors, partly shedding light on the spoken language, but partly also on the poor knowledge of Latin of either the person who commissioned the tombstone or the stone-cutter.

# Bibliography


Abascal Palazón, Juan Manuel 1994. *Los nombres personales en las inscripciones latinas de Hispania.* Murcia: Universidad de Murcia.

Abramenko, Andrik 1993. *Die munizipale Mittelschicht im kaiserzeitlichen Italien. Zu einem neuen Verständnis von Sevirat und Augustalität* (Europäische Hochschulschriften, Reihe III, 547). Frankfurt am Main: Peter Lang.

Acrudoae, Ionuţ 2012. Ala I Pannoniorum Tampiana: history and mobility. *Classica et Christiana* 7/1: 9–16.

Acrudoae, Ionuţ 2013. The prosopography of the militaries from Pannonia in *legio XV Apollinaris. Classica et Christiana* 8/2: 377–393.

Adam, Jean-Pierre 2007. *Roman Building. Materials and Techniques* (Translated by Anthony Mathews). London – New York: Routledge.

Alexandrescu, Cristina-Georgeta (ed.) 2015. *Cult and Votive Monuments in the Roman Provinces. Proceedings of the 13th International Colloquium on Roman Provincial Art (CSIR).* Cluj-Napoca: Mega Publishing House.

Alföldi, András 1948. I Varciani della Pannonia meridionale ed i loro vicini. *L'antiquité classique* 17/1: 13–18.

Alföldy, Géza 1961. Die Sklaverei in Dalmatien zur Zeit des Prinzipats. *Acta Antiqua Academiae Scientiarum Hungaricae* 9: 121–151.

Alföldy, Géza 1962. Die Auxiliartruppen der Provinz Dalmatien. *Acta Archaeologica Academiae Scientiarum Hungaricae* 14: 259–296.

Alföldy, Géza 1964. Municipium Iasorum. *Archaeologiai Értesítő* 91: 218–221.

Alföldy, Géza 1965a. Die *Valerii* in Poetovio. *Arheološki vestnik* XV–XVI: 137–144.

Alföldy, Géza 1965b. Epigraphica, in *Arheološke študije* II / *Varia archaeologica* II (*Situla* 8): 93–112.

Alföldy, Géza 1968. *Die Hilfstruppen der römischen Provinz Germania Inferior.* (Epigraphische Studien, Band 6). Düsseldorf: Rheinland-Verlag.

Alföldy, Géza 1969. *Die Personnennamen in der römischen Provinz Dalmatien* (Beiträge zur Namenforschung, Bh. 4). Heidelberg: Universitätsverlag Carl Winter.

Alföldy, Géza 1977. Die Personennamen in der römischen Provinz Noricum, in N. Duval (ed.). *L'onomastique latine* (Actes du Colloque International sur l'onomastique latine organisée à Paris du 13 au 15 octobre 1975): 249–265. Paris: CNRS.

Alföldy, Géza 1993. Die Personennamen auf den Bleietiketten von Kalsdorf (Steiermark) in Noricum. *Festschrift für Jürgen Untermann* (Innsbrucker Beiträge zur Sprachwissenschaft 78): 1–32. Innsbruck: Institut für Sprachwissenschaft der Universität Innsbruck.

Alföldy, Géza 2001. Eine *clarissima femina* in Lucus Augusti. *Zeitschrift für Papyrologie und Epigraphik* 136: 233–238.

Anreiter, Peter 2001. *Die vorrömischen Namen Pannoniens* (Archaeolingua, Ser. Minor 16). Budapest: Archaeolingua Alapítvány.

von Artner, Therese 1830. *Briefe über einen Theil von Croatien und Italien an Caroline Pichler.* Halberstadt: Carl Brüggemann.

Agusta-Boularot, Sandrine and Bousbaa, Mohamed 1993–1995 (1997). Inscriptions récemment découvertes à Cherchell (Caesarea de Maurétanie). *Bulletin archéologique du Comité des Travaux historiques et scientifiques, Afrique du Nord,* n.s. 24: 101–128.

Baćani, Iskra 2018. Nekropole / Necropolises, in I. Baćani and T. Tomaš Barišić (eds) 2018: 4–20.

Baćani, Iskra and Tomaš Barišić, Tea (eds) 2018. *Nekropole Siscije / Necropolises of Siscia* (exhibition catalogue). Sisak: Gradski muzej Sisak.

Barbieri, Guido 1952. *L'albo senatorio da Settimio Severo a Carino (193-285).* Roma: A. Signorelli.

von Barcley Lloyd, Joan E. 1990. Das goldene Gewand der Muttergottes in der Bildersprache mittelalterlicher und frühchristlicher Mosaiken in Rom. *Römische Quartalschrift für christliche Altertumskunde und Kirchengeschicte* 85/1–2: 66–85.

Barker, Simon J. 2012. Roman marble salviging, in A. Gutiérrez Garcia-M, P. Lapuente Mercadal and I. Rodá de Llanza (eds) 2012: 22–30.

Barkóczi, *László* 1964. The Population of Pannonia from Marcus Aurelius to Diocletian. *Acta Archaeologica Academiae Scientiarum Hungaricae* XVI: 257-356.

Barta, Andrea 2017. The Siscia Curse Tablet from a Linguistic Point of View. A New Autopsy. *Graeco-Latina Brunensia* 22/2: 23–41.

Bauchhenß, Gerhard (ed.) 1996. *Akten des 3. Internationalen Kolloquiums über Probleme des provinzialrömischen Kunstschaffens.* Köln – Bonn: Rheinland-Verlag, Dr. Rudolf Habelt.

Bekić, Luka and Lovenjak, Milan 2002. Novi natpis vojnog obilježja iz okolice Zagreba. *Vjesnik Arheološkog muzeja u Zagrebu* XXXV: 197–204.

Belaj, Juraj and Migotti, Branka 2011. Rimski grobni spomenik s lokaliteta Gornji grad u Ivancu / Roman tombstone from the Stari grad site in Ivanec. *Prilozi Instituta za arheologiju u Zagrebu* 28: 147–168.

Beltrán, José, Ontiveros, Esther, Loza Azuaga, Maria L., and Romero, Manuel (eds) 2012. Roman use, petrography and elemental geochemistry of the Surco Intrabético limestones (western region of

Málaga province, Spain), in A. Gutiérrez Garcia-Moreno, P. Lapuente, and I. Rodà de Llanza (eds) 2012: 500–510.

Beltrán Lloris, Francisco 2015. Towards a Typology of Latin Inscriptions, in Ch. Bruun and J. Edmondson (eds) 2015: 89–110.

Ben Abdallah, Zeïneb and Ladjimi Sebai, Leïla 1983. *Index onomastique des inscriptions latines de la Tunisie suivi de l'Index onomastique des inscriptions latines d'Afrique* (*Etudes d'antiquités africaines*). Paris: CNRS.

Beneš, Jan 1978. *Auxilia Romana in Moesia atque in Dacia.* Prague: Academia.

Bennett, Julian 2006. New evidence from Ankara for the *collegia veteranorum* and the *albata decursio*. *Anatolian Studies* 56: 93–99.

Bergmann, Birgit 2010. Der Kranz der Kaisers. Genese und Bedeutung einer römischen Insignie. Berlin – New York: Walter de Gruyter.

Bergmann, Marianne 1977. *Studien zum römischen Porträt des 3. Jahrhunderts n. Chr.* Bonn: Rudolf Habelt Verlag GMBH.

Bergmann, Marianne 2007. Bildnisse der Tetrarchenzeit, in A. Demandt and J. Engemann (eds). *Konstantin der Grosse: Imperator Caesar Flavius Constantinus* (exhibition catalogue): 58–70. Mainz am Rhein: Philipp von Zabern.

Berti, Fede 2006. *Le stele dei* Fadieni, in F. Berti (ed.). *Mors inmatura. I Fadieni e il loro sepolcreto* (Quaderni di Archeologia dell'Emilia Romagna 16): 9–19. Borgo San Lorenzo (FI): Edizioni all'Insegna del Giglio.

Birk, Stine 2013. *Depicting the Dead. Self-Representation and Commemoration on Roman Sarcophagi with Portraits* (Aarhus Studies in Mediterranean Antiquity XI). Aarhus: Aarhus University Press.

Birley, Eric 1969. Septimius Severus and the Roman Army. *Epigraphische Studien* 8: 63–82.

Bishop, Mike C. 2012. *Handbook to Roman Legionary Fortresses*. Barnsley: Pen and Sword Books.

Blanc, Nicole and Gury, Françoise 1986. Amor, Cupido. *LIMC* III: 952–1049.

Bogaers, Julianus Egidius 1969. Cohortes Breucorum. *Berichte van de Riksdienst voor het Oudheidkundig Bodemonderzoek* 19: 25–50.

Bojanovski, Ivo 1982. Antičko rudarstvo u unutrašnjosti provincije Dalmacije u svjetlu epigrafskih i numizmatičkih izvora. *Arheološki radovi i rasprave* VIII–IX: 89–120.

Bojanovski, Ivo 1974. *Dolabelin sistem cesta u rimskoj provinciji Dalmaciji (Dolabellae systema viarum in provincia Romana Dalmatia).* Djela XLVII. Sarajevo: Akademija nauka i umjetnosti Bosne i Hercegovine.

Bojanovski, Ivo 1988. *Bosna i Hercegovina u antičko doba (Bosnie et Herzegovine a l'*époque antique). Djela LXVI. Sarajevo: Akademija nauka i umjetnosti Bosne i Hercegovine.

Bojović, Dragoljub 1983. *Rimske fibule Singidunuma.* Beograd: Muzej grada Beograda.

Bonetto, Jacopo and Previato, Caterina 2013. Trasformazioni del paesaggio e trasformazioni della città: le cave di pietra per Aquileia, in. G. Cuscito (ed.). *Le modificazioni del paesaggio nell'Altoadriatico tra pre-protostoria ed altomedioevo* (*Antichità Altoadriatiche* LXXVI): 141–162. Trieste: Editreg.

Bonfante, Larissa 1994. Introduction, in J. L. Sebesta and L. Bonfante (eds). *The World of Roman Costume*: 3–10. Madison: The University of Wisconsin Press.

Borhy, László 2003. Überlegungen zur Änderung des Begriffes *civitas* in der Spätantike, in Á. Szabo and E. Tóth (eds). *Pannonica provincialia et archaeologia. Studia sollemnia Eugenio Fitz octogenario dedicata*: 415–424. Budapest: Magyar Nemzeti Múzeum.

Borhy, László 2010. *COREG, Legio VII Claudia, Ala I Contariorum milliaria civium Romanorum* – Neue Angaben zur Militärgeschichte von Brigetio: Spolien eines Steinkistengrabs aus dem Gerhát-Gräberfeld, in L. Borhy (ed.). *Studia Celtica Classica et Romana Nicolae Szabó septuagesimo dedicata*: 65–77. Budapest: Pytheas.

Boschung, Dietrich 1987. *Antike Grabaltäre aus den Nekropolen Roms* (Acta Bernensia, Band X). Bern: Verlag Stämpfli.

Bozzo, E., Merlanti, Federica and Rocca, Giuseppe 1992. Building stones and their provenance in the Etruscan township of Marzabotto near Bologna (Northern Italy), in M. Waelkens, N. Herz and L. Moens (eds) 1992: 105–114.

Braemer, François 1959. Les stèles funéraires a personnages de Bordeaux, Ier–IIIe siècles. Paris: Éditions A. et J. Picard et Cie.

Braemer, François 2005. Les steles funéraires a personnages. Origine des thèmes, modèles et dates à travers l'Empire romain, in M. Sanader, A. Rendić-Miočević and D. Tončinić (eds) 2005: 31–51.

Brunšmid, Josip 1898a. Arheološke bilješke iz Dalmacije i Panonije. II. *Vjesnik Hrvatskoga arheološkoga društva* III: 150–205.

Brunšmid, Josip 1898b. Rimski vojnički diplom iz Krnješevaca. *Vjesnik Hrvatskog arheološkog društva* 3: 144–149.

Brunšmid, Josip 1904. Kameni spomenici hrvatskog narodnog muzeja u Zagrebu. *Vjesnik Hrvatskoga arheološkoga društva* VII (1903-1904): 209–240.

Brunšmid, Josip 1905. Kameni spomenici hrvatskog narodnog muzeja u Zagrebu. *Vjesnik Hrvatskoga arheološkoga društva* VIII: 35–106.

Brunšmid, Josip 1907. Kameni spomenici hrvatskog narodnog muzeja u Zagrebu. *Vjesnik Hrvatskoga arheološkoga društva* IX (1906-1907): 81–134.

Brunšmid, Josip 1909. Spomenici hrvatskog narodnog muzeja u Zagrebu. *Vjesnik Hrvatskoga arheološkoga društva* X (1908–1909): 149–222.

Brunšmid, Josip 1911. Spomenici hrvatskog narodnog muzeja u Zagrebu. *Vjesnik Hrvatskoga arheološkoga društva* XI (1910-1911): 61–144.

Bruun, Christer 2015. Senators and Equites: Prosopography, in Ch. Bruun and J. Edmondson (eds) 2015: 202–226.

Bruun, Christer and Edmondson, Jonathan (eds) 2015. *The Oxford Handbook of Roman Epigraphy*. Oxford: Oxford University Press.

Bulat, Mirko 1977. Stanje istraživanja antičkih naselja u Slavoniji, in B. Vikić-Belančić (ed.) 1977. *Antički gradovi i naselja u južnoj Panoniji i graničnim područjima.* Materijali XIII: 63–88. Beograd: Savez arheoloških društava Jugoslavije.

Bulat, Mirko 2001. Novi rimski vojnički natpis iz Novog naselja kod Pakraca. *Osječki zbornik* 24-25: 61–66.

Bulat, Mirko 2007. Tko su bili Ateriksi, navodno, iz Osijeka i iz Odre kraj Zagreba. *Osječki zbornik* 28: 9–14.

Burkowsky, Zdenko 1996. Nekropole antičke Siscije. *Prilozi Instituta za arheologiju u Zagrebu* 10 (1993): 69-80.

Burkowsky, Zdenko 1999. *Sisak u prapovijesti, antici i starohrvatskom dobu.* Sisak: Gradski muzej Sisak.

Burkowsky, Zdenko 2000. Antički kameni spomenici Gradskog muzeja Sisak. *Godišnjak Gradskog muzeja Sisak* I: 71-88.

Burkowsky, Zdenko 2007. Arheološko istraživanje na mjestu nove zgrade Državnog arhiva u Sisku, in N. Kušanić (ed.) 2007. *Od latenskih nastambi do suvremene arhivske zgrade* (exhibition catalogue): 6–13. Sisak: Gradski muzej Sisak - Državni arhiv u Sisku.

Burnand, Yves 2002. I legionari originari dalla regione transpadana: un ceto medio della società?, in A. Sartori and A. Valvo (eds). *Ceti medi in Cisalpina* (Atti del colloquio internazionale, 14-16 settembre 2000, Milano): 215–228. Milano: Comune di Milano, Settore Musei e Mostre.

Buzov, Marija 1996. Segestika i Siscija - topografija i povijesni razvoj. *Prilozi Instituta za arheologiju u Zagrebu* 10 (1993): 47–68.

Buzov, Marija 2002. Grad mrtvih uz grad živih - nekropole Siscije. *Histria antiqua* 8:175–191.

Buzov, Marija 2012. Cults in Roman Siscia. *Classica et Christiana* 7/1: 57–83.

Buzov, Marija 2015. Roman sculpture in Pannonia between imports and local production, in P. Pensabene and E. Gasparini (eds) 2015: 995–967.

Calabi Limentani, Ida 1968. *Epigrafia latina. Con un'appendice bibliografico di Attilio Degrassi.* Milano: Istituto editoriale cisalpino.

Cambi, Nenad 1975. Antički sarkofazi iz Like, in I. Marović and Ž. Rapanić (eds). *Arheološka problematika Like.* Posebna izdanja Hrvatskog arheološkog društva 1: 75–81. Split: Hrvatsko arheološko društvo.

Cambi, Nenad 1982. Sarkofag iz Šipova. *Godišnjak Centra za balkanološka ispitivanja* XX (18): 91–110.

Cambi, Nenad 1989. Bilješke uz dvije panonske nadgrobne stele. *Vjesnik Arheološkog muzeja u Zagrebu* XXII: 59–76.

Cambi, Nenad 2002. *Antika.* Zagreb: Naklada Ljevak d.o.o.

Cambi, Nenad 2003. Stela iz župne crkve u Bjelovaru. *Radovi Instituta za povijest umjetnosti* 27: 19–25.

Cambi, Nenad 2005a. *Kiparstvo rimske Dalmacije / The Sculpture of the Roman Province of Dalmatia.* Split: Književni krug.

Cambi, Nenad 2005b. Funerary Monuments from Dalmatia, Istria and the Croatian Part of Pannonia. A Comparative Study, in M. Sanader, A. Rendić-Miočević, and D. Tončinić (eds) 2005: 13-30.

Cambi, Nenad 2010. *Sarkofazi lokalne produkcije u rimskoj Dalmaciji / Die Sarkophage der lokalen Werkstätten im römischen Dalmatien.* Split: Književni krug.

Cambi, Nenad 2017. Sarajevski medaljon 'Tračkog konjanika'. *Arheološki radovi i rasprave* 17: 145-162.

Cambi, Nenad and Koch, Guntram (eds) 2013. *Funerary Sculpture of the Western Illyricum and the Neighbouring Regions of the Roman Empire.* Split: Književni krug.

Campbell, Brian 2000. *The Writings of the Roman Land Surveyors. Introduction, Text, Translation and Commentary.* London: Society for the Promotion of Roman Studies.

Campbell, Brian 2012. *Rivers and the Power of Ancient Rome.* Chape Hill: The University of North Carolina Press.

Carroll, Maureen 2006. *Spirits of the Dead. Roman Funerary Commemoration in Western Europe.* Oxford: Oxford University Press.

Carroll, Maureen 2015. Commemorating Military and Civilian Families on the Danube limes, in L. Vagalinski and N. Sharankov (eds) 2015: 501–509.

Cermanović, Aleksandrina 1965. Die dekorierten Sarkophage in den römischen Provinzen von Jugoslawien. *Archaeologia Iugoslavica* VI: 89–103.

Charbonneaux, Jean, Martin, Roland and Villard, François 1969. *La Grecia arcaica (620 - 480 a. C)* (translated by M. Lenzini and L. Sosio). Milano: Rizzoli Editore.

Chioffi, Laura 2015. Death and Burial, in Ch. Bruun and J. Edmondson (eds) 2015: 627–648.

Chiriac, Costel, Mihailescu-Bîrliba, Lucretiu and Matei, Ionel 2004. Ein neues Militärdiplom aus Moesien. *Zeitschrift für Papyrologie und Epigraphik* 150: 265–269.

Christol, Michel and Demougin, Ségolène 1986. Notes de prosopographie equestre. IV – Un chevalier romain juge des cinq decuries originaire de Siscia. *Zeitschrift für Papyrologie und Epigraphik* 64: 185–194.

Cichorius, Conrad 1894. Ala. *RE* I: 1224–1270.

Cichorius, Conrad 1900. Cohors. *RE* IV/1: 231–356.

Ciliberto, Fulvia 2007. Un singolare monumento funerario a San Canzian d'Isonzo presso Aquileia, in E. Walde and B. Kainrath (eds) 2007: 325–334.

Ciliberto, Fulvia and Mainardis, Fulvia 2015. Der Sarkophag der Magia Marcellina in Venafro. Überlegungen im antiken Samnium, in B. Porod and G. Koiner (eds) 2015: 62–72.

Clauss, Manfred 1973. *Untersuchungen zu den principales des römischen Heeres von Augustus bis Diokletian -*

*Cornicularii, speculatores, frumentarii*. Bochum: Ruhr Universität.

Clauss, Manfred 1999. *Lexikon lateinischer militärischer Fachausdrücke*. Stuttgart: Konrad Theiss.

Cleland, Liza, Davies, Glenys and Llewellyn-Jones, Lloyd 2012. *Greek and Roman Dress from A to Z*. London – New York: Routledge.

Croom, Alexandra 2002. *Roman Clothing and Fashion*. Stroud: Tempus Publishing Ltd.

Curbera, Jaime B. and Jordan, David R. 1996. A Curse Tablet from Pannonia Superior. *Tyche* 11: 45–50.

Čače, Slobodan and Milivojević, Feđa 2017. Rimski Ilirik u 1. st. pr. Kr.: nekoliko zapažanja / Roman Illyricum in the first century BC: A few remarks. *Vjesnik za arheologiju i historiju dalmatinsku* 110/2: 425–450.

Čremošnik, Irma 1963. Nošnja na rimskim spomenicima u Bosni i Hercegovini. *Glasnik Zemaljskog muzeja Bosne i Hercegovine* XVIII: 103–125.

Čremošnik, Irma 1964. Die einheimische Tracht Noricums, Pannoniens und Illyricums und ihre Vorbilder. *Latomus* 23/4: 760–773.

Čučković, Lazo 1986. Arheološka topografija karlovačke regije, in N. Majnarić-Pandžić (ed.). *Arheološka istraživanja na karlovačkom i sisačkom području*. Posebna izdanja Hrvatskog arheološkog društva 10: 9–18. Zagreb: Hrvatsko arheološko društvo.

Čučković, Lazo 2007. *Toplice - Topusko*. Topusko: Turistička zajednica općine Topusko.

Čučković, Lazo 2008. Sjeverna granica Japoda, in T. Kolak (ed.). *Arheološka istraživanja u Lici i arheologija pećina i kraša*. Posebna izdanja hrvatskog arheološkog društva 23: 79–83. Zagreb - Gospić: Hrvatsko arheološko društvo – Muzej Like Gospić.

Dally, Ortwin 2000. Die Diskussion des Kulturbegriffes in der Klassischen Archäologie – ein Desiderat?, in S. Frölich (ed.). Kultur – ein interdisciplinäres Kolloquium zur Begrifflichkeit: 81–115. Halle: http://archiv.ub.uni- heidelberg.de/propylaeumdok/3152/1/Dally_Die_Diskussion_des_Kulturbegriffs_2000.pdf

Dana, Dan 2014. *Onomasticon Thracicum (OnomThrac). Répertoire des noms indigènes de Thrace, Macédoine orientale, Mésies, Dacie et Bithynie* (Meletemata 70). Athènes: KERA.

Dautova-Ruševljan, Velika 1983. *Rimska kamena plastika u jugoslovenskom delu provincije Donje Panonije*. Novi Sad: Savez arheoloških društava Jugoslavije.

Davies, Roy William. 1969. The medici of the Roman armed forces. Epigraphische Studien 8: 83–99. Köln: Rheinland Verlag, Bonn: Habelt in Komm.

Dean, Lindley Richard 1916. *A Study of the Cognomina of Soldiers in the Roman Legions*. Princeton: Princeton University.

Degavre, Jean 1998. *Lexique gaulois, Recueil de mots attestés, transmis ou restitués et de leurs interprétations* (Mémoires de la Société belge d'Études celtiques 10). *Supplément au lexique*, tome III, 2004 (Mémoires de la Société belge d'Études celtiques 20: 523–561). Bruxelles: Société belge d'Études celtiques.

Degmedžić, Ivica 1957. Sadržaj antiknih kamenih spomenika nađenih u Zagrebu i okolici. *Iz starog i novog Zagreba* I: 91–117.

Degmedžić, Ivica 1958. Prvi arheološki spomen liječničkog zvanja u okolici Zagreba. *Večernji vjesnik*, 15. travnja 1958: 6. Zagreb.

Delamarre, Xavier 2003. *Dictionnaire de la langue gauloise*, 2ème édition. Paris: Editions Errance.

Delamarre, Xavier 2007. *Nomina Celtica Antiqua Selecta Inscriptionum. Noms de personnes celtiques dans l'épigraphie classique*. Paris: Editions Errance.

Demicheli, Dino, Radman-Livaja, Ivan and Wiewegh, Zoran 2012. List of personal names on a ceramic bowl fragment from *Siscia*, in M. E. Fuchs, R. Sylvestre and C. Schmidt Heidenreich (eds). *Inscriptions mineures : nouveautés et réflexions*. Actes du premier colloque Ductus (19-20 juin 2008, Université de Lausanne): 185–192. Berne – Berlin – Bruxelles – Frankfurt am Mein, New York – Oxford – Wien: Peter Lang Gmbh.

Demo, Željko 1982. Prilog topografiji križevačke regije u antičko doba s osvrtom na numizmatičke nalaze. *Križevački zbornik* 2: 75–92.

Demo, Željko (ed.) 1994. *Od nepobjedivog sunca do sunca pravde / From the Invincible Sun to the Sun of Justice* (exhibition catalogue). Zagreb: Arheološki muzej u Zagrebu.

Demougin, Ségolène 1975. Les juges des cinq décuries originaires de l'Italie. *Ancient Society* 6: 143–202.

Demougin, Ségolène 1992. *Prosopographie des chevaliers romains Julio-claudiens*. Rome: L'Ecole française de Rome.

Dészpa, Mihály L. 2012. *Peripherie-Denken. Transformation und Adaptation des Gottes Silvanus in den Donauprovinzen (1.-4. Jahrhundert n. Chr.)* (Potsdamer Altertumswiss. Beiträge 35). Stuttgart: Franz Steiner.

Dexheimer, Dagmar 1998. *Oberitalienische Grabaltäre. Ein Beitrag zur Sepulkralkunst der römischen Keiserzeit*. British Archaeological Reports International Series 741. Oxford: Hadrian Books.

De Stefano, Alessandra. 2015. The limestone quarries of the Karaburum peninsula (southern Albania), in P. Pensabene and E. Gasparini (eds) 2015: 513–518.

Dise, Robert L. Jr. 1997. Variation in Roman administrative practice: the assignements of beneficiarii consularis. *Zeitschrift für Papyrologie und Epigraphik* 116: 284–299.

Djurić, Bojan 1991. Noriško-panonska proizvodnja nagrobnih spomenikov in trgovina z marmornimi izdelki. Unpublished PhD dissertation, University of Ljubljana.

Djurić, Bojan 1997. Eastern Alpine Marble and Pannonian Trade, in B. Djurić and I. Lazar (eds). *Akten des IV. internationalen Kolloquiums über Probleme des provinzialrömischen Kunstschaffens* (Situla 36): 73–86. Ljubljana: Narodni muzej Slovenije.

Djurić, Bojan 2001a. Production of Marble Sarcophagi in Poetovio, in P. Zsidi and E. Hanny (eds) 2001: 47–62.

Djurić, Bojan 2001b. Ossuaria Poetoviensia: Iconography and Structure, in T. A. S. M. Panhuysen (ed.). *Die Maastrichter Akten des 5. Internationalen Kolloquiums über das provinzialrömische Kunstschaffen*: 117–129. Maastricht: Stichting Willem Goossens.

Djurić, Bojan 2004. V Saksanovem svetu / In Saxanus' Welt, in I. Lazar (ed.). *Rimljani - steklo, glina, kamen / Die Römer - Glas, Ton, Stein* (exhibition catalogue): 150–167. Celje: Pokrajinski muzej Celje.

Djurić, Bojan 2005. Poetovio and the Danube marble trade, in M. Mirković (ed.). *Römische Städte und Festungen an der Donau*: 75–82. Beograd: Filozofski fakultet.

Djurić, Bojan 2007. Predrimske forme na nagrobniku iz Mrzlega Polja pri Ivančni Gorici, in M. Blečić *et al.* (eds). *Scripta praehistorica in honorem Biba Teržan* (*Situla* 44): 895–902. Ljubljana: Narodni muzej Slovenije.

Djurić, Bojan 2008. Early stelae from Poetovio and the marble studies, in Ch. Franek, S. Lamm, T. Neuhauser, B. Porod and K. Zöhrer (eds). *Thiasos - Festschrift für Erwin Pochmarski zum 65 Geburtstag*: 159–165. Wien: Institut für Archäologie der Karl-Franzens-Universität Graz.

Djurić, Bojan 2010. Nagrobna stela P(ublija) Aelija Viatorja (UEL 3783) iz Murske Sobote. *Zbornik soboškega muzeja* 15: 201–211.

Djurić, Bojan 2013. Zagreb, Arheološki muzej: Proben der Steinmonumente (unpublished report). Ljubljana.

Djurić, Bojan 2016. The *Ma]sculus* stela in Poetovio and the Flavia Solva workshops, in R. Lafer (ed.) 2016: 173–185.

Djurić, Bojan. The logistic behind ancient art: the case of Noricum and Pannoniae. Papers of the 15th International Colloquium on Roman Provincial Art, 14 - 20 June 2017, Graz/Austria (in press).

Djurić, Bojan, Hebert, Bernhard *et al.* 2005. Marmore römischer Brüche und Steindenkmäler in der Steiermark und in Štajerska. Ergebnisse eines Forschungsprojektes. *Fundberichte aus Österreich* 43(2004): 365–431.

Djurić, Bojan, Kele, Sándor and Rižnar, Igor 2015. The Budakalász Travertine Production in *Pannonia Inferior* and *Moesia Superior*, in K. Marasović (ed.) 2015: 102–103.

Djurić, Bojan and Müller, Harald 2009. *White Marbles in Noricum and Pannonia: an Outline of the Roman Quarries and Their Products, in Ph. Jockey (ed.).* Leukos Lithos. *Marbres et autres roches de la Méditerranée antique - Interdisciplinary Studies on Mediterranean Ancient Marble and Stones.* Proceedings of the VIIIth International Conference of the Association for the Study of Marble and Other Stones Used in Antiquity VIII: 111-127. Paris: Maisonneuve & Larose.

Djurić, Bojan, Müller, Harald and Filipović, Slavica 2010. Karakteriziranje mramornih spomenika Murse. *Osječki zbornik* 29: 9–18.

Djurić, Bojan and Rižnar, Igor 2016. Kamen Emone / The rocks for Emona, in B. Vičić and B. Županek (eds). *Emona MM. Urbanizacija prostora - nastanek mesta / Urbanisation of space - beginning of a town*: 121–144. Ljubljana: Zavod za varstvo kulturne dediščine Slovenije – Mestni muzej, Muzej in galerije mesta Ljubljane.

Dobó, Árpad 1975[4]. *Inscriptiones extra fines Pannoniae Daciaeque repertae ad res earundem provinciarum pertinentes.* Budapest: Magyar Nemzeti Múzeum.

von Domaszewski, Alfred and Dobson, Brian 1967. *Die Rangordnung des römischen Heeres*, 2. Auflage (Beihefte der Bonner Jahrbücher, Band 14). Köln-Graz: Böhlau Verlag.

Domić Kunić, Alka 1988. Augzilijari ilirskog i panonskog porijekla u natpisima i diplomama (od Augusta do Karakale). *Arheološki radovi i rasprave* 11: 83–114.

Domić Kunić, Alka 1995. Rimske provincijalne flotile, s posebnim obzirom na udjel Panonaca. *Arheološki radovi i rasprave* 12: 83–100.

Domić Kunić, Alka 2006. Bellum Pannonicum (12.–11. pr. Kr.), posljednja faza osvajanja južne Panonije. *Vjesnik Arheološkog muzeja u Zagrebu* 39: 59–164.

Domić Kunić, Alka 2012: Literary Sources before the Marcomannic Wars, in B. Migotti (ed.) 2012b: 29–69.

Donati, Angela and Gentili, Giovanni (eds) 2005. *Costantino il Grande. La civiltà antica al bivio tra Occidente e Oriente.* Milano: SilvanaEditoriale.

Dondin-Payre, Monique 2001a. L'onomastique dans les cités de Gaule centrale (Bituriges Cubes, Éduens, Senons, Carnutes, Turons, Parisii), in M. Dondin-Payre and M. T. Raepsaet-Charlier (eds) 2001: 193–341.

Dondin-Payre, Monique 2001b. Secundus et ses dérivés en Narbonnaise, dans les Trois Gaules et les Germanies, in M. Dondin-Payre and M. T. Raepsaet-Charlier (eds) 2001 : 537–595.

Dondin-Payre, Monique and Raepsaet-Charlier, Marie-Thérèse (eds) 2001. *Noms, identités culturelles et romanisation sous le Haut-Empire*: Bruxelles: Le livre Timperman.

Dorcey, Peter F. 1992. *The Cult of Silvanus. A Study in Roman Folk Religion* (Columbia Studies in the Classical Tradition XX). Leiden - New York - Köln: Brill.

Dresken-Weiland, Jutta 1998. *Repertorium der christlich-antiken Sarkophage*, Bd. 2. Mainz: Philipp von Zabern.

Dross-Krüpe, Kirsten 2012. Purchase orders of Military Garments from Papyri of Roman Egypt, in M.-L. Nosch (ed.). *Wearing the Cloak. Dressing the soldier in Roman Times* (Ancient Textiles Series, vol. 10): 13–18. Oxford: Oxbow Books.

Durman, Aleksandar 1992. O geostrateškom položaju Siscije. *Opuscula archaeologica* 16: 117–131.

Dušanić, Slobodan 1978. A Military Diploma of A.D. 65. *Germania* 56/1: 461–475.

Dušanić, Slobodan 1983. Moesia and Pannonia in Domitian's Last War on the Danube. Živa antika XXXIII: 13–21.

Dušanić, Slobodan 2010. *Selected Essays in Roman History and Epigraphy*. Beograd: Zavod za udžbenike.

Duthoy, Robert 1976. Recherches sur la répartition géographique et chronologique des termes sevir Augustalis, Augustalis et sevir dans l'Empire romain. Epigraphische Studien 11: 143–124. Köln - Bonn: Rheinland-Verlag.

Đurić, Bojan, Davidović, Jasmina, Maver, Andreja and Müller, Harald 2008. Stone Use in Roman Towns. Resources, Transport, Products and Clients. Case Study Sirmium. First Report. *Starinar* LVI (2006): 103–137.

Eck, Werner 1985. *Die Statthalter der germanischen Provinzen vom 1.-3. Jahrhundert* (Epigraphische Studien 14). Köln - Bonn: Rheinland-Verlag.

Edmondson, Jonathan 2015. Roman Family History, in Ch. Bruun and J. Edmondson (eds) 2015: 559–581.

Elton, Hugh 1996. *Frontiers of the Roman Empire*. London: Batsford.

Ertel, Christine 2010. *Bestandteile von römischen Grabbauten aus Aquincum und dem Limesabschnitt im Stadtgebiet von Budapest. CSIR, Ungarn*, Band IX. Budapest: Archaeolingua.

Evans, David Ellis 1967. *Gaulish Personal Names. A Study of some Continental Celtic Formations*. Oxford: Clarendon Press.

Facsády, Annamária 2014. Les traits iconographiques des stèles funéraires familiales à Aquincum, in I. Koncani Uhač (ed.) 2014: 194–202.

Falileyev, Alexander and Radman-Livaja, Ivan 2016. More Celtic names from Roman Pannonia. *Zeitschrift für celtische Philologie* 63: 49–67.

Farnum, Jerome H. 2005. *The Positioning of the Roman Imperial Legions*. British Archaeological Reports, International Series 1458. Oxford: Archaeopress.

Faure, Patrice 2013. Entre épigraphie et archéologie: la IIe légion parthique et les structures des légions romaines, in F. Bertholet and C. Schmidt Heidenreich (eds). *Entre archéologie et épigraphie. Nouvelles perspectives sur l'armée romaine* (Echo. Collection de l'Institut d'Archéologie et des Sciences de l'Antiquité de l'Université de Lausanne): 17–77. Bern etc.: Peter Lang.

Feig Vishnia, Rachel 2002. The Shadow Army - The Lixae and the Roman Legions. *Zeitschrift für Papyrologie und Epigraphik* 139: 265–272.

Feissel, Denis 1983. *Recueil des inscriptions chrétiennes de Macédoine du IIIe au VIe siècle* (BCH, Suppl. 8). Athènes: École Française d'Athènes.

Fejfer, Jane 2008. *Roman Portraits in Context*. Berlin - New York: Walter de Gruyter.

Fellmann, Rudolf 2000. Die 11. Legion Claudia Pia Fidelis, in Y. Le Bohec and C. Wolff (eds). *Les légions de Rome sous le Haut-Empire* (Actes du Congrès de Lyon 17-19 sept. 1998): 127–131. Lyon: De Boccard.

Ferjančić, Snežana 2002. *Naseljavanje legijskih veterana u balkanskim provincijama, I - III vek n.e.* (Balkanološki institut, Posebna izdanja 79). Beograd: Srpska akademija nauka i umetnosti.

Ferjančić, Snežana 2015. The Ala Pannoniorum in the army of Illyricum. *Istraživanja - Journal of Historical Researches* 26: 37–45.

Filipec, Krešimir 2017. Antički nadgrobni spomenici Lobora - prilog poznavanju antičke skulpture panonsko-noričkog žiteljstva i njegove sredine. *Arheološki radovi i rasprave* 18: 95–144.

Filippini, Alister and Gregori, Gian Luca 2014. *Procuratores Augusti et praepositi vexillationibus ab Imperatore missi*: le missioni speciali di L. Iulius Iulianus e di M. Valerius Maximianus a confronto, in S. Demougin and M. Navarro Caballero (eds). *Se déplacer dans l'Empire romain. Approches épigraphiques* (Ausonius Scripta Antiqua 59): 85–120. Bordeaux: Ausonius Éditions.

Fishwick, Duncan 2002. *The Imperial Cult in the Latin West. Studies in the Ruler Cult of the Western Provinces of the Roman Empire.* Vol. III: *Provincial Cult*. Part 2: *The Provincial Priesthood*. Leiden - Boston - Köln: Brill.

Fitz, Jenő 1962. A Military history of Pannonia from the Marcomann Wars to the death of Alexander Severus (180–235). *Acta Archaeologica Academiae Scientiarum Hungaricae* 14: 25–112.

Fitz, Jenő 1980a. Administration and army, in A. Lengyel and G.T.B. Radan (eds) 1980: 125–140.

Fitz, Jenő 1980b. Population, in A. Lengyel and G. Radan (eds) 1980: 141–159.

Fitz, Jenő 1993–1995. *Die Verwaltung Pannoniens in der Römerzeit, I-IV*. Budapest: Encyclopedia.

Fitz, Jenő 2003. Die Städte Pannoniens, in M. Šašel Kos and P. Scherrer (eds) 2003: 47–52.

Fodorean, Florin-Gheorghe 2017. *Praetorium* and the Emona–Siscia–Sirmium–Tauruno road in the ancient geographical and epigraphic sources. *Arheološki vestnik* 68: 337–348.

Forier, Sarah 2001. Les anthroponymes formés sur des noms d'animaux en Gaule Narbonnaise et dans les provinces alpines, in M. Dondin-Payre and M. T. Raepsaet- Charlier (eds) 2001: 473–535.

Forni, Giovani 1982. Revisione di epigrafi da Apulum (Dacia). *Zeitschrift für Papyrologie und Epigraphik* 46: 188–190.

Forni, Giovani 1985. *Le tribù romane III, 1. Le pseudo-tribù*. Roma: G. Bretschneider.

France, Jérôme and Nelis-Clément, Jocelyne 2014. *Tout en bas de l'empire. Les* stationes, *lieux de contrôle et de représentation du pouvoir*, in J. France and J. Nelis-Clément (eds). *La statio. Archéologie d'un lieu de pouvoir dans l'empire romain*: 117-246. Bordeaux: Ausonius Éditions.

Frank, Tenney 1916. Race Mixture in the Roman Empire. *The American Historical Review* 21/4: 689–708.

Franke, Thomas 2000. *Legio XIV Gemina*, in Y. Le Bohec and C. Wolff (eds). *Les légions de Rome sous le Haut-Empire*, Tome I: 191-202. Lyon: Diffusion De Boccard.

Freber, Philipp-Stephan G. 1993. *Der hellenistische Osten und das Illyricum unter Caesar* (Palingenesia 42). Stuttgart: Steiner.

Fulir, Miroslav 1969. Topografska istraživanja rimskih cesta na varaždinskom i međimurskom području (1960–1967). *Razprave* (Slovenska akademija znanosti in umetnosti, Ljubljana) VI: 363–431.

Gabelmann, Hanns 1972. Die Typen der römischen Grabstelen am Rhein. *Bonner Jahrbücher* 172: 65–140.

Gabelmann, Hanns 1973. *Die Werkstattgruppen der oberitalischen Sarkophage.* Bonn: Rheinland-Verlag GMBH.

Gabelmann, Hanns 1977. Zur Tektonik oberitalischer Sarkophage, Altäre und Stelen. *Bonner Jahrbücher* 177: 199–244.

Gabler, Dénes 2007. Der gesellschaftliche Hintergrund der mythologischen Darstellungen in Pannonien, in E. Walde and B. Kainrath (eds) 2007: 13–26.

Gaggadis-Robin, Vassiliki, Hermary, Antoine, Reddé, Michael and Sintés, Claude (eds) 2009. *Les ateliers de sculpture régionaux: techniques, styles et iconographie.* Actes du X[é] colloque international sur l'art provincial romain. Aix-en-Provance: Centre Camille- Julian, Musée departmental de l'Arles antique.

Galić, Miljenka and Radman-Livaja, Ivan 2002. Zaštitno istraživanje u Samoborskim Novakima 2000. godine. *Vjesnik Arheološkog muzeja u Zagrebu* XXXV: 239–248.

Garbsch, Jochen 1965. Die norisch-pannonische Frauentracht im 1. und 2. Jahrhundert (Münchner Beiträge zur Vor- und Frühgeschichte 11). München: Beck.

Garbsch, Jochen 1985. Die norisch-pannonische Tracht, in *Aufstieg un Niedergang der römischen Welt*, Bd II., 12.3: 546–577. Berlin – New York: Walter de Gruyter.

Gáspár, Dorottya 2002. *Christianity in Roman Pannonia. An evaluation of Early Christian finds and sites from Hungary.* British Archaeological Reports International Series 1010. Oxford: Archaeopress.

Gawroński, Radosław A. 2016. The origin of *cornicularii*. *Archaeologia* 65: 7–24.

Gilliver, Catherine M. 1999. *The Roman Art of War.* Stroud: Tempus Publishing.

Girardi Jurkić, Vesna 2012. The Cavae Romanae quarry. Properties and use of the stone for the amphitheatre in Pula (Croatia), in A. Gutiérrez Garcia-Moreno, P. Lapuente, and I. Rodà de Llanza (eds) 2012: 640–644.

Glaser, Franz 1997. Isisverehrerinnen in Noricum, in B. Djurić and I. Lazar (eds). *Akten des IV. internationalen Kolloquiums über Probleme des provinzialrömischen Kunstschaffens* (*Situla* 36): 127–130. Ljubljana: Narodni muzej Slovenije.

Goette, Hans Rupprecht 1989. *Studien zu römischen Togadarstellungen.* Mainz am Rhein: Verlag Philipp von Zabern.

Gómez-Pantoja, Joaquín 2000. Legio X Gemina, in Y. Le Bohec and C. Wolff (eds). *Les légions de Rome sous le Haut-Empire* (Collection du Centre d'Etudes Romaines et Gallo- Romaines, Nouvelle série, N° 20): 169–190. Lyon: De Boccard.

Gordon, Mary L. 1924. The Nationality of Slaves under the early Roman Empire. *Journal of Roman Studies* 14: 93–111.

Gorenc, Marcel 1960. Ostaci antiknog groblja u Držićevoj ulici i počeci urbanizacije užeg područja grada Zagreba. *Iz starog i novog Zagreba* II: 9–28.

Gorenc, Marcel 1971. Antičko kiparstvo jugoistočne Štajerske i rimska umjetnost Norika i Panonije. *Vjesnik Arheološkog muzeja u Zagrebu* V: 15–46.

Goss, Vladimir 2007. 'Kamen iz Križovljana' nakon 'Kamena iz Belca' – pitanje konteksta / 'Stone from Križovljan' following the 'Stone from Belec': Some Questions concerning the Context. *Prilozi Instituta za arheologiju u Zagrebu* 22: 187–198.

Goss, Vladimir P. (ed.) 2007. *Stotinu kamenčića izgubljenog raja. Romanička skulptura u muzejima i zbirkama između Save i Drave / A Hundred Stones from a Lost Paradise. Romanesque Sculpture in Museums and Collections between the Sava and the Drava Rivers* (exhibition catalogue). Zagreb: Arheološki muzej u Zagrebu.

Gračanin, Hrvoje 2004. Svijet antičke žene u dva južnopanonska pokrajinska središta, Sisciji i Sirmiju. *Scrinia Slavonica* 4: 15–106.

Gračanin, Hrvoje 2010. Rimske prometnice i komunikacije u kasnoantičkoj južnoj Panoniji. *Scrinia Slavonica* 10: 9–69.

Graf, Andreas 1936. Übersicht der antiken Geographie von Pannonien (Diss. Pann. ser. l, 5). Budapest: Institut für Münzkunde und Arch. der Peter Pázmány-Universität.

Graßl, Herbert 1991. Die wirtschaftlichen Grundlagen für das Kunstschaffen in Noricum, in M. Hainzmann, D. Kramer and E. Pochmarski (eds) 1991/II: 3–13.

Grbić, Dragana 2014. *Tribal Communities in Illyricum. Pre-urban Administrative Structures in the Roman Provinces between the Adriatic and the Danube (First-Third Centuries).* Belgrade: Institut for Balkan Studies of the Serbian Academy of Sciences and Arts.

Greene, Kevin 1990. *The Archaeology of the Roman Economy.* Berkeley – Los Angeles: University of California Press.

Gregl, Zoran 1990. Noričko-panonski tumuli u Hrvatskoj, in N. Majnarić-Pandžić (ed.). *Arheološka istraživanja u Podravini i kalničko-bilogorskoj regiji.* Posebna izdanja Hrvatskog arheološkog društva 14: 101–106. Zagreb: Hrvatsko arheološko društvo.

Gregl, Zoran 1991. *Rimljani u Zagrebu. Tragovi Rimskog Carstva u gradu i okolici.* Zagreb: Latina et Graeca.

Gregl, Zoran 1996. Sadržaj antiknih kamenih spomenika nađenih u Zagrebu i okolici II. *Iz starog i novog Zagreba* VII: 9–17.

Gregl, Zoran 1997. *Rimske nekropole sjeverne Hrvatske*. Zagreb: Arheološki muzej u Zagrebu.

Gregl, Zoran 2000. Antičke kružne grobnce s dromosom u Hrvatskoj. *Opuscula archaeologica* 23–24 (1999–2000): 165–168.

Gregl, Zoran and Migotti, Branka 2000. Nadgrobna stela iz Siska (*CIL* III 3985). *Vjesnik Arheološkog muzeja u Zagrebu* XXXII–XXXIII (1999–2000): 119–164.

Gregl, Zoran and Migotti, Branka 2004. Civitas Iovia (Botivo), in M. Šašel Kos and P. Scherrer (eds) 2004: 131–143.

Groh, Stefan 2014. Neues zur Urbanistik des Munizipiums Andautonia-Ščitarjevo (Pannonia Superior, Kroatien): Auswertung und archäologisch-historische Interpretation der geophysikalischen Messungen 2012 / Novosti u urbanizmu municipija Andautonije- Ščitarjevo (Gornja Panonija, Hrvatska): analiza i arheološko-povijesna interpretacija geofizičkih mjerenja 2012. godine. *Vjesnik Arheološkog muzeja u Zagrebu* XLVI: 89–113.

Gummerus, Herman 1932. *Der Ärztestand im römischen Reiche nach den Inschriften*. Helsingfors: Societas Scientiarum Fennica.

Guštin, Mitja 2015. Roman camps following the route to Segestica and the western Balkans / Rimski vojaški tabori v smeri proti Segestiki in zahodnemu delu Balkanskega polotoka, in J. Istenič, B. Laharnar and J. Horvat (eds). *Evidence of the Roman army in Slovenia / Sledovi rimske vojske na Slovenskem*. Katalogi in monografije / Catalogi et monographiae 41: 221–233. Ljubljana: Narodni muzej Slovenije.

Gutiérrez Garcia-Moreno, Anna, Lapuente, Pilar and Rodá de Llanza, Isabel (eds) 2012. *Interdiscplinary Studies on Ancient Stone*. Proceedings of the IX the Association for the Study of Marble and Other Stones in Antiquity (ASMOSIA) Conference. Tarragona: Institut Català d'Arqueologia Clàssica.

Haensch, Rudolf 1997. *Capita provinciarum. Statthaltersitze und Provinzialverwaltung in der römischen Kaiserzeit* (Kölner Forschungen 7). Mainz: De Gruyter.

Hagenweiler, Patrizia E. 2001. Drei Reliefplatten von einem römischen Grabmonument von Verona und dionysische Themen in der Grabkunst Oberitaliens, in G. Brands, J-N. Andrikopoulou-Strack, D. Dexheimer and G. Bauchhenß (eds). *Rom und die Provinzen. Gedenkschrift für Hanns Gabelmann* (Beihefte der Bonner Jahrbücher, Band 53): 75–81. Mainz: Verlag Philipp von Zabern.

Hainzmann, Manfred 2011. *Maromogios* und die Gauvorsteher (*pagi magistri*), in S. Groh and H. Sedlmayer (eds). *Forschungen im römischen Heiligtum am Burgstall bei St. Margarethen im Lavanttal (Noricum)* (Zentraleurop. Arch. 2): 207–218. Wien: Verlag VÖAW.

Hainzmann, Manfred 2015. *Sagum* – zu problematischen Datierungsansätzen eines keltisch- römischen Kleidungsstückes, in C.-G. Alexandrescu (ed.) 2015: 263–274.

Hainzmann, Manfred, Kramer, Dietreich and Pochmarski, Erwin (eds) 1991. *Akten des 1. Internationalen Kolloquiums über Probleme des Provinzialrömischen Kunstschaffens*. (*Mitteilungen der Archaeologischen Gesellschaft Steiermark* 3/4 [1989/90]). Wien: VWGÖ.

Handy, Markus and Pochmarski Erwin 2014. Grabaltäre im südlichen Noricum. *Jahreshefte des Össtereichischen Archäologischen Institutes in Wien* 85: 107–130.

Handy, Markus and Pochmarski Erwin 2016. Das *sagum* in Norricum. *Jahreshefte des Österreichischen Archäologischen Institutes in Wien* 83: 57–99.

Harding, Merete and Jacobsen, Gurli 1988. Die Bedeutung der zivilen Zuwanderung aus Norditalien für die Entwicklung der Städte in Noricum und Pannonia. *Classica et Mediaevalia* 39: 117–206.

Hardy, Ernest G. 1912. *Roman Laws and Charters*. Clark, New Jersey: The Lawbook Exchange, Ltd.

Harl, Friederike 2016. Ein distanziertes Verhältnis, in R. Lafer (ed.) 2016: 17–27.

Harl, Ortolf 1991a. Zu den Voraussetzungen für das Entstehen einer römischen Steinskulptur im Ostalpenraum, in M. Hainzmann, D. Kramer and E. Pochmarski (eds) 1991/I: 3–32.

Harl, Ortolf 1991b. Historische Selektion und Datierung römischer Steindenkmäler im Ostalpenraum, in M. Praznovszky (ed.) 1991: 15–51.

Harl, Ortolf 2003. Die Donauarmee als Träger der norisch-pannonischen Kunst – der Fall Norikum, in P. Noelke, F. Naumann-Steckner and B. Schneider (eds) 2003: 337–361.

Häussler, Ralph 2013. *Becoming Roman? Diverging Identities and Experiences in Ancient Northwest Italy*. Walnut Creek, California: Left Coast Press.

Heger, Norbert 2007. Eine Randbemerkung zum pannonisch-norischen Rahmenornament, in V. Höck, F. Lang and W. Wohlmayr (eds). Akten zum 2. Österreichischen 'Römerstein-Treffen' 2006 in Salzburg: 185–187. Wien: Phoibos Verlag.

Hemelrijk, Emily A. 2015. *Hidden Lives, Public Personae. Women and Civic Life in the Roman West*. Oxford: Oxford University Press.

Hermary, Antoine, Cassimatis, Hélène and Vollkommer, Rainer 1986. Eros. *LIMC* III: 850–942.

Hirt, Alfred M. 2010. *Imperial Mines and Quarries in the Roman World. Organizational Aspects 27 BC – AD 235*. Oxford: Oxford University Press.

Hoffiller, Viktor 1937a. Nove rimske vojničke kacige. *Časopis za zgodovino in narodnopisje* (Kovačićev zbornik) 32/1–4: 29–32.

Hoffiller, Viktor 1937b. Arheološki nalazi u Novigradu Podravskom, in B. Madjer. *Časti i dobru zavičaja*: 295–298. Novigrad: Blaž Madjer.

Holder, Alfred 1896–1913. *Alt-Celtischer Sprachschatz*, 3 vols. Leipzig: Teubner.

Holder, Paul A. 1980. *Studies in the Auxilia of the Roman Army from Augustus to Trajan*. British Archaeological Reports International Series 70, Oxford: Archaeopress.

Holtheide, Bernard 1980. Matrona stolata - femina stolata. *Zeitschrift für Papyrologie und Epigraphik* 38: 127–134.

Hope, Valerie M. 2001. *Constructing Identity: The Roman Funerary Monuments of Aquileia, Mainz and Nimes.* British Archaeological Reports International Series 960. Oxford: John and Erica Hedges Ltd.

Horvat, Anđela 1962. Rudine u Požeškoj kotlini – ključni problem romanike u Slavoniji. *Peristil* 5/1: 11–27.

Horvat, Jana, Lovenjak, Milan, Dolenc Vičič, Andreja, Lubšina-Tušek, Marija, Tomanič-Jevremov, Marjana and Šubic, Zorka 2003. Poetovio. Development and Topography, in M. Šašel Kos and P. Scherrer (eds) 2003: 153–189.

Hoti, Marina 1992. Sisak u antičkim izvorima. *Opuscula archaeologica* XVI: 133–163.

Howells, Daniel Thomas 2015. *A Catalogue of the Late Antique Gold Glass in the British Museum.* London: The British Museum.

Hudák, Krisztina and Nagy, Levente 2009. *A Fine and Private Place. Discovering the Early Christian Cemetery of Sopianae/Pécs.* Pécs: Pécs/Sopianae Örökség Nonprofit Kft.

Hudeczek, Erich 2008. *Die Rundskulpturen des Stadtgebietes von Flavia Solva. CSIR, Österreich*, Band IV, 1. Wien: Österreichische Akademie der Wissenschaften.

Humer, Franz, Kremer, Gabrielle, Pollhammer, Eduard and Pülz, Andreas (eds) 2016. Akten der 3.Österreichischen Römersteintagung in Carnuntum. Wien: Amt der NÖ Landesregierung, Abteilung Kunst und Kultur.

Ilić, Luka Oriovčanin 1852. Starožitnosti kraljevstva Slavonije (unpublished manuscript). Zagreb: The National and University Library.

Ilijanić, Mira 1961. Sarkofag Julija Viktorija Quadrata. *Bulletin Zavoda za likovne umjetnosti Jugoslavenske akademije znanosti i umjetnosti* IX/1–2: 43–47.

Jaeger, Bettina 2003. Zur Porträtrepresentation auf norischen, pannonischen und gallischen Grabdenkmälern in der römischen Kaiserzeit, in P. Noelke, F. Naumann-Steckner and B. Schneider (eds) 2003: 475–480.

Jaeger, Bettina 2007. Römische Grabdenkmäler mit Porträtdarstellungen aus Aquincum (Budapest), in E. Walde and B. Kainrath (eds) 2007: 27–36.

Jakovljević, Goran 2012. *Registar arheoloških nalaza i nalazišta Bjelovarsko-bilogorske županije.* Bjelovar: Gradski muzej Bjelovar.

Jarak, Mirja 1994. Povijest starokršćanskih zajednica na tlu kontinentalne Hrvatske / The history of early Christian communities in continental Croatia (exhibition catalogue), in Ž. Demo (ed.) 1994: 15–39 / 154–179.

Joshel, Sandra R. 2013. *Slavery in the Roman World.* Cambridge: Cambridge University Press.

Kajanto, Iiro 1963. *Onomastic studies in the early Christian inscriptions of Rome and Carthage.* Rome: Institutum Romanum Finlandiae.

Kajanto, Iiro 1965. *The Latin Cognomina.* Commentationes Humanarum Litterarum 36. Helsinki-Helsingfors: Keskurkirjapaino.

Kakoschke, Andreas 2012. *Die Personennamen in der römischen Provinz Noricum* (Alpha-Omega, Reihe A, 262). Hildesheim - Zürich - New York: Olms - Weidmann.

Kane, Susan and Carrier, Sam C. 1992. Relationships between style and size of statuary and the availability of marble in the Eastern Roman Empire, in M. Waelkens, N. Herz and L. Moens (eds) 1992: 121–127.

Karlović, Tomislav 2006. The Legal Status of *municipium Andautonia*, in G. Thür and Z. Lučić (eds). *'Imperium und Provinzen' (Zentrale und Regionen)*: 51–74. Sarajevo: Pravni fakultet Univerziteta.

Karlović, Tomislav, Milotić, Ivan and Petrak, Marko 2015. Andautonia. An Example of Local Self-Government in Pannonia. *Lex localis - Journal of Local Self-Government* 13/1: 35–48.

Kastelic, Jože 1988. *Simbolika mitov na rimskih nagrobnih spomenikih. Šempeter v Savinjski dolini.* Ljubljana: Slovenska matica v Ljubljani.

Kaszab-Olschewski, Tünde 2007. Ländliche Alltagsszenen auf Grabreliefs der Nordwestprovinzen, in E. Walde and B. Kainrath (eds) 2007: 173–182.

Katičić, Radoslav 1963. Das mitteldalmatinische Namengebiet. *Živa antika* 12/2: 255–292.

Kehne, Peter 2007. War - and Peacetime Logistics: Supplying Imperial Armies in East and West, in P. Erdkamp (ed.). *A Companion to the Roman Army*: 323–338. London: Blackwell.

Keppie, Lawrence 1984. *The Making of the Roman Army, from Republic to Empire.* London: Batsford.

von Kersauson, Kate 1996. *Catalogue des portraits romanis. Tome II: De l'année de la guerre civile (68-69 après J.-C.) à la fin de l'Empire.* Paris: Réunion des Musées Nationaux.

Kleiner, Diana E. E. 1977. *Roman Group Portraiture. The Funerary Reliefs of the Late Republic and Early Empire.* New York - London: Garland Pub.

Kleiner, Diana 1992. *Roman Sculpture.* New Haven - London: Yale University Press.

Klemenc, Josip 1935. Ranorimski napis iz Siska. *Vjesnik Arheološkog muzeja u Zagrebu* XVI: 67-68.

Klemenc, Josip 1961. Limes u Donjoj Panoniji, in M. Grbić (ed.). *Limes u Jugoslaviji.* Zbornik radova sa simposiuma o limesu (Zagreb, 1960): 5–34. Beograd: Societas arch. Jugoslaviae

Knezović, Ivan 2010. The Worship of Savus and Nemesis in Andautonia. *Arheološki vestnik* 61: 187–202.

Knezović, Ivan 2012. Construction Materials and Techniques, in B. Migotti (ed.) 2012b: 225–249.

Koch, Guntram 1993. *Sarkophage der römischen Kaiserzeit.* Darmstadt: Wissenschaftliche Buchgesellschaft.

Koch, Guntram 2015. Einige Überlegungen zum Verhältnis der erhaltenen kaiserzeitlichen Sarkophage zur ursprünglichen Produktion, in B. Porod and G. Koiner (eds) 2015: 144–162.

Koch, Guntram and Sichtermann, Helmut 1982. *Römische Sarkophage.* München: C. H. Bek'sche Verlagsbuchhandlung.

Kolar-Sušanj, Sonja 1971. Gola, Koprivnica - rimskodobni grobni humci. *Arheološki pregled* 13: 40–42.

Kolšek, Vera 1991. Die Todesgenien im Stadtbereich von Celeia, in M. Praznovszky (ed.) 1991: 139–146.

Koncani Uhač, Ida (ed.) 2014. *Akti XII. međunarodnog kolokvija o rimskoj provincijalnoj umjetnosti. Datiranje kamenih spomenika i kriteriji za određivanje kronologije / Proceedings of the 12th International Colloquium on Problems of Roman Provincial Art. The dating of stone monuments and criteria for determination of chronology.* Pula: Arheološki muzej Istre / Archaeological Museum of Istria.

Konrad, Sven 2004. *Die Grabstelen aus Moesia Inferior: Untersuchungen zu Chronologie, Typologie und Ikonographie.* Leipzig: Casa Libri.

Koščević, Remza 2013. *Siscia, Pannonia Superior, Old and New Finds.* British Archaeological Reports, International Series 2461. Oxford: Archaeopress.

Kovachich, Márton György 1786. Anmerkungen ueber Prabners Vertheidigung der Sylvestrinischen Bulle (eingesendet). *Merkur von Ungarn* 1, VI.8., 7,12: 894–909.

Kovács, Péter 2007. *Corpus inscriptionum Graecarum Pannonicarum. Editio III. aucta* (Hungarian Polis Studies 15). Debrecen: University, Dept. of Ancient History; Budapest: Pan.

Kovács, Péter 2008. Some Notes on the Division of Illyricum, in J. Piso (ed.). *Die Römischen Provinzen. Begriff und Gründung*: 237–248. Cluj-Napoca: Editura Mega.

Kovács, Péter 2014. *A History of Pannonia during the Principate.* Bonn: Dr. Rudolf Habelt GMBH.

Kovács, Péter 2017. Neue römische Inschriften aus Komitat Komárom-Esztergom III. *Zeitschrift für Papyrologie und Epigraphik* 201: 261–269.

Kovács, Péter 2018. Northern Pannonia and the Roman Conquest, in M. Milićević Bradač and D. Demicheli (eds). *The Century of the Brave. Roman Conquest and Indigenous Resistance in Illyricum during the time of Augustus and His Heirs / Stoljeće hrabrih. Rimsko osvajanje i otpor starosjedilaca u Iliriku za vrijeme Augusta i njegovih nasljednika*: 163–174. Zagreb: FF press.

Koželj, Tony and Wurch-Koželj, Manuela 1992. The military protection of the quarries of the Aliki area during the Byzantine period, in M. Waelkens, N. Herz and L. Moens (eds) 1992: 43–57.

Kraft, Konrad 1951. *Zur Rekrutierung der Alen und Kohorten an Rhein und Donau.* Bern: A. Francke Verlag.

Krahe, Hans 1929. *Lexikon altillyrischen Personennamen.* Heidelberg: Carl Winter.

Krahe, Hans 1955. *Die Sprache der Illyrier, Erster Teil: Die Quellen.* Wiesbaden: Harrassowitz.

Kranz, Peter 1977. Zu den Anfängen der stadtrömische Säulensarkophage. *Mitteilungen des Deutschen Archäologischen Instituts, Römische Abteilung* 84: 349–380.

Kranz, Peter 1986. Die Grabmonumente von Šempeter. Die Beobachtungen zur Entwicklung der Bildhauer Kunst in Noricum während der mittleren und späten römischen Kaiserzeit. *Bonner Jahrbücher* 186: 193–239.

Kremer, Gabrielle 2001. *Antike Grabbauten in Noricum. Katalog und Auswertung von Werkstücken als Beitrag zur Rekonstruktion und Typologie.* Wien: Österreichisches Archäologisches Institut.

Kremer, Gabrielle and Kitz, Isabelle 2016. Steindenkmäler und Steingewinnung - Neue interdisciplinäre Forschungen im Rahmen des CSIR Carnuntum, in F. Humer, G. Kremer, E. Pollhammer and A. Pülz (eds) 2016: 71-85.

Križman, Mate 1991. *Rimska imena u Istri. Osobna imena na istarskim natpisima iz rimskog doba.* Zagreb: Latina et Graeca.

Kronberger, Michaela, Mosser, Martin and Insulander, Sophie 2016. Gesteinbestimmung an Römersteinen aus Vindobona - Lösungsansätze, erste Ergebnisse und Perspektiven aus archäologisher Sicht, in F. Humer, G. Kremer, E. Pollhammer and A. Pülz (eds) 2016: 87–99.

Kubitschek, Jos Wilhelm 1889. *Imperium Romanum tributim discriptum.* Pragae - Vindobonae - Lipsiae: Pragae F. Tempsky.

Kulenović, Igor and Muštra, Miona 2002. Novi nalaz kruništa s lavljim parom s lokaliteta Gabajeva greda kod Koprivnice. *Opuscula archaeologica* 26: 143–152.

Kuntić-Makvić, Bruna 1993. Sisciensia Kercselichiana. *Opuscula archaeologica* XVII: 263–271.

Kušan Špalj, Dora 2014. Aquae Iasae - zdraviliško, kultno in preroško svetišče / Aquae Iasae - a Centre of Health, Cult and Oracle, in N. Pirnat Spahić (ed.) 2014. *Aquae Iasae. Najnovejše rimske najdbe na območju Varaždinskih Toplic / Recent Discoveries of Roman Remains in the Region of Varaždinske Toplice*: 56–106. Ljubljana: Cankarjev dom.

Kušan Špalj, Dora. 2017. *Aquae Iasae* - nova otkrića u rimskom svetištu - s posebnim osvrtom na kultove Apolona, Eskulapa i Serapisa / *Aquae Iasae* - new discoveries in the Roman sanctuary - with special regard to the cults of Apollo, Aesculapius and Serapis. *Vjesnik Arheološkog muzeja u Zagrebu* L: 255–308.

Lafer, Renate (ed.) 2016. *Römische Steindenkmäler im Alpen-Adria-Raum.* Klagenfurt/Celovec - Ljubljana/Laibach - Wien/Dunaj: Hermagoras Verlag/Mohorjeva založba.

Landskron, Alice 2007. Zur Gewandstatue im Museum Carnuntinum, in E. Walde and B. Kainrath (eds) 2007: 349–355.

Langner, Martin 2003. Attribute auf gallo-römischen Grabreliefs als Ausdruck einer gesteigerten Wertschätzung materieller Güter, in P. Noelke, F.

Naumann-Steckner and B. Schneider (eds) 2003: 191–202.

Lapuente, Pilar, Rojo, Hernando, Cuchi, José Antonio, Justes, Julia and Preite-Martinez, Maria 2012. Local stones and marbles found in the territory of 'Alto Aragon' (Hispania) in Roman times, in P. Pensabene and E. Gasparini (eds) 2012: 191–200.

Lazar, Irena 2013. The Tombs of the Nobles in Celeia and its Territory, in N. Cambi and G. Koch (eds) 2013: 253–279.

Le Bohec, Yann 1989. *Les unités auxiliaires de l'armée romaine en Afrique Proconsulaire et Numidie sous le Haut-Empire*. Études d'antiquités africaines. Paris: Editions du CNRS.

Le Bohec, Yann and Wolff, Catherine 2000. Legiones Moesiae Superioris, in Y. Le Bohec and C. Wolff (eds). *Les légions de Rome sous le Haut-*Empire (Collection du Centre d'Etudes Romaines et Gallo-Romaines, Nouvelle série, N° 20): 239–245. Lyon: De Boccard.

Lefebvre, Sabine 2001. À propos de la répartition du nom Verecundus en Gaule et en Germanie, in M. Dondin-Payre and M. T. Raepsaet-Charlier (eds) 2001 : 597–647.

Lehner, Manfred 2012. Die 'Grüne Grenze' zwischen Noricum und Obepannonien: Alte Meinungen und neue Indizien. *Studia Universitatis Babeş Bolyai, Series Historica* 57/1: 38–51.

Leleković, Tino 2011. Antičke nekropole Siscije i Murse (Roman Cemeteries of Siscia and Mursa ). Unpublished PhD dissertation, University of Zagreb.

Leleković, Tino 2012. Cemeteries, in B. Migotti (ed.) 2012b: 313–357.

Leleković, Tino and Rendić-Miočević, Ante 2012. Rural Settlements, in B. Migotti (ed.) 2012b: 279–311.

Lengyel, Alfonz and Radan, George T. (eds) 1980. *The Archaeology of Roman Pannonia*. Budapest: The University Press of Kentucky – Akademiai Kiado.

Le Roux, Patrick 2000. Legio VII Gemina (pia) felix, in Y. Le Bohec and C. Wolff (eds). *Les légions de Rome sous le Haut-Empire* (Collection du Centre d'Etudes Romaines et Gallo-Romaines, Nouvelle série, N° 20) : 383–396. Lyon: De Boccard.

Leveau, Philippe 1993. *Territorium urbis*. Le territorie de la cité romaine et ses divisions: du vocabulaire aux réalités administratives. *Revue des Études Anciennes* 95/3-4: 459–471.

Leventopoulou, Maria 1997. Die Kentauren in der griechischen Kunst. *LIMC* VIII: 674–710.

Lindgren, Claire K. 2014. Provincial Ladies in Sculpture: Criteria for Determining Chronology, in I. Koncani Uhač (ed.) 2014: 55-58.

Liu, Jinyu 2009. Collegia Centonariorum*: The Guilds of Textile Dealers in the Roman West* (Columbia Studies in the Classical Tradition 34). Leiden – Boston: Brill.

Liu, Jinyu 2012. Clothing supply for the military. A look at the inscriptional evidence, in M.- L. Nosch (ed.). *Wearing the Cloak. Dressing the soldier in Roman Times* (Ancient Textiles Series, vol. 10): 19–28. Oxford: Oxbow Books.

Lochner-Hüttenbach, Fritz 1965. Die Antiken Personennamen aus Ig bei Ljubljana. *Situla* 8: 15–45.

Lochner von Hüttenbach, Fritz 1989. *Die römerzeitlichen Personennamen der Steiermark*. Graz: Leykam.

Lolić, Tatjana 2003. Colonia Flavia Siscia, in M. Šašel Kos and P. Scherrer (eds) 2003: 131–152.

Lolić, Tatjana 2014. Urbanizan antičke Siscije na temelju analize i interpretacije povijesne i moderne nacrtne dokumentacije (Layout of Roman Siscia Based on Analysis and Interpretation of Historic and Modern Drawings and Plans). Unpublished PhD dissertation, University of Zagreb.

Lolić, Tatjana and Wiewegh, Zoran 2012. Urbanism and Architecture, in Migotti (ed.) 2012b: 191–224.

Lo Monaco, Annalisa 2011. Algide e belle come dee. Immagini private e apoteosi a Roma in età medio-imperiale, in E. La Rocca and C. Parisi Presicce (eds). *Ritratti, le tante facce del potere*: 335–357. Roma: MondoMostre.

Lőrincz, Barnabás 1985. Zu den Kaiserbeinamen der römischen Truppen im 3. Jahrhundert. *Acta Archaeologica Academiae Scientiarum Hungaricae* 37: 177–189.

Lőrincz, Barnabás 2001. *Die römischen Hilfstruppen in Pannonien während der Prinzipatszeit*. Wien: Forschungsgesellschaft Wiener Stadtarchäologie.

Lőrincz, Barnabás and Redő, Ferenc 1994. *Onomasticon Provinciarum Europae Latinarum, Vol. I : ABA – BYSANVS*. Budapest: Archaeolingua.

Lovenjak, Milan 1998. *Inscriptiones Latinae Sloveniae* 1: *Neviodunum* (*Situla* 37). Ljubljana: Narodni muzej Slovenije.

Lovenjak, Milan 2003. Municipium Flavium Latobicorum Neviodunum, in M. Šašel Kos and P. Scherrer (eds) 2003: 93–105.

Lovrenčević, Zvonko 1981. Rimske ceste i naselja u bilogorsko-podravskoj regiji (II.). *Arheološki pregled* 22: 195-208.

Lučić, Lovorka, 2013. Rimski natpisi iz Varaždinskih Toplica / The Roman inscriptions from Varaždinske Toplice. *Vjesnik Arheološkog muzeja u Zagrebu* XLVI: 185–255.

MacMullen, Ramsay 1960. Inscriptions on arms and the supply of arms in the Roman Empire. *American Journal of Archaeology* 64/1: 23–40.

Mainardis, Fulvia. Epigrafia e iconografia, in R. Lafer (ed.) 2016: 283–307.

Mainecke, Katharina 2015. Inschriften mit Angaben zum sozialen Status in Grabanlagen mit Sarkophagen in Rom und Umgebung, in B. Porod and G. Koiner (eds) 2015: 180–194.

Malone, Stephen James 2006. *Legio XX Valeria Victrix, Prosopography, archaeology and history*. British Archaeological Reports, International Series 1491. Oxford: Archaeopress.

Mander, Jason 2013. *Portraits of Children on Roman Funerary Monuments*. Cambridge: Cambridge University Press.

Marasović, Katja (ed.) 2015. ASMOSIA XI (XI International Conference, Split, Croatia, 18–22 May 2015). Abstracts. Split: Faculty of Civil Engineering, Architecture and Geodesy.

Marco Simón, Francisco and Rodà de Llanza, Isabel 2008. A Latin *defixio* (Sisak, Croatia) to the river god Savus mentioning L. Licinius Sura, Hispanus. *Vjesnik Arheološkog muzeja u Zagrebu* XLI: 167–198.

Marsigli, Luigi Ferdinando 1726. *Danubius Pannonico-mysicus observationibus geographicis, historicis, physicis perlustrarus et in sex tomos digestus* (tomus secundus). Hagae–Amstelodami: Comitum, apud P. Gosse, R.C. Alberts, P. de Hondt.

Maršić, Dražen 2014. Female Fashion on Late Antique Monuments from Salona (3rd – 4th c. A.D), in Koncani Uhač (ed.) 2014: 203–210.

Marvin, Miranda 1983. Freestanding Sculptures from the Baths of Caracalla. *American Journal of Archaeology* 87/3: 347–384.

Matasović, Ranko 2003. Jezični tragovi Kelta u Iliriku. *Latina et Graeca*, n. s. 3: 5–23.

Matijašić, Robert and Tassaux, Francis 2000. Liber et Silvanus, in C. Delplace and F. Tassaux (eds). *Les cultes polythéistes dans l'Adriatique romaine* (Études 4): 65–117. Bordeaux: Ausonius.

Maver, Andreja 2008. Funerary Aediculae in Poetovio and her Ager, in Ch. Franek, S. Lamm, T. Neuhauser, B. Porod and K. Zöhrer (eds). *Thiasos - Festschrift für Erwin Pochmarski zum 65 Geburtstag*: 611–622. Wien: Institut für Archäologie der Karl- Franzens-Universität Graz.

Mayer, Anton 1957 and 1959. *Die Sprache der alten Illyrier* I and II. Wien: Rohrer in Komm.

Medar, Mladen 1986. O otkriću antičkog reljefa nad portalom bjelovarske župne crkve. *Muzejski vjesnik* 9: 11–13.

Medar, Mladen 2012. *Informativni vodič*. Bjelovar: Gradski muzej Bjelovar.

Meid, Wolfgang 2005. *Keltische Personennamen in Pannonien*. Budapest: Archaeolingua.

Meinecke, Katharina 2015. Inschriften mit Angaben zum sozialen Status in Grabanlagen mit Sarkophagen in Rom und Umgebung, in B. Porod and G. Koiner (eds) 2015: 180–194.

Meischner, Jutta 1964. *Das Frauenporträt der Severerzeit*. Berlin: Wasmuth.

Mercando, Liliana and Paci, Gianfranco 1998. Stele *romane in Piemonte* (Monumenti antichi, Serie miscellanea 5). Roma: G. Bretschneider.

Mesihović, Salmedin 2011. *Antiqui homines Bosnae*. Sarajevo: Filozofski fakultet u Sarajevu.

Migotti, Branka 1996. Ranokršćanski grobni nalaz iz Velikih Bastaja kod Daruvara. *Vjesnik Arheološkog muzeja u Zagrebu* XXVIII–XXIX (1995–1996): 127–157.

Migotti, Branka 1997. *Evidence for Christianity in Roman Southern Pannonia (Northern Croatia). A catalogue of finds and sites*. British Archaeological Reports International Series 684. Oxford: Archaeopress.

Migotti, Branka 2001. Sarkofag Romanije Nevije iz Siska. *Godišnjak Gradskog muzeja Sisak* II: 37–88.

Migotti, Branka 2002. *Two Gold-Sandwich Glasses from Štrbinci (Đakovo, Northern Croatia)*. Zagreb: Hrvatska akademija znanosti i umjetnosti.

Migotti, Branka 2005a. The iconography of the Dioscuri on a sarcophagus from Siscia. *Histria antiqua* 13: 277–285

Migotti, Branka 2005b. The Ash-chest of Marcus Aurelius Glabrio from Siscia reconsidered, in M. Sanader (ed.) 2005. *Illyrica Antiqua. Ob honorem Duje Rendić-Miočević*: 367–384. Zagreb: Filozofski fakultet – Arheološki muzej u Zagrebu.

Migotti, Branka 2007a. Rimski sarkofazi Siscije. *Godišnjak Gradskog muzeja Sisak* 7: 7–30.

Migotti, Branka 2007b. The Social/Gender Context of the Sarcophagus of a *togata clarissima femina* from Siscia (Pannonia Superior). *Archaeologia Bulgarica* XI/1: 61–78.

Migotti, Branka 2008a. Nadgrobni spomenik robovske obitelji iz Donjih Čeha kod Zagreba. *Archaeologia Adriatica* 2: 453–465.

Migotti, Branka 2008b. *The Crosssbow Brooches with Portraits in the Roman Empire / Lukovičaste fibule s portretima na području Rimskog Carstva*. Zagreb: Hrvatska akademija znanosti i umjetnosti.

Migotti, Branka 2010a. Rimska vojnička stela iz Svojića na Kordunu, in J. Dukić, A. Milošević and Ž. Rapanić (eds). *Scripta Branimiro Gabričević dicata*: 93–114. Trilj: Kulturno društvo Trilj.

Migotti, Branka 2010b. Weaving in Roman antiquity: between a female virtue and economic reality. *Instrumentum (Bulletin du Groupe de travail européen surl'artisanat et les productions manufacturées dans l'Antiquité)* 31: 30–31.

Migotti, Branka 2011a. Vojnička nadgrobna stela severskog razdoblja iz Lobora. *Archaeologia Adriatica* 3 (2009): 155–171.

Migotti, Branka 2011b. Prostor ranokršćanske biskupije Siscije i njegova arheološka ostavština, in D. Tepert and S. Jurić (eds). *Antiquam fidem* (Radovi sa znanstvenog skupa Sisak, 3. – 5. prosinca 2010.): 37–68. Zagreb: Glas Koncila.

Migotti, Branka 2012a. Rimski nadgrobni reljef iz Križovljana u sjeverozapadnoj Hrvatskoj, in B. Migotti, Ph. Mason, B. Nadbath and T. Mulh (eds). *Scripta in Honorem Bojan Djurić*: 387–398. Ljubljana: Zavod za varstvo kulturne dediščine Slovenije.

Migotti, Branka (ed.) 2012b. *The Archaeology of Roman Southern Pannonia*. British Archaeological Reports International Series 2393. Oxford: Archaeopress.

Migotti, Branka 2012c. Introduction and Commentaries, in B. Migotti (ed.) 2012b: 1–27.

Migotti, Branka 2012d. Dvije rimske stele iz Savarije, in J. Balen and M. Šimek (eds). *Arheologija varaždinskog kraja i srednjeg Podravlja*. Izdanja Hrvatskog arheološkog društva 28: 141–161. Varaždin: Hrvatsko arheološko društvo - Gradski muzej Varaždin.

Migotti, Branka 2013a. *The Stelae of Northern* Croatia, in N. Cambi and G. Koch 2013: 303–341.

Migotti, Branka 2013b. The Roman Sarcophagi of Siscia. *Arheološki radovi i rasprave* 17: 181–226.

Migotti, Branka 2013c. Can soldiers on 3rd-century stelae in Pannonia be recognized only by the sagum?, in M. Sanader, A. Rendić-Miočević, D. Tončinić and I. Radman-Livaja (eds). *Proceedings of the XVIIth Roman Military Conference: Weapons and Military Equipment in a Funerary Context*: 259–271. Zagreb: Filozofski fakultet Sveučilišta u Zagrebu - Arheološki muzej u Zagrebu.

Migotti, Branka 2014. Urbanizam i arhitektura rimske Panonije, in M. Šegvić and D. Marković (eds). *Klasični Rim na tlu Hrvatske. Arhitektura, urbanizam, skulptura* (exhibition catalogue): 104–117. Zagreb: Galerija Klovićevi dvori.

Migotti, Branka 2016a. The military stelae of northern Croatia, in J. Horvat (ed.). *The Roman army between the Alps and the Adriatic*: 171–191. Ljubljana: Inštitut za arheologijo ZRC SAZU.

Migotti, Branka 2016b. An architectonic sarcophagus from Siscia (Sisak, NW Croatia), in R. Lafer (ed.) 2016: 223–240.

Migotti, Branka 2017a. Saxa loquuntur. *Roman Epitaphs from North-Western Croatia*. Oxford: Archaeopress.

Migotti, Branka 2017b. Roman Sarcophagi of Northern Croatia Featuring Peculiar Iconographies, in D. Demicheli (ed.). *Illyrica Antiqua II. In honorem Duje Rendić- Miočević* (Proceedings of the International Conference Šibenik 12th - 15th September 2013): 499–513. Zagreb: Department of Archaeology, Faculty of Humanities and Social Sciences, University of Zagreb.

Migotti, Branka 2017c. The population of Aquae Balissae (Pannonia Superior). *Studia Antiqua et Archaeologica* 23/1: 83–124.

Migotti, Branka 2018. Grobni spomenici Siscije u kontekstu novijih geoarheoloških istraživanja. *Godišnjak Gradskog muzeja Sisak* 12: 18–38.

Migotti, Branka. The Interdependence of the Epitaph and Portraiture in Interpreting Inscribed Funerary Monuments (in press: Papers of the 15th International Colloquium on Roman Provincial Art, 14 - 20 June 2017, Graz/Austria).

Migotti, Branka. Aspects of Characterisation of Stone Monuments from southern Pannonia (in press: Papers of ASMOSIA XI, Split, Croatia, 18–22 May 2015).

Mihailescu-Bîrliba, Lucretiu 2006. Les patrons des affranchis privés dans les provinces balkano-danubiennes (Dalmatie, Pannononies, Dacies et Mésies). *Studia Antiqua et Archaeologica* 12: 65–76.

Minkova, Milena 2000. *The Personal Names of the Latin Inscriptions in Bulgaria*. Frankfurt am Main - Berlin - Bern - Bruxelles - New York - Wien: Peter Lang GmbH.

Mirković, Miroslava and Dušanić, Slobodan 1976. *Inscriptions de la Mésie Supériere*, vol. I: *Singidunum et le nord-ouest de la province*. Beograd: Faculté de philosophie de l'Université de Beograd.

Mitthof, Fritz and Cenati, Chiara 2016. *Miles in urbe* - Soldati di area danubiana e balcanica a Roma tra Settimio Severo e Costantino (193–312 d.C.), in F. Humer, G. Kremer, E. Pollhammer and A. Pülz (eds) 2016: 117-124.

Mócsy, András 1959. *Die Bevölkerung von Pannonien bis zu den Markomannenkriegen*. Budapest: Verlag der Ungarischen Akademie der Wissenschaften.

Mócsy, András 1962. Pannonia. *RE Suppl*. IX: 516–776.

Mócsy, Andreas 1968a. Municipium Iasorum. *RE Suppl.* XI: 1003–1004.

Mócsy, Andreas 1968b. Incerum. *RE Suppl.* XI: 849–850.

Mócsy, András 1974. *Pannonia and Upper Moesia*. London: Routledge and Kegan Paul Books.

Mócsy, András 1983. *Nomenclator provinciarum Europae Latinarum et Gallia Cisalpinae* (Dissertationes Pannonicae III/1). Budapest: Népművelési Propaganda Iroda.

Mócsy, András 1984a. Lateinische Cognomina als Geschichtsquelle: zwei Typen der provinzialrömsichen Kultur. *Acta Archaeologica Academiae Scientiarum Hungaricae* 36: 197–222.

Mócsy, András 1984b. Bemerkungen zu den negotiatores von Colijnsplaat. *Münstersche Beiträge zur Antiken Handelsgeschichte* 3/2: 43–58

Mócsy, András 1985a. *Beiträge zur Namenstatistik* (Dissertationes Pannonicae III/ 3). Budapest: ELTE Régészeti Tanszék.

Mócsy, András 1985b. Zum Gentiliz der Neubürger unter Septimius Severus, in E. Weber and G. Dobesch (eds). *Festschrift für Artur Betz zur Vollendung seines 80.* Lebensjahres (Archäologisch-Epigraphische Studien, Band 1. Römische Geschichte, Altertumskunde und Epigraphik): 403–415. Wien. Österreichische Gesellschaft für Archäologie.

Morris, John and Roxan, Margaret 1977. The witnesses to Roman military diplomata. *Arheološki vestnik* 28: 299–333.

Mosser, Martin 2002. C. Atius und die legio XV Apollinaris in Vindobona. *Fundort Wien. Berichte zur Archäologie* 5: 102–126.

Mosser, Martin 2003. *Die Steindenkmäler der legio XV Apollinaris* (Wiener Archäologische Studien 5). Wien: Forschungsgesellschaft Wiener Stadtarchäologie.

Mouritsen, Henrik 2011. *The Freedman in the Roman World*. Cambridge: Cambridge University Press.

Mrozewicz, Leszek 2010. I Flavi e il bacino danubiano, in L. Zerbini (ed.). *Roma e le province del Danubio*: 35–45. Soveria Mannelli: Rubbettino Edittore.

Mrozewicz, Leszek 2015. Flavische Städtegründungen auf dem Balkan, in G. von Bülow (ed.). *Kontaktzone Balkan. Beiträge des intern. Kolloquiums 'Die Donau-Balkan-Region als Kontaktzone zwischen Ost-West und Nord-Süd' vom 16.–18. Mai 2012 in Frankfurt a. M.* (RGK Kolloquien zur Vor- und Frühgeschichte 20): 151–163. Bonn: Rudolf Habelt.

Müller, Harald 2002. Provenance determination of Roman marble sculptures from Pannonia, in E. Jerem and K. Biró (eds). *Archaeometry 98.* Proceedings of the 31st International Symposium on Archaeometry: 767–775. Oxford: British Archaeological Reports International Series 1043.

Müller, Harald 2007. Die Herkunft der römischen Marmorobjekte im Vorarlberger Landesmuseum in Bregenz. *Jahrbuch des Vorarlberger Landesmuseumsvereines - Funde der Landeskunde 2006/07*: 361–369.

Nagy, Mihály 2004. Mogetiana, in M. Šašel Kos and P. Scherrer (eds) 2004: 75–83.

Nelis-Clément, Jocelyne 2000. *Les* beneficiarii*: militaires et administrateurs au service de l'Empire (I^er s. a.C. - VI^e s. p.C.)* (Études 5). Bordeaux: Ausonius.

Nemeth-Ehrlich, Dorica and Kušan Špalj, Dora 2003. Municipium Andautonia, in M. Šašel Kos and P. Scherrer (eds) 2003: 107–129.

Nemeth-Ehrlich, Dorica and Kušan Špalj, Dora 2007. *2000 godina Andautonije. Od rimskog grada do arheološkog parka / 2000 Years of Andautonia. From the Roman Town to the Archaeological Park.* Zagreb: Arheološki muzej u Zagrebu.

Nervi, Cristina 2015. Life of Nora (Province of Cagliari - South Sardinia). Roman quarries and their organization in the rural landscape, in P. Pensabene and E. Gasparini (eds) 2015: 585–591.

Newby, Zahra 2007. Art at the crossroads? Themes and styles in Severan art, in S. Swain, S. Harrison and J. Elsner (eds). *Severan culture*: 201–249. Cambridge: Cambridge University Press.

Nicole, Georges 1873. Sphinx, in Ch. Daremberg and E. Saglio (eds). *Dictionnaire des antiquités grecques et romaines* 4/2: 1431–1439. Paris: Librarie Hachette.

Nikoloska, Aleksandra 2007. *Aspects of the Cult of Cybele and Attis on the Monuments from the Republic of Croatia.* British Archaeological Reports, International Series 2010. Oxford: Archaeopress.

Nikoloska, Aleksandra and Burmaz, Josip 2008. Novi nalazi kultnih spomenika iz Siska / New Discoveries of Cult Monuments in Sisak. *Opuscula archaeologica* 31 (2007): 205–220.

Nodelman, Sheldon 1983. A Portrait of the Empress Plautilla. *The Journal of J. P. Getty Museum* 1: 105–120.

Noelke, Peter, Naumann-Steckner, Friederike and Schneider, Beate (eds) 2003. *Romanisation und Resistenz in Plastik, Architektur und Inschriften der Provinzen des Imperium Romanum.* Akten des VII. internationalen Colloquiums über Probleme des provinzialrömischen Kunstschaffens. Mainz am Rhein: Philipp von Zabern.

Noll, Rudolf 1986. *Griechische und lateinische Inschriften der Wiener Antikensammlung.* Wien: Notring Verlag (1st ed. 1962).

Nosch, Marie-Louise (ed.) 2012. *Wearing the Cloak. Dressing the Soldier in Roman Times.* Oxford: Oxbowbooks.

Njegovan Stárek, Siniša 2009. Rimski kameni nadgrobni spomenik iz Kusonja. *Zbornik Povijesnog društva Pakrac-Lipik* 6: 120–133.

Öğüş, Esen 2014. The Rise and Fall of the Sarcophagus Production at Aphrodisias. *Phoenix* LXVIII: 137–156.

Okoń, Danuta 2017. *Album senatorum vol. I. Senatores ab Septimii Severi aetate usque ad Alexandrum Severum (193–235 AD).* Szczecin: Wydawnictwo Naukowe Uniw. Szczecińskiego.

Olujić, Boris 2007. Sustavno arheološko istraživanje lokaliteta Viničica kod Josipdola. *Modruški zbornik* 1: 53–69.

Onuk, Peter and Prochaska, Walter 2015. How in Situ Laser Ablation ICP-MS Analysis Can Help in tthe Provenance Analysis of Isotopically Very Similar Marbles, in K. Marasović (ed.) 2015: 191.

Ortalli, Jacopo 2005. 'Simbolo e ornato nei monumenti sepolcrali romani: il caso aquileiese'. *Antichità altoadriatiche* LXI: 245–286.

Osgood, Josiah 2014. Making Romans in the Family, in M. Peachin (ed.). *The Oxford Handbook of Social Relations in the Roman World*: 69–83. Oxford: Oxford University Press.

Ott, Joachim 1995. *Die Beneficiarer: Untersuchungen zu ihrer Stellung innerhalb der Rangordnung des römischen Heeres und zu ihrer Funktion* (Historia, Einzelschriften, Band 92). Stuttgart: Franz Steiner Verlag.

Palagia, Olga 1986. Daphne. *LIMC* III: 345–346.

Palágyi, Sylvia 1991. Zwei Grabmedaillons aus dem Balácaer Hügelgrab, in M. Praznovszky (ed.) 1991: 287–294.

Palágyi, Sylvia 2003. Landschaftsbilder auf pannonischen Grabstelen?, in P. Noelke, F. Naumann-Steckner and B. Schneider (eds) 2003: 481–488.

Palmentieri, Angela 2015. Local workshops of the Roman Imperial age. A contribution to the study of the production of Campanian sarcophagi, in P. Pensabene and E. Gasparini (eds) 2015: 283–293.

Pape, Wilhelm and Benseler, Gustav E. 1911. *Wörterbuch der griechischen Eigennamen.* 2 vols. Graz 1950 (reprint of the 3rd ed.: Vieweg & Sohn, Braunschweig).

Parisi Presicce, Claudio 2005. L'abandono della moderazione. I ritratti di Costantino e della sua progenie, in A. Donati and G. Gentili (eds). *Costantino il Grande. La civilità antica al bivio tra Occidente e Oriente.* Milano: SilvanaEditoriale.

Paškvalin, Veljko 2012. *Antički sepulkralni spomenici s područja Bosne i Hercegovine.* Sarajevo: Akademija nauka i umjetnosti Bosne i Hercegovine.

Patsch, Carl 1897. Bolentium. *RE* III/1: 673–674.

Patsch, Karl (Carl) 1900. Archäologisch-epigraphische Untersuchungen zur Geschichte der römischen

Provinz Dalmatien, Vierter Teil. *Wissenschaftliche Mittheilungen aus Bosnien und der Hercegovina* VII: 33–166.

Peachin, Michael 2014. Introduction, in M. Peachin (ed.). *The Oxford Handbook of Social Relations in the Roman World*: 3–36. Oxford: Oxford University Press.

Pearson, Andrew 2006. *The Work of Giants. Stone and Quarrying in Roman Britain*. Stroud: Tempus Publishing Limited.

Pensabene, Patrizio and Gasparini, Eleonora (eds) 2015. *ASMOSIA X: Interdisciplinary Studies on Ancient Stone*. Proceedings of the Tenth Association for the Study of Marble and Other Stones in Antiquity International Conference. Rome: L'Erma di Bretschneider.

Perdrizet, Paul 1900. Jupiter, Ζεύς, in Ch. Daremberg and E. Saglio (eds). *Dictionnaire des antiquités grecques et romaines* 3/1: 691–713. Paris: Librarie Hachette.

Perinić Muratović, Ljubica and Vulić, Hrvoje 2009. Razmatranja o Silvanovom kultu u Panoniji povodom nalaza žrtvenika u Vinkovcima / Observations on the Cult of Silvanus in Pannonia in the Light of the Find of an Altar in Vinkovci. *Prilozi Instituta za arheologiju u Zagrebu* 26: 165–180.

Perkić, Domagoj 2012. Antičke nekropole u speleološkim objektima u kontekstu ostalih antičkih nalazišta na području Korduna. Unpublished PhD dissertation, University of Zagreb.

von Petrikovits, Harald 1980. *Lixae*, in W.S. Hanson and L.J.F. Keppie (eds). *Roman Frontier Studies 1979*. British Archaeological Reports, International Series 71: 1027–1035. Oxford: Archaeopress.

Pflaum, Hans-Georg 1960 and 1961. *Les carrières procuratoriennes équestres sous le Haut-Empire romain*, I–III. Paris: P. Geuthner.

Pflaum, Hans-Georg *et al.* 1983. *Index onomastique des Inscriptions latines d'Afrique* (Etudes d'antiquités africaines): 51–90. Paris: CNRS.

Pflug, Hermann 1989. *Römische Porträtstelen in Oberitalien. Untersuchungen zur Chronologie, Typologie und Ikonographie*. Mainz am Rhein: Verlag Philipp von Zabern.

Piccottini, Gernot 1972. *Die Rundmedaillons und Nischenporträts des Stadtgebietes von Virunum. CSIR, Österreich*, Band II, 2. Wien: Österreichische Akademie der Wissenschaften.

Piccottini, Gernot 1977. *Die Dienerinnen- und Dienerreliefs des Stadtgebietes von Virunum. CSIR, Österreich*, Band II, 3. Wien: Österreichische Akademie der Wissenschaften.

Piganiol, André 1892. Venus, in Ch. Daremberg and E. Saglio (eds). *Dictionnaire des antiquités grecques et romaines* 5/2: 721–736. Paris: Librarie Hachette.

Pintarić, Tatjana 1998. *Nekropole i groblja na području Ščitarjeva* (exhibition catalogue). Velika Gorica: Muzej Turopolja.

Pinterović, Danica 1975. Nepoznata Slavonija. *Osječki zbornik* 14-15: 123-166.

Pochmarski Erwin 1986. Nachtrag zu den 'Girlandenhaltenden Eroten'. *Römisches Österreich* 13-14 (1985-1986): 243-262.

Pochmarski, Erwin 1991. Zum Porträtmedaillon in Noricum. Herkunft und Ausformung einer Gattung von Sepulkralreliefs, in M. Praznovszky (ed.) 1991: 123–134.

Pochmarski, Erwin 1996. Severische Reliefs aus Stadtterritorium von Virunum, in G. Bauchhenß (ed.) 1996: 127–139.

Pochmarski, Erwin 1998. Überlegungen zur Chronologie der pannonischen Sarkophage, in G. Koch (ed.). *Akten des Symposiums '125 Jahre Sarkophag-Corpus'*: 182–200. Mainz am Rhein: Verlag Philipp von Zabern.

Pochmarski, Erwin 2001. Zur Ikonographie und Chronologie der römischen Sarkophage aus Brigetio, in P. Zsidi and E. Hanny (eds) 2001: 201-221.

Pochmarski, Erwin 2005. Norische Mädchen in besonderer Tracht, in G. Grabher, B. Kainrath, A. Larcher and B. Welte (eds). *Vis Imaginum. Festschrift für Elisabeth Walde*: 350–366. Innsbruck: Institut für Klassische und Provinzialrömische Archäologie der Universität Innsbruck.

Pochmarski, Erwin 2009. Die girlandentragenden Eroten vom Forum in Aquileia. Reliefs zwischen der stadtrömischen und der provinzialrömischen Kunst, in T. Nogales and I. Rodà (eds). *Roma y las provincias; modelo y diffusion*. XI Coloquio Internacional de Arte Romano Provincial: 111–119. Roma: L'Erma di Bretschneider.

Pochmarski, Erwin 2011. *Die Porträtmedaillons und Porträtnischen des Stadtgebietes von Flavia Solva. CSIR, Österreich*, Band IV/ 2. Wien: Österreichische Akademie der Wissenschaften.

Pochmarski, Erwin 2015. Aschenkisten und Sarkophage auds Poetovio, in B. Porod and G. Koiner (eds) 2015: 212–225.

Pochmarski, Erwin 2016a. Die typologischen, ikonographischen und chronologischen Probleme der Grabstelen und Grabaltäre im Territorium von Flavia Solva, in F. Humer, G. Kremer, E. Pollhammer and A. Pülz (eds) 2016: 143-161.

Pochmarski, Erwin 2016b. Hunde in der Reliefkunst der Provinz Noricum. *Schield von Steier* 27 (2015–2016): 252–261.

Pochmarski, Erwin 2016c. Grabaltäre mit dionysischen Darstellungen und ihre Vorbilder in Rom, Oberitalien und Noricum. *ANODOS. Studies of the Ancient World* 12 (2012): 205–218.

Pochmarski, Erwin and Pochmarski-Nagele, Margaretha 2005. Die dionysischen Dreifigurenreliefs von Hartberg und Bad Waltersdorf (Steiermark), in M. Sanader, A. Rendić-Miočević, and D. Tončinić (eds) 2005: 141-155.

Pochmarski, Erwin and Pochmarski-Nagele, Margaretha 2013. Die Sarkophage von Intercisa, in N. Cambi and G. Koch (eds) 2013: 281-382.

Pochmarski, Erwin and Weber-Hiden, Ingrid 2016. *Die Grabstelen und Grabaltäre im Stadtgebietes von Flavia Solva. CSIR, Österreich*, Band IV, 3. Wien: Österreichische Akademie der Wissenschaften.

Póczi, Klara 1980. Pannonian Cities, in A. Lengyel and G. Radan (eds) 1980: 239-274.

Porod, Barbara and Koiner, Gabriele (eds) 2015. *Römische Sarkophage. Veröffentlichungen des Instituts für Archäologie der Karl-Franzens-Universität Graz* 12. Graz: Karl-Franzens-Universität.

Praznovszky, Mihály (ed.) 1991. *Akten des 2. Internationalen Kolloquiums über Probleme des Provinzialrömischen Kunstschaffens*. Veszprém: Laczkó Múzeum.

Previato, Caterina and Mareso, Nicolo 2015. Marbles from the *Domus* of 'Bestie ferrite' and from the *Domus* of 'Tito Macro' in Aquileia (UD), Italy, in P. Pensabene and E. Gasparini (eds) 2015: 299–309.

Prochaska, Walter 2015. Fine-grained dolomitic marble of high sculptural quality used in antiquity, in P. Pensabene and E. Gasparini (eds) 2015: 661–669.

Prochaska, Walter 2017. Provenance analysis of 25 white marble samples. Unpublished report, Leoben.

Prochaska, Walter and Attanasio Donato 2012. Tracing the origin of marbles by inclusion fluid chemistry, in A. Gutiérrez Garcia-Moreno, P. Lapuente Mercadal and I. Rodá de Llanza (eds) 2012: 230–237.

Prochaska, Walter, Donato, Attanasio and Bruno, Mattihias 2015. Unravelling the Carrara – Göktepe Entanglement, in K. Marasović (ed.) 2015: 219–220.

Prohászka, Péter 2013. Eeinige römische Inschriften aus Siscia (Sisak, Kroatien) nach einem Brief des Kaufmanns Paul Bitroff, in *Studia epigraphica in memoriam Géza Alföldy* (Antiquitas Reihe 1, Bd. 61): 285–294. Bonn: Dr. Rudolf Habelt GMBH.

Radman-Livaja, Ivan 2004. *Militaria Sisciensia. Nalazi rimske vojne opreme iz Siska u fundusu Arheološkog muzeja u Zagrebu / Finds of the Roman Military Equipment in Sisak in the Holding oft he Archaeological Museum in Zagreb (*Musei Archaeologici Zagrabiensis Catalogi et Monographiae, Vol. I). Zagreb: Arheološki muzej u Zagrebu.

Radman-Livaja, Ivan 2007. In Segestica... *Prilozi Instituta za arheologiju u Zagrebu* 24: 153–172.

Radman-Livaja, Ivan 2010. *Siscia* kao rimsko vojno uporište / *Siscia* as a Roman Military Stronghold, in I. Radman-Livaja (ed.). *Nalazi rimske vojne opreme u Hrvatskoj / Finds of the Roman Military Equipment in Croatia*: 179–201. Zagreb: Arheološki muzej u Zagrebu.

Radman-Livaja, Ivan 2012. *The Roman Army*, in B. Migotti (ed.) 2012b: 159–189.

Radman-Livaja, Ivan 2013. Rimski trnokop iz Siska s urezanim natpisom. *Vjesnik Arheološkog muzeja u Zagrebu* XLV: 483–497.

Radman-Livaja, Ivan 2014. *Tesere iz Siska. Olovne tesere iz Siscije / Plombs de Siscia*. Zagreb: Arheološki muzej u Zagrebu.

Radman-Livaja, Ivan 2015. Rimska vojska u Sisciji od Augusta do Klaudija /, in R. Škrgulja and T. Tomaš Barišić (eds). *35. prije Krista / 35 Before Christ* (exhibition catalogue): 24–44. Sisak: Gradski muzej Sisak.

Radman-Livaja, Ivan 2016. Immigrants from Other Areas of the Roman Empire Documented on Siscia Lead Tags, in D. Davison *et al.* (eds). *Croatia at the Crossroads. A consideration of archaeological and historical connectivity*: 169–191. Oxford: Archaeopress.

Radman-Livaja, Ivan and Ivezić, Hana 2012. A Review of South-Pannonian Indigenous Anthroponymy, in B. Migotti (ed.) 2012b: 137–158.

Radman-Livaja, Ivan and Vukelić, Vlatka 2015. Roman Military Inscriptions from Siscia: An Overview, in L. Vagalinski and N. Sharankov (eds) 2015: 399–405.

Radman-Livaja, Ivan and Vukelić, Vlatka 2016. Not exactly new epigraphic testimonies for the *beneficiarii* in Siscia, in J. Horvat (ed.) 2016. *The Roman army between the Alps and the Adriatic*: 205–212. Ljubljana: Inštitut za arheologijo ZRC SAZU.

Radman-Livaja, Ivan and Vukelić, Vlatka 2018. The whereabouts of Tiberius' ditch in Siscia, in M. Milićević Bradač, D. Demicheli (eds). *The Century of the Brave. Roman Conquest and Indigenous Resistance in Illyricum during the time of Augustus and His Heirs / Stoljeće hrabrih. Rimsko osvajanje i otpor starosjedilaca u Iliriku za vrijeme Augusta i njegovih nasljednika*: 407–421. Zagreb: FF Press.

Radnóti, Aladár and Barkóczi, László 1951. The Distribution of Troops in Pannonia Inferior during the 2nd Century A.D. *Acta Archaeologica Academiae Scientiarum Hungaricae* 1/3–4: 191–230.

Raepsaet-Charlier, Marie-Thérèse 2001. Onomastique et romanisation: éléments d'une comparaison entre les provinces de Gaule Belgique et de Germanie inférieure, in M. Dondin-Payre and M. T. Raepsaet-Charlier (eds) 2001 : 399–470.

Raepsaet-Charlier, Marie-Thérèse 2012. 'Decknamen', Homophony, Assonance: an Appraisal of Consonance Phenomena in Onomastics of the Roman Empire, in T. Meißner (ed.). *Personal Names in the Western Roman World* (Studies in Classical and Comparative Onomastics 1): 11–23. Berlin: Curach Bhán Publications.

Ragolič, Anja 2014. The territory of Poetovio and the boundary between Noricum and Pannonia / Upravno območje Petovione in meja med Norikom in Panonijo. *Arheološki vestnik* 65: 323–351.

Rankov, Boris 1986. The beneficiarii consularis in the western provinces of the Roman Empire. Unpublished PhD dissertation, University of Oxford.

Raybould, Marilynne E. and Sims-Williams, Patrick 2007. *A Corpus of Latin Inscriptions of the Roman Empire containing Celtic personal Names*. Aberystwyth: CMCS Publications.

Reali, Mauro 2000. Macro-storie di legioni e micro-storie di legionari: i *Mediolanenses* sotto le armi, in Y.

Le Bohec and C. Wolff (eds). *Les légions de Rome sous le Haut- Empire* (Actes du Congrès de Lyon 17-19 sept. 1998): 655–661. Lyon: Diffusion De Boccard.

Rebecchi, Fernando 1977. Sarcophagi cispadani di età imperiale romana. Richerce sulla decorazione figurate, sulla produzione e sul loro commercio. *Mitteilungen des Deutschen Archäologischen Instituts, Römische Abteilung* 84: 107–158.

Rebecchi, Fernando 1978. I sarcophagi romani dell'arco Adriatico. *Antichità Altoadriatiche* XIII: 201–258.

Reddé, Michel 1986. *Mare Nostrum, Les infrastructures, le dispositif et l'histoire de la marine militaire sous l'empire romain* (Bibliothèque des Ecoles françaises d'Athènes et de Rome 260). Rome – Paris: Ecoles françaises d'Athènes et de Rome.

Reed, Nicholas 1978. Pattern and Purpose in the Antonine Itinerary. *The American Journal of Philology* 99/2: 228–254.

Rémy, Bernard 2001. La dénomination des Viennois à l'époque impériale, in M. Dondin- Payre and M. T. Raepsaet-Charlier (eds) 2001 : 55–174.

Rendić-Miočević, Ante 1993. Rimska nadgrobna stela iz Odre nadomak Zagreba. *Obavijesti Hrvatskog arheološkog društva* 25–2: 28–31.

Rendić-Miočević, Ante 1994. Catalogue entry 147, in A. Rendić-Miočević (ed.). *Zagreb prije Zagreba. Arheološka baština Zagreba od pretpovijesti do osnutka biskupije 1094. godine / Zagreb Before Zagreb. The Archaeological Heritage of Zagreb from Prehistory to the Foundation of the Bishopric in 1904* (exhibition catalogue): 104. Zagreb: Arheološki muzej u Zagrebu.

Rendić-Miočević, Ante 2013. The Marble Funerary Stele of the *Cohors II Varcianorum Equitata*'s Veteran, *Titus Flavius Ateboduus* from the Odra Village near Zagreb, in N. Cambi and G. Koch (eds): 343–381.

Rendić-Miočević, Duje 1980 (1989). Neki ikonografski i onomastički aspekti Silvanove kultne zajednice u Panoniji i Iliriku. *Vjesnik Arheološkog muzeja u Zagrebu* XII-XIII: 105–124 (= *Iliri i antički svijet / The Illyrians and the Classical World*: 507–522. Split: Književni krug).

Reuter, Marcus 2003. Die 'keltische Renaissance' in den Nordwestprovinzen des Römischen Reiches, in P. Noelke, F. Naumann-Steckner and B. Schneider (eds) 2003: 21–26.

Ricci, Cecilia 1993. Balcanici e Danubiani a Roma. Attestazioni epigrafiche di abitanti delle province Rezia, Norico, Pannonia, Dacia, Dalmazia, Mesia, Macedonia, Tracia (I–III sec.), in L. Mrozewicz and K. Ilski (eds). *Prosopographica*: 175–208. Poznań: WERS.

Riha, Emilie 1979. *Die römischen Fibeln aus Augst und Kaiseraugs* (Forschungen in Augst 3). Basel: Amt für Museen und Archäologie des Kantons Basel-Landschaft.

Rinaldi Tufi, Sergio 1983. *Yorkshire. CSIR, Great Britain* I, 3. Oxford: Oxford Univeresity Press (for the British Academy).

Ritterling, Emil 1925. Legio. *RE* XII (Halbband XXIII–XXIV): 1186–1829.

Rižnar, Igor and Jovanović, Divna 2008. Stone Material of Regional Provenance from Sirmium. *Starinar* 56 (2006): 139–152.

Rodà, Isabel de Llanza 2014. Lucius Licinius Sura, Hispanus, in I. Piso and R. Varga (eds). *Trajan und seine Städte*: 21–35. Cluj-Napoca: Editura Mega.

Rohatsch, Andreas, Moshammer, Beatrix, Hodits, Barbara, Draganits, Erich and Heinrich, Maria 2016. Steindenkmäler und Steingewinnung im Raum Carnuntum-Vindobona – Vorstellung des geologischen Teils eines interdisziplinären Projektes, in F. Humer, G. Kremer, E. Pollhammer, and A. Pülz (eds) 2016: 177–183.

Rose, Hannelore 2007. Privatheit als öffentlicher Wert – Zur Bedeutung der Familie auf Grabmonumenten der Gallia Belgica, in E. Walde and B. Kainrath (eds) 2007: 207–224.

Roth, Jonathan P. 1999. *The Logistics of the Roman Army at War (264 B.C. – A.D. 235)* (Columbia Studies in the Classical Tradition, vol. 23). Leiden – Boston – Köln: Brill.

Rothe, Ursula 2009. *Dress and Cultural Identity in the Rhine-Moselle Region of the Roman Empire*. British Archaeological Reports International Series 2038. Oxford: Archaeopress.

Rothe, Ursula 2012. Clothing in the Middle Danube provinces. The garments, their origins and their distribution. *Jahreshefte des Österreichischen Archäologischen Institutes in Wien* 81: 137–231.

Russell, Ben 2011. The Roman Sarcophagus 'Industry': a Reconsideration, in J. Elsner and J. Huskinson (eds). *Life, Death and Representation: Some New Work on Roman Sarcophagi. Millennium-Studien / Millenium Studies* 29: 119–147. Berlin – New York: Walter de Gruyter GmbH.

Russell, Ben 2013. *The Economics of the Roman Stone Trade*. Oxford: Oxford University Press.

Russell, Miles and Laycock, Stuart 2011. *Unroman Britain. Exposing the Great Myth of Britannia*. Stroud: The History Press.

Salajić, Silvija 2001. *Arheologija virovitičkog kraja* (exhibition catalogue).Virovitica: Gradski muzej Virovitica.

Salomies, Olli 1987. *Die römischen Vornamen. Studien zur römischen Namengebung* (Commentationes Humanarum Litterarum 82). Helsinki: Societas Scientiarum Fennica.

Salway, Benet 2015. Late Antiquity, in C. Bruun and J. Edmondson (eds) 2015: 364–393.

Sanader, Mirjana 1994. O kultu Herkula u Hrvatskoj. *Opuscula archaeologica* 18: 87–114.

Sanader, Mirjana 2003. Grabsteine der *legio VII* aus Tilurium – Versuch einer Typologie, in P. Noelke, F. Naumann-Steckner and B. Schneider (eds) 2003: 501–510.

Sanader, Mirjana 2012. Uz skulpturu Ikara iz Daruvara / Contribution to the Sculpture of Icarus from Daruvar. *Opuscula archaeologica* 36: 106–142.

Sanader, Mirjana, Rendić-Miočević, Ante and Tončinić, Domagoj (eds) 2005. *Akti VIII. međunarodnog kolokvija o problemima rimskog provincijalnog umjetničkog stvaralaštva. Religija i mit kao poticaj rimskoj provincijalnoj plastici / The Proceedings of the 8th International Colloquium on Problems of Roman Provincial Art. Religion and myth as an impetus for the Roman provincial sculpture.* Zagreb: Golden marketing- Tehnička knjiga.

Santangelo, Federico 2016. Caesar's Aims in Northeast Italy. *Papers of the British School at Rome* 84: 101–129.

Saria, Balduin 1955. Varciani. *RE* VIII A,1: 363–365.

Sarnowski, Tadeusz 1988. Quellenkritische Bemerkungen zu den Polizeikräften in Niedermoesien. *Eos* 76: 99–104.

Saxer, Robert 1967. *Untersuchungen zu den Vexillationen des römischen Kaiserheeres von Augustus bis Diokletian* (Epigraphische Studien 1). Köln – Graz: Böhlau Verlag.

Scardozzi, Giuseppe 2012. Ancient marble and alabaster quarries near Hierapolis in Phrygia (Turkey): new data from archaeological surveys, in A. Gutiérrez Garcia-Moreno, P. Lapuente Mercadal and I. Rodá de Llanza (eds) 2012: 573–583.

Schejbal, Berislav 2003. Nova razmatranja o Aquae Balissae i narodu Jaza: Pejzaž – Vode – Etimologija – Kultovi – Mitologija, pitanje atribucije i kontinuiteta. *Opuscula archaeologica* 27: 393–416.

Schejbal, Berislav 2004. Municipium Iasorum (Aquae Balissae), in M. Šašel Kos and P. Scherrer (eds) 2004: 99–129.

Schejbal, Berislav 2013. Daruvarska muzejska zavičajna zbirka: jučer – danas – sutra. *Vrela* 38: 16–22. Daruvar: Ogranak Matice hrvatske za znanost, kulturu i umjetnost.

Scherrer, Peter 2003. Savaria, in M. Šašel Kos and P. Scherrer (eds) 2003: 53–80.

Scherrer, Peter 2017. Gallische Römer oder römische Gallier? Menschen und Gottheiten in ausgewählten Bild-Schrift-Denkmälern der Nordprovinzen im 1. Jh. n. Chr.: Der Pilier des Nautes in Paris und ein Iuppiter-Taranis-Altar in Noricum. *Acta Paleohispanica XII (Paleohispanica* 17): 425–436. Zaragoza: Fritz Thyssen Stiftung, AELAW, COST.

Schmidt, Evamaria 1997. Venus. *LIMC* 8: 192–230.

Schmidt, Karl-Horst 1957. Die Komposition in gallischen Personennamen. *Zeitschrift für celtische Philologie* 26: 33–301.

von Schnurbein, Siegmar 1982. Die kulturgeschichtliche Stellung des nördlichen Rätien. *Bericht der Römisch-Germanischen Kommission* 63: 5–16.

Schober, Arnold 1923. *Die römischen Grabsteine von Noricum und Pannonien.* Wien: Österreichische Verlagsgesellschaft.

Scholz, Birgit 1996. Zu einigen Frauenporträts des 2. und frühen 3. Jahrhunderts aus den Balkanprovinzen, in G. Bauchhenß (ed.) 1996: 159–171.

Scholz, Markus 2012. *Grabbauten des 1.-3. Jahrhunderts in den nördlichen Grenzprovinzen des Römischen Reiches* (Teil 1, 2). Mainz: Römisch-Germanisches Zentralmuseum.

Schuler, Markus 2007. Vindonissa (Windisch Kt. Aargau, Schweiz), geologische Bestimmung epigraphischer Objekte, in E. Walde and B. Kainrath (eds) 2007: 107–116.

Schulze, Wilhelm 1904. *Zur Geschichte Lateinischer Eigennamen.* Abhandlungen der Königlichen Gesellschaft der Wissenschaften zu Göttingen, Phil.-hist. Klasse 5, 5, 1904 (2. unveränderte Auflage, Berlin/Zürich/Dublin, 1966; Neuausgabe mit einer Berichtigungsliste zur Neauausgabe von Olli Salomies, Zürich, Hildesheim, 1991).

Selem, Petar 1980. *Les religions orientales dans la Pannonie romaine: partie en Yougoslavie* (Études préliminaires aux religions orientales dans l'Empire romain 85). Leiden: Brill.

Selem, Petar 1997. *Izidin trag – Egipatski kultni spomenici u rimskom Iliriku (Sur les traces d'Isis – Les monuments cultuels égyptiens sur le territoire de l'Illyricum romain)* (Biblioteka znanstvenih djela 96). Split: Književni krug.

Sengelin, Thomas 1997. Die Kentauren in der römischen Kunst. *LIMC* VIII: 710–721.

Sergejevski, Dimitrije 1940. Arheološki nalazi kod Bihaća i Bos. Novog. *Glasnik Zemaljskog muzeja Kraljevine Jugoslavije* LI (1939): 7–14.

Sergejevski, Dimitrije 1957. Epigrafski nalazi iz Bosne. *Glasnik Zemaljskog muzeja Bosne i Hercegovine* XII: 109–125.

Sinn, Friederike 2003. Die Grabkammer in Köln-Weiden. Formen stadtrömischer Privatgräber für den lokalen Grabbrauch, in P. Noelke, F. Naumann-Steckner and B. Schneider (eds): 307–325.

Sinobad, Marko 2010. Jupiter i njegovi štovatelji u svjetlu epigrafskih izvora na području Hrvatske. *Opuscula archaeologica* 34: 145–228.

Skupińska-Løvset, Ilona 2014. Roman Fashion as a Chronological Criterion for the Dating of Palestinian Portraits, in Koncani Uhač (ed.) 2014: 40–45.

Slapšak, Božidar 2001. Poetovio as a supra-regional center, in M. Vomer Gojkovič and N. Kolar (eds). *Ptuj v rimskem cesarstvu. Mitraizem in njegova doba / Ptuj im Römischen Reich. Mithraskult und seine Zeit / Ptuj in the Roman Empire. Mithraism and its Era*, (Archaeologia Poetovionensis 2): 11–25. Ptuj: Pokrajinski muzej Ptuj.

Slukan Altić, Mirela 2012. *Povijesni atlas gradova: Sisak* (drugo dopunjeno izdanje). Zagreb: Institut društvenih znanosti 'Ivo Pilar' – Sisak: Državni arhiv u Sisku.

Sokač-Štimac, Dubravka 2004. Arheološki nalazi u požeškom kraju, in N. Čerti (ed.) 2004. *Kulturna baština Požege i Požeštine*: 14-35. Požega: Spin Valis.

Sokač-Štimac, Dubravka 2005. *Tekić, rimsko groblje i naselje* (exhibition catalogue). Požega: Gradski muzej Požega.

Sokač-Štimac, Dubravka and Korać, Zrinka 2005. *Rimske kupke u Tekiću. Rezultati arheoloških istraživanja 2006.* (exhibition catalogue). Požega: Gradski muzej Požega.

Solin, Heikki 1996. *Die Stadtrömischen Sklavennamen. Ein Namenbuch.* I.Teil: *Lateinische Namen* (Forschungen zur antiken Sklaverei 2/1). Stuttgart: Franz Steiner Verlag.

Solin, Heikki 2001. Un aspetto dell'onomastica plebea e municipale. La ripresa di nomi illustri da parte di comuni cittadini, in G. Angeli Bertinelli and A. Donati (eds). *Varia Epigraphica (Atti del Colloquio Intern. di Epigrafia, Bertinoro, 8-10 giugno 2000* (Epigrafia e Antichità 17): 411–427. Faenza: Fratelli Lega.

Solin, Heikki 2003a. *Die griechischen Personennamen in Rom: ein Namenbuch*, Band 1–3. Berlin – New York: Walter de Gruyter.

Solin, Heikki 2003b. Le trasformazioni dei nomi personali tra antichità e medioevo, in F. De Rubeis, W. Pohl (eds). *Le scritture dai monasteri.* Atti del II° seminario internazionale di studio 'I monasteri nell'alto medioevo' (AIRF 29): 15–45. Roma: Institutum Romanum Finlandiae.

Solin, Heikki 2015a. Analecta epigraphica. *Arctos* 49: 195–267.

Solin, Heikki 2015b. *Nomen omen.* Ripresa di nomi illustri nella società romana, in *In amicitia per Renato Badalì* (Una giornata di studi lunedì 8 giugno 2015): 16–40. Viterbo: Casa editrice Serena.

Solin, Heikki and Salomies, Olli 1994. *Repertorium nominum gentilium et cognominum Latinorum. Editio nova addendis corrigendisque augmentata.* Hildesheim – Zürich –New York: Olms.

Spaul, John 1995. Ala I Pannoniorum – one or many? *Zeitschrift für Papyrologie und Epigraphik* 105: 63–73.

Spaul, John 2000. *Cohors². The evidence for and a short history of the auxiliary infantry units of the Imperial Roman army.* British Archaeological Reports International Series 841. Oxford: Archaeopress.

Spaul, John E. H. 1994. *Ala².* Andover: Nectoreca Press.

Spaul, J. E. H. 1995. Ala I Pannoniorum, one or many? *Zeitschrift für Papyrologie und Epigraphik* 105: 63–73.

Speidel, Michael A. 2009. Roman Army Pay scales, in M. A. Speidel 2009. *Heer und Heerschaft im Römischen Reich der Hohen Kaiserzeit*: 349–380. Stuttgart: Franz Steiner Verlag.

Speidel, Michael A. 2015. The Roman Army, in Ch. Bruun and J. Edmondson (eds) 2015: 319–344.

Speidel, Michael P. 1983 (1992). The Centurions' Titles. *Epigraphische Studien* 13: 43–61 (= 1992: 21–39).

Speidel, Michael P. 1984 (1992). Regionarii in Lower Moesia. *Zeitschrift für Papyrologie und Epigraphik* 57: 185–188 (= 1992: 140–144).

Speidel, Michael P. 1989. The Soldiers' Servants. *Ancient Society* 20: 239–248.

Speidel, Michael P. 1992. *Roman Army Studies.* Vol. Two (*MAVORS* 8). Stuttgart: Franz Steiner.

Speidel, Michael P. 1994. *Die Denkmäler der Kaiserreiter – Equites singulares Augusti* (Beihefte der Bonner Jahrbücher 50). Köln: Rheinland-Verlag.

Spieß, Annette 1988. Studien zu den römischen Reliefsarkophagen aus den Provinzen Germania inferior und superior, Belgica und Raetia. *Kölner Jahrbuch für Vor- und Frühgeschichte* 21: 253–324.

Starac, Alka 1999. *Rimsko vladanje u Histriji i Liburniji. Društveno i pravno uređenje prema literarnoj, natpisnoj i arheološkoj građi – Histrija*, I. Pula: Arheološki muzej Istre.

Starac, Alka 2002. Rimske stele u Histriji. *Histria arhcaeologica* 31 (2000): 61–132.

Steiner, Alexandra 2007. Römerzeitlichen Denkmäler aus heimischen marmor in Kärnten, in E. Walde and B. Kainrath (eds) 2007: 393–400.

Stout, Ann M. 1994. Jewelry as a Symbol of Status in the Roman Empire, in J.L. Sebesta and L. Bonfante (eds). *The World of Roman Costume*: 77–100. Madison: The University of Wisconsin Press.

Strobel, Karl 2010. *Kaiser Traian. Eine Epoche der Weltgeschichte.* Regensburg: Friedrich Pustet.

Sumner, Graham 2009. *Roman Military Dress.* Stroud: The History Press.

Szabó, Ádám 2006. *Pannoniciani sacerdotes. A szervezett vallási élet principatuskori vezetői.* Pécs: Bornus Nyomda Kft.

Szabó, Gjuro 1934. Iz prošlosti Daruvara i okolice. *Narodna starina* 28 (1932): 79–97.

Szirmai, Krisztina 1991. Die Lupa Capitolina-Darstellungen auf den pannonischen Steindenkmälern, in M. Hainzmann, D. Kramer and E. Pochmarski (eds) 1991/I: 162–167.

Šačić Beća, Amra 2017. Rimske administrativne jedinice na pretpostavljenoj teritoriji panonskog naroda Oserijatâ: municipium Faustinianum i Servitium / Roman administrative units on the presumed territory of the Pannonian community of the Oseriates: municipium Faustinianum and Servitium. *Glasnik Zemaljskog muzeja Bosne i Hercegovine* 54 (2012): 109–166.

Šajn, Metka 2013. Clothing of the Deceased Portrayed on the Roman Gravestones Found on the Territory of Slovenia, in N. Cambi and G. Koch (eds) 2013: 223–252.

Šarić, Ivan 1978. Antičko naselje u Petrijancu, in: Ž. Rapanić (ed.). *Arheološka istraživanja u sjeverozapadnoj Hrvatsko.* Posebna izdanja Hrvatskog arheološkog društva 2: 177–195. Zagreb: Hrvatsko arheološko društvo.

Šarić, Ivan 1980. Antički kamenolom u Prozoru (tehnologija vađenja kamena), in M. Suić and M. Zaninović (eds). *Materijali, tehnike i strukture predantičkog i antičkog graditeljstva na istočnom jadranskom prostoru*: 115–123. Zagreb: Odjel za arheologiju, Centar za povijesne znanosti.

Šarić, Ivan 1992. Dva zanimljiva rimska kamena spomenika na području Korduna. *Vjesnik Arheološkog muzeja u Zagrebu* XXIV-XXV (1991-1992): 111-120.

Šašel, Jaroslav 1955. C. Vindonius Successus. *Živa antika* V: 127–139 (English translation in Šašel 1992: 44–53).

Šašel, Jaroslav 1960. Caesernii. Živa antika X: 201–221 (= Šašel 1992: 54–74).

Šašel, Jaroslav 1970. Andautonia. *RE Suppl.* XII: 71–75 (= Šašel, 1992: 580–582).

Šašel, Jaroslav 1974. Siscia. *RE, Suppl.* XIV: 702-741 (= Šašel, 1992: 600–620).

Šašel, Jaroslav 1981. Ancora un Caesernius Aquileiese. *Aquileia Nostra* 52: 165–168 (= Šašel 1992: 160–161).

Šašel, Jaroslav 1983 (1992). La fondazione delle città Flavie quale espressione di gratitudine politica. in *La città antica come fatto di cultura*: 79–91. Como: Società arch. comense (= 1992: 332–344).

Šašel, Jaroslav 1985. K rimskim napisom v Beli krajini. *Arheološki vestnik* XXXVI: 325 - 336.

Šašel, Jaroslav 1989. Die regionale Gliederung in Pannonien, in G. Gottlieb (ed.). *Raumordnung im Römischen Reich. Zur regionalen Gliederung in den gallischen Provinzen, in Rätien, Noricum und Pannonien*: 57–73 (= Šašel 1992: 690–706).

Šašel, Jaroslav 1992. *Opera selecta* (*Situla* 30). Ljubljana: Narodni muzej Slovenije.

Šašel Kos, Marjeta 1995. The 15th Legion at Emona - Some Thoughts. *Zeitschrift für Papyrologie und Epigraphik* 109: 227–244.

Šašel Kos, Marjeta 1997. The end of the Norican Kingdom and the formation of the provinces of Noricum and Pannonia, in B. Djurić and I. Lazar (eds). *Akten des IV. internationalen Kolloquiums über Probleme des provinzialrömischen Kunstschaffens* (*Situla* 36): 21–42. Ljubljana: Narodni muzej Slovenije.

Šašel Kos, Marjeta 1998. Autochthonous Cults between Emona and Poetovio, in J. Fitz (ed.). *Religions and Cults in Pannonia*: 17–22. Székesfehérvár: Fejér Megyei Múzeumok Igazgatósága.

Šašel Kos, Marjeta 1999. Aspects of the Sevirate and *Augustalitas* in the Northeastern Adriatic Area. *Histria antiqua* 5: 173–181.

Šašel Kos, Marjeta 2000. Caesar, Illyricum, and the Hinterland of Aquileia, in G. Urso (ed.). *L'ultimo Cesare. Scritti, Riforme, Progetti, Poteri, Congiure* (Monografie / *Centro ricerche e documentazione sull'ant. class.* 20): 277–304. Roma: 'L'Erma' di Bretschneider.

Šašel Kos, Marjeta 2005. *Appian and Illyricum* (*Situla* 43). Ljubljana: Narodni muzej Slovenije.

Šašel Kos, Marjeta 2006. Centenarians in the Emona area and the adjacent Norican and Pannonian regions, in M. G. Angeli Bertinelli and A. Donati (eds). *Misurare il tempo, misurare lo spazio* (Epigrafia e antichità 25): 175–197. Faenza: Fratelli Lega.

Šašel Kos, Marjeta 2010. The early urbanization of Noricum and Pannonia, in L. Zerbini (ed.). *Roma e le province del Danubio* (Pubblicazioni del LAD 1): 209–230. Soveria Mannelli: Rubbettino.

Šašel Kos, Marjeta 2014a. Poetovio before the Marcomannic Wars: from Legionary Camp to *Colonia Ulpia*, in I. Piso and R. Varga (eds). *Trajan und seine Städte*: 139–165. Cluj-Napoca: Editura Mega.

Šašel Kos, Marjeta 2014b. Kaj se je leta 14/15 dogajalo v Emoni - cesarski napis in upor panonskih legij / What Was Happening in Emona in AD 14/15? An Imperial Inscription and the Mutiny of the Pannonian Legions, in: M. Ferle (ed.). *Emona. Mesto v imperiju / A city of the Empire*: 79–93. Ljubljana: Mestni muzej.

Šašel Kos, Marjeta 2017a. Emona and its pre-Roman population: epigraphic evidence / Predrimsko prebivalstvo Emone v luči rimskih napisov. *Arheološki vestnik* 68: 439–458.

Šašel Kos, Marjeta 2017b. The Sisciani in the Roman Empire. *Studia Europaea Gnesnensia* 16: 173–204.

Šašel Kos, Marjeta 2017c. A sacred river landscape with a sanctuary - the worship of rivers in the south-eastern Alpine area, in R. Haeussler and A. King (eds). *Celtic Religions in the Roman Period. Personal, Local, and Global* (Celtic Studies Publications XX): 441–459. Aberystwyth: Celtic Studies Publications.

Šašel Kos, Marjeta 2018. The residents of Siscia: a population study. *Epigraphica* LXXX/1–2: 265–321.

Šašel Kos, Marjeta and Scherrer, Peter (eds) 2003. *The Autonomous Towns of Noricum and Pannonia, Pannonia I* (*Situla* 41). Ljubljana: Narodni muzej Slovenije.

Šašel Kos, Marjeta and Scherrer, Peter (eds) 2004. *The Autonomous Towns of Noricum and Pannonia, Pannonia II* (*Situla* 42). Ljubljana: Narodni muzej Slovenije.

Šegvić, Marina 1988. Natpis veterana XV. Apolinarske legije u Topuskom. *Vjesnik Arheološkog muzeja u Zagrebu* XXI: 57–66.

Šegvić, Marina 1996. Croatiae schedae epigraphicae Latinae (CSEL). Inscriptiones quae in Croatia ab anno MCMXCI usque ad annum MCMXCV repertae et editae sunt. *Opuscula archaeologica* 20: 131–139.

Šegvić, Marina 2006. Topusko, in S. Durman (ed.). *Stotinu hrvatskih arheoloških nalazišta*: 270–271. Zagreb: Leksikografski zavod Miroslav Krleža.

Šimunić, Antun 1986. Geološka građa okolice Lepoglave i osvrt na pojave mineralnih sirovina. *Radovi Zavoda za znanstveni rad Varaždin* 1: 19–32.

Škegro, Ante 1997. Inscriptiones Latinae et Graecae Bosniae et Hercegovinae. *Opuscula archaeologica* 21: 85–116.

Škegro, Ante 2006. The economy of Roman Dalmatia, in D. Davison, V. Gaffney and E. Marin (eds). *Dalmatia. Research in the Roman Province 1970-2001.* British Archaeological Reports International Series 1576: 149–173. Oxford: Archaeopress.

Škrgulja, Rosana 2018. Grobni spomenici od kamena / Funerary Stone Monuments, in I. Baćani and T. Tomaš Barišić (eds) 2018: 232–242.

Škrgulja, Rosana and Tomaš Barišić, Tea (eds) 2015. *35. prije Krista / 35 Before Christ.* Sisak: Gradski muzej Sisak.

Škrgulja, Rosana and Migotti, Branka 2015. Stela veterana 14. Legije Gaja Antonija Sentina iz Siscije / Stele of the Veteran of the 14th Legion *Gaius Antonius Sentinus* from *Siscia*. *Vjesnik Arheološkog muzeja u Zagrebu* XLVIII: 19–49.

Talbert, Richard J. A. (ed.) 2000. *Barrington Atlas of the Greek and Roman World*. Princeton – Oxford: Princeton University Press.

Tarpin, Michel 2002. Vici et pagi *dans l'Occident romain* (CEFR 299). Rome: École française.

Tataki, Argyro B. 2006. *The Roman presence in Macedonia, Evidence from personal names*, ΜΕΛΕΤΗΜΑΤΑ 46. Athens: KEPA – Diffusion de Boccard.

Thomasson, Bengt E. 1984. *Laterculi praesidum*. Vol. 1. Gothoburgi: Editiones Radii.

Thylander, Hilding 1952. *Étude sur l'épigraphie latine*. Lund: CWK Gleerup.

Tiussi, Cristiano 2003. *La collezione di Franco Marinotti a Torviscosa (Udine). Materiali scultorei di età romana. CSIR, Italia, Regio X, Friuli - Venezia Giulia*. Roma: Quasar.

Todisco, Elisabetta 2000. I veterani italici nelle province: l'integrazione sociale, in Y. Le Bohec and C. Wolff (eds). *Les légions de Rome sous le Haut-Empire* (Collection du Centre d'Etudes Romaines et Gallo-Romaines, Nouvelle série, N° 20): 663–673. Lyon: De Boccard.

Tomaš Barišić, Tea 2018. Jugoistočna nekropola / Southeastern Necropolis, in I. Baćani and T. Tomaš Barišić (eds) 2018: 60–135.

Tomasevic Buck, Teodora 2007. Vindonissa, die Steinmetzen, in E. Walde and B. Kainrath (eds) 2007: 83–106.

Tončinić, Domagoj 2009. Ziegelstempel römischer Miltäreinheiten in der Provinz Dalmatien, in Á. Morillo, N. Hanel and E. Martín (eds). *Limes XX. Estudios sobre la frontera romana / Roman Frontier Studies (2006). Anejos de Gladius* 13: 1447–1459. Madrid: Consejo Superior de Investigaciones Cientificas; Polifemo.

Ubl, Hannsjörg 2013. *Waffen und Uniform des römischen Heeres der Prinzipatsepoche nach den Grabreliefs Noricums und Pannoniens* (Austria Antiqua 3). Wien: Österr. Gsellschaft für Archäologie.

Vacinová, Lenka 2017. The Punishment of Tarpeia and its Possible Iconographic Inspiration. *Acta Universitatis Carolinae Philologica 2 / Graecolatina Pragensia*: 43–55.

Vagalinski, Lyudmil and Sharankov, Nicolay (eds) 2015. *Limes XXII. Proceedings of the 22nd International Congress of Roman Frontier Studies* (*Bulletin of the National Archaeological Institute* XLII). Sofia: National Archaeological Institute with Museum – Bulgarian Academy of Science.

Valentić, Mirko and Prister, Lada 2002. *Zbirka kamenih spomenika*. Zagreb: Hrvatski povijesni muzej.

Van Haeperen, Françoise 2012. Collèges de dendrophores et autorités locales et romaines, in M. Dondin-Payre and N. Tran (eds). Collegia. *Le phénomène associatif dans l'Occident romain* (Scripta Antiqua 41): 47–62. Bordeaux: Ausonius Éditions.

Verzár-Bass, Monika 1996. Arte funerarioa lungo la via dell'ambra, in M. Buora (ed.). *Lungo la via dell'Ambra. Apporti altoadriatici alla romanizzazione dei territori del Medio Danubio (I sec. a.C - I sec. d.C)*: 245–271. Udine: Civici Musei di Udine.

Vidović, Josip 2003. Pogrebni ritus u grobnim humkama Međimurja, in: B. Čečuk (ed.). *Arheološka istraživanja u Bjelovarsko-bilogorskoj županiji i Pogrebni ritusi na teritoriju Hrvatske*. Posebna izdanja Hrvatskog arheološkog društva 21: 79–86. Zagreb: Hrvatsko arheološko društvo.

Vikić-Belančić, Branka 1978. Tipovi naselja u sjeverozapadnoj Hrvatskoj (u svjetlu novijih istraživanja), in Ž. Rapanić (ed.). *Arheološka istraživanja u sjeverozapadnoj Hrvatskoj*. Posebna izdanja Hrvatskog arheološkog društva 2: 177–195. Zagreb: Hrvatsko arheološko društvo.

Visy, Zsolt 1986. Die kryptotopographische Truppenaufzählung in den Auxiliardiplomen von Pannonien, in W. Eck and H. Wolff (eds). *Heer und Integrationspolitik, Die römische Militärdiplome als historische Quelle* (Passauer historische Forschungen, Band 2): 482- 517. Cologne – Vienna: Böhlau.

Visy, Zsolt 2016. The Late Roman cemeteries of Sopianae, in E. Tóth, T. Vida and I. Takács (eds). *Saint Martin and Pannonia. Christianity on the Frontiers oft he Roman World* (exhibition catalogue): 68–76. Pannonhalma: Pannonhalmi Föapátság.

Visočnik, Julijana 2017. *The Roman Inscriptions from Celeia and its Ager*. Celje: Celjska Mohorjeva družba.

Vittinghoff, Friedrich 1953. Portorium. *RE* 22/1: 346–399.

Vukelić, Vlatka 2006. Prilog istraživanju antičke Siscije. Prvi pisani spomen Severillina sarkofaga u ranom novovjekovlju i pokušaj rekonstrukcije njegovog izvornog nalazišta. *Vjesnik Arheološkog muzeja u Zagrebu* XXXIX: 201–216.

Vukelić, Vlatka 2009. Sarkofag Maksimijane Emilije iz Siska. *Vjesnik Arheološkog muzeja u Zagrebu* XLII: 393–420.

Vukelić, Vlatka 2011. Povijest sustavnih arheoloških istrživanja u Sisku od 16. stoljeća do 1941. godine (The History of Systematic Archaeological Research in Sisak from the 16th Century until 1941). Unpublished PhD dissertation, University of Zagreb, Croatian Studies.

Vukelić, Vlatka and Glazer, Eva Katarina 2015. Siscian Villa at the site of 'Mrcinište'. Example of an Economic Production Complex in a Roman Period Pannonian Suburb. *Radovi - Zavod za hrvatsku povijest* 47: 567–580.

Vuković, Domagoj 1994. *Siscija - vizija rimskog grada u Panoniji*. Sisak: Centar za kulturu 'Vladimir Nazor'.

Waelkens, Marc, Herz, Norman and Moens, Luc (eds) 1992. *Ancient Stones: Quarrying, Trade and Provenance.*

*Interdisciplinary Studies on Stone and Stone Technology in Europe and Near East from the Prehistoric to the Early Christian Period*. Leuven: Leuven University Press.

Wagner, Walter 1938. *Die Dislokation der römischen Auxiliarformationen in den Provinzen Noricum, Pannonien, Moesien und Dakien von Augustus bis Gallienus*. Berlin: Junker and Dünnhaupt.

Walde, Elisabeth and Kainrath, Barbara (eds) 2007. *Die Selbstdarstellung der römischen Gesellschaft in den Provinzen im Spiegel der Steindenkmäler. IX Internationales Kolloquiums über Probleme des provinzialrömischen Kunstschaffens* (Ikarus - Innsbrucker klassisch-archäologische Universitäts-Schriften 2). Innsbruck: University Press.

Weaver, Paul R. C. 1972. *Familia Caesaris. A Social Study of the Emperor's Freedmen and Slaves*. Cambridge: Cambridge University Press.

Weber, Ekkehard. 2007. Grabsteine von Freigelassenen, in E. Walde and B. Kainrath (eds) 2007: 377–379.

Weber, Ekkehard. 2012. Stadtterritorien im römischen Österreich - Möglichkeiten und Grenzenihrer Bestimmung am Beispiel von Iuvavum. *Römisches Österreich* 34/35 (2011- 2012), 219-226.

Weber-Hiden, Ingrid 2014. Versuche zur Datierung des Inschriftenmaterials von Carnuntum anhand der Grabstelen, in Koncani Uhač (ed.) 2014: 70–73.

Wedenig, Reinhold 1997. *Epigraphische Quellen zur städtischen Administration in Noricum* (Aus Forschung und Kunst 31). Klagenfurt: Geschichtsvereins für Kärnten.

Wegner, Max 1939. *Die Herrscherbildnisse in antoninischer Zeit*. Berlin: Verlag Gebr. Mann.

Wenner, David B. and Herz, Norman 1992. Provenance signatures for classical limestone, in M. Waelkens, N. Herz and L. Moens (eds) 1992: 199–202.

Wessel, Klaus 1949. Römische Frauenfrisuren von der severischen bis zur konstantinischen Zeit. *Jahrbuch des Deutschen Archäologischen Instituts* 61–62 (1946/47): 62–76.

Wheeler, Everett L. 2000. Legio XV Apollinaris, From Carnuntum to Satala and beyond, in Y. Le Bohec and C. Wolff (eds). *Les légions de Rome sous le Haut-Empire* (Collection du Centre d'Etudes Romaines et Gallo-Romaines, Nouvelle série, N° 20): 259–308. Lyon: De Boccard.

Whittaker, Charles R. 1994. *Roman Empire. A Social and Economic Study*. Baltimore - London: The Johns Hopkins Uniersity Press.

Whittaker, Charles R. 2004. *Rome and its frontiers: the dynamics of Empire*. London: The Johns Hopkins Press.

Wiewegh, Zoran 2003. *Jugoistočna nekropola Siscije (The Southeast Necropolis of Siscia)*. Sisak: Gradski muzej Sisak.

Wilkes, John Joseph 1969. *Dalmatia*. London: Routledge.

Wilkes, John Joseph 2000. Roman Legions and their Fortresses in the Danube Lands (first to third centuries AD), in R. J. Brewer (ed.). *Roman Fortresses and Their Legions*: 101–119. London and Cardiff: Society of Antiquaries of London - National Museums and Galleries of Wales.

Winkler, Gerhard 1970. Flavius Verus Metrobalanus. *RE Suppl*. XII: 380 no. 204a.

Wolff, Catherine 2000. La legio XIII Gemina au 1er siècle, in Y. Le Bohec and C. Wolff (eds). *Les légions de Rome sous le Haut-Empire* (Collection du Centre d'Etudes Romaines et Gallo-Romaines, Nouvelle série, N° 20): 203–204. Lyon: De Boccard.

Zaccaria, Claudio 1994. Testimonianze epigrafiche di spettacoli teatrali e di attori nella Cisalpina romana, in *Spettacolo in Aquileia e nella Cisalpina romana* (*Antichità Altoadriatiche* 41): 69–98. Udine: Arti Graphice Friulane.

Zaccaria, Claudio 2006. *Palatina tribus*. Cavalieri e senatori di origine libertina certa o probabile ad Aquileia. I. - I *Caesernii*, in M. Faraguna and V. Vedaldi Iasbez (eds). Δύνασθαι διδάσκειν. *Studi in onore di Filippo Càssola per il suo ottantesimo compleano* (Fonti e studi per la storia della Venezia Giulia. Studi 11): 439–455. Trieste: Editreg.

Zaccaria, Claudio 2015. Tergestini nell'Impero Romano: Affari e carriere. Le testimonianze delle iscrizioni. *Archeografo Triestino* 75: 283–308.

Zachos, Giorgios and Leka, Evridiki 2012. The ancient quarry at Pagani-Alyfanta, Lesbos, Greece, in A. Gutiérrez Garcia-Moreno, P. Lapuente, and I. Rodà de Llanza (eds) 2012: 592–603.

Zaninović, Marin 1981. Siscia u svojim natpisima, in Ž. Rapanić (ed.). *Arheološka istraživanja u Zagrebu i njegovoj okolici* (*Izdanja Hrvatskog arheološkog društva* 6): 201–208. Zagreb: Hrvatsko arheološko društvo.

Zaninović, Marin 1993. Classis Flavia Pannonica, in B. Čečuk (ed.). *Arheološka istraživanja u Slavonskom Brodu i Brodskom Posavlju*. Posebna izdanja Hrvatskog arheološkog društva 16: 53–58. Zagreb: Hrvatsko arheološko društvo.

Zgusta, Ladislav 1964. *Kleinasiatische Personennamen*. Prag: Tschechoslowakische Akademie der Wissenschaften.

Ziegler, Daniela 2000. *Frauenfrisuren der römischen Antike: Abbild und Realität*. Berlin: *Weißensee Verlag*.

Zimmer, Gerhard 1982. *Römische Berufdarstellungen*. Berlin: Gebr. Mann Verlag.

Zsidi, Paula and Hanny, Erzsébet (eds) 2001. *Akten des 6. internationalen Kolloquiums über Probleme des provinzialrömischen Kunstschaffens* (*Budapest Régiségei* XXXIV). Budapest: Budapesti Történeti Múzeum.

Zorković, Anđelko 1995. *Povjesnica općine Petrijanec*. Petrijanec: Općina Petrijanec.

Zupčić, Krunoslav 2007. Lokalitet: Rijeka Mrežnica. *Hrvatski arheološki godišnjak* 3(2006): 217–218.

# Abbreviations of epigraphic and onomastic corpora and Internet databases

*AE* *L'Année épigraphique*

*AIJ* Hoffiller, Viktor and Saria, Balduin (eds) 1938. *Antike Inschriften aus Jugoslavien, Heft 1: Noricum und Pannonia Superior.* Zagreb: F. Pelikan - St. Kugli.

*CBI* Schallmayer, Egon, Eibl, Kordula, Ott, Joachim, Preuss, Gerhard and Wittkopf, Ester 1990. *Der römische Weihbezirk von Osterburken I. Corpus der griechischen und lateinischen Beneficiarer-Inschriften des Römischen Reiches* (Forschungen und Berichte zu Vor- und Frühgeschichte in Baden-Würtenberg 40). Stuttgart.

*CIL* *Corpus Inscriptionum Latinarum*

*CLE* *Carmina Latina epigraphica*

*CLEPann* Cugusi, Paolo and Sblendorio Cugusi, Maria Teresa 2007. *Studi sui carmi epigrafici. Carmina Latina epigraphica Pannonica.* Bologna, Pàtron Editore

*CSIR* *Corpus Signorum Imperii Romani*

*EDCS* *Epigraphik-Datenbank Clauss / Slaby.*

*EDR* *Epigraphic Database Roma.*

*EE* *Ephemeris Epigraphica. Corporis Inscriptionum Latinarum Supplementum*, Vols. I-IX. 1872-1913. Berlin: Deutsches Archäologisches Institut, Römische Abteilung.

*HD* *Epigraphic Database Heidelberg.*

*ILCV* Ernst Diehl, *Inscriptiones Latinae Christianae veteres* I-III, Berolini 1961[2] (1st ed. 1925-1931).

ILJug Šašel, Ana and Šašel, Jaro. *Inscriptiones Latinae quae in Iugoslavia inter annos MCMXL et MCMLX repertae et editae* sunt, *Situla* 5, 1963; *iidem*, *Inscriptiones Latinae quae in Iugoslavia inter annos MCMLX et MCMLX X repertae et editae sunt*, *Situla* 19, 1978; *iidem*, *Inscriptiones Latinae quae in Iugoslavia inter annos MCMLXII et MCMXL repertae et editae sunt*, *Situla* 25, 1986.

*ILN* *Inscriptions latines de Narbonnaise*

*ILS* *Inscriptiones Latinae selectae*, ed. Hermann Dessau, Berlin 1892-1916.

*ILSl* Lovenjak, Milan 1998. *Inscriptiones Latinae Sloveniae 1. Neviodunum* (*Situla* 37). Ljubljana: Narodni muzej Slovenije.

*Inscr. Aquil.* Giovanni Battista Brusin, *Inscriptiones Aquileiae*, I-III, Udine 1991-1993.

*LIMC* *Lexicon Iconographicum Mithologiae Classicae.* Zürich - München - Düsseldorf.

*lupa* Friederike und Ortolf Harl, http://lupa.at (Bilddatenbank zu antiken Steindenkmälern).

*OPEL* Lőrincz, Barnabás. *Onomasticon provinciarum Europae Latinarum.* Vol. I : *Aba - Bysanus.* Budapest 2005[2]: M. Opitz; Vol. II: *Cabalicius - Ixus.* Wien 1999: Forschungsgesellschaft Wiener Stadtarchäologie; Vol. III: *Labareus - Pythea.* Wien 2000: Forschungsgesellschaft Wiener Stadtarchäologie; Vol. IV: *Quadratia - Zures.* Wien 2002: Forschungsgesellschaft Wiener Stadtarchäologie.

*PME* *Prosopographia militiarum equestrium quae fuerunt ab Augusto ad Gallienum.* Hubert Devijver 1976-1980. Vols I-III. Louvain: Symbolae Facultatis Litterarum et Philosophiae Lovaniensis, Katholieke Universiteit Faculteit Wijsbegeerte en Letteren.

*RE* *Realencyclopädie der Classischen Altertumswissenschaft*, Stuttgart.

*RIU* *Römische Inschriften Ungarns*, Budapest.

*RMD* *Roman Military Diplomas.* Margaret Roxan and Paul Holder1978-2006. Vols. I-III. London: Institute of Archaeology. Vols. IV-V. London: Institute of Classical Studies.

*Tit. Aq.* II *Tituli sepulcrales et al. Budapestini reperti*, eds. P. Kovács et Á. Szabó, Budapest 2010.

*TLL* *Thesaurus Linguae Latinae*, 1894-. München: Bayerische Akademie der Wissenschaften.